Hipparcos, the New Reduction of the Raw Data

Astrophysics and Space Science Library

Recently Published in the ASSL series

Volume 350: *Hipparcos, the New Reduction of the Raw Data*, by Floor van Leeuwen. Hardbound ISBN: 978-1-4020-6341-1

Volume 349: *Lasers, Clocks and Drag-Free Control, Exploration of Relativistic Gravity in Space*, edited by Dittus, Hansjörg; Lämmerzahl, Claus; Turyshev, Slava . Hardbound ISBN: 978-3-540-34376-9

Volume 348: *The Paraboloidal Reflector Antenna in Radio Astronomy and Communication, Theory and Practice*, edited by Baars, Jacob W.M. Hardbound ISBN: 978-0-387-69733-8

Volume 347: *The Sun and Space Weather*, edited by Hanslmeier, Arnold. Hardbound ISBN: 978-1-4020-5603-1

Volume 346: *Exploring the Secrets of the Aurora,* edited by Akasofu, Syun-Ichi. Softcover ISBN: 978-0-387-45094-0

Volume 345: *Canonical Perturbation Theories, Degenerate Systems and Resonance*, edited by Ferraz-Mello, Sylvio. Hardbound ISBN: 978-0-387-38900-4

Volume 344: *Space Weather: Research Toward Applications in Europe*, edited by Jean Lilensten. Hardbound 1-4020-5445-9

Volume 343: *Organizations and Strategies in Astronomy: volume 7*, edited by A. Heck. Hardbound 1-4020-5300-2

Volume 342: *The Astrophysics of Emission Line Stars*, edited by Tomokazu Kogure, Kam-Ching Leung. Hardbound ISBN: 0-387-34500-0

Volume 341: *Plasma Astrophysics, Part II: Reconnection and Flares*, edited by Boris V. Somov. Hardbound ISBN: 0-387-34948-0

Volume 340: *Plasma Astrophysics, Part I: Fundamentals and Practice,* by Boris V. Somov. Hardbound ISBN 0-387-34916-9, September 2006

Volume 339: *Cosmic Ray Interactions, Propagation, and Acceleration in Space Plasmas,* by Lev Dorman. Hardbound ISBN 1-4020-5100-X, August 2006

Volume 338: *Solar Journey: The Significance of Our Galactic Environment for the Heliosphere and the Earth,* edited by Priscilla C. Frisch. Hardbound ISBN 1-4020-4397-0, September 2006

Volume 337: *Astrophysical Disks,* edited by A. M. Fridman, M. Y. Marov, I. G. Kovalenko. Hardbound ISBN 1-4020-4347-3, June 2006

Volume 336: *Scientific Detectors for Astronomy 2005,* edited by J. E. Beletic, J. W. Beletic, P. Amico. Hardbound ISBN 1-4020-4329-5, December 2005

For other titles see www.springer.com/astronomy

Hipparcos, the New Reduction of the Raw Data

Floor van Leeuwen
Institute of Astronomy, Cambridge University
Cambridge, UK

Springer

Editor

Floor van Leeuwen
Institute of Astronomy, Cambridge University
Cambridge, UK

ISBN: 978-1-4020-6341-1 e-ISBN: 978-1-4020-6342-8

Library of Congress Control Number: 2007936112

Printed on acid-free paper.

9 8 7 6 5 4 3 2 1

springer.com

This book is dedicated to all those who made the Hipparcos mission a success, and successfully negotiated, together, the many obstacles along the way.

Contents

Part II Exploring the Hipparcos Astrometric Data

Part III A Description of the Contents and Peculiarities of the Hipparcos Photometric Data

Part IV Hipparcos Attitude Modelling

Part V Summary of Selected Spacecraft and Payload Calibration Results

List of Figures

List of Tables

Acronyms

ABM Apogee Boost Motor

AC Alternating Current (used here for the modulated signal)

ACV Alpha-2 CVn (Rotational modulation variable)

AF Astrometric Field

AOV Analysis of Variance

AU Astronomical Unit

BP Blue Photometry (Gaia, dispersion spectra)

CCD Charge-Coupled Device

CHARA Center for High Angular Resolution Astronomy

CNES Centre National des Etudes Spatial

COG Centre of Gravity

CTI Charge Transfer Inefficiency

DC Direct Current (used here for the flat signal)

DFA Discrete Fourier Analysis

DPAC Data Processing and Analysis Consortium

ESA the European Space Agency

ESAC European Space Astronomy Centre

ESOC European Space Operations Centre

ESRO European Space Research Organisation

ESTEC the European Space Technoly Centre

FAST Fundmental Astrometry with Space Techniques

FDA Fully Dynamic Attitude

FFOV Following Field of View

FOV Field of View

FWHM Full Width Half Maximum

GASS Gaia System Simulator

GIBIS Gaia Instrument and Basic Image Simulator

GOG Gaia Object Generator

GSC Guide Star Catalogue

HR Hertzsprung-Russell

ICRS International Celestial Reference System

IDT Image Dissector Tube

IFOV Instantaneous Field of View

IGRF International Geomagnetic Reference Field

INCA Input Catalogue Consortium

LKE Lutz Kelker Effect

LMC Large Magellanic Cloud

LSF Line Spread Function

LSR Local Standard of Rest

ML Maximum Likelihood

MOC Mission Operation Centre

NDAC Northern Data Analysis Consortium

OBC On-Board Clock

OTF Optical Transfer Function

PFOV Preceding Field of View

PDF Probability Density Function

PPN Parameterized Post Newtonian

PSF Point Spread Function

QSO Quasi Stellar Object

RA Right Ascension

RGB the Red Giant Branch

RGC Reference Great Circle

RGO Royal Greenwich Observatory

RP Red Photometry (Gaia, dsipersion spectra)

RTAD Real Time Attitude Determination

RVS Radial Velocity Spectrograph

SC Scan Circle

SM Star Mapper

SMC Small Magellanic Cloud

SPB Slowly-pulsating B stars

SSRF Single-Slit Response Function

TCE Thermal Control Electronics

TDI Time-Delayed Integration

UV Ultra Violet

VIM Variability Induced Mover

VLBI Very Large Baseline Interferometry

Preface

The publication of the *Hipparcos* and *Tycho* Catalogues in 1997 transformed astrometry, and as a consequence astronomers' perception of astrometry. What had before often been regarded as a somewhat quaint specialty of limited relevance to modern astrophysics, was suddenly seen to produce a wealth of data of immediate practical use. The ready availability of many thousand precise trigonometric stellar distances and the access to an accurate and dense optical reference frame have changed the way astronomers think about certain problems and plan their experiments. Inevitably, the exploitation of so much new data not only solved some old problems, tidied up several confused areas and sharpened many observational constraints, but it also generated new questions about established theory – and about the data themselves. The author of this book has taken a radical approach to answer some of these questions: a complete re-examination of the satellite data and the models used to represent them, in particular the attitude modelling. Eventually this resulted in the new and very significantly improved *Hipparcos* reduction described in this book. This remarkable achievement was made possible by a combination of many factors, including time and the exponential growth of computing power, but mainly an incredible amount of detailed, tedious and ingenious work by the author and the resulting insight into what really went on with the satellite in its unhappy orbit.

Dr. Floor van Leeuwen has been involved in, or rather been at the centre of, the data analysis effort for *Hipparcos* since its beginning around 1980. A key person in NDAC (one of the *Hipparcos* data analysis consortia), he was the main person responsible, within that consortium, for the reduction of the detector signals, the instrument calibration, attitude determination, attitude control and torque determination, parts of the photometric analysis, and several other tasks. He was a member of the *Hipparcos* Science Team 1986–1997 and made very considerable contributions to the documentation and publication of the results. From his long experience, he is extremely well acquainted with all aspects of the *Hipparcos* raw data set, including its many defects and idiosyncrasies. Dr. van Leeuwen is currently a member of the Gaia Science Team and leads the photometric processing task for that mission.

In this book Dr. van Leeuwen describes the improved concepts, models and calibrations on which his new reduction of the *Hipparcos* data is founded. However, a considerable part of the book is devoted to discussing the accuracies of the new reduction, including practical applications permitting an objective assessment of the data quality. The result is persuasive evidence for the superiority of the new reduction. Many of the applications are however also text-book examples of how to use the data. This has great pedagogical value, as most modern investigations using this kind of data depend on sophisticated statistical procedures. An example is when marginally significant parallaxes are used for luminosity calibration. It is then critically important to understand how the data were obtained, their statistical characteristics and proper use. The book thus provides indispensible background information for anyone interested in getting the most out of the *Hipparcos* data, as well as an excellent introduction to space astrometry in general, including the future Gaia mission.

LENNART LINDEGREN, LUND OBSERVATORY

Introduction

In the ten years since the publication of the *Hipparcos* data a very substantial number of papers have been produced based, at least partly, on the wealth of astrometric and photometric data provided by this mission. The *Hipparcos* astrometric data formed an entirely new concept in observational astronomy: a full sky survey of one-dimensional positions obtained simultaneously in two fields of view over a 3.2 year period was processed to produce a rigid reference frame of positions and proper motions for 118 300 stars. Within this reference frame were measured absolute parallaxes to accuracies of up to 0.5 milli arcseconds (mas), providing significant measures of distance for nearly 40 000 stars. Further analysis of the data provided details on 8 000 double stars and 7 000 variable stars, as well as positional measurements for 48 minor planets and the moons Europa and Titan. The data had been reduced by two independent reduction consortia, FAST and NDAC, and numerous verifications of intermediate results had taken place before the mission, using simulated data, and during the mission on real data. The final catalogue (ESA 1997) has been obtained by carefully merging the results from the two groups. The merged results were subjected to a variety of tests to establish the validity of the published results.

Why, one may ask, would there be the need for a new reduction, and why a book on the *Hipparcos* data now, 14 years after the completion of the mission and 10 years after the publication of its results? Since the publication of the *Hipparcos* data in 1997, further investigations have been going on involving the raw data as well as some intermediate data products, such as the satellite attitude. The reconstruction accuracy of the satellite attitude is a fundamental parameter in defining the noise on the astrometric measurements. Attitude modelling errors will appear as systematic noise, showing through correlations between measurement errors for stars measured nearly simultaneously. In the post-publication analysis the main source for modelling errors in the along-scan attitude reconstruction was traced down to scan-phase discontinuities and small external hits of the spacecraft. The attitude model built to accommodate these effects, once identified, is presented in part IV of this book. The new attitude model and the identification of around 1500 scan-phase jumps and some 80 hits, led to a reduction in the attitude noise from 3 mas for the published data to around 0.6 mas in the new reduction. At the same time, a lot of attention was paid to the weight ratios between data used from the two fields of view for the attitude reconstruction, this to avoid the potential "Pleiades problem". Also, for very red variables, epoch-dependent colour indices, as presented in the literature after the publication of the *Hipparcos* data, were implemented throughout the new reduction. All these improvements reflect in the astrometric accuracies: while in the published data these accuracies were dominated by Poisson noise from the photon counts only for stars fainter than magnitude 9.5 to 10, in the new reduction photon statistics are the dominant noise down to magnitude 3 to 4. As a result, formal errors on the astrometric parameters for the brightest stars have been decreased by a factor 2 to 4, with the smallest formal errors down to 0.1 mas. At the same time, correlations between measurement errors had dropped by a factor 40 to an insignificant level. When these results became clear, there was no other option left than to prepare and present a new reduction of the *Hipparcos* astrometric data. Some of the methods explored in the new reduction have provided guidance for the similar Astrometric Global Iterative Solution as currently being developed for the Gaia mission (Part VI).

Apart from the new reduction, there was a further reason for preparing a book on the *Hipparcos* data. Over the years various papers had been produced showing more or less successful methods of how one may interpret these data for various applications. There is no single reference point where methods for analysing the *Hipparcos* data can be found. Part II of this book therefore provides an overview of some of the most relevant methods for the analysis of these data, using examples drawn from the application to the new reduction results, and as such providing at the same time some interesting astrophysical information: open cluster distances, period-luminosity calibrations, local galactic kinematics, etc. Part III summarizes the processing and results of the photometric data as a reference for their use. Part IV, as described above, presents the modelling of the satellite's attitude.

Finally, in part V, calibration data are presented that provide some further background information on the mission, payload as well as hardware, some of which may be useful for understanding peculiarities in the data.

At the start of the *Hipparcos* project, the measurement process, the kind of data produced and many other aspects of this mission were all new concepts, for which it takes time to become familiar with, and for which some peculiarities may remain hidden for a long time. This book is not intent on showing that the original reductions had been unsuccessful or inadequate in handling these peculiarities. On the contrary, the original reductions have been highly successful, and performed, under very difficult circumstances, significantly better than the mission targets. Improvements in the new reduction could only be obtained starting from the published results and methods. The new reduction, in that sense, constitutes a final set of iterations, for which there had been no time or computing power within the tight schedule that led to the data release in May 1997. These final iterations allowed at the same time correcting a few problems recognised, after the publication, to exist in the data, and to perform an additional series of tests on intermediate and final results. It also provides more background data which may be of interest for other studies. And last, but not least, this is an exercise that should benefit the processing and use of the forthcoming Gaia mission data, a mission based on a very similar measuring principle, but producing a three to four orders of magnitude larger data volume.

A study like this involves more than just the author. It started with the Doctoral Thesis of Elena Fantino and the Master Thesis of Andrea Dalla Torre, both from the University of Padova. Their enthusiasm and dedication to the project provided its essential kick-off. From thereon, however, it has been a matter of finding time in evenings and weekends, reducing and examining data many times (15 iterations were finally made through the entire data reduction), and gradually getting a sufficient understanding of the raw data and in particular the workings of the attitude reconstruction to justify the publication of a complete new reduction. Along the way errors were made and (hopefully all) corrected. For those inadvertently left I apologise. A few people led me back to the right track when I was lost, in particular Lennart Lindegren and Rudolf LePoole. My family, and in particular my wife, have been a huge support, and in particular during the final stages of the preparation of the data and the manuscript. Staff at the Cambridge Institute of Astronomy and other colleagues, and especially Dafydd Wyn Evans, Robin Catchpole, Michael Feast, Gerry Gilmore, Anthony Brown, Michael Perryman, Ulrich Bastian and Derek Jones provided over the years the encouragement to continue, for which I am grateful. Some of them, as well as my wife Gośka, Elena Fantino, Stephan Theil, Ralf Keil and Francesca De Angeli proof-read several chapters of the book, which was much appreciated. I am also grateful to Springer (and former Kluwer) publishers, who stood right behind me in preparing this book.

This is the kind of project that at the start is rather poorly defined: "we have this very-high accuracy attitude reconstruction for the *Hipparcos* satellite, let's see what we can learn from it concerning the dynamics of the satellite". As such, it defined the doctoral thesis of Elena Fantino. The next step, preparing for a new reduction of the *Hipparcos* data, was only decided upon at a much later stage, and was initially not considered as a likely outcome of this study at all. Only afterwards, when it became clear what wat could be learned from such study, and what impact it can make, could we have possibly phrased the second part of the project in a way that it may possibly have attracted some funding and a student, but by that time this was no longer needed.

PART I

INTRODUCTION TO THE HIPPARCOS MISSION

The *Hipparcos* mission concept originated in astrometric techniques developed over the past 150 to 200 years, explored in a new environment, free from the constraints and limitations of ground-based observations. Working with one-dimensional observations, and aspiring to provide a distortion-free full sky catalogue of positions and proper motions as well as absolute parallaxes is, however, far from simple. Operating a satellite at the precision level required by an astrometric mission exposes unexpected instrumental features that have to be accounted for, as well as very specific requirements on the data processing. Here we outline some of the developments that led to the Hipparcos mission, and show the main concepts of global astrometry based on one-dimensional data.

Chapter 1

THE HIPPARCOS MISSION

1.1 Overture

Astrometry in the 1960s and 1970s was still entirely ground-based, and all too often seen (and experienced by very few) as a rather tedious, time-consuming technique that no longer could create much of an impact on the study of astrophysics. Parallax measurements seemed to have reached their limit and were hampered by calibration uncertainties, and proper motion studies were only in a few cases still breaking new ground. The maintenance of the optical reference frame, the Fundamental Katalog, was a lengthy process based on accumulating, cleaning and calibrating observations spread over two centuries. Most meridian telescopes were closed down and dismantled, the resulting space being used for library extensions, offices or a canteen. Limitations to measuring accuracy by the Earth's atmosphere, and the inability to observe all the sky with the same instrument under the same conditions made it difficult for astrometrists to improve significantly upon already achieved results. There remained a need, however, for improved astrometric data, and in particular for improvements in the quality and quantity of parallax measurements, to support ever more detailed astrophysical theories and models.

Around the same time, and fuelled by those needs, new ideas on obtaining more, and more accurate, astrometric data were developing, such as an astrometric satellite proposal to the French Space Agency by Pierre Lacroute, modernisation of meridian telescope technology by Erik Høg in Hamburg, and later in Copenhagen, and automatic measuring machines for photographic plates like the Lick-Gaertner automatic measuring system (Klemola et al. 1974), Galaxy at the Royal Greenwich Observatory (RGO) (Murray and Nicholson 1975) and Astroscan at Leiden Observatory (De Vries 1986). Computing facilities were rapidly developing way beyond the capacity of the largely manual calculators of the preceding era. Amalgamation of these and several more developments led in the late 1970s to a proposal for an astrometric satellite at the European Space Agency (ESA). It was approved as a project in 1981, and launched in 1989. It operated for 3.5 years until its on-board computer failed in May 1993, probably caused by radiation damage sustained through frequent (unplanned) passages through the Van Allen belts. The data, after careful reductions by two independent consortia, and overseen by the invaluable Michael Perryman as ESA project scientist, were presented in 1997. Since then it has become one of the most frequently queried astronomical-data catalogues. The intensive use of the data also exposed some potential weaknesses in the data as published in 1997, which could not be addressed without a complete new reduction of the original, raw data.

To understand the origin of the *Hipparcos* mission, we briefly follow some of the developments that preceded and contributed to the *Hipparcos* mission: from determining time and longitude (which led to the meridian telescope developments), through mapping the sky (a crucial input to the mission) to measuring parallaxes (one of the main aims of the mission).

1.1.1 Seasons, Time and Longitude

Mapping the sky must be among the oldest scientific pursuits. It may have started with the attempt to understand the daily, yearly and longer-term motions of the Sun, the Moon, the major planets (Venus, Mars, Jupiter and Saturn) and the stars. How important this was to a number of civilisations in the Old as well as the New World is shown by the accuracies of their calendars and for example predictions and observations of eclipses. Its development has been attributed to the need for an agricultural society to keep track of the seasons, to indicate suitable times for sowing and harvesting. For many of the early seafaring peoples this knowledge must have been crucial for navigation. The sheer awe of a night sky, not hampered in visibility by today's air and light pollution, must also have played a role. A most impressive example of the early use of astrometry is shown by the complex of Central American civilisations. The accuracy of their calendar and calculations on the periodic appearances of the Moon and Venus are all well documented. Most of this information is available through written records, carved in stone by the Maya (see for example Drew 2000), but the traditions are likely to be older still.

What was measured, and how? The times and directions of rising and setting can be measured with relatively high accuracy, without the need of a telescope. For measuring these, details along the horizon are needed, and a fixed reference point for the observations. There are indications of such observatories among the Maya ruins (at Chichen Itza for example), and indications that at least some temples were explicitly positioned to serve as reference points for observations.

This technique of measuring positions relative to some reference position(s) has been crucial in all astrometry.

Mapping the sky and keeping track of the bodies of the solar system was also practised extensively in the Ancient Near East. The Greeks, after the conquest of Alexander the Great, built on that knowledge base and extended it. One invention in particular, the Astrolabe, enabled the measurement of large angles on the sky. The invention of the Astrolabe is attributed to Hipparchus of Biythynia (150 B.C.), who used it for what may have been the first large-scale and relatively accurate mapping of the sky. Here the reference 'point' is the horizon, and the Astrolabe provides a means of accurately measuring the altitude of an object above the horizon. A further developed version of the Astrolabe arrived in Spain in mediaeval Europe at the start of the second millennium, via the Moors. In the 14th century this amazing tool was widely used for the determination of local latitude and time from just a single measurement of the Sun or a bright star. The Astrolabe had become a very complex tool, as becomes clear from a contemporary description by Geoffrey Chaucer in 1391. It required 21 points to just explain the meaning and working of all its principal elements. His last point referred to a map of the fixed stars on the sky, incorporated in the "rete" (the movable disk) of the Astrolabe: this map of the sky allows the determination of local time at night as does the position of the Sun during the day. To measure time, the altitude of the Sun or one of the stars on the rete is measured, then knowing whether it is rising or setting, this can be set in the Astrolabe using a latitude-specific grid (for which various discs were incorporated) to determine the time of day. A modern description of the way the mediaeval Astrolabe worked can, for example, be found in Hoskin (1997).

A much simplified hand-held Astrolabe was used intensively for navigation by seafarers. A much larger form used by astronomers developed into the quadrant and was used, up to the end of the 18th century, for mapping the sky and time keeping. However, Astrolabes, and the quadrant which followed it, could not reach the precision needed for the ever increasing requirements of positional astronomy through later centuries.

A major driving force behind 16th to 19th century developments in astronomy was the need for accurate maps of the sky, to be used in navigation. King Charles II of England, recognising the strategic importance of good navigation tools, founded a Royal Observatory in Greenwich in 1675 for their further development. A combination of reliable clocks and an ephemeris is needed to determine the relative longitudes of different locations on Earth. The observatory, which after moving out of London in the early 1950s (to escape from the very poor sky conditions) changed its name to the Royal Greenwich Observatory (RGO), remained part of the Royal Navy until the 1960s. During its 323-year history the observatory has contributed greatly to the construction of accurate catalogues of stellar positions, and has stimulated the development of time-keeping instruments that are insensitive to the variable conditions on an ocean-going ship. It was closed down in 1998 by the UK government of Prime Minister Tony Blair following a recommendation by the Particle Physics and Astronomy Research Council. The time-keeping activities of the Royal Observatory at Greenwich led to the globally accepted concept of a reference longitude, the Greenwich Meridian, which was the reference line of the Airy Greenwich Meridian Telescope. Earlier transit telescopes all lay to the west. Greenwich Mean Time (GMT) still measures the position of the Greenwich Meridian with respect to the mean direction of the Sun. A mean position is used rather than an exact position, avoiding the modulating effects of the Earth's orbital ellipticity.

Time keeping used to be the equivalent of defining the rotation phase of the Earth with respect to the sky above, and was thus entirely Earth-bound (McCarthy 2005). Today this is still partly the case: the "Universal Time" used to set clocks of our daily life is still tuned to the rotation phase of the Earth (as we wouldn't want the time of mid-day to drift away from 12 noon). Since 1967, the length of the second is fixed as the duration of 9,192,631,770 periods of the radiation corresponding to the transition between two hyperfine levels of the ground state of the caesium 133 atom. This is equivalent to an earlier definition, which defined the second as the fraction 1/31,556,925.9747 of the tropical year for 1900 January 0 at 12 hours ephemeris time. Back in 1820 the length of the day was exactly 86,400 seconds, but since then it has increased by roughly 2 millisecond, as the Earth slowly loses angular momentum due to mainly tidal friction of the oceans (Stephenson and Morrison 1982; Stephenson 1997). This tidal friction effectively transfers angular momentum from the Earth rotation to the orbital motion of the Moon, causing an increase in the mean distance between the Earth and the Moon of about 1 cm per year. An Earth-bound clock is therefore not very good to describe for example the motions of the planets in our solar system. For this purpose we use "Dynamical Time", which over the years drifts away from Universal time. This drift is accounted for in the form of "Leap Seconds", a second of time added to the clock usually at midnight GMT on New Years Eve, but which can also be applied mid-year.

1.1.2 Mapping the Sky

The oldest catalogue of stellar positions we know of is the compilation made around 129 BC by Hipparchus, a catalogue that is still being investigated (Schaefer 2004). Its only surviving copy appears to be a map of the sky on a late Roman statue, and is known as the Farnese Atlas.

During the renaissance understanding the orbits of the Moon and planets required ever more accuracy for the stellar reference frame against which their motions were (and still are) measured. This reference frame was constructed from measurements of angular separations of stars using quadrants and similar instruments, and measurements of stellar transits using meridian telescopes. Measuring stellar transits through a meridian telescope served two purposes that were closely interlinked: establishing a reference frame, and measuring the Earth's rotation phase with respect to this reference frame. This duplicity in the use of the data, where the data provide their own reference frame, is a very common situation is astrometric studies, and one that needs to be understood well to appreciate the limitations of some of these studies. Another typical example is the measurement of differential proper motions from a collection of photographic plates of a small part of the sky. In this case the stars for which the proper motions are determined are the same as those used to calibrate the plates. As a result is not possible to determine the mean translation, rotation or expansion of these stars from the same measurements.

Photographic astronomy was introduced during the second half of the 19th century, stimulated amongst others by Gill's observations of star images on his pioneering photographs of the great comet of 1882 (Glass 1989). A major initiative from the Observatoire de Paris led to the Astrographic Catalogue project, in which observatories spread over the world collaborated to create an all-sky stellar catalogue down to about magnitude 12, based on photographic plates. This project was later followed by the "Carte du Ciel", providing a photographic sky survey down to about magnitude 14. The Carte du Ciel was never completed, but the astrographic catalogue was, and the photographic plates were re-measured in recent times. Those measurements were used in combination with the *Tycho* survey, which was part of the *Hipparcos* mission, to provide a stellar catalogue down to about magnitude 11, with well determined proper motions (Hoeg et al. 2000b,a). The positions, as measured on the standard 2×2 degrees fields (on photographic plates of 16×16 cm), were fixed on the reference frame provided by meridian-telescope measurements from all over the world. The main centre for accumulating these data has been, and still is, the Astronomisches Rechen Institut, currently in Heidelberg. The publications of the various versions of the Fundamental Katalog by this institute, served as the optical reference frame in astronomy until it was superseded by the *Hipparcos* Catalogue in 1997.

During the 1960s ways to automate the meridian-telescope readouts were being investigated by van Herk and van Woerkom (1961) and Hoeg (1968). Instead of an observer timing the passage of a star at a cross-wire, slit systems and photon counting devices were developed (see for example Hoeg 1970, 1972), initially for the Perth/Bergendorf Meridian Telescope. These developments culminated in the Carlsberg Meridian Telescope (Fig. 1.1), which operated as a Meridian telescope on the Island of La Palma over a period of 13 years (Helmer and Morrison 1985), for the final two years effectively as a robotic telescope (Evans 2001). It could be activated and checked by computer over a telephone line from any of the three institutes involved in these

Figure 1.1. The Carlsberg Meridian Telescope, which operated from 1984 till 1997 on the island of La Palma, formed the culmination of meridian telescope technology

projects: Copenhagen Observatory in Denmark, the RGO in the UK and San Fernando Observatory in Spain.

These days we can get more accurate information on the rotation phase of the Earth through the Global Positioning System or GPS, and maps of the sky are now fixed on the *Hipparcos* catalogue as published by ESA (1997). Meridian telescopes have therefore become largely obsolete. In recent years the Carlsberg Telescope has instead been used for the systematic mapping of the sky extending towards fainter magnitudes than reached by the *Hipparcos* mission (Evans et al. 2002; Evans 2003), using CCDs and an observational technique very similar to what is now being developed for the next generation of astrometric satellites.

1.1.3 Measuring Distances Through Parallaxes

Imagine we are standing on the side of a lake and observe a tall, lonely tree on the other side. Without being able to cross the lake, we can determine the distance of this tree, provided we can measure angles accurately. The principle is one of basic geometry: from our side of the lake we draw a line in the direction of the tree, to which we refer as the 'direction line'. Subsequently, we draw a line perpendicular to this direction, which we call the 'baseline'. The baseline is drawn such that it is cut in equal halves by the 'direction line'. We measure the length of the baseline and call it $2l$. In the middle of the baseline the direction of the tree will be perpendicular to it, but at the ends it will no longer be so. The difference from the perpendicular direction at those points is measured by the angle ϖ, which we call the parallax. The distance of the tree is now measured as $l/\tan\varpi$, without ever crossing the lake. When we estimate distances using our two eyes, we employ the same mechanism, as the brain registers the difference in orientation between the two eyes for which the images in both eyes coincide. The baseline for our eyes is about 60 mm. With a pixel spacing in the eyes of about 0.3 arcmin, distances may in principle be perceived up to several hundred meters away (effectively only through comparison with objects still further away), though only with low resolution.

How accurate we will measure the distance to the tree depends mostly on the *relative* accuracy of the angle ϖ. When the distance is large, the angle is small, and it will require a high *absolute* accuracy to obtain a good *relative* accuracy. Say, the distance to the tree is 10 km (if the tree is 50 m tall, it's image would be about half the size of that of the Moon). We use a baseline of 10 m for our experiment. The parallax angle then has an expected value of $\varpi = \arctan(10/10000) = 3.4$ arcmin, about one tenth of the apparent diameter of the Moon as seen from the Earth. To get the distance correct to about 10 per cent, the accuracy of the measurement needs to be about 20 arcsec, about equivalent to the thickness of a hair as seen from a distance of one meter.

The simplest way to measure such a small angle is with respect to a background object, much further away than the tree itself, for example a mountain in the far distance. Assuming the mountain is very much further away than the tree, the apparent displacement of the tree with respect to the mountain in the background as seen from two positions on the baseline can provide a measurement of ϖ. This is a relative rather than an absolute measurement, but as long as the mountain is far enough away, this will only give a small underestimate of the parallax, and a small overestimate of the distance. But, for example, if the mountain is at a distance of 50 km, the value measured relative to the mountain for ϖ is only 2.75 arcmin, giving the impression that the distance of the tree is nearly 12.5 km. Thus, the distance ratio of the mountain versus the tree needs to be quite large to obtain a reliable estimate of the tree's distance.

Distances in the solar system were determined during the 18th century in a very similar manner. From Kepler's laws the relative distances of the planets as measured from the Sun were known, but in order to transform these in absolute distances, the measurement of at least one distance in absolute terms was required. The opportunity to do so was provided by transits of the planet Venus in front of the Sun. Captain Cook's journey to the Pacific Ocean in 1769 was partly aimed at observing such an event (Orchiston 2005). By accurately timing the start and end of the transit seen from different positions on Earth, the actual distance to Venus could in principle be calculated. These, and parallax measurements for Mars and some minor planets, and later still radar measurements of Venus, gave the actual dimensions of planetary orbits in the solar system (see for example Standish 2005).

Once the scale of the solar system, and thus the radius of the Earth's orbit around the Sun was known, it became theoretically possible to measure the distances of nearby stars. The Earth-orbital radius provides a baseline generally referred to as the Astronomical Unit (AU) (1 AU $\approx 1.49 \cdot 10^{11}$ m). The background is provided by stars that are assumed to be at much larger distances than the object under investigation. Since the early 19th century, all ground-based measurements of stellar parallaxes have been performed by measuring the positional displacements of a target star, over a period of several years, with respect to a reference frame of background stars. A measurement over several years is needed because, unlike our tree, the stars are moving, and we need to be able to distinguish the parallax from the movements or proper motions of the stars. Furthermore, many observations are needed to reduce the error on the measured parallax.

As the typical size of the first measured parallaxes was just less than 1 arcsecond, it became the custom to measure parallaxes in arcseconds, and distances in parsec, where an object at a distance of 1 parsec shows a parallax of 1 arcsec. Expressed in the average radius of the Earth's orbit, 1 parsec or "pc" is equivalent to 206 264.8 AU or $3.07 \cdot 10^{16}$ m. It takes 3.4 years for light from an object at a distance of 1 pc to reach us.

Determining parallaxes thus depends on measuring angular distances between stars. A most ingenious way of measuring such angles, before photography was introduced in astronomy, was used by Bessel in the early 19th century in his determination for the parallax of the star 61 Cyg (HIP 104214). The star 61 Cyg was selected by Bessel on the basis of its large proper motion: he argued that a nearby star was likely to show a larger proper motion than a star much further away.

Bessel made his measurements using a heliometer with optics made by Fraunhofer (see for example Hirshfeld 2001). The heliometer is a refractor for which the objective lens cut in two halves (see Fig. 1.2). To measure the angular distance on the sky between two stars the cut-direction of the objective is aligned with the images of both stars. Each half of the objective produces images of both stars, and by shifting one half with respect to the other, using a micrometer, the image of one star can be made to coincide with an image of the other star. The shift applied to the objective provides an accurate measure

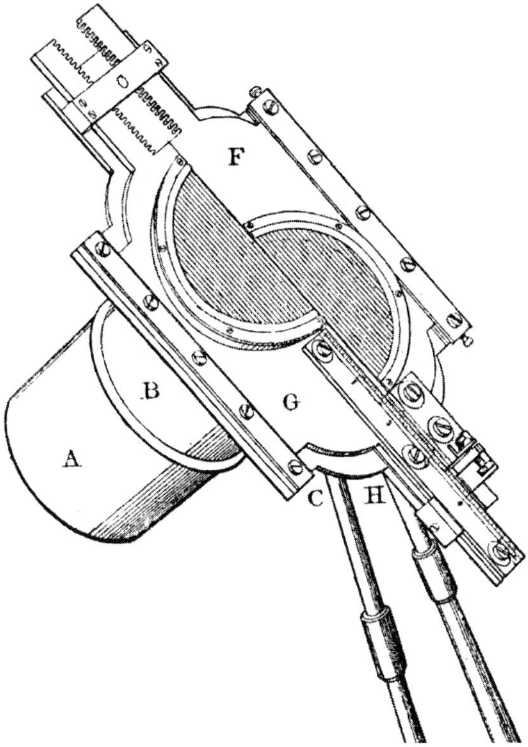

Figure 1.2. Bessel used a lens made and cut in two by Fraunhofer as part of a heliometer to very precisely measure the angular distances between two stars, by making the image of one star as seen through one half of the objective coincide with the image of the other star seen through the other half of the objective (from Ambronn, 1900)

of the angular distance between these stars as projected on the sky. Splitting an optical element to combine the light from two different directions was also a crucial step in the design of the *Hipparcos* satellite.

Through the years ever more sophisticated methods have been developed to correct for the residual parallax of the reference stars, so that the relative parallaxes have become nearly as reliable as absolute measurements (see for example Gatewood et al. 2000). The measurement of parallaxes using relative displacements in a small field on the sky (and this includes the Hubble Space Telescope) remains, however, a time consuming and very tedious procedure, for which the accuracies will never quite reflect the precision of the measurements (Jenkins 1952, 1963; Upgren 1977; Gliese and Upgren 1990; van Altena et al. 1995).

So how can we solve the small-field problem? The main problem within a small field on the sky is that the parallax-displacement direction for each object, as the Earth moves around the Sun, is the same, and only the amounts differ according to the distance of each individual object. This may be solved by using extra-galactic objects as reference points, but these tend to be faint and are in almost all cases extended rather than point sources, causing problems in defining an image centre. The problem can also be solved if we can measure large angles on the sky with very high precision. In the example of the tree, there would be no residual parallax for the background object, if that background object was situated along, rather than perpendicular to, our baseline. But in that case we need to measure an angle very close to 90 degrees with an accuracy of a few arcsec, a relative accuracy of 1 part in 55 thousand, which is much more difficult than measuring the relative accuracy of 1 part in 10. For the measurement of stellar parallaxes down to a milli arcsec (mas), the accuracy requirement for such angles would be 1 part in 300 million.

1.1.4 Space Astrometry

Space astrometry merged and extended the techniques that had been developed for ground-based astrometry over more than 200 years: measuring large angular distances between stars to triangularize the sky, using an optical element cut in two and combining the light on a single detector to make those measurements resemble those of small angular displacements, rotating the satellite instead of relying on the rotation of the Earth to measure transit times, and measuring these transit times by means of light modulation through slit systems. It took a while, however, before all these ideas were fitted together properly, and the first ever astrometric mission, the European Space Agency (ESA) *Hipparcos* satellite, became a reality, worthy of honouring through its name one of the great ancient astronomers. The name *Hipparcos* was a compromize between an acronym (HIgh-Precission Parallax COllecting Satellite) and the famous Greek astronomer Hipparchus.

1.1.4.1 Early developments and basic principles

During the 1960s ideas were developed to use the much more favourable conditions outside the Earth's atmosphere for astrometric observations (Lacroute 1982). These ideas evolved from a relatively small mission targeting a few thousand selected objects, to a proposal with the European Space Research Organisation (ESRO), which later became ESA, for a scanning satellite with two fields of view, separated by a fixed and very stable basic angle. This creates the possibility of measuring the angular distances between objects in parts of the sky that are sufficiently well separated to have very different parallax displacement directions, and to measure those distances to very high precision. Thus, it becomes possible for the first time to measure absolute rather than relative parallaxes. This requires, however, that the distribution of measured stars on the sky is more or less homogeneous (more in terms of integrated light than in actual numbers of stars), and the angle between the two viewing directions is stable to a level better than the accuracy to which the parallaxes are to be obtained. Such conditions can only be reached outside the Earth's atmosphere, and are best utilized by an instrument for which all the sky is visible.

Measurements from such instrument can provide a very rigid reference system of relative positions and proper motions. Through linking this frame with the absolute (Earth-bound) equatorial reference frame of radio sources, for which the large radio interferometers can obtain positions at the mas level, the relative positions and proper motions can be transformed to absolute (i.e. related to an Earth-bound reference frame) values. This reference frame is referred to as the optical realisation of the International Celestial Reference Frame (Arias et al. 1995; Kovalevsky et al. 1997; Hemenway et al. 1997; Feissel and Mignard 1998).

1.1.4.2 The double telescope

An astrometric satellite for measuring positional information for stars over the entire sky thus becomes in a way equivalent to a meridian telescope with two entrance pupils, the light of which is combined into a single optical path. Instead of the rotation of the Earth, it spins around an axis such that the two fields of view describe one and the same great circle on the sky. A modulating grid in the focal plane (Fig. 2.4) transforms the light of a transiting stellar image into a time-modulated signal, the phase of which provides the positional information.

The telescope design for *Hipparcos* is shown schematically in Fig.1.3. The light from two entrance pupils arrives at the beam combiner, which combines it to a single optical path. The beam combiner is a mirror, cut in two halves and glued together such that the angle between the optical axes for the two mirror halves was $\gamma/2$, where γ is the angle on the sky between a fiducial reference point in the focal plane, as projected through the two fields of view on the sky, and is referred to as the *basic angle*. The flight spare of this mirror can be seen at the National Maritime Museum at Greenwich. The beam combiner is part of an all-reflective Schmidt telescope with a focal length of 1400 mm. The "cut and glued" solution for the beam combiner was one of the measures to ensure a high stability of the basic angle, which was a crucial parameter for the success of the *Hipparcos* mission. Further stability was ensured by combining the optical paths of the light from the two fields of view, and by a tight thermal control of the payload. Furthermore, the telescope had to be protected from stray light to allow it to observe fields of view quite close to the Earth or the Moon. Long uninterrupted stretches of data have been an essential element in the construction of the *Hipparcos* catalogue, as these provide the most accurate calibration of the basic angle. The reduction of straylight also set a minimum for the angular separation of the scan circle and the direction of the Sun, and thus between

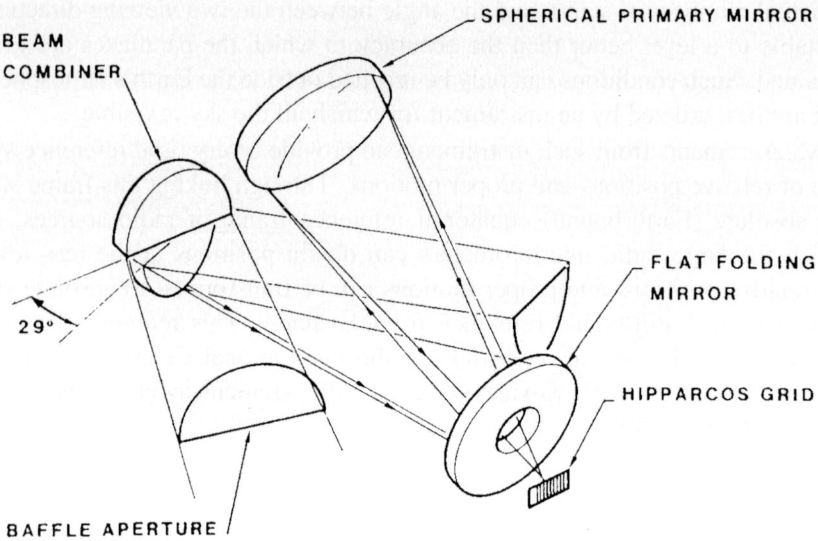

Figure 1.3. Configuration of the Schmidt telescope that formed the payload of the *Hipparcos* Satellite. Light entered from the two baffle directions, and was brought together at the beam combiner, which was configured as a Schmidt corrector. The combined light was reflected from the flat folding and spherical primary mirrors onto the focal surface where the modulating grids were located

the directions of the Sun and the satellite's spin axis. This angle is referred to as the solar aspect angle ξ, and was fixed at 43° for the *Hipparcos* mission.

The two areas of sky seen by the telescope are referred to as the Preceding Field of View (PFOV) and the Following Field of View (FFOV). The satellite was spinning such that most stars observed in the PFOV are seen about 20 minutes later in the FFOV, while many of these stars appear again 108 minutes later in the PFOV. The repeated observation of the same stars in the two Field of Views (FOVs) is important for the construction of an all-sky astrometric catalogue, allowing an intermediate step of linking of positions on a great circle (and making it possible to calibrate the basic angle) before linking data all over the sky.

1.1.4.3 *The basic angle choice*

The basic angle for the *Hipparcos* telescope is 58 degrees. Why was this value and not any other value chosen? It is mainly a question of resonances. What has to be avoided is a situation where an integer, multiplied by the basic angle, is (almost) equal to the time interval, times the mean rotation velocity, between any two field-of-view transits of a star in a 12-hour period of

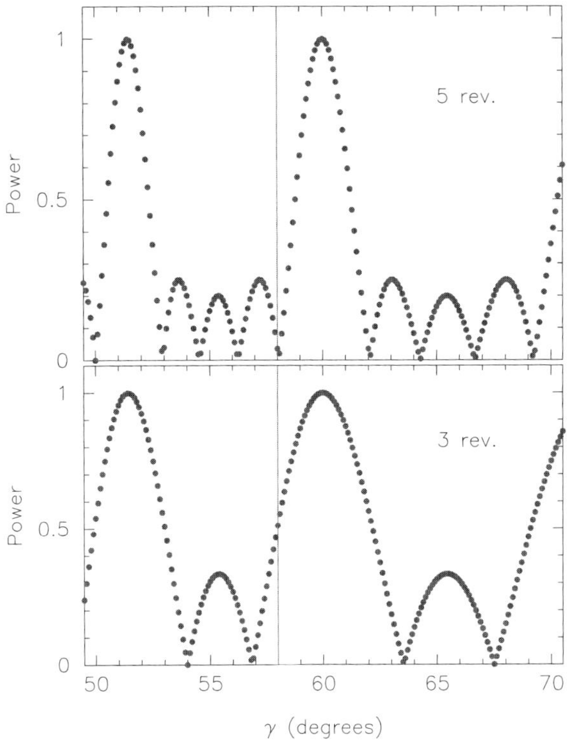

Figure 1.4. Resonance tests on basic angle values when combining data for 5 rotations (top) and for 3 rotations (bottom) of the satellite. The basic angle value was optimized at 58 degrees for the nominal mission, where on average units of 5 rotations of the satellite would have been used in the accumulation of the data. Instead, average unit lengths were more like 3 to 4 rotations

observations. The three-fold symmetry in the overall design of the satellite required in addition, an angle close to 60 degrees.

If a star is observed in, say, the preceding field of view, it may again be observed after the telescope has rotated by γ, 2π, $2\pi + \gamma$ etc. A resonance is reached when $n\gamma$ fits exactly between any two of these observations, where n has to be an integer number larger than one. The larger value we find for n, the lower is the sensitivity to resonances. For example, for $\gamma = 60^\circ$, we find $n = 3$, and resonances are very likely. For $\gamma = 58^\circ$, we find $n = 90$, and resonances are very unlikely. In a more detailed evaluation, the actual width of the field of view needs to be accounted for too in these calculations.

The possibility of resonances has been particularly important for the so-called great-circle reduction used in the preparation of the published *Hipparcos* catalogue (van Leeuwen 1997). That type of solution, which was necessitated by the limited accuracies of the *a priori* positions in the *Hipparcos* Input Catalogue (Turon et al. 1992; Jahreiss et al. 1992), provided a simultaneous fitting of the along scan attitude (the scan phase of the satellite, Section 2.3) and corrections to the catalogue positions. A resonance as described above may create a situation where the measurements of a star are linked to only a small subset of other stars rather than with all other measurements incorporated in the solution. This subset can then create its "own" reference phase, which is only weakly linked to reference phases for other subsets of measurements.

The basic angle value was therefore optimized to collect, on average, data from 5 rotations of the satellite for each great-circle reduction. A total of 5 rotations was chosen to provide sufficient average number of measurements for each star (to be able to separate the two contributions to the differences between the observed and predicted along-scan coordinates), while the interval covered had to be short enough to allow a projection on a single reference great circle.

Due to a problem with the satellite's orbit (see Section 1.2.3), observations were interrupted every 10.7 hours by a 1 to 3 hour gap, when the satellite went through perigee. These gaps naturally divided the data into data sets, referred to by the sequentially counted orbits of the satellite (see Chapter 11 for more details). The maximum number of rotations that could be incorporated in a great-circle solution for the revised mission was therefore only just over 4, while the actual number available was often smaller. For small numbers of rotations the angle of 58°is a little too close to 60°(Fig. 1.4). This has led to the spontaneous appearance of 6th order harmonics in the along-scan attitude solution for short data sets, and thus in the abscissa residuals, which are measured relative to the reconstructed attitude. This could occur for data sets with less than two full rotations or data sets with large gaps. These harmonics were filtered out later in the sphere solution, although one orbit has slipped through the net: orbit 59 as processed by the FAST consortium (Fig. 1.5).

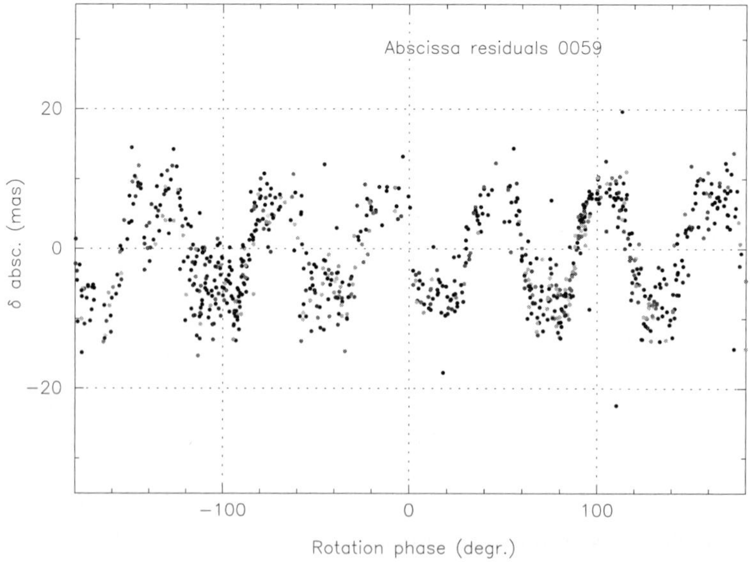

Figure 1.5. The only data set in the published data for which the 6th harmonic resonance in the abscissa residuals had not been removed: orbit 59 as processed by FAST

Also, for very short data sets the 12th harmonic could be in resonance (van Leeuwen 2005a), but these data sets were generally given such low weight that they didn't contribute significantly to the published catalogue.

1.2 The mission

1.2.1 The Spacecraft

For a telescope to be able to operate in space, it requires a supporting structure, which is referred to as the spacecraft. In a scanning satellite the spacecraft occupies the lower platform of the satellite, and the telescope or payload the upper platform. The spacecraft provides all the hardware to "manage" the mission, for example: computers for regulating and recording the observations and the communications with the ground station, rate-integrating gyroscopes which assist in the attitude control, cold-gas tanks and piping to thrusters to keep the scan on course. The spacecraft also takes care of the power supplies to all the equipment on board the satellite: regulators for the power from the solar panels and batteries for times of solar eclipse. Everything happening on-board the satellite (spacecraft and payload) is ultimately regulated by signals from the on-board clock, which is also situated in the spacecraft.

A major part of the spacecraft is occupied by the Apogee Boost Motor (ABM) and its solid-fuel tank. As the motor failed to fire (see below), the tank remained full during the mission. The result of this was to increase in the mass of the satellite and shift its centre of gravity during its operational phase. This shift can be observed in the calibration of the cold-gas thrusters used to control the rotation rate around the spin-axis. These thrusters are situated in the plane which was supposed to contain the centre of gravity for the operational phase. If this were the case, then activation of those thrusters would only affect the rotation around the spin axis. However, with a shift of the centre of mass, the activation of these thrusters caused additional small torques on the other two axes. Calibrations of these torques showed that the centre of gravity of *Hipparcos* was situated 133 mm below its intended position (See Section 13.3 for more detail).

1.2.2 Launch and Early Orbit Phases

Hipparcos was launched on 8 August 1989 by flight 21 of Ariane 4 from Kourou in French Guyana. The launch vehicle put the satellite in a geostationary transfer orbit, from where it had to inject itself into a geostationary position using the ABM. This would provide continuous contact with the satellite from the Odenwald ground station. Unfortunately, things didn't go quite according to plan, and the ABM refused to fire. This left the satellite stranded in the transfer orbit, requiring three ground stations to maintain contact.

When it became clear that the ABM would not fire, the perigee height of the geostationary transfer orbit was increased from around 200 km to an average of 500 km above the Earth's surface. The thrusters and fuel intended for station keeping were used to increase the velocity at apogee. The final orbital characteristics were chosen so that approximately 5 rotations of the satellite would fit in one orbit, and 9 orbits fit in 4 days. Both aspects are important in the interpretation of the satellite dynamics (Part IV).

1.2.3 The Operational Orbit

The scientific aims of the mission put stringent requirements on the accuracy with which the orbital parameters had to be known during observations. A positional accuracy requirement of 1.5 km ensured that no significant noise would be added to the interpretation of observations of solar-system objects, in particular minor planets. To keep the noise on the aberration corrections below 0.1 mas, the velocity vector had to be known to better than $0.2 \cdot 10^{-3}$ km s^{-1}. Thus, the orbital parameters and their evolution over the mission are known to a high level of accuracy, generally to about 3 parts in 10^5 (Table 1.1).

Semi-periodic variations in the orbital parameters are caused by the gravitational fields of the Sun and the Moon, while the long-term evolution is mainly the result of loss of orbital energy during perigee passages, and can be linked directly to the varying height of perigee during the mission. The variations in perigee height are the result of changes in the orientation of the major axis of the orbit with respect to the direction of the Sun: when perigee is on the mid-day side of the Earth, the perigee height is lowest (450 to 460 km). At the mid-night side it is slightly higher (480 to 490 km), and it is highest when the orientation of the major axis of the orbit is perpendicular to the direction of the Sun as seen from Earth (570 to 580 km).

The observed orbital energy is the sum of the two-body energy h and the potential $V(\boldsymbol{r},t)$ (Bond and Allman 1996):

$$E_{\mathrm{T}} = h + V(\boldsymbol{r},t), \tag{1.1}$$

and can for example be directly observed in the orbital period of the satellite. After accounting for all conservative forces acting on the satellite, the

Table 1.1. The main orbital parameters and their evolution

Element	Mean value	Change/yr	Variation	Units
Period	38340	-20.4	±4.5	s
Semi-major axis	24582	-8.2	±2.5	km
Eccentricity	0.7196	0.000	±0.005	-
Perigee	6890	-3.0	±140	km
Precession	0.37		±0.07	°/day

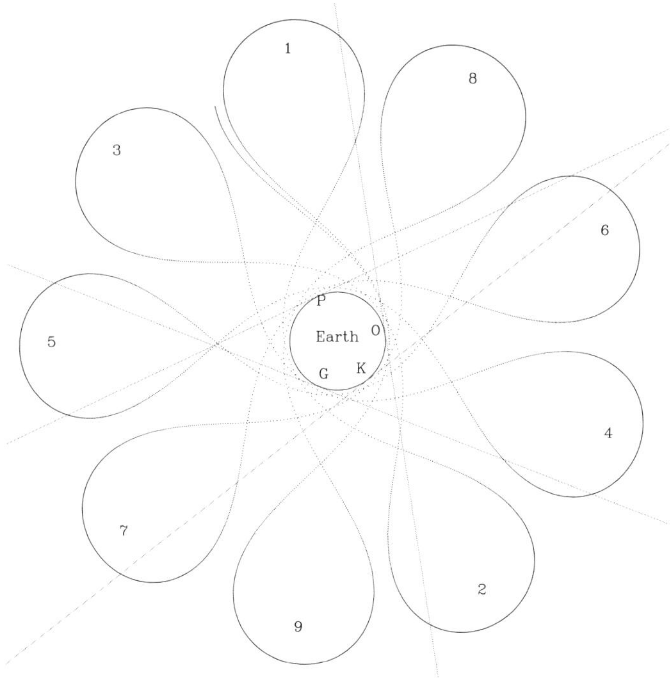

Figure 1.6. The position of the satellite in the final orbit plotted relative to a stationary Earth. Points in the orbit are drawn at equal time intervals. The just over 9 successive orbits shown are numbered. Orbit 10 almost coincides with orbit 1

remaining loss in orbital energy provides information on the non-conservative forces, which are, during perigee passages, dominated by the friction in the outer layers of the Earth's atmosphere. The loss of energy is largest during periods of low perigee height, and in particular when the perigee passage takes place on the mid-day side of the Earth. These calculations, shown in Fig. 1.7, ultimately provide some information on models of the outer atmosphere, as was shown by Dalla Torre and van Leeuwen (2003).

1.2.4 Communications and Ground Stations

The on-board computing power and memory was very limited. Observations could only be made while the satellite was in contact with a ground station. Scientific, spacecraft and housekeeping data, were relayed to the ground station where it was time-tagged. This process required the ground station to lock onto a one-second block signal from the satellite in order to synchronize with the on-board clock (Section 13.1). This did not always work successfully, and there are a few cases where a one-second error is made, and one case with a two-second error. These errors, as well as a number of inconsistencies

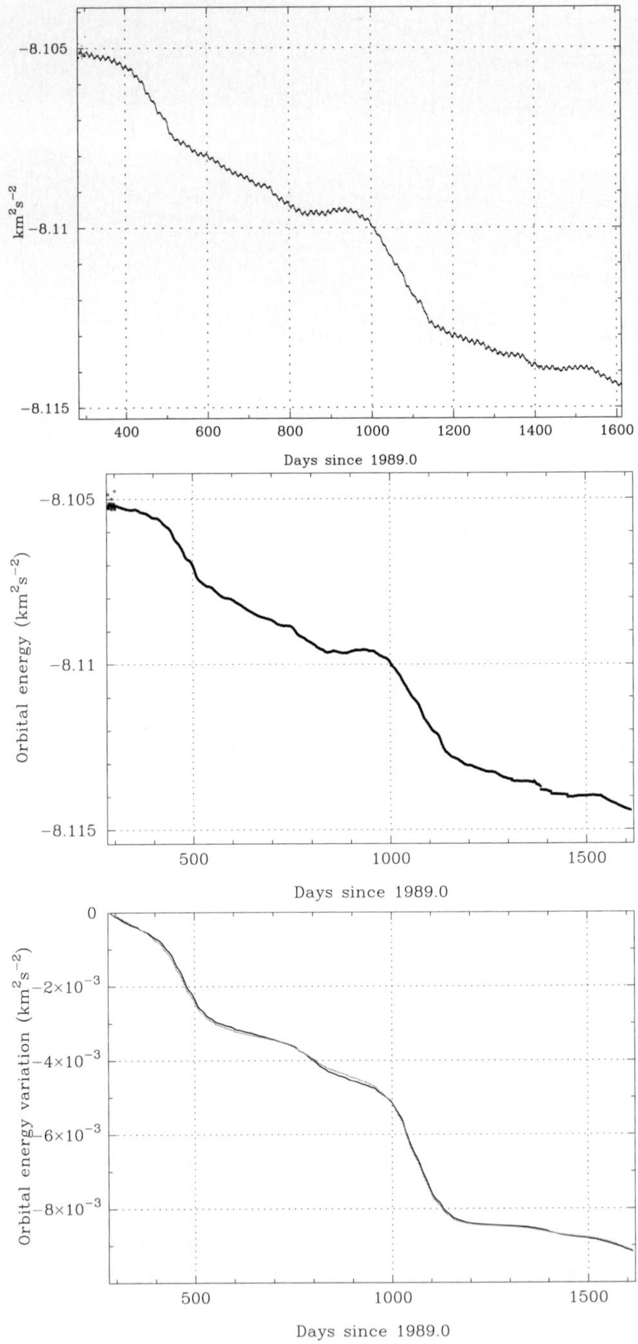

Figure 1.7. The orbital energy of *Hipparcos* as measured and reconstructed over the mission. Top graph: the total orbital energy. The little wobbles refer to the gravitational attraction of the Moon. Middle graph: after removal of the effects of varying gravitational potentials. Bottom graph: Predicted loss based on atmosphere models, with satellite parameters adjusted to obtain the closest fit with the observed curve

in the ground-station delay times (the time between the receiver end and the time-tagging electronics), are now all corrected in the data (van Leeuwen and Penston 2003). The study of the behaviour of the on-board clock provided the most detailed information on the thermal environment for the spacecraft (see Section 13.1).

The mission was served from ground stations in Odenwald and Perth (entire mission), Kourou (start and end), and Goldstone (from April 1990 till end). Only the configuration of Odenwald, Perth and Goldstone provided a nearly complete visibility of the satellite, with the exception of perigee passages, (Fig. 1.6). Only occasionally Kourou was able to observe the satellite around perigee passage. At regular intervals, the ground station sent the satellite a list of stars (for the epoch of observation and corrected for aberration) that could be observed over a period of about 20 minutes. In some cases only bright stars were selected, which allowed for a longer time span to be covered. It is still not always clear why this was done.

1.2.5 On-board Time Units

The data collected and sent down by the spacecraft was organized in fixed-length units of nominally 32/3 s, referred to as telemetry formats. All observational data was synchronized with these telemetry formats, which were sub-divided into 256 telemetry frames. One telemetry frame covered 50 sampling periods (at 1200 Hz) for the main detector, the Image Dissector Tube (IDT), and 25 sampling periods for the Star Mapper (600 Hz) detectors. The IDT data were collected and processed in units of 2560 samples, referred to as an observing frame (32/15 s). A telemetry format covers exactly five observing frames. Extracts of star mapper data, as used in the attitude reconstruction, covered 10 telemetry frames, or 250 star mapper samples. One rotation of the satellite covered nominally 720 telemetry formats or 3600 observing frames. An orbit covered nominally five rotations of the satellite or 18000 observing frames. All telemetry and observing was regulated and tagged by the on-board clock, and observations were sent to the ground station almost immediately after being obtained.

1.2.6 The Scanning Law or Nominal Attitude

The *Hipparcos* survey was defined by a scanning law, which describes the *a priori* coverage of the sky during the mission. A scanning law is a compromise between several requirements and constraints. Coverage should in principle be homogeneous, but that is very difficult to achieve when scanning the sky along great circles. Thermal stability of the spacecraft required a fixed *solar aspect angle*, the angle ξ between the spin axis and the direction to the Sun (see Fig. 1.8). This condition is fulfilled by letting the spin-axis precess around the

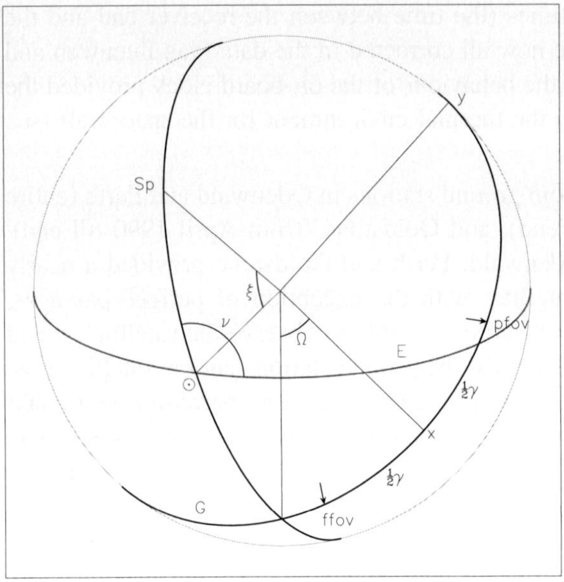

Figure 1.8. The heliotropic angles (ν, ξ, and Ω), used to describe the satellite's nominal attitude, and their relation to the satellite axes (**x**, **y**, **z**) (the **z** axis coincides with the spin axis, indicated by "Sp"). The plane of the ecliptic is indicated by 'E', the scanning circle by 'G'. The two fields of view (PFOV and FFOV) are positioned at $\pm\gamma/2$ (half the basic angle) from the x axis

direction of the Sun. For the duration of the *Hipparcos* mission, the nominal ecliptic longitude of the Sun is described by:

$$\begin{aligned}\lambda_\odot &= L + 2e\sin g + 1.2e^2\sin 2g,\\ L &\equiv -1.38691 + 0.0172021240d,\\ g &\equiv -0.04144 + 0.0172019696d,\\ e &\equiv +0.016714,\end{aligned}\tag{1.2}$$

where d is measured in mean solar days from the reference epoch used in the description of the nominal attitude: 1988 January 1, $12^h00^m00^s$ UT, and e is the ellipticity of the Earth's orbit. All angles are measured in radians.

A further requirement is that the scan directions across the sky vary sufficiently to ultimately determine two-dimensional positions for all selected stars. This condition becomes better fulfilled when ξ gets larger. For small values of ξ, stars positioned near the ecliptic plane will only be resolved in ecliptic latitude, as all scans tend to be nearly perpendicular to the ecliptic plane. However, ξ can't be chosen too large either, as it would bring the scan circle too close to the direction of the Sun (creating unacceptable levels of stray light), and it would be difficult and inefficient to employ the solar panels for energy provision. A value of $\xi = 43°$ was chosen for *Hipparcos* as a reasonable compromise. As a result there are no transits in a pure East-West (or opposite) direction for stars with ecliptic latitudes between $\pm 47°$.

The next requirement states that in every 6 months, (Fig. 1.9) period stars in any part of the sky should be scanned in at least two different directions.

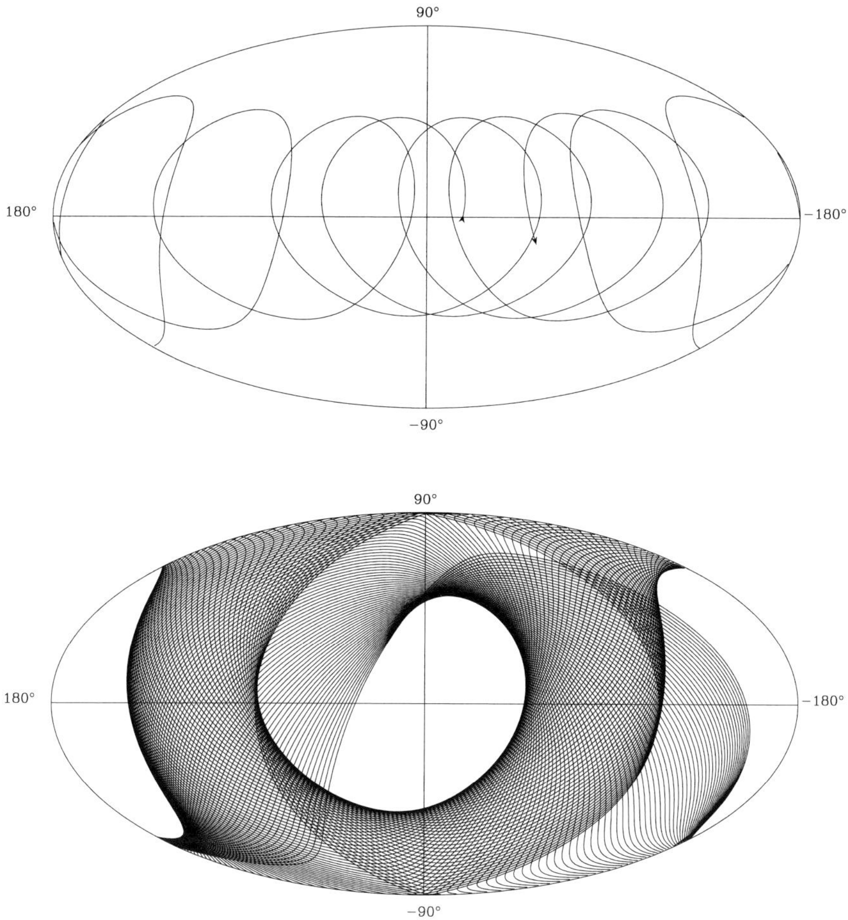

Figure 1.9. The scanning motion of the *Hipparcos* satellite on the celestial sphere, shown in ecliptic coordinates. The top figure shows the path of the spin axis between 22 May 1990 and 23 September 1991. The scan direction is indicated by the arrows. The bottom figure shows one reference great circle for each orbit (5 rotations of the satellite) for the period between 22 May and 14 July 1990. The actual scanning was 5 times denser

Thus, at least one 2D reference point is obtained for each star every 6 months, (Fig. 1.9). This sets the rate of change for the precession phase $\bar{\nu}$ of the spin axis position. An average of 3 to 4 cycles of $\bar{\nu}$ in half a year fulfil the requirement, and for *Hipparcos*:

$$\bar{\nu} = \bar{\nu}_0 + K\lambda_\odot, \tag{1.3}$$

where $K \equiv 6.4$. Further complications are caused by the precessional motion of the spin axis, which, if left uncorrected, would cause a significant difference in the effective scan speed depending on whether the spin axis moves in the same direction, or in opposite direction from the Sun. If this were to reflect in the scanning of the sky, the scan density would show significant and undesired systematic variations. This is compensated by describing the actual precession phase ν as a modulated signal of $\bar{\nu}$, such that the first time derivative of ν is constant:

$$\nu = \bar{\nu} + a_1 \cos(\bar{\nu}) + a_2 \sin(2\bar{\nu}) + a_3 \cos(3\bar{\nu}) + a_4 \sin(4\bar{\nu}), \tag{1.4}$$

with the following values for the coefficients:

$$\begin{aligned} a_1 &= -0.16378459 \\ a_2 &= -0.01307777 \\ a_3 &= +0.00123243 \\ a_4 &= +0.00012341. \end{aligned} \tag{1.5}$$

This modulation, however, leads to variations in the actual scanning speed on the sky if the scan velocity of the satellite is kept constant. This was not allowed, and the evolution of the scan phase Ω is modulated to ensure that the actual velocity on the sky is constant:

$$\Omega(t) = \Omega_0 + 2\pi R(t - t_0) - \nu \cos\xi + (b_1\bar{\nu} + b_2 \cos(\bar{\nu}) + b_3 \sin(\bar{\nu}))\frac{\sin\xi}{K}, \tag{1.6}$$

where $R = 11.25$ rotations per day, and the coefficients have the following values:

$$\begin{aligned} b_1 &= 0.08215269 \\ b_2 &= 0.99006117 \\ b_3 &= 0.04045213, \end{aligned} \tag{1.7}$$

where t is a time unit of the on-board clock.

The zero point of the scan phase, Ω_0, was kept constant over most of the first two years of the mission, but later it was adjusted on an almost daily basis to lose less time in the convergence of the on-board attitude control: instead of bringing the satellite to where it was supposed to be, this scanning-law parameter was adjusted to where the satellite was actually pointing at. During the final few months of the mission this was also applied to the zero point of the precession phase, ν_0. The positive effect of the adjustments on the amount of observing time available, more than compensated the very small negative effect they had on the sky coverage.

The actual attitude of the satellite is described by a '1-2-3' set of *Cardan* or *Tait-Bryant* angles with respect to the nominal attitude, the details of which are explained in Appendix A.

1.2.7 Scanning Anomalies

At a few instances during the mission the scanning law had to be abandoned, either for a few days to observe in Sun-pointing mode (with the spin axis pointing in the direction of the Sun), or spun-up in save-keeping mode, when no observations took place. Any data obtained in Sun-pointing mode contains no information on the stellar parallaxes, because the great circle described by the two FOVs is perpendicular to the direction to the Sun. Thus, in that case all parallax displacements are perpendicular to the scan circle and do not affect the stellar transit times (see also Chapter 2). The most serious of these events is a gap of nearly 2 months at the end of 1992 when no observations took place due to problems with two of the gyros. At that time a strategy was being developed to make it possible to observe with only two active gyros. This strategy operated during the last four months of the mission, until the breakdown of the on-board computer in May 1993 brought to an end all further observations.

1.3 The published data

The results of the *Hipparcos* mission were published in 1997 in the form of 17 Volumes (ESA 1997), providing details of the data products, the mission, the data analysis for *Hipparcos* and *Tycho*, the astrometric catalogues (5 volumes), light curves and tables of variable stars, maps and details of double stars, cross references with other identifiers etc. This book can in no way, and does not intend to, replace that publication. What it will do is provide a critical look at some aspects of the data analysis, and the results of the new reduction, which may be regarded as a final set of iterations through the *Hipparcos* data. At various occasions the published data will be referenced, and this section provides a brief summary of the activities that led to those data. A more extensive review is presented by van Leeuwen (1997) and a detailed description of all the original data products is presented in Volume 1 of ESA (1997).

1.3.1 The Two Data-reduction Consortia

The *Hipparcos* data reductions were done by two independent consortia, the Fundmental Astrometry with Space Techniques (FAST) (Kovalevsky et al. 1992) and the Northern Data Analysis Consortium (NDAC) (Lindegren et al. 1992). The reason for this approach was to provide some quality assurance, as there is insufficient independent data to do an overall verification of the quality of the *Hipparcos* astrometric data. This was also the reason why the mission took place! Independent approaches by two groups should bring to light problems in the data analysis at an early stage through comparisons of intermediate results. Such comparisons were carried out at several stages of the data analysis, as well as using simulated data before the mission. Although a few minor problems were discovered and taken care off, the overall results were satisfactory, and consistently indicated a higher than expected level of accuracy in the data. This was confirmed by the final data quality verifications (see below). Comparisons were carried out on the following (intermediate) data products (with the person responsible in brackets):

- Star mapper reductions (van Leeuwen, NDAC)
- Attitude reconstruction (Donati, FAST)
- Main detector photon-count analysis (Perryman, ESA)
- Great-circle reduction (van der Marel, FAST)
- Sphere reconstruction (Lindegren, NDAC)
- Photometry, main mission (Evans, NDAC)
- Astrometric parameters (Arenou, INCA)

For the first 4 items only a relatively small amount of data was compared, while for the last three items all the data was used. Comparison results were presented and discussed in the *Hipparcos* Science Team, and any significant discrepancies were identified and had to be explained and corrected. Despite this rigorous control, some of the main processes of the data reduction remained fully independent simply because of the different techniques employed. This was most clearly the case in the attitude reconstruction, but also applied to details of the great-circle reduction and the sphere reconstruction.

1.3.2 The Basic Data

The basic data produced by the *Hipparcos* mission are transit times, which, through the reconstruction of the along-scan attitude, are translated to transit positions: one-dimensional positional measurements on the sky. These measurements were compared with predicted positions, based on the best available astrometric data for each object, and the differences (observed minus predicted) are referred to as the abscissa residuals. These residuals potentially contain information on corrections required to the astrometric parameters and corrections needed for the along-scan attitude.

1.3.3 The Three-step Reduction

One of the major obstacles in getting *Hipparcos* accepted as a mission was the ability to analyse the data. Ideally one would like to do a global solution, in which the along-scan attitude and astrometric parameters for the whole mission are solved together. This was well out of reach for 1980s, or even 1990s computing power. The solution to this problem was provided by L. Lindegren. It splits the reduction in three discrete steps:

1. the "great-circle reduction", which reduces data obtained over up to 5 rotations of the satellite to a single great circle with abscissae;
2. the "sphere reconstruction", which combines the abscissae from the great circles in a single common solution;
3. the "astrometric parameter solution", which analyses the abscissae on the sphere to provide the 5 astrometric parameters for all observed single stars.

In practice there was a lot of processing to be done before the data was ready for step 1, and steps 2 and 3 were in the end combined in one solution. It is important to notice some features of the great-circle reduction that have a bearing on the new reduction presented here.

In the great-circle reductions three parameter sets were solved simultaneously: the along-scan attitude, the instrument parameters and the abscissa corrections. The reason was the noise on the abscissa residuals, most of which was due to poor *a priori* knowledge of positions. This noise was therefore non-white, in particular due to many observations of the same stars that could contribute to the solution. The great-circle reduction provided single, combined abscissae for all observations of each individual star on a reference great circle. This implied that the actual observations had to be projected on this reference circle, using the reconstruction of the satellite's attitude from the star mapper data. The error level for the reconstructed attitude was generally low enough for this not to be a problem, but at times of high activity in the Van Allen belts (causing high to very high background readings in the Star Mapper detectors) and at times of penumbra phases of solar eclipses, the errors are likely to have been larger than the strict requirement, adding noise to the abscissae. Due to the use of the abscissae in both the along-scan attitude and the abscissa-correction determinations only stars with multiple transits of the field of view could contribute safely to the solution.

1.3.4 Merging the Catalogues

The two catalogues produced by the FAST and NDAC reduction chains were investigated individually and compared at the Meudon observatory by Frédéric Arenou. This investigation focused largely on formal errors and correlations between the underlying abscissa measurements produced by the two groups. Comparisons revealed relatively high correlation levels for faint stars. This would be expected as the main noise contribution for those measurements is the photon noise, which is the same in both reductions. Correlations decreased towards brighter stars, as here the attitude-fitting noise became dominant. Thus, for brighter stars the combination of the two reduction results should improve the accuracy. This was shown to indeed be the case from a histogram of observed parallax values. In the combined solution, the peak in the histogram is higher than in either the FAST or the NDAC solutions, and the tail of negative parallaxes is slightly smaller.

The normalisation of the formal errors per reduction chain was not straightforward. The transit data were obtained from a combination of all frame transits over an orbit, with a formal error estimated in the great-circle reduction (Fig. 1.10). These error estimates did not contain the attitude errors, or information on the integrated intensity of the transit, given by the total photon count of the observation. An empirical fit of the formal errors to the observed noise levels was applied in the normalisation, which was hiding in some cases poor solutions behind large assigned errors (see also van Leeuwen 2005a). The merging of the catalogues is described in some detail in Chapter 17 of Volume 3 of (ESA 1997).

The published data are obtained from the abscissa measurements of both reduction chains, with weights and correlation coefficients as determined in the comparison process. The abscissa residuals, with respect to the accepted solution are presented in the intermediate astrometric data file on one of the CDROMs contained in Volume 17 of ESA (1997). Both accepted and rejected data are included (and flagged accordingly), as well as any other information required to check or change the accepted solution. How to use those data is described by van Leeuwen and Evans (1998).

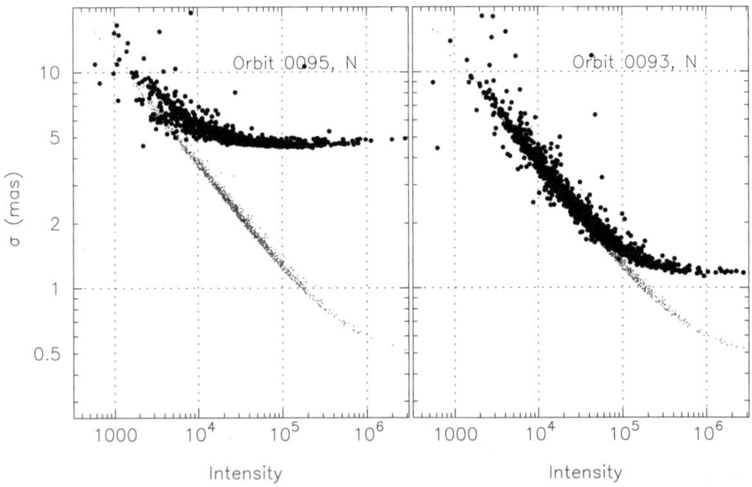

Figure 1.10. Two examples of formal errors on abscissae as assigned in the final merging of the FAST and NDAC data (large dots). On the left, an orbit with large systematic errors left in the abscissa residuals, on the right an orbit which performed better in the reductions. Both graphs show in the background, as small dots, the formal errors that were assigned to data from these same orbits in the new reduction

1.3.5 Data Quality Verification

The verification of the accuracy and lack of systematic errors in the *Hipparcos* astrometric data is non trivial. Two studies on the global accuracy levels were performed: Lindegren (1995) examined the distributions of parallaxes observed in bins of different formal errors, using the Richardson-Lucy deconvolution algorithm and a fitting procedure to derive simultaneously the distribution of true parallaxes and external errors (see further Section 3.1.2). The ratio of external over internal error thus found is 0.992 ± 0.019, in other words, no significant under or over estimate of the internal errors was detected. Much of this study relies on the presence and distribution of small and negative parallaxes.

In a second study, Arenou et al. (1995) examined the sample for a zero point in the parallaxes as well as the reliability of the formal errors. Parallaxes were compared with several externally determined values, in particular for objects of which the expected parallax value would be very close to zero as observed by *Hipparcos*. The observed parallaxes for a sample of 700 distant stars, with parallax errors below 2 mas, were investigated as functions of position on the sky, proper motion and magnitude. Only marginally significant trends were found, in particular with respect to magnitude and proper motion in ecliptic longitude. Overall the internal errors were found to be within 5 per cent of the external errors. The global zero point in the parallaxes was found to be smaller than 0.1 mas.

With the results of these global studies in mind it becomes very difficult to assume the widespread presence of systematic errors in the *Hipparcos* data: such errors would have contributed to the observed external errors, and would cause the external errors to be systematically higher than the internal errors, which is contradicted by the two studies mentioned above. The results of these studies also effectively exclude the introduction of systematic errors in the data that result from the instrument itself. One type of error in particular, a systematic modulation of the basic angle with the rotation of the satellite, would reflect in a global zero-point offset for the parallaxes, contrary to what is observed, as explained in Section 1.3.6.

1.3.6 Basic-angle Modulation and Parallax Zero Point

While the satellite rotates, different parts of it are exposed to solar radiation. A temperature fluctuation develops, which, in terms of satellite coordinates, is spin-synchronous, i.e. periodic with the rotation period of the satellite. These periodic temperature variations can be observed in the drift of the on-board clock, situated in the spacecraft. The payload, however, was subject to very tight thermal control, and such effects are not expected to affect the instrument. The low limit observed for the parallax zero point confirms this, and the reason for this is as follows.

A modulation of the basic angle by $\delta h_0 = A{\cdot}\cos\Omega_H$ (where h_0 is defined as the correction to half the basic angle and Ω_H the rotation phase of the satellite's x axis) causes different offsets in the abscissae $\mathrm{d}a$ of a star as observed in the preceding or following field of view:

$$\begin{aligned} \mathrm{d}a_p &= A \cdot \cos(\Omega_* + \gamma/2), \\ \mathrm{d}a_f &= -A \cdot \cos(\Omega_* - \gamma/2), \end{aligned} \tag{1.8}$$

where $\gamma/2$ is half the basic angle between the two fields of view ($\gamma/2 \approx 29$ degrees). The combined, average effect from two different stars observed simultaneously in the two fields of view is:

$$\mathrm{d}a = -A \cdot \sin\Omega_* \sin(\gamma/2). \tag{1.9}$$

The parallax factor for a star is given by (see Chapter 2):

$$p = -\sin\xi \sin\Omega_*, \tag{1.10}$$

where ξ is the solar aspect angle of 43 degrees.

Thus, a modulation of the basic angle as described above would on average produce a systematic offset on the measured parallaxes: the parallax contributions for all stars would be offset by a fixed amount of $\varpi_0 = -A\sin(\gamma/2)/\sin\xi$. A more detailed calculation, incorporating the effect the modulation has on the along-scan attitude reconstruction, shows that the full equation for the parallax zero point under these conditions is given by van Leeuwen (2005a) as:

$$\varpi_0 = A\frac{\sin(\gamma/2)}{(1 + (\cos(\gamma/2))^2)\sin\xi} \approx 0.40A. \tag{1.11}$$

With the observed systematic offset smaller than 0.1 mas (Arenou et al. 1995), an upper limit of $|A| < 0.25$ mas is derived. Any other modulation has no systematic effect on the parallax zero point, and contributes mainly noise to the abscissa measurements.

1.3.7 Possible Local Systematics

The methods described above deal with the catalogue data on a global basis and at accuracy levels representative for individual stars. There is, however, a possibility that at local level systematic errors did enter the catalogue. To understand how this may have happened, we have to look again at the two basic conditions required for constructing absolute parallaxes from the *Hipparcos* measurements:

1 The basic angle needs to be constant preferably to within a range of a few tenths of a mas over a period of 10 hours, and in particular no modulation with the rotation period of the satellite is allowed (Fig. 1.11);

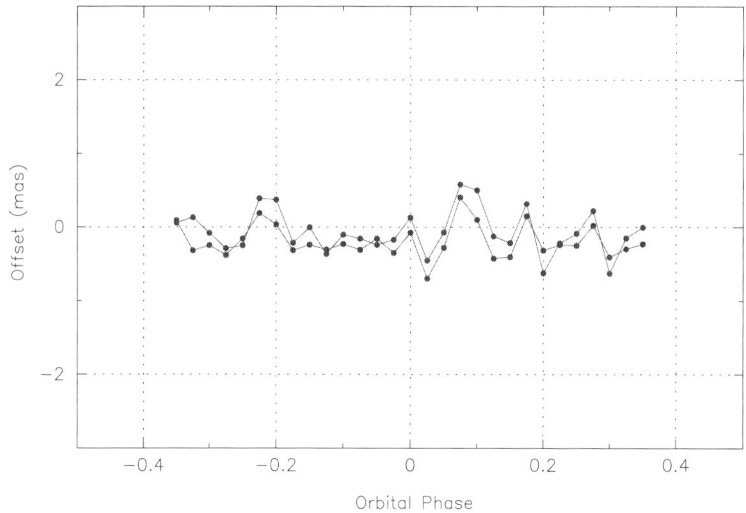

Figure 1.11. Small variations in the observed basic angle over orbit 505. Such variations are not corrected, as this may have led to instabilities in the construction of the catalogue

2 The reconstruction of the along-scan rotation phase has to be based in more or less equal amounts on contributions from both fields of view.

The first condition is found to be well fulfilled for all but 16 (0.7 per cent) of the orbits (see Table 11.3, page 290). In nearly all cases the cause of the drift was identified as a thermal-control anomaly of the payload. Only a few of those orbits were recognized in the reductions of the published data. The second condition is far more difficult, and may get violated for the cores of rich, young open clusters. These systems can cause a violation of condition (2) by contributing excessive weight from a small area on the sky in the attitude solution. Such area may not, or only very slowly, get properly connected to the catalogue, as it can force the reconstruction of the along-scan attitude to adapt itself to predicted positions based on earlier estimates of the astrometric parameters. The area can become partially detached, and the parallaxes measured are no longer absolute, in other words, a locally significant zero point is allowed. In these areas there is no direct test for this effect because there are unlikely to be any stars observed with very small parallaxes. However, the two independent reduction chains are unlikely to produce the same local zero-point deviation, and comparisons between the results from the two chains do not indicate that this has been a serious problem in the *Hipparcos* data reductions.

1.4 Concepts of the new reduction

The new reduction of the raw *Hipparcos* data, as presented and used in this book, differs in a fundamental way from the old reduction. The processes that used to be linked in the great-circle reduction (GCR) are now decoupled. The noise on the *a priori* stellar positions is now assumed white, which could not be the case for the initial stages of the old reduction. Furthermore, basic angle drifts are now recognized and corrected for 16 orbits, while elsewhere the basic-angle is found to be highly stable, with only slow drifts over the mission. After experimenting with the attitude reconstruction, a physical rather than a mathematical model is used. In the process of applying this model it was (re-)discovered that discrete jumps in the scan phase could occur under a wide range of conditions. Some 1500 jumps have been identified (and taken care off in the attitude modelling) in several hundreds of orbits. Similarly, over 80 external hits were identified and incorporated in the attitude modelling. Finally, an attempt was made to prevent potential "soft spots" in the catalogue by assigning appropriate weights to the two fields of view in the along-scan attitude reconstruction. Each of these issues will now be described in more detail.

1.4.1 Decoupling of the Great-circle Reduction Processes

The traditional great-circle reduction process as it was applied in the production of the published catalogue is described by van der Marel (1988) and van der Marel and Petersen (1992). It uses the field-of-view transit times as measured for single stars to derive three parameter sets:

- The along-scan attitude;
- The instrument parameters;
- The stellar position corrections in the scan direction.

The accuracy aimed at is of the order of 2-3 mas, but the initial positional errors in the *Hipparcos* Input Catalogue (ESA 1992) showed a dispersion of the order of 300 mas. Thus, in order to use the same stellar abscissae for solving all three parameter sets it was necessary to combine their solutions in one. Combined with the sometimes quite short intervals of observations this led occasionally to minor instabilities in the solution, manifesting themselves mainly in the form of a sixth-order harmonic modulation in the abscissa residuals. These modulations were solved for and removed in the sphere solution. They were the result of the basic angle of 58 degrees being close enough to 60 degrees to cause these problems in short data sets (see Fig. 1.4). The harmonic could be triggered also by a few outlying measurements with high weights.

In the new reduction the *Hipparcos* catalogue (ESA 1997) is used for the initial values of the astrometric parameters. The errors on the predicted positions are therefore at a level of 0.6 to 2.0 mas, and can be considered as white noise. This means that the along-scan attitude can be solved for separately, using a time resolution appropriate for that parameter set. As will be explained in Part IV, this means that abscissa residuals, per field of view, can be averaged over periods of 10 s, allowing more visibility of attitude variations. It also allows the implementing of a dynamical model for the attitude, and modelling the underlying torques rather than the positional variations. In the new solution, the basic angle determination is solved as part of the along-scan attitude determination. It also still has to enter the instrument-parameter solution, to account for the field-of-view dependence of first- and second-order along-scan positional dependencies in the instrument model.

The instrument parameters are solved at the full resolution of the frame transits. However, third-order parameters have been fixed according to their observed evolution over the mission, which has been approximated by spline functions, interrupted when necessary at instances of refocusing. This is justified by the observation that the variation in the fitted third-order parameters is mostly due to fitting noise and not to actual variations. Constraining the third-order parameters should provide more reliable calibrations for stars near the edge of the field of view and for very red stars (colour indices $\mathrm{B} - \mathrm{V} > 2.5$). Constraining the third-order parameters also takes away the natural correlations with the first-order parameters, the determination of which becomes less noisy and easier to monitor.

Finally, the abscissa residuals are collected per transit of the FOV. This too is a potential improvement on the old reduction, as there is no longer any need to combine all frame transits over an orbit, for which it was required to project these transits on a reference great circle. The accuracy of this projection depends on the accuracy of the star-mapper based reconstruction of the instantaneous position of the satellite's spin axis. At times of high background signal (activity in the Van Allen belts) this accuracy could be less than required. Keeping the field transits separated also allows a better detection of corrupted measurements. The most likely reason for problems with a field transit is the accidental superposition of an image from the other field of view, but this will always be specific to (part of) a single field-of-view transit only. Due to the generally different across-scan rotation rates in the two fields of view, such an accidental superpositions will not remain the same over a complete field-of-view transit.

1.4.2 Dynamic Approach to Attitude Reconstruction

A dynamic approach to the *Hipparcos* attitude reconstruction led to the Fully Dynamic Attitude (FDA) modelling as first presented by van Leeuwen and Fantino (2003). The FDA makes two main assumptions:

1. The satellite can be considered a rigid body;
2. The torques to which the satellite is subjected are a continuous function of time.

Even though there are a few minor violations of these criteria, experiments with the FDA showed encouraging results. The first of the violations is fully accommodated in the FDA: the actions of the gyros on board the spacecraft. These actions are modelled as additional torques. The underlying torques are also likely to experience small discontinuities as a result of changes in the magnetic moment of the satellite. Such changes may be the result of a heater being switched on or off. However, these effects are generally very small and ignoring them doesn't seem to have a significant negative effect on the quality of the reconstructed attitude. Details of the FDA are presented in Part IV.

1.4.3 Discontinuities in the Scan Phase

Discontinuities in the scan phase (Fig. 1.12) had first been recognized in the NDAC reductions by Carsten Skovmand Petersen (Copenhagen University Observatory). A few events were noted to occur shortly after the end of an eclipse. In the preparations for the new reduction, these jumps have been found to be far more frequent than was recognized in the preparation of the published data. They occur also frequently after the start of an eclipse, and for some stretches of time around certain orbital phases (possibly related to temperature

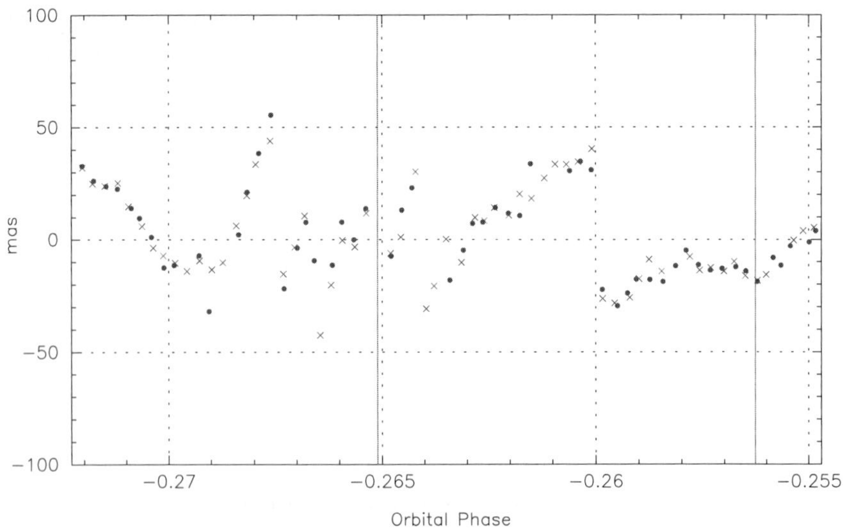

Figure 1.12. Scan-phase discontinuities are observed when the satellite is warming up rapidly, for example after an eclipse as shown here for orbit 253 (February 1990). The discontinuities reflect discrete adjustments of the inertia tensor to the changes in the temperature. Here 4 jumps of around 40 to 60 mas in scan phase are observed from the abscissa residuals with respect to the star-mapper based reconstruction of the along-scan attitude. The crosses and circles represent data from the two fields of view. The data points are weighted averages over 10.6 s of observations. The vertical lines represent thruster firings

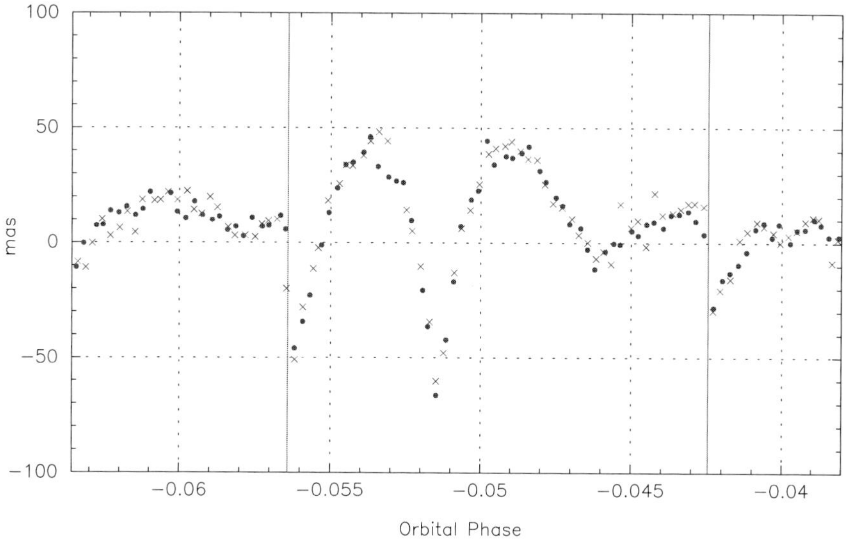

Figure 1.13. An external hit of the satellite as shown in the abscissa residuals relative to the star-mapper based scan-phase reconstruction. The data are for orbit 715 (October 1990), for which the scan velocity at orbital phase -0.05148 changed abruptly by 4.7 mas s^{-1}. The crosses and circles represent data from the two fields of view. The data points are weighted averages over 10.6 s of observations. The vertical lines represent thruster firings

adjustments following low perigee passages). These events have now been manually (visually) identified and are accounted for in the data analysis. In the original reductions it was left to the node adjustments and outlier-removal procedures to remove the worst consequences of these events. A detailed description of the scan-phase discontinuities is presented in Section 10.14.

1.4.4 Discontinuities in the Scan Velocity

There are two types of scan-velocity discontinuities: controlled ones caused by firings of the cold-gas thrusters for attitude control (see also Section 13.3.2), and those caused by external hits of the satellite (micro meteoroids, space debris). The first type is easy to account for, as these events are marked in the satellite telemetry. The second type had to be identified in the data first. The largest hits had been identified from examinations of the gyro data, they constitute rate changes at the level of a few arcsec s^{-1}. Only three of these have been observed over the mission. The smaller hits can be identified from the abscissa residuals, as is shown in Fig. 1.13. The reason that these hits appear differently from the scan-velocity discontinuities caused by thruster firings is because only the latter are taken into account in the preceding phase of the attitude modelling, as based on the star mapper data. The discontinuities that show at the thruster firings reflect the inaccuracies in the star mapper based attitude model. Hits at this level took place roughly once every two weeks, and were not accounted for in the published data, but are, as much as possible, taken care of in the new reduction by adding artificial thruster-firing instances at the times of hits. A detailed description of the properties of the detected hits is presented in Section 10.5.5.

1.4.5 Stability of the Basic Angle

The new reduction allowed for a more systematic monitoring of the stability of the basic angle. The main tool for this is the trend analysis of the instrument parameters. Whenever the basic angle shows a significant drift, this also shows in the scale and rotation of the field of view. The reason for this is that a change to the basic angle tends to be caused by a temperature-control failure, which also affects the focus of the telescope and thus the scale in the focal plane. Therefore, any data set for which one or more of the instrument parameters showed results deviating from the general trend, has been further examined by, for example, showing the effective basic angle as a function of time over the time interval of that data set. This way, 16 orbits have been found where a significant (10 to 20 mas) basic-angle drift took place (see for example orbit 1150 in Fig. 1.14). All of these have been linked to features in the events log for the mission. The most important and frequent of these, which had not been noted

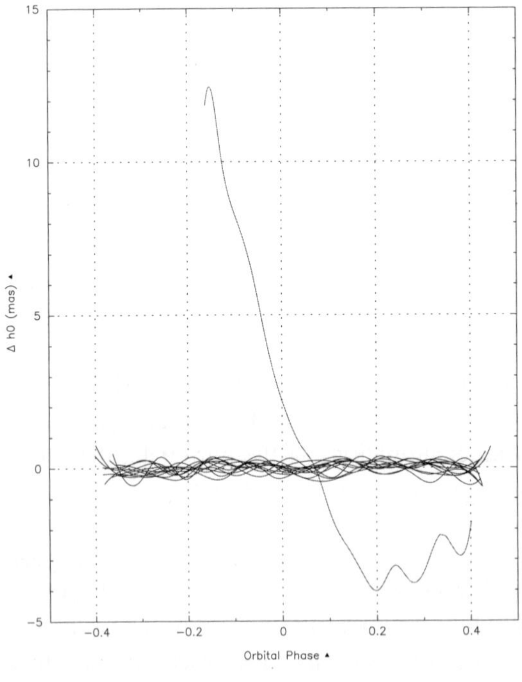

Figure 1.14. Drift and stability of the basic angle. The group of curves with values around 0 show the observed fluctuations in the fitted basic angle over 10 orbits. The other curve shows the drift in the basic angle for orbit 1150 (30 March 1991), following an anomaly in the power supply on the spacecraft (Anomaly Report 48)

as an "anomaly" and had slipped through in the NDAC and FAST reductions, was a restart of the payload after a telemetry problem. In the new reductions these drifts are detected and fitted, though in some cases there is still some drift left in the field-of-view rotation. Apart from these cases, the basic angle shows a remarkable level of stability (better than a few tenths of a mas, see Fig. 1.14), as well as a rapid convergence to a stable configuration (generally within 2 to 4 hours). This is an important criterion for the potential accuracy of the final catalogue, for which a stable basic angle is a crucial parameter. As was described in Section 1.3.6, the stability of the basic angle has also been shown through tests on the parallax zero-point in the *Hipparcos* catalogue: the most likely systematic modulation of the basic angle would have caused an offset of that zero point.

1.4.6 Weight Adjustments Between the Two Fields of View

The weight adjustments between the two fields of view in the along-scan attitude reconstruction is an issue that was triggered by possible problems in the Pleiades parallaxes for members in the centre of the cluster. In order to create a rigid sphere from the *Hipparcos* data it is essential that connections are made between the two fields of view. These connections consist of the very stable basic angle and the reconstruction of the along-scan attitude as based on input from both fields of view. In the along-scan attitude fitting the interval between two nodes of the spline function (or, in other words, the number of measurements per estimated parameter) is crucial, and in particular how this translates into the length of a strip of sky scanned. In the old solution this was equivalent to 5 to 6 degrees, in the new solution it is 4 degrees. If an area smaller, but not much smaller, than about 3 degrees in diameter, imposes an excessive weight on the attitude reconstruction, then this area can become semi detached from the catalogue, and the astrometry will no longer be absolute, but relative instead. The reason for this is that the along-scan attitude can adjust itself to a faulty set of astrometric parameters if dominated by only one field of view. The weight adjustments prevent this from happening by never allowing the weight ratio between the two fields of view (for all data contributing to an interval between two spline nodes) to be above a factor 2.72 (sigma ratio <1.65). Further details can be found in Section 10.5.3.

Chapter 2

HIPPARCOS ASTROMETRY

This chapter presents the basic concepts of astrometric measurements and how astrometric information is extracted from the *Hipparcos* data. The equations relating space positions and velocities to six observed parameters are derived, followed by an overview of the *Hipparcos* technique of measuring five of these six parameters, and an assessment of the formal errors on the measurements. This is followed by brief descriptions of two calibrations affecting the measurements: the reconstruction of the along-scan rotation phase of the satellite at the time of observation, and the reconstruction of grid distortions. Finally, the actual derivation of the astrometric parameters is described, with an assessment of the accuracies obtained.

2.1 From positions and velocities to astrometric data

Of the six parameters that define the phase space of a star relative to the Sun, five are derived from astrometric measurements: the three components of the position and the two components of the relative velocity of the star that are perpendicular to the direction in which the star is observed. The final, sixth parameter measures the velocity along the line of sight (referred to as the radial velocity), and is derived from measuring the Doppler shift of absorption lines in the stellar spectrum. In this Section we recall the relations between astrometric measurements and the five astrometric parameters.

2.1.1 Parallax Displacements in Ecliptic Coordinates

Take an inertial Cartesian reference system in which the position of a star is given by the vector $\boldsymbol{R} = (X, Y, Z)$, relative to the position of the Sun as fixed at the centre $(0, 0, 0)$. We use, for simplicity later on, ecliptic coordinates (λ, β), with (X, Y) in the ecliptic plane as defined by the orbit of the Earth around the Sun, and positive (Z) towards the Ecliptic North Pole. Within this system we assume that a star has a linear motion. Given that we will be looking primarily within a relatively small radius around the Sun and over a time interval which, on a Galactic scale, is very short, that seems to be a reasonable assumption. Issues of orbital motions in double and multiple systems will be discussed in Chapters 3 and 4. Thus, the motion of the star relative to the Sun can be given by a fixed vector $(\dot{X}, \dot{Y}, \dot{Z})$. With the distance of the star given by $R = |\boldsymbol{R}|$, the relation between the its Cartesian and ecliptic coordinates is described by:

$$\begin{aligned} X &= R\cos\lambda\cos\beta, \\ Y &= R\sin\lambda\cos\beta, \\ Z &= R\sin\beta. \end{aligned} \tag{2.1}$$

For an observation made from Earth, with position:

$$(X, Y, Z)_E = (r(t)\cos\lambda_E(t), r(t)\sin\lambda_E(t), 0), \tag{2.2}$$

the Geocentric position of the star is a function of time:

$$\begin{aligned} X - r(t)\cos\lambda_E(t) &= R'\cos\lambda'\cos\beta', \\ Y - r(t)\sin\lambda_E(t) &= R'\sin\lambda'\cos\beta', \\ Z &= R'\sin\beta'. \end{aligned} \tag{2.3}$$

where $\lambda_E(t)$ is the ecliptic longitude of the Earth as seen from the Sun. We ignore for simplicity the ellipticity of the Earth orbit ($r(t) = r$). Let $R' = R + \Delta R(t)$, $\lambda' = \lambda + \Delta\lambda(t)$, and $\beta' = \beta + \Delta\beta(t)$. We assume, in first instance, the time derivative of (X, Y, Z) to be zero. Equating the third lines of Eq. 2.1 and Eq. 2.3 gives the relation:

$$\Delta R = -R\frac{\cos\beta}{\sin\beta}\Delta\beta, \tag{2.4}$$

with which the first two lines of Eq. 2.3 simplify in first approximation to:

$$\begin{aligned} r\cos\lambda_E(t) &= R\sin\lambda\cos\beta\cdot\Delta\lambda(t) + R(\cos\lambda/\sin\beta)\cdot\Delta\beta(t), \\ r\sin\lambda_E(t) &= -R\cos\lambda\cos\beta\cdot\Delta\lambda(t) + R(\sin\lambda/\sin\beta)\cdot\Delta\beta(t). \end{aligned} \tag{2.5}$$

These equations represent a circle on the sky at the ecliptic poles, becoming an ellipse away from the pole and a displacement in longitude only in the ecliptic plane. We now normalize Eq. 2.5 with respect to the distance R, and introduce the parallax: $\varpi \equiv r/R$. Furthermore, we define $\Delta\lambda_*(t) \equiv \Delta\lambda(t)\cos\beta$:

$$\begin{aligned} \varpi\cos\lambda_E(t) &= \sin\lambda\cdot\Delta\lambda_*(t) + \frac{\cos\lambda}{\sin\beta}\Delta\beta(t), \\ \varpi\sin\lambda_E(t) &= -\cos\lambda\cdot\Delta\lambda_*(t) + \frac{\sin\lambda}{\sin\beta}\Delta\beta(t), \end{aligned} \tag{2.6}$$

from which we derive:

$$\varpi = \sin(\lambda - \lambda_E(t))\Delta\lambda_*(t) + \frac{\cos(\lambda - \lambda_E(t))}{\sin\beta}\Delta\beta(t), \tag{2.7}$$

and:

$$\begin{aligned} \Delta\lambda_*(t) &= \varpi \sin(\lambda - \lambda_E(t)), \\ \Delta\beta(t) &= \varpi \cos(\lambda - \lambda_E(t)) \sin\beta. \end{aligned} \tag{2.8}$$

When using the ecliptic longitude of the Sun as seen from the Earth ($\lambda_S = \lambda_E + \pi$), the above equations change sign. Equations 2.7 and 2.8 thus show how the parallax is reflected in the observed yearly variations in ecliptic longitude and latitude for a given position on the sky. In calculating the parallax corrections for the position of a star, Eq. 2.8 is generally more practical, but to derive the parallax from a set of observations, Eq. 2.7 can be used.

2.1.2 Parallax Displacements in Heliotropic Coordinates

The parallax displacement for a star is restricted to the great circle through the direction to this star and the direction of the Sun. We can see this when we transform the coordinate system from ecliptic to heliotropic (centred on the Earth, and with the Sun in a fixed direction). Because of the movement of the Earth around the Sun, this system is not a fixed inertial reference frame. However, it is useful for deriving the relations for parallax displacements as measured by *Hipparcos* on great circles. The heliotropic angles ξ measures the angular separation between a star and the direction of the Sun (the equivalent of a co-latitude), while ν measures the equivalent of a longitude. Both angles are generally time dependent. We can obtain positions in this reference system by applying first a rotation $-\lambda_E(t)$ around the Z axis, followed by a rotation of -90° around the Y axis to Eq. 2.1:

$$\begin{aligned} \cos\nu(t)\sin\xi(t) &= -\sin\beta, \\ \sin\nu(t)\sin\xi(t) &= \sin(\lambda - \lambda_E(t))\cos\beta, \\ \cos\xi(t) &= \cos(\lambda - \lambda_E(t))\cos\beta. \end{aligned} \tag{2.9}$$

With the Earth positioned at $(0, 0, 0)$ and the position of the Sun at $(0, 0, -r)$, the following relations, equivalent to Eq. 2.3 above, are obtained:

$$\begin{aligned} R\cos\nu(t)\sin\xi(t) &= R'\cos\nu(t)'\sin\xi(t)', \\ R\sin\nu(t)\sin\xi(t) &= R'\sin\nu(t)'\sin\xi(t)', \\ R\cos\xi(t) - r &= R'\cos\xi(t)'. \end{aligned} \tag{2.10}$$

Here the first two equations give:

$$\begin{aligned}\Delta R(t) &= -R\frac{\cos\xi(t)}{\sin\xi(t)}\Delta\xi(t),\\ \Delta\nu(t) &= 0.\end{aligned} \tag{2.11}$$

Combined with the third equation, the parallax displacement is derived as:

$$\Delta\xi(t) = \varpi \sin\xi(t), \tag{2.12}$$

which reaches, as we would expect, a maximum at $\xi(t) = 90°$, i.e. when the baseline is perpendicular to the direction of the star. Also as expected, there is no displacement in ν: all parallax displacements are a function of the co-latitude ξ only, and are thus restricted to great circles through the directions of the Sun and the star.

The polar axis of a great circle as described by the two FOVs of *Hipparcos* is tilted with respect to the direction of the Sun by a nearly fixed value of $\xi_H = 43°$, also referred to as the solar-aspect angle. The parallax displacement for a star as measured along such a great circle depends on the local angle between the directions of the Sun and the normal to the circle (the direction of the spin axis). Figure 2.1 shows this inclination as the angle ψ, which is related to the two angles defining the abscissa Ω_* and heliotropic co-latitude ξ_* of a star for a great circle:

$$\sin\psi = \frac{\sin\Omega_*}{\sin\xi_*}\sin\xi_{\rm H}, \tag{2.13}$$

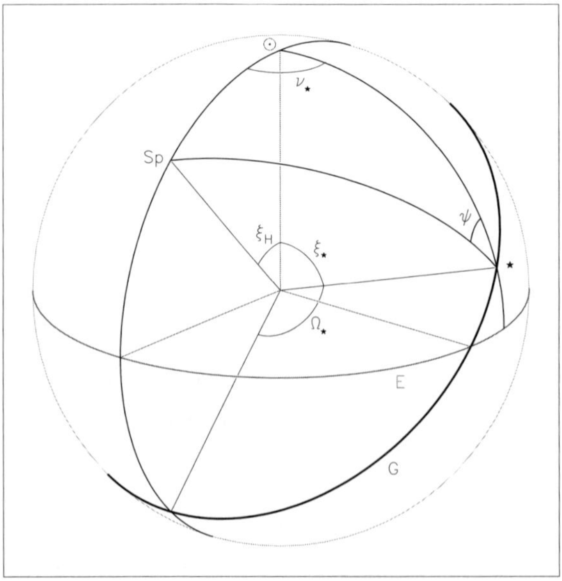

Figure 2.1. A great circle (G) in the heliotropic reference frame, showing the relation between the angle ψ, the coordinates of the star (ν_*, ξ_*), the stellar abscissa Ω_* and the solar aspect angle of the satellite's spin axis (Sp) $\xi_{\rm H}$, used for the calculation of the stars' parallax displacements along the great circle

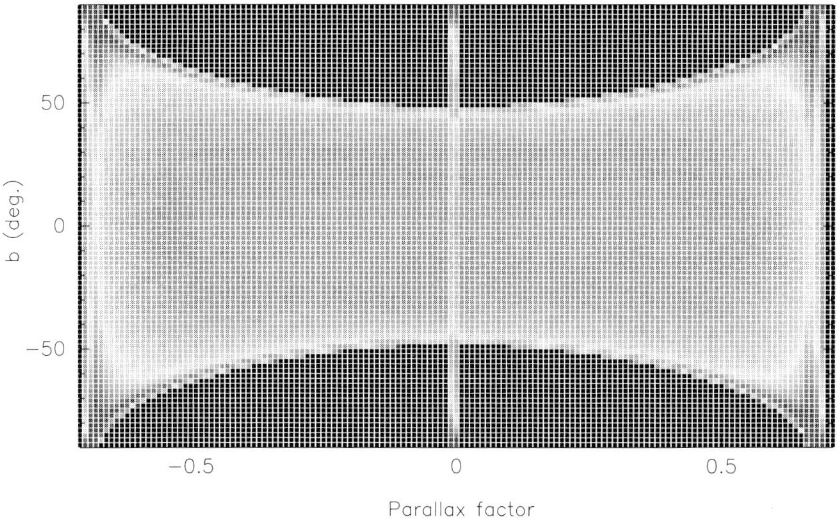

Figure 2.2. The distribution of the parallax factor $\sin\Omega_* \sin\xi_{\rm H}$ over ecliptic latitude. Near the ecliptic poles the parallax factor is always close to the maximum or minimum value, while zero values can be reached for b between $\pm 47°$(values in the black areas of the diagram can and do not occur in the data). The zero values that cover the entire range of latitudes are due to Sun-pointing mode data sets

where ξ_* is the value of $\xi(t)$ for the star observed at the time of observation. The parallax displacement along the great circle is given by (remember that the co-latitude ξ is measured from the direction of the Sun):

$$da = -\Delta\xi \sin\psi, \tag{2.14}$$

which, together with Eqs. 2.12 and 2.13 gives:

$$da = -\varpi \sin\Omega_* \sin\xi_{\rm H}. \tag{2.15}$$

Thus, the parallax displacement along the great circle only depends on the abscissa Ω_* of the star and the instantaneous solar aspect angle $\xi_{\rm H}$ of the satellite. For a star positioned not exactly on the great circle, at an ordinate υ, Eq. 2.15 becomes:

$$da \cos\upsilon = -\varpi \sin\Omega_* \sin\xi_{\rm H}. \tag{2.16}$$

The effect of the $\cos\upsilon$ coefficient is a correction of less than 0.003 per cent over the width of the *Hipparcos* focal plane, and thus of little significance. The distribution of parallax factors as observed over the mission for 13.6 million field transits is shown in Fig. 2.2.

The angle Ω_* is directly related to the rotation phase Ω_H of the satellite:

$$\Omega_* = \Omega_H \pm \gamma/2, \tag{2.17}$$

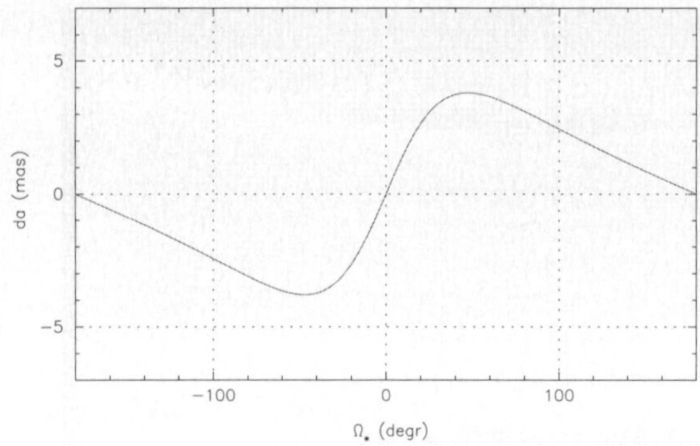

Figure 2.3. The effects of light bending due to the Sun as projected along the scan circle

where γ is the basic angle between the two fields of view. The angle Ω_H also defines the direction of exposure to the solar radiation. The direct relation between the parallax displacements and the rotation phase (and exposure) of the satellite is a potential weakness of the mission (see Chapter 1, Section 1.3.6).

2.1.3 Correction for Natural Direction

The isotropic coordinate direction of an object is given by its instantaneous direction in space for the time and position of observation. The transformation from isotropic coordinate direction to natural direction takes into account the light bending by the Sun and by the Earth. Considering here only the component for the Sun, as given in Volume 3, Chapter 12 of ESA (1997), gives the following relation:

$$\hat{\boldsymbol{u}} = \left\langle \bar{\boldsymbol{u}} + \boldsymbol{h}\frac{(1+\gamma)GS}{c^2 h(h + \bar{\boldsymbol{u}}'\boldsymbol{h})} \right\rangle, \tag{2.18}$$

where $\bar{\boldsymbol{u}}$ is the isotropic, and $\hat{\boldsymbol{u}}$ the natural direction for a star, c is the velocity of light, and γ is the PPN parameter, equal to unity in general relativity. The vector $\boldsymbol{h}$ is the heliocentric position of the satellite, like the heliocentric position of the Earth as defined in Eq. 2.2. As for the parallax displacements, the displacements due to light bending are in ξ only. Expressed in the heliotropic coordinates defined above, we have:

$$\hat{\boldsymbol{u}} = \left\langle \bar{\boldsymbol{u}} + \langle \boldsymbol{h} \rangle \frac{2GS}{c^2 h(1 - \cos\xi_*)} \right\rangle, \tag{2.19}$$

where the constant $2GS/c^2h \approx 1.974 \cdot 10^{-8}$ for a distance of $h = 1$ AU, and $h\cos\xi_* \equiv -\bar{\boldsymbol{u}}'\boldsymbol{h}$.

What we are interested in is the change in ξ_* as a result of the light bending:

$$\cos\Delta\xi_* = \bar{\boldsymbol{u}}'\hat{\boldsymbol{u}}. \tag{2.20}$$

Application to Eq. 2.19 gives:

$$\cos\Delta\xi = \frac{1 - q\cos\xi_*}{|\bar{\boldsymbol{u}} + q\langle\boldsymbol{h}\rangle|} = \frac{1 - q\cos\xi_*}{\sqrt{1 - 2q\cos\xi_* + q^2}}, \tag{2.21}$$

where $q \equiv (2GS/c^2h)(1/(1-\cos\xi_*))$. The square root in Eq. 2.21 is approximated by:

$$\sqrt{1 - 2q\cos\xi_* + q^2} \approx 1 - q\cos\xi_* + \frac{1}{2}q^2\sin^2\xi_*. \tag{2.22}$$

Applied to Eq. 2.21 gives:

$$\cos\Delta\xi_* \approx 1 - \frac{1}{2}\Delta\xi_*^2 \approx 1 - \frac{1}{2}q^2\sin^2\xi_*, \tag{2.23}$$

which leads to:

$$\Delta\xi_* = q\sin\xi_* = \frac{2GS}{c^2h}\,\frac{\sin\xi_*}{1-\cos\xi_*}. \tag{2.24}$$

This relation is usually expressed as:

$$\Delta\xi_* = \frac{2GS}{c^2h}\,\frac{1}{\tan\frac{\xi_*}{2}}. \tag{2.25}$$

Combining Eq. 2.24 with Eq. 2.13 and Eq. 2.14 we obtain:

$$da = -\frac{2GS}{c^2h}\,\frac{\sin\Omega_*\sin\xi_H}{1-\cos\xi_*}, \tag{2.26}$$

and with $\cos\xi_* = \cos\Omega_*\sin\xi_H$:

$$da = -\frac{2GS}{c^2h}\,\frac{\sin\Omega_*\sin\xi_H}{1-\cos\Omega_*\sin\xi_H}. \tag{2.27}$$

The maximum effect of solar light bending as projected along the great circle, and thus affecting the abscissae, is found at $\cos\Omega_* = \sin\xi_H$, for which:

$$da = \mp\frac{2GS}{c^2h}\,\tan\xi_H, \tag{2.28}$$

and just within ± 4 mas. The effects along the scan circle are shown in Fig. 2.3.

A check on the correlation between the abscissa residuals after the astrometric-parameter fitting and the corrections applied for the light-bending (including Earth and Sun), provides a limit for the PPN parameter γ. Using the abscissa residuals for nearly 11.7 million field-of-view transits of stars with simple 5-parameter solutions, the correlation between the abscissa residuals and the correction for light-bending gives the correction to γ as

$$\Delta\gamma = 0.00072 \pm 0.00076, \tag{2.29}$$

not significantly different from zero, and thus confirming $\gamma = 1$ within the measurement error (see also Froeschle et al. 1997).

2.1.4 Proper Motions and Radial Velocities

We now return to Eq. 2.1 to see how changes in the position of a star, $(\dot{X}, \dot{Y}, \dot{Z})$ reflect in changes in the coordinates (λ, β) and distance R:

$$\begin{aligned}\dot{X} &= \dot{R}\cos\lambda\cos\beta - \dot{\lambda}R\sin\lambda\cos\beta - \dot{\beta}R\cos\lambda\sin\beta,\\ \dot{Y} &= \dot{R}\sin\lambda\cos\beta + \dot{\lambda}R\cos\lambda\cos\beta - \dot{\beta}R\sin\lambda\sin\beta,\\ \dot{Z} &= \dot{R}\sin\beta + \dot{\beta}R\cos\beta. \end{aligned} \tag{2.30}$$

The radial velocity is defined as $V_R \equiv \dot{R}$, and is measured in km s^{-1}. The displacements on the sky are given by the proper motion components:

$$\begin{aligned}\mu_{\lambda*} &\equiv \dot{\lambda}\cos\beta,\\ \mu_\beta &\equiv \dot{\beta}, \end{aligned} \tag{2.31}$$

which are measured in, for example, mas yr^{-1}. We already defined the parallax as $\varpi \equiv r/R$. We further define $r = 1$ AU and use the parsec as unit for R, so that an object at a distance of $R = 1000$ parsec produces a parallax of $\varpi = 1$ mas. Furthermore, a proper motion of 1 mas yr^{-1} at 1000 parsec is equivalent to $\kappa = 4.74047$ km s^{-1}. Then, expressing the parallax in mas, we find:

$$\begin{aligned}\dot{X} &= V_R\cos\lambda\cos\beta - \kappa\left[\frac{\mu_{\lambda*}}{\varpi}\sin\lambda + \frac{\mu_\beta}{\varpi}\cos\lambda\sin\beta\right],\\ \dot{Y} &= V_R\sin\lambda\cos\beta + \kappa\left[\frac{\mu_{\lambda*}}{\varpi}\cos\lambda - \frac{\mu_\beta}{\varpi}\sin\lambda\sin\beta\right],\\ \dot{Z} &= V_R\sin\beta + \kappa\frac{\mu_\beta}{\varpi}\cos\beta. \end{aligned} \tag{2.32}$$

The position, parallax and proper motion of a star are referred to as the five astrometric parameters, which are sufficient to represent the positions on the sky as a function of time for most single stars. There are small complications for very nearby stars with large proper motions, where second order effects may become relevant. One of these secondary effects is the change in proper motion and parallax resulting from the radial velocity of the star. In a few extreme cases it may even become possible to use these effects for an independent determination of the star's radial velocity (Dravins et al. 1997; Lindegren et al. 2000).

2.2 The Hipparcos astrometric data

2.2.1 The Grid and the Image Dissector Tube

Hipparcos measured positions by means of, and relative to, a modulating grid, the details of which can be seen in Fig. 2.4. The entire main grid measured just over 2 by 2 cm. The optical block on which the grid was engraphed

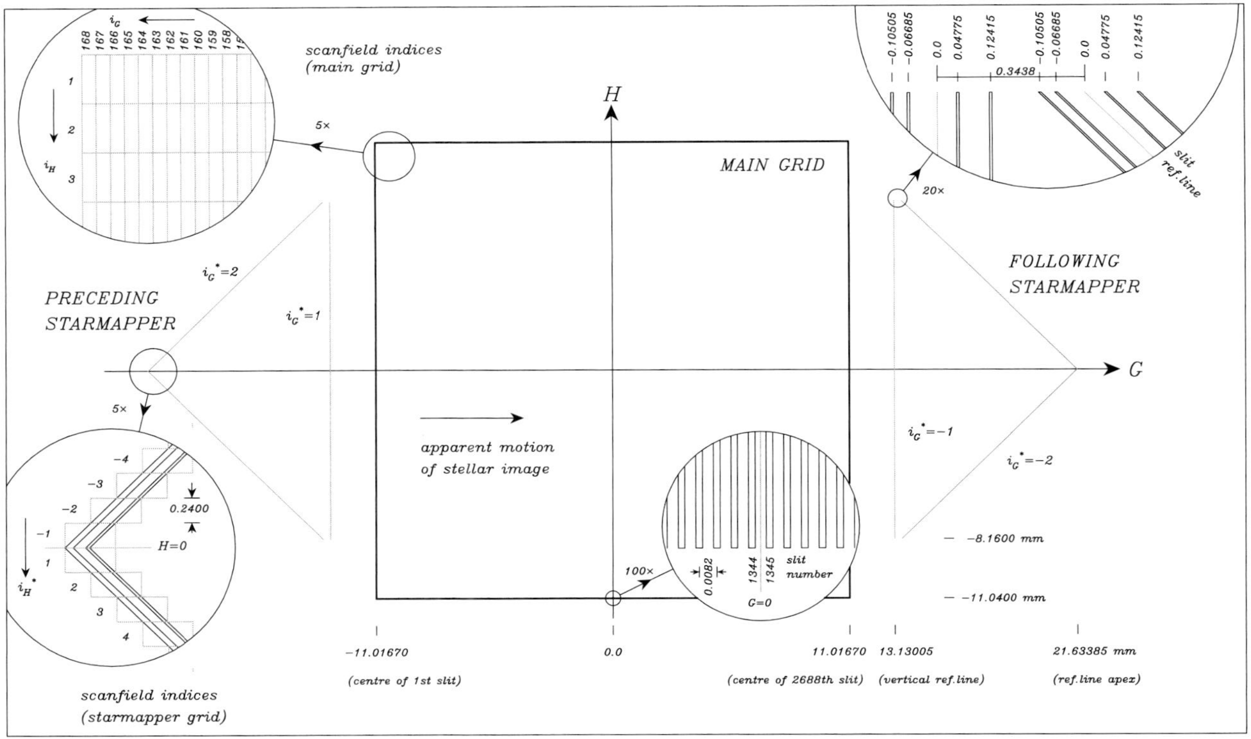

Figure 2.4. The grid assembly characteristics. The actual main grid measured about 2 × 2 cm. Grid lines were engraved in separate scan fields, 7728 for the main grid and 136 for each star mapper, the geometry of which is shown on the left of the diagram. Each scan field for the main grid contained 16 equally spaced grid lines

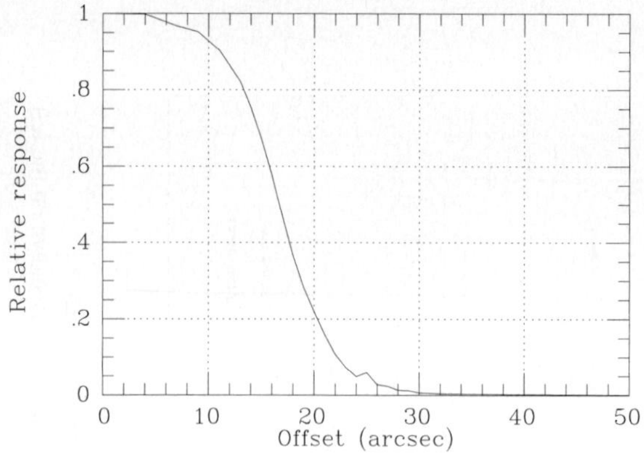

Figure 2.5. The sensitivity curve for the Instantaneous Field of View of the image dissector tube. This curve varied with position on the grid and colour of the star

consists of a piece of Suprasil 1, a high-quality silica. The convex surface of the optical block matches the curved focal surface of the telescope. Its rear side is flat there where star mapper grids are situated, and is ground with a radius of curvature of 213.8 mm where the main grid is, for which it also serves as a field lens.

The main grid and the star mapper grids were engraved in discrete "scan fields", with 46×168 of these scan fields forming the main-detector grid. The method used to engrave these grids was such that when projected orthogonally on a plane normal to the optical axis of the optical block, the grids are strictly rectangular, but when projected on the curved surface of the optical block, they are slightly distorted. The spacing of the grid lines was 8.20 μm, and their width 3.13 μm. These values were optimized such that a significant second harmonic modulation signal could be obtained, which is essential for disentangling the signals of double stars (Chapter 4). The accuracy of the grid line positions reached a standard deviation level of around 10 nm (see also Fig. 2.14).

The Image Dissector Tube (IDT) directs the light from a small area behind the grid onto a photo-multiplier tube. This area, with a diameter of about 15 arcsec, is referred to as the Instantaneous Field of View (IFOV). The response profile of the IFOV is shown in Fig. 2.5. The IDT provided the mechanism for the IFOV to follow a stellar image as it crossed the grid, and made it possible to switch in very short time between the images of different stars. The switching time could not be entirely ignored, and the first sampling period after a repositioning is therefore always rejected. The positioning of the IFOV was regulated through coil currents and based on three inputs:

- The on-board Real Time Attitude Determination (RTAD), providing the instantaneous estimate of the pointing directions of the satellite axes;

- The pre-calculated apparent star positions, which together with the RTAD provide a predicted grid position for the stellar image;

- The coil-current calibration matrix, allowing the IDT to direct the IFOV to the predicted position.

Thus, when the on-board attitude was poorly or not known at the time of observation, no sensible IDT measurements could be obtained, as the IFOV pointing would be offset from the actual star position. In some extreme cases this situation has been used to learn more about the background signal for the main grid measurements, examining the contributions from, zodiacal light and the effects of radiation during transits of the Van Allen belts (see also Fig. 12.9, page 313). This was important for the analysis of the photometric data.

The coil-current values used for each observation are part of the IDT data stream, and have been preserved. The relation between the coil currents and positions on the grid was calibrated about once every day for a grid of 11 by 11 reference points. These calibration values are fitted with a two-dimensional third order polynomial in the grid coordinates to provide an interpolated response at every position on the grid. The differences between the reconstructed positions of the IFOV and the target star are now preserved in the reduced field-transit data, which may in some cases be relevant for the photometric and double-star analysis.

2.2.2 Abscissae or One-dimensional Positions

While for most ground-based observations positional information is obtained from both coordinates on the sky, this is not the case for the *Hipparcos* data. As is explained in Chapter 1, *Hipparcos* is in many ways equivalent to a meridian telescope, timing stellar transits. *Hipparcos* therefore obtained only accurate positional measurements along the scan direction. These positions are referred to as abscissae, and are derived from the phase of the modulated signal created by the passage of a star over the grid.

The grid modulates the light of a passing star into a regular sinusoidal signal of which only the first and second harmonics are significant (Fig. 2.6). The light is sampled at 1200 Hz, and data collected over a period of 32/15 s, an (observing) frame, is processed together (see also Section 1.2.5). Within a frame up to 10 stars could be sharing the observing time, the average being between 4 and 5. Faint stars had generally more observing time assigned to them than bright stars. The distribution of time among stars was regulated on-board by the observing-strategy program.

Following the description of the modulated signal as defined by Lindegren for the data processing by NDAC, we express the intensity I_k as a function of

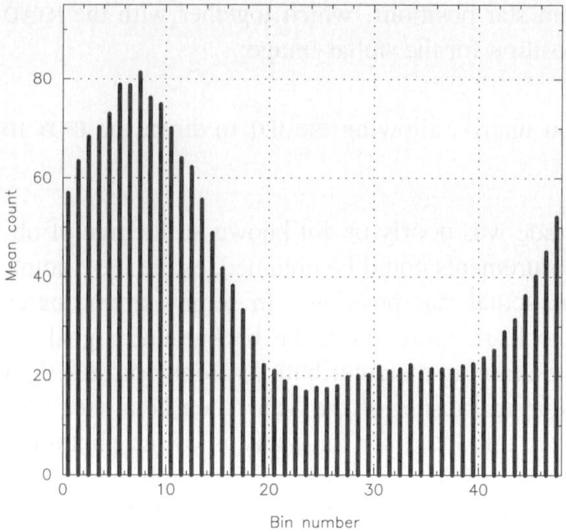

Figure 2.6. The modulated signal obtained from a transit of a star over the grid. The signal has been binned at 48 rather than 12 bins (which is sufficient for the signal analysis, see text) for increased resolution on the graph. The data are for a transit of HIP 26093 (Hp=5.565) in orbit 384 (24 April 1990)

the phase p_k at sample k:

$$I_k = \beta_1 + \beta_2 \big[\cos(p_k + \beta_3) + \beta_4 \cos 2(p_k + \beta_3) + \beta_5 \sin 2(p_k + \beta_3)\big], \quad (2.33)$$

where β_1 and β_2 are referred to as the Direct Current (DC) (flat signal) and Alternating Current (AC) (modulated signal) intensities respectively and are input to the photometric reductions, β_3 is the modulation phase and will be used to determine the stellar abscissa for the observation. The parameters β_4 and β_5 describe the relative properties of the second harmonic amplitude and phase. These model parameters can be expressed as a function of the Field of View, position in the Field of View, and colour of the measured star. This process is referred to as the calibration of the Optical Transfer Function (OTF) (see Chapter 12). We will come back to this equation when describing the double star treatment and the photometry; at this stage what is important is the phase of the modulated signal, β_3, which defines the relative position of the star on the modulating grid at a given reference time.

2.2.3 Solving for the Modulation Parameters

2.2.3.1 The modulated signal

Equation 2.33 does not lend itself to a least squares solution. We rewrite it in the simpler form:

$$I_k = b_1 + b_2 \cos(p_k) + b_3 \sin(p_k) + b_4 \cos(2p_k) + b_5 \sin(2p_k). \quad (2.34)$$

The relations between the β_i in Eq. 2.33 and b_i in Eq. 2.34 are given by:

$$\begin{aligned}
\beta_1 &= b_1, \\
\beta_2 &= \sqrt{b_2^2 + b_3^2}, \\
\beta_3 &= \arctan(-b_3/b_2), \\
\beta_4 &= \frac{(b_2^2 - b_3^2) \cdot b_4 - 2 \cdot b_2 b_3 b_5}{\sqrt{(b_2^2 + b_3^2)^3}}, \\
\beta_5 &= \frac{(b_2^2 - b_3^2) \cdot b_5 + 2 \cdot b_2 b_3 b_4}{\sqrt{(b_2^2 + b_3^2)^3}},
\end{aligned} \tag{2.35}$$

and the reverse relations by:

$$\begin{aligned}
b_1 &= \beta_1, \\
b_2 &= \beta_2 \cos \beta_3, \\
b_3 &= -\beta_2 \sin \beta_3, \\
b_4 &= \beta_2 \big(\beta_4 \cos 2\beta_3 + \beta_5 \sin 2\beta_3 \big), \\
b_5 &= -\beta_2 \big(\beta_4 \sin 2\beta_3 - \beta_5 \cos 2\beta_3 \big).
\end{aligned} \tag{2.36}$$

Each observation of I_k represents an integration over 1/1200 s (Section 1.2.5), during which a stellar image moved across the grid by nominally 0.14 arcsec. Data was collected in slots of 8 successive sampling periods, i.e. 1/150 s or 1.125 arcsec on the sky. This compares with an average value of 1.2074 arcsec for the grid period as projected on the sky. Thus, a slot covered nearly one complete modulation cycle. The observing strategy sub-divided a 32/15 s frame into 16 interlacing periods (2/15 s), each covering 20 slots of 1/150 s. An interlacing period was associated with an observing pattern, which required defining the stars to be observed, their sequence of observation, and the integration time (in slots) assigned to each star. Up to three different strategies could be employed over that period. This made it possible to partially observe stars that were either entering or leaving the field of view during the frame. Per frame there could only be a maximum of one entering and one leaving star included in the strategy, and only brighter stars were observed this way. Partially observed stars were always assigned two slots per interlacing period, either at the end (entering) or the beginning (leaving).

2.2.3.2 *Phase-binning the data*

To process the data in an efficient manner, we phase-bin the samples such that the first and second order phase information is preserved. This procedure allows the number of bins to remain small. Experiments with simulated data have shown that 6 bins per full cycle of the highest harmonic fitted is sufficient. In this case the second harmonic is the highest, and 12 (equally-spaced) bins

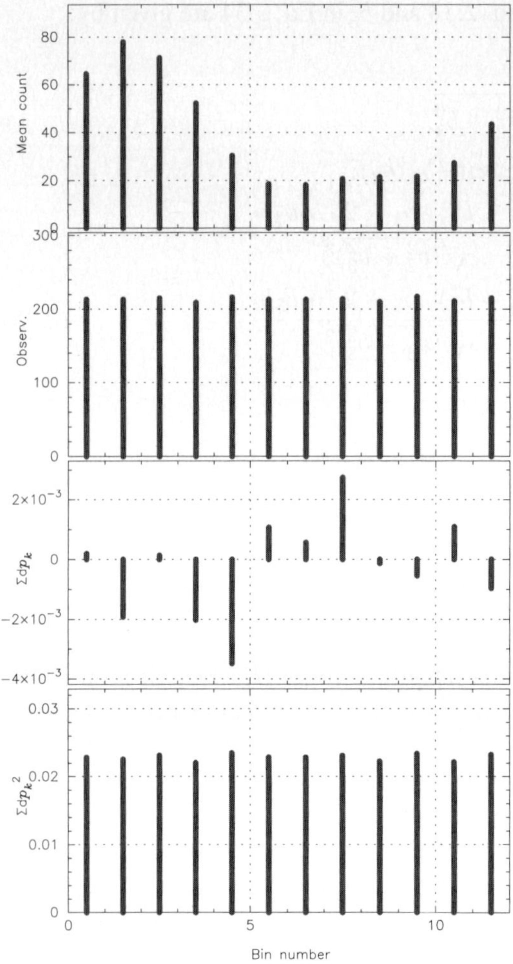

Figure 2.7. Phase-binning data for the same transit as shown in the previous figure, now binned at the normal processing resolution of 12 bins. This star is the only star observed in the frame. From top to bottom: the mean counts; the total number of samples per bin; the mean phase offset; the phase-offset variance

are used. A nominal reference phase p_r is assigned to each bin. A sample is added to the bin with the reference phase p_r nearest to the sample phase p_k. The phase difference is given by:

$$p_k - p_r = \mathrm{d}p_{k,r}. \tag{2.37}$$

We can now rewrite Eq. 2.34 for a sample in bin r:

$$\begin{aligned} I_{k,r} &= b_1 + b_2 \cos(p_r + \mathrm{d}p_{k,r}) + b_3 \sin(p_r + \mathrm{d}p_{k,r}) + \\ &\quad b_4 \cos(2(p_r + \mathrm{d}p_{k,r})) + b_5 \sin(2(p_r + \mathrm{d}p_{k,r})). \end{aligned} \tag{2.38}$$

Up to second order in $\mathrm{d}p_{k,r}$, Eq. 2.38 can be written as:

$$\begin{aligned} I_{k,r} &= b_1 + \\ &\quad b_2(\cos(p_r)\cdot(1-\mathrm{d}p_{k,r}^2/2) - \sin(p_r)\cdot \mathrm{d}p_{k,r}) + \\ &\quad b_3(\sin(p_r)\cdot(1-\mathrm{d}p_{k,r}^2/2) + \cos(p_r)\cdot \mathrm{d}p_{k,r}) + \\ &\quad b_4(\cos(2p_r)\cdot(1-2\mathrm{d}p_{k,r}^2) - \sin(2p_r)\cdot 2\mathrm{d}p_{k,r}) + \\ &\quad b_5(\sin(2p_r)\cdot(1-2\mathrm{d}p_{k,r}^2) + \cos(2p_r)\cdot 2\mathrm{d}p_{k,r}). \end{aligned} \tag{2.39}$$

Summing over all n_r samples in a bin, and defining for each bin r: $I_r \equiv \Sigma_k I_{k,r}$, $T_r \equiv \Sigma_k \mathrm{d}p_{k,r}$ and $S_r \equiv \Sigma_k \mathrm{d}p_{k,r}^2$ then gives:

$$\begin{aligned} I_r/n_r &= b_1 + \\ &\quad b_2(\cos(p_r)\cdot(1-\frac{S_r}{2n_r}) - \sin(p_r)\cdot\frac{T_r}{n_r}) + \\ &\quad b_3(\sin(p_r)\cdot(1-\frac{S_r}{2n_r}) + \cos(p_r)\cdot\frac{T_r}{n_r}) + \\ &\quad b_4(\cos(2p_r)\cdot(1-2\frac{S_r}{n_r}) - \sin(2p_r)\cdot 2\frac{T_r}{n_r}) + \\ &\quad b_5(\sin(2p_r)\cdot(1-2\frac{S_r}{n_r}) + \cos(2p_r)\cdot 2\frac{T_r}{n_r}). \end{aligned} \tag{2.40}$$

Thus, the only trigonometric terms to be calculated are for the fixed reference phases of the bins. The most significant correction in Eq. 2.40 comes from S_r, which, unlike T_r, does not average out. The third-order terms will usually also average out, and the fourth order terms are so small that their accumulation is of no significance. There is a small risk of resonances between the sampling and the binning frequencies, but these are generally of very short duration due to the constantly changing scan velocity. An example of phase binning is shown in Fig. 2.7. Assuming Poisson statistics for the counts, the formal error for the contribution from bin r is $\sqrt{I_r}/n_r$. Based on the observed counts, this formal error will generally be more accurate than the formal errors on the individual samplings.

2.2.3.3 *Transformation to β parameter solution*

The least-squares solution for Eq. 2.40 provides the information for the parameters b_i. We call the matrix A the 12×5 matrix representing the model parameters for the binned observations. Then the information matrix for the solution of the parameters b_i is given by:

$$\mathsf{R}_b \equiv \mathsf{A}^T\mathsf{A}. \tag{2.41}$$

However, for further processing we need the information matrix for the equivalent β_i values of Eq. 2.33. This is obtained through the Jacobian matrix product:

$$\mathsf{R}_\beta = \mathsf{J}\mathsf{R}_b\mathsf{J}^T, \tag{2.42}$$

where:

$$J_{ij} = \frac{\partial b_i}{\partial \beta_j}. \tag{2.43}$$

Using the relations given in Eq. 2.36, the matrix J is determined as:

$$\mathsf{J} = \begin{bmatrix} 1 & 0 & 0 & 0 & 0 \\ 0 & q \cdot b_2 & q \cdot b_3 & q \cdot b_4 & q \cdot b_5 \\ 0 & b_3 & -b_2 & 2b_5 & -2b_4 \\ 0 & 0 & 0 & q(b_2^2 - b_3^2) & -q \cdot 2b_2b_3 \\ 0 & 0 & 0 & q \cdot 2b_2b_3 & q(b_2^2 - b_3^2) \end{bmatrix}, \tag{2.44}$$

where $q \equiv 1/\sqrt{b_2^2 + b_3^2}$. The inverse of the matrix R_β thus obtained forms the basis of the formal error estimates and correlations of the β parameters.

2.2.4 The Formal Errors on the Abscissa Measurements

Assuming Poisson statistics for the intensity measurements I_k, and applying weights accordingly, we derive estimated formal errors for b_2 and b_3:

$$\sigma_{b_2} \approx \sigma_{b_3} \approx b_1 \cdot s/\sqrt{I_{\text{tot}}}, \tag{2.45}$$

where $s \approx 1.32$, with a slight non-linear dependence on the relative amplitude of the first-order modulation. The accuracy of $\beta_3 = \arctan(-b_3/b_2)$ is approximately given by:

$$\sigma_{\beta_3} \approx \frac{\sigma_{b_2}}{\sqrt{b_2^2 + b_3^2}}. \tag{2.46}$$

This approximate relation can also be derived from the inverse of the matrix R_β as defined above. The denominator in Eq. 2.46 is the amplitude β_2 of the second harmonic, while b_1 in Eq. 2.45 is identical to β_1 in Eq. 2.33. The modulation coefficient of the first harmonic is defined as $M_1 \equiv \beta_2/\beta_1$, and varies around a value of 0.7 (see Fig. 2.8), with a dependence on the colour of the observed star and the position on the grid. With these substitutions, the expected accuracy of β_3 becomes:

$$\sigma_{\beta_3} \approx \frac{s}{M_1\sqrt{I_{\text{tot}}}} \times \frac{1.2074}{2\pi} \approx \frac{0.25}{M_1\sqrt{I_{\text{tot}}}} \text{ arcsec}, \tag{2.47}$$

where the extra factor $\frac{1.2074}{2\pi}$ translates the phase accuracy from radians to arcseconds on the grid, given a grid period of 1.2074 arcsec. The final constant (0.25) in the above equation represents an average over the mission, over

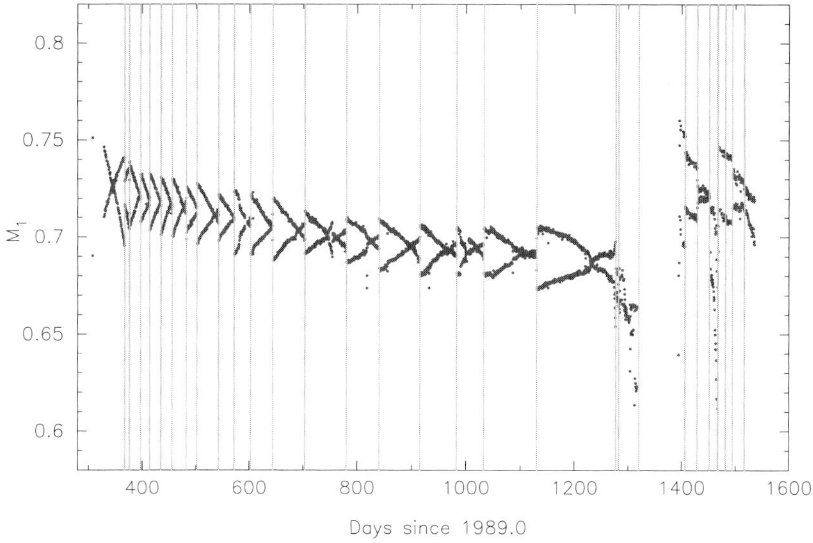

Figure 2.8. The modulation coefficient M_1 as observed over the mission, for the two fields of view. Most of the discontinuities are the result of refocusing (indicated by the grey lines), others are caused by temporary thermal-control problems. The rising curves relate to the preceding, the descending curves to the following field of view. These data are for the centre of each field of view, and for a colour index of $(\mathrm{V}-\mathrm{I})_\mathrm{C} = 0.5$

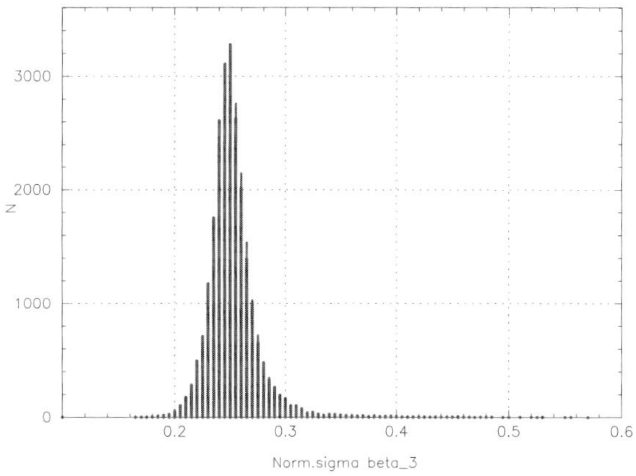

Figure 2.9. Distribution of the formal errors on the abscissa measurements in orbit 75, multiplied by $M_1\sqrt{I_{\mathrm{tot}}}$ to show the constant in Eq. 2.47. The long tails of the distribution result from double or multiple images

which it varies by less than 1 per cent (Fig. 3.1, page 74). An example of the observed distribution of formal errors, multiplied by the factor $M_1\sqrt{I_{\mathrm{tot}}}$, is shown in Fig. 2.9. The total counts observed varied from around 30 to nearly one million, giving precisions for the brightest single-frame transits at a level of 0.5 mas, and for the faintest at around 50 mas.

The modulation phase provides a position with respect to the grid pattern. In order to place the transit properly on the grid, in other words to identify the relevant grid line at the reference time for the transit, *a priori* positions are required to an accuracy of around 0.2 to 0.3 arcsec. These were originally provided by the *Hipparcos* Input Catalogue Consortium (INCA), led by Catherine Turon (ESA 1992; Turon et al. 1992; Jahreiss et al. 1992; Grenon et al. 1992). With the publication of the *Hipparcos* catalogue (ESA 1997; Perryman et al. 1997; van Leeuwen et al. 1997a; Lindegren et al. 1997), almost all of these positions (at the mean epoch of the *Hipparcos* mission, 1991.25) are now known to better than 2 mas. At any other epoch the limited accuracies of the proper motions will contribute further, though small, uncertainties.

2.3 Reconstruction of the along-scan rotation phase

Hipparcos was not only designed to accurately measure positions on a grid, it was also designed to relate these positions to a suitable reference frame, which describes the scan phase of the satellite as a function of time. Using the observations from the two fields of view and a stable basic angle, it is possible to reconstruct the along-scan rotation phase of the satellite to a high degree of accuracy. This process is usually referred to as the along-scan attitude reconstruction, a detailed description of which is presented in Chapters 8 to 10. The stellar reference positions for this reconstruction are the same positions we try to derive from the accumulated data over the mission (see Section 2.5). It is through iterations between the along-scan attitude reconstruction and the astrometric parameter determination (including a variety of calibrations) that the reconstructed along-scan rotation phase of the satellite ultimately becomes unaffected by parallactic displacements. This is the essential condition for deriving absolute parallaxes.

An important aspect of the attitude reconstruction is the power spectrum of the positional variations (see also Section 9.2.3 and Fig. 9.6). From this power spectrum the requirements for the degrees of freedom in the fitting model can be derived, as well as a unit of time within which the attitude variations can be averaged. If we consider variations in position down to a level of 0.3 mas, then one degree of freedom per minute of observations is most of the time sufficient, and data can be averaged (per field of view) over 10 s intervals. The simultaneous solutions for the abscissae and attitude in the original reduction would not allow such binning. There are on average between 5 and 15 observations per field of view for each 10 s interval. This averaging or binning of the data, improves the chance of early detection and removal of poor quality observations, and exposes more clearly, the hits and scan phase discontinuities, that from time to time disturb the satellite's attitude.

For the proper operation of the iteration between the along-scan attitude reconstruction and the astrometric parameter determination, data from both fields

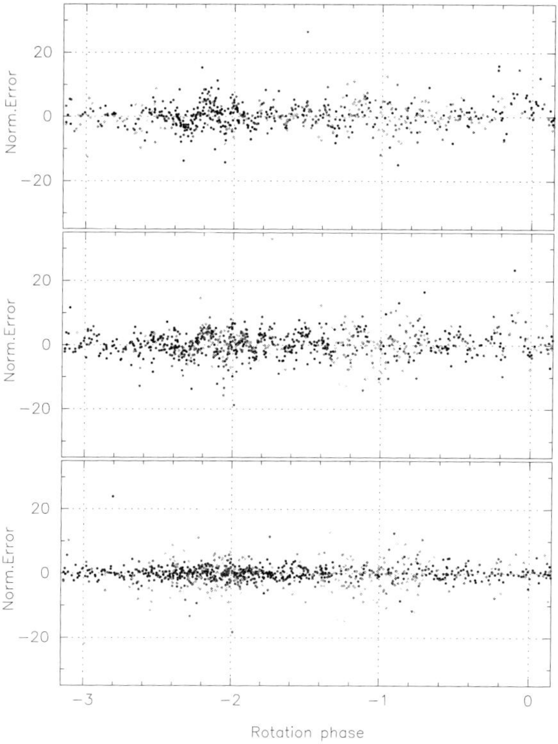

Figure 2.10. Abscissa residuals for part of orbit 237. From top to bottom: results from the FAST, NDAC, and new reductions. The residuals in the FAST and NDAC solutions show clearly very similar systematic and correlated variations, which may be the result of not having corrected for the scan-phase discontinuities that exist for this orbit. The rotation phase is given in radians

of view need to contribute significantly. Only when the determination of every degree of freedom in the along-scan attitude reconstruction is significantly affected by data from both fields of view, is the overall solution likely to converge properly to a reference frame, free of parallactic displacements. This is referred to as the *connectivity condition*, as it provides the essential connection between measurements in the two fields of view that allows for the elimination of the parallax factors. Together with the stability requirement for the basic angle this makes it possible to obtain absolute parallaxes all over the sky.

The possibilities of reconstructing the along-scan attitude are limited, and will depend on a range of circumstances such as the proximity of an eclipse and the orbital phase at the time of observation. The modelling noise, as introduced by the along-scan attitude reconstruction, is therefore not entirely random or homogeneous, but more or less systematic (Fig. 2.10). In the published data the systematic noise from the modelling errors is noticeably larger than in the new reduction, and shows in the correlations between abscissa errors for brighter stars. In the new reduction the average modelling noise has been reduced from around 3 mas to 0.6 mas, leaving Poisson noise as the only significant noise contribution for all stars fainter than magnitude 3 to 4. This has significantly reduced the abscissa-error correlations, as is shown in Fig. 2.11. Reduction

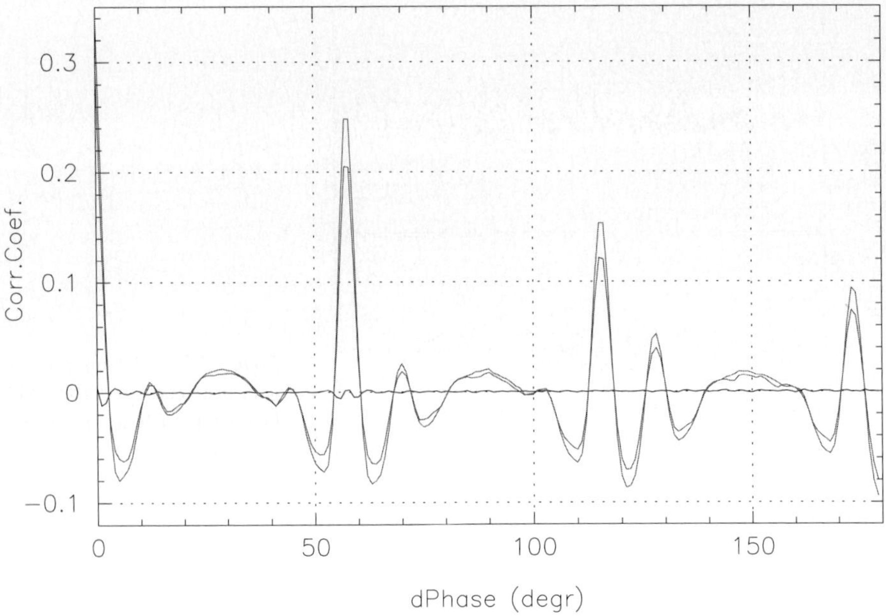

Figure 2.11. Average levels of abscissa-error correlations for stars brighter than magnitude 7 in the published data (shown by the large-amplitude curves, representing the FAST and NDAC data) and the new reduction (shown by the much smaller-amplitude curve). In the new reduction correlation levels have been reduced by a factor 30 to 40 with respect to the published data. The peaks visible in the curves for the published data show the correlations at intervals close to an integer times 58°, the basic-angle. They fold back at 180° and again at 0°, and in total at least 9 such resonances can be distinguished. In the new reduction only one resonance peak can be observed. The correlations have been determined for measurements in the same orbit as a function of their separation in phase on the reference great circle

of error-correlation levels is crucial for determining astrometric parameters in open clusters, as it removes a major uncertainty in data correlations arising from a solution where data for several stars in close proximity on the sky, are combined.

2.4 Grid distortions

The grid (see Fig. 2.4) as well as the projection on the grid are subject to distortions at the mas level. These distortions are taken care of in three stages:

1 Large-scale distortions or instrument parameters;

2 Medium-scale or scan-field level distortions;

3 Small-scale or sub-scan-field distortions.

The first of these is strongly affected by the focal adjustments of the optics, and also covers the orientation of the grid in the focal plane. The second represents the mean positions of the grid scan fields and projection characteristics of the telescopes. The third covers the systematic distortions of the grid lines in the scan fields and any systematic tilting of scan fields along the scan direction. When properly corrected, the systematic errors introduced by the grid can be reduced to an insignificant level, below 0.1 mas. Some characteristics of the three distortions are briefly summarized below. More detailed descriptions are given in Chapter 12.

2.4.1 Large-scale Grid Distortions

The instrument parameters relate the large-scale properties of the actual positions of stellar images on the modulating grid, to an idealized grid with simple projection properties. These simple projection properties state that the grid is projected on a spherical surface, fitting the focal plane of the telescope. This in principle is also the way the grid had been manufactured, and in fact the substrate on which it was printed formed the final element in the optical system (see Section 2.2.1).

By far the largest effects described by the instrument parameters are the rotation of the grid by approximately 5 arcminutes and changes in the scale, resulting from focus corrections and adjustment of the satellite to the space environment. The value found for the rotation was considerably larger than expected, and was not fully recognized until a month into the mission. As a result, measurements over the first month can be affected by relatively large offsets between the expected (IFOV) and actual star positions on the grid during observations, in particular towards the corners of the grid (see also Fig. 11.6, page 296).

The full set of instrument parameters describe the large-scale distortions as a function of position (up to 3$^{\rm rd}$ order) and colour, a total of 21 parameters. As was also done in the NDAC reductions, all parameters (except for the basic angle correction) come in pairs, g and h, such that $g+h$ and $g-h$ give the correction for the preceding and following fields of view respectively. Not all parameters are estimated in the final reductions. Third-order parameters were fitted in the initial stages of the reductions, and for large data sets only. The formal errors on these parameter estimates for single orbits are much larger than the features observed in their long-term evolution. The evolution over the mission of each of these parameters was therefore fitted with a cubic spline, allowing for discrete steps in all scale-related parameters at times of refocusing. Constraining the third-order parameters decreases the formal errors on the first-order parameters, with which they are naturally correlated. A similar effect can be observed from the instrument parameters as derived by FAST, where the third-order parameters had been given fixed values over short in-

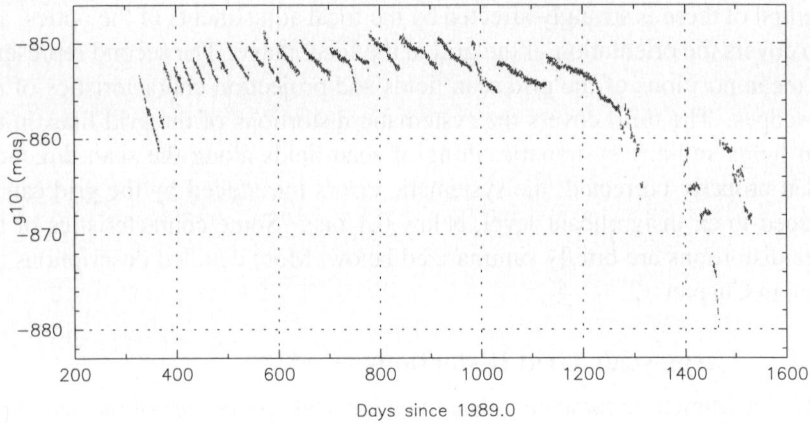

Figure 2.12. The evolution of the mean scale correction, one of the large-scale distortion parameters, shows the effects of refocusing and a few thermal control anomalies over the mission. Notice the similarity with the evolution of the modulation amplitude M_1 as shown in Fig. 2.8

tervals of time (Volume 3 of ESA 1997). As an example of the instrument parameters, Fig. 2.12 shows the changes that took place in the mean scale over the mission, as recorded in the new solution.

A colour term has been included in the instrument model. This was not possible in the original reductions due to the linking of the instrument parameter solution with the abscissa corrections and along-scan attitude solution. There the colour or chromaticity term was instead solved as part of the sphere solution. The colour term was also fitted with a spline function as a function of time, and not solved for individual orbits in the final reductions. A residual colour term is recovered from an accumulation of residuals in the astrometric-parameter determination. Further details on the large-scale distortions can be found in Section 12.2, page 304.

2.4.2 Medium-scale Grid Distortions

Abscissa residuals obtained after fitting the along-scan attitude and instrument parameters have been collected on a 42×46 points grid. This resolution was chosen to coincide with the printing process of the grid: the modulating grid was assembled from 168×46 individually engraved prints or scan fields (see Fig. 2.4). To account for the smearing along the scan over the 2.133 s integration period and to have similar resolutions along and across scan, the resolution in the along-scan direction was set at four scan fields. Residuals have been collected for all orbits over the mission, in such a way that they can be added afterwards, to give properly weighted mean values and dispersions. Solutions for different time intervals over the mission, show no significant evo-

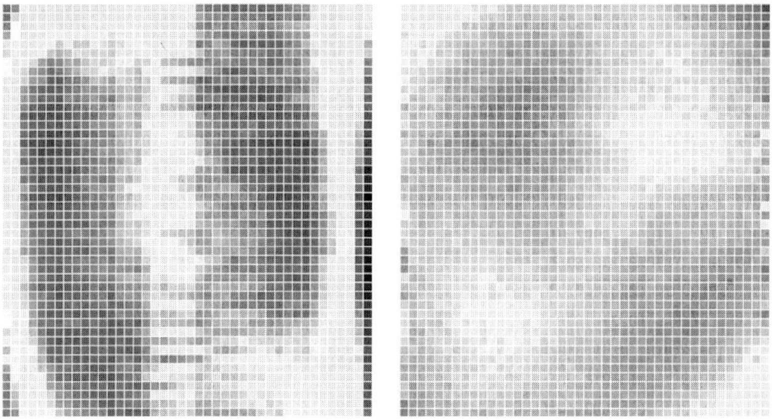

Figure 2.13. The medium-scale distortion over the two fields of view. The mean distortion is shown on the left, the differences between the two fields of view on the right. The total range of the medium-scale distortions is about 2 mas, and the details are accurately reproduced over different intervals of the mission. The scan direction is from right to left

lution in these distortions and in the final reductions one set of corrections, as displayed in Fig. 2.13, has been applied to all mission data.

2.4.3 Small-scale Distortions

The small-scale distortions take care of systematic abscissa residuals on a scale smaller than a scan field. From an accumulation of abscissa residuals in the astrometric parameter solutions, these residuals are collected as a function of the mean ordinates of the transits at a resolution of 1/24 scan field, about 1.5 arcsec. Due to the transverse velocities of the satellite, most transits took place over wider bands. A systematic pattern is observed for these distortions, showing two main contributions: a distortion with a total amplitude of around 0.6 mas, which affected all scan fields in the same way, and variations in the mean tilt of the scan fields for each row. After removal of these two effects, the remaining noise is at a level of less than 0.1 mas (see Fig. 2.14).

The transverse velocity will cause a suppression of the underlying modulation, and the actual distortion per scan field is likely to be at a higher level than observed over an entire row. These effects will therefore be dependent on the transverse scan velocity for a transit. One other interesting aspect is that the systematic modulation permits an accurate calibration of the scale and positioning of the scan fields in the across-scan direction, from which a slightly different scale is derived for the across-scan position than found along-scan. It appears that the *Hipparcos* optical system was slightly astigmatic.

The removal of the small-scale distortions only significantly affects the astrometric parameter solutions for the brightest stars. Most importantly, they

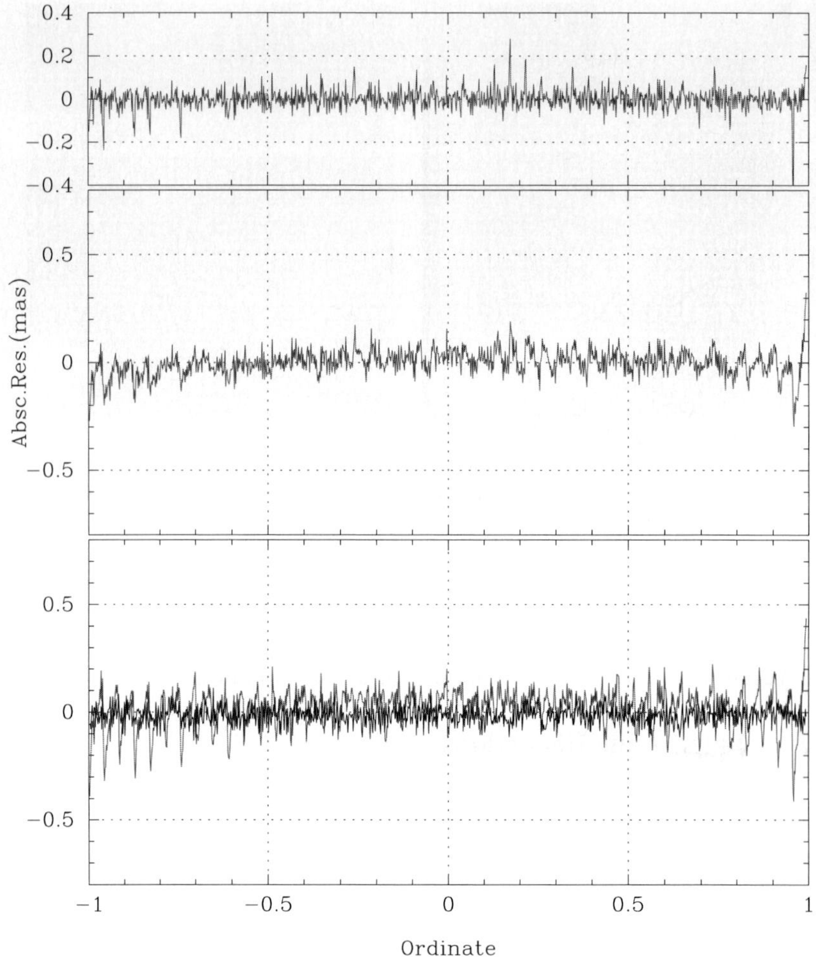

Figure 2.14. The small-scale distortions as observed across the grid for field transits. Bottom: the actual distortions, showing the 46 individual rows of scan fields as a regular pattern. The large fluctuations represent the mean over the two fields of view, the much smaller fluctuations (darker) represent half the difference. Middle: after correcting for a systematic non-linearity of the grid lines. Top: after removing the mean scan-field tilt for each row

show down to what level systematics can now be recognized in the *Hipparcos* data. These data also confirm that the modulating grid more than satisfied the original specifications, and that in principle one can remove systematics from grid distortions down to a level of 0.1 mas, and possibly even better. The equivalent of 0.1 mas on the sky is 16 nm on the grid.

2.4.4 Colour Dependencies

The V − I colour index, as presented in the *Hipparcos* catalogue, was used in modelling the colour dependencies of calibrations. In some cases the catalogue provides an improved value from that used in the NDAC and FAST reductions. Further improvements were obtained for very red and variable objects, for which epoch-resolved colour indices were used, as supplied by Dimitri Pourbaix (Pourbaix and Jorissen 2000; Pourbaix and Boffin 2003; Knapp et al. 2001, 2003). This affects the data in two ways: the colour-dependency of instrument models becomes more stable towards very red stars, and the application to those stars is more accurate. Noise levels on abscissa residuals for very red stars have as a result been significantly reduced.

2.5 Astrometric-parameter solutions

The astrometric-parameter solutions are derived as corrections to the assumed parameters, which at the start of the mission often consisted of no more than the approximate positions, without prior knowledge of proper motion or parallax values. The predicted positions are used to calculate the assumed abscissae for the observations. The differences with the observed positions then form the input for the astrometric-parameter solutions.

2.5.1 Position and Proper Motions Displacements on a Great Circle

The projection of positional corrections into components along and perpendicular to a great circle, is defined by the position of the pole of the circle at the time of observation and the position of the star. Using ecliptic coordinates (λ, β), the positions of the star and the instantaneous pole (or spin axis) of the great circle are given by (λ_*, β_*) and (λ_s, β_s) respectively. As seen from the position of the star on the great circle, the angle between the directions of the ecliptic pole and the pole of the great circle is given by ψ. Using a variant of Fig. 2.1, where $\nu_* = \lambda_* - \lambda_s$, the following relations can be derived:

$$\begin{aligned} \sin\psi &= -\cos\beta_s \sin(\lambda_* - \lambda_s), \\ \cos\psi &= \frac{\sin\beta_s}{\cos\beta_*}. \end{aligned} \tag{2.48}$$

A star will generally have a position ζ relative to the scan circle, or in other words it will be positioned at a distance of $\pi/2 - \zeta$ from the direction of the spin axis at the time of measurement. The transit ordinate ζ relative to the scan circle is derived from:

$$\sin\zeta = \sin\beta_s \sin\beta_* + \cos\beta_s \cos\beta_* \cos(\lambda_* - \lambda_s). \tag{2.49}$$

The transit ordinate will always be between ±0.45 deg, set by the width of the field of view. Taking into account the transit ordinate, the orientation ψ is given by:

$$\begin{aligned} \sin\psi &= -\cos\beta_s \sin(\lambda_* - \lambda_s)/\cos\zeta, \\ \cos\psi &= \frac{\sin\beta_s - \sin\beta_* \sin\zeta}{\cos\beta_* \cos\zeta}. \end{aligned} \tag{2.50}$$

The correction to the stellar coordinates as projected along the great circle is now given by:

$$da = \mathrm{d}\lambda \cos\beta_* \cos\psi + \mathrm{d}\beta \sin\psi. \tag{2.51}$$

The proper motion correction simply describes the time dependence of the positional corrections. Thus, the distribution of the angle ψ for a given star, over the entire mission, will determine how well the two coordinates and their time dependence can be determined. Very similar equations are obtained in equatorial coordinates.

The description of the scanning law in ecliptic coordinates imposes a limit of ±43° on β_s, which imposes restrictions on the range of values obtainable for ψ, depending on the ecliptic latitude β_* of a star. Figure 2.15 shows the actual distribution of ψ for all abscissa measurements of the *Hipparcos* mission, as a function of ecliptic latitude. The complete lack of values for ψ near 0° and 180° for stars near the ecliptic plane reflects in generally larger errors in

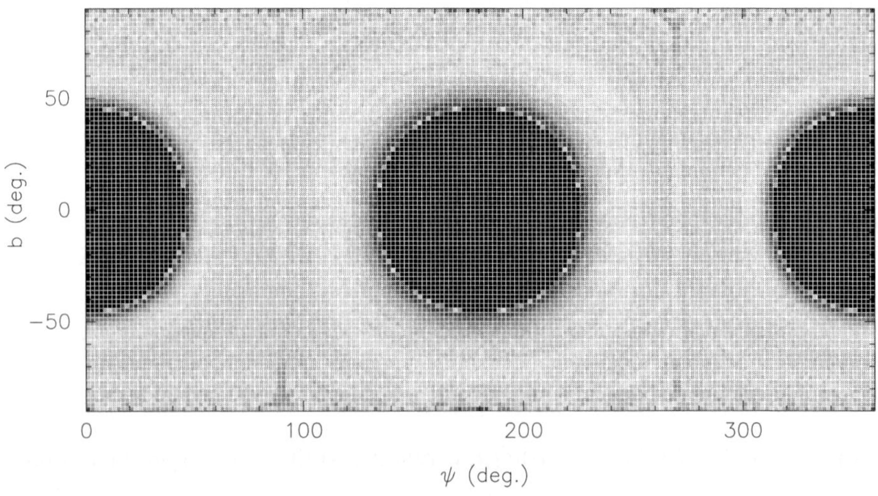

Figure 2.15. Distribution of the scan-orientation angle ψ for stars as a function of ecliptic latitude. The black circles show the areas that can't be reached. The peaks at 90° and 270° are due to observations in the Sun-pointing mode, where $\xi_H = 0°$

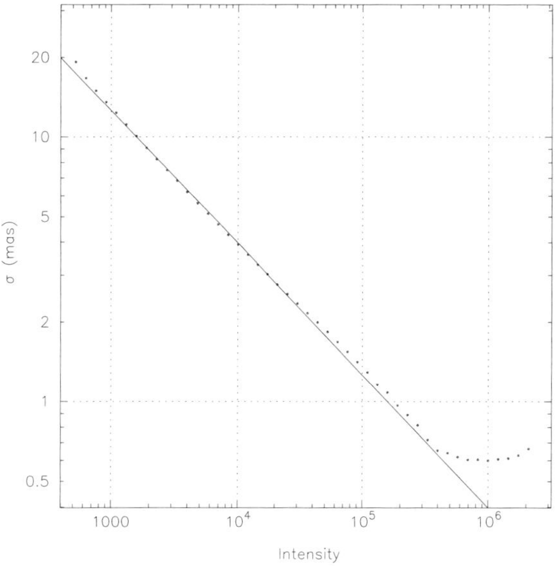

Figure 2.16. The standard deviation for field-transit abscissa residuals as a function of the transit intensity. The relation follows the expected relation for Poisson statistics, indicated by the diagonal line, over more than three orders of magnitude in observed intensity (equivalent to > 7.5 magnitudes). The flattening towards the bright end is due to the noise contributions from, primarily, the attitude reconstruction, at a level of 0.6 mas

positions and proper motions in ecliptic longitude than ecliptic latitude, and, when rotated to equatorial coordinates, in quite significant error correlations between right ascension and declination.

2.5.2 Formal Errors and Error Correlations

The formal error on an abscissa measurement is obtained from the distribution of the abscissa residuals after fitting the astrometric-parameter corrections, and is shown as function of the integrated intensity for field transits in Fig. 2.16. These residuals are dominated by two contributions: the photon noise (Section 2.2.4), which is proportional to the inverse square root of the intensity, and a near-constant noise introduced by reconstructing the along-scan attitude. Noise contributions from inaccuracies in the instrument parameter calibration, including the basic-angle reconstruction, are at a level of 0.3 mas or less, and play no significant role. The photon noise contribution derives from the formal error estimate obtained in fitting the modulated signal. Not only is the photon noise relation clearly preserved, as is shown by the solid line in Fig. 2.16, but also the factor introduced in Eq. 2.47. Thus, the same errors that are observed at frame-transit level, are also observed to be the dominating error for all but the very brightest stars ($Hp < 4$) in the astrometric solutions. The average attitude error contribution over the mission is observed to be around 0.6 mas for the new reduction, considerably less than the value of 2 to 3 mas obtained for the published data. Intermediate reduction results showed that systematic deviations from the Poisson relation were the result of unresolved problems in the reconstruction of the along-scan attitude, such as scan-phase discontinuities, external hits, and attitude convergence problems. Only after all these effects had been taken care off, were proper statistical relations recovered. In the published data, where they were not recognized, and therefore not removed, these problems are the main reason behind the correlated errors in the abscissa residuals for the published data. After removal of these effects, no significant abscissa-error correlations are expected except for the very few areas on the sky, where stars brighter than 3$^{\rm rd}$ magnitude appear with separations less than about 2 degrees (see also Fig. 2.11). The scarcity of these configurations makes it impossible to obtain a reliable empirical estimate of the correlation levels for such cases.

Small corrections to the formal errors were needed to incorporate variations with the colour of the star and the ordinate of the transit (Fig. 2.17). The observed unit-weight standard deviations are applied as a correction factor to the formal errors of the field-transit abscissae, before determining the astrometric parameters. In determining these correction factors, the ordinate dependence was established first, and then the colour dependence, after applying the ordinate dependence. As only few stars have colour indices beyond $\mathrm{V} - \mathrm{I} = 2.0$, the corrections for the reddest stars play almost no role in determining the initial colour dependence, which was done for each orbit as part of the large-scale

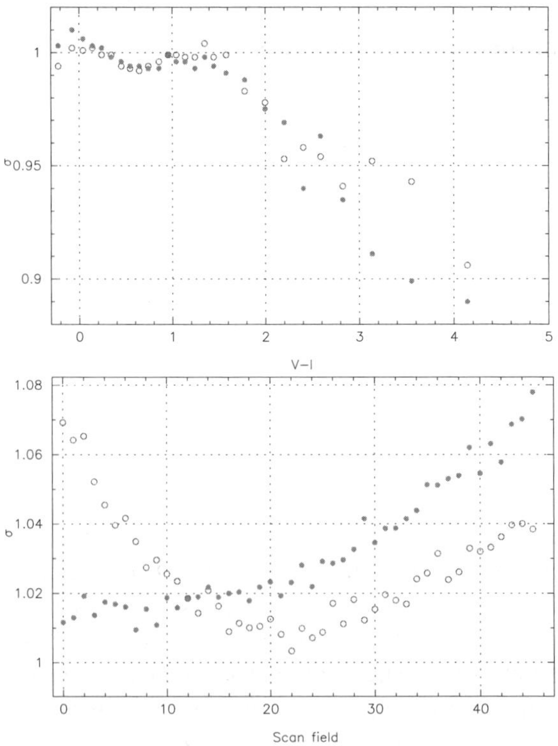

Figure 2.17. The unit-weight standard errors of the abscissa residuals for single star five-parameter solutions, as a function of star colour index $\mathrm{V} - \mathrm{I}$ (top) and as a function of the across-scan scan field (bottom). Open and filled symbols represent the preceding and following fields of view respectively

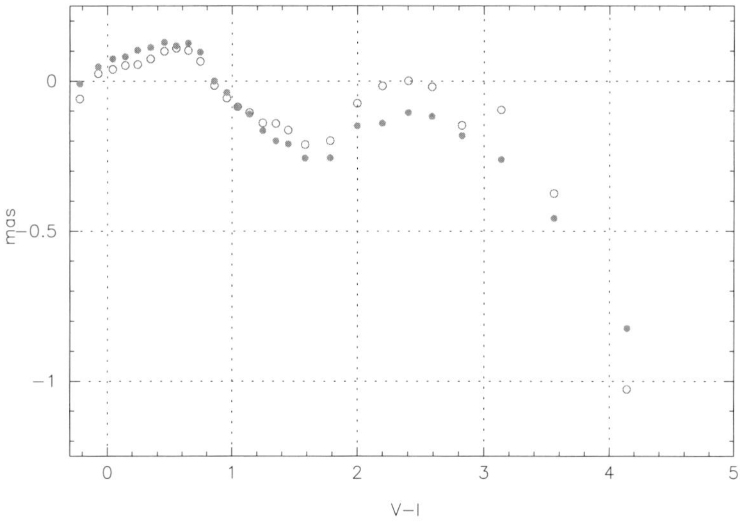

Figure 2.18. The final chromaticity corrections as derived from the accumulation of abscissa residuals in the astrometric parameters solutions. The closed and open symbols refer to data from the preceding and following fields of view respectively. A linear term (variable over the mission) had already been subtracted as part of the instrument-parameters solution

distortion (see above). The entire determination has been iterated several times in order to reach convergence on these corrections. The results shown in the figures are the converged results. The abscissa corrections as a function of ordinate (Fig. 2.14) and as a function of colour index (Fig. 2.18) were determined in the same iteration.

2.5.3 Astrometric Parameters and the Sphere Solution

Astrometric parameters derived here are only based on the new solution, that is, on field transits from a single reduction. Details on the processing of the solutions for the published data can be found in van Leeuwen and Evans (1998).

The solutions for the astrometric parameters are always differential. We start from an assumed set of parameters (which may be no more than a preliminary position). For each measurement (a combined abscissa measurement for all observations in an orbit for the published solution, and a mean abscissa per field transit for the new solution), the predicted apparent position is calculated from the epoch of observation, the assumed astrometric parameters, the velocity vector of the spacecraft (aberration correction) and the positions of the Sun and Earth (light-bending corrections, Section 2.1.3). Together with the instantaneous directions of the satellite axes in the observational reference frame, predicted abscissae are obtained. The differences $\mathrm{d}a$ between these predicted and the observed abscissae are referred to as the abscissa residuals. These residuals are expressed as a (linear) function of the corrections to the astrometric parameters. For this we use the expressions derived above for the projection of the parallax (Eq. 2.15) and positional displacements (Eq. 2.51) along the great circle:

$$\begin{aligned} da &= \cos\psi \mathrm{d}\alpha\cos\delta + \sin\psi \mathrm{d}\delta + \sin\Omega\sin\xi_H \mathrm{d}\varpi \\ &\quad + \cos\psi\Delta t \mathrm{d}\mu_{\alpha\cos\delta} + \sin\psi\Delta t \mathrm{d}\mu_\delta. \end{aligned} \tag{2.52}$$

The values for ψ, $\sin\Omega\sin\xi_H$ and Δt used in Eq. 2.52 are contained in the intermediate astrometric data files for each field transit observation, together with the residuals relative to the accepted solution, the formal error on the observation and other auxiliary information.

The sphere solution has been reduced to collecting residuals from the astrometric-parameter solutions for each orbit, as well as a function of colour (Fig. 2.18) and ordinate (Fig. 2.14), for all orbits together. These accumulations are applied as final corrections to the abscissae, but are generally so small that they have little effect on accuracies. Small corrections for the formal errors are also applied. The formal errors for very red stars, for example, tend to be up to 10 per cent smaller than for stars of more average colour index, at the same total photon count (Fig. 2.17).

2.5.4 Parallax Accuracies in the New Solution

The distribution of the final parallax errors as a function of magnitude shown in Fig. 2.19, nicely summarizes the various aspects described in this Chapter:

- The general trend follows the Poisson statistics down to magnitude 3 to 4, while in ESA (1997) this trend was only maintained down to magnitude 9. This reflects the much lower noise contribution from the attitude reconstruction, as well as a good understanding and representation of the grid distortions at various levels.

- The effects of the scanning strategy can be clearly recognized in the bimodal structure of the distribution: the higher accuracies are found for stars near the ecliptic poles, where the scanning is well distributed in time; the lower accuracies are found near the ecliptic plane, where scanning directions are limited, as shown in Fig. 2.15.

How to verify accuracies at the level of 0.10 to 0.5 mas, is described in Section 3.1.

2.5.5 Summary Information on Disturbed Solutions

A star, for which the data does not fit the simple 5-parameter model, shows positional disturbances that can originate from a variety of causes. These cases

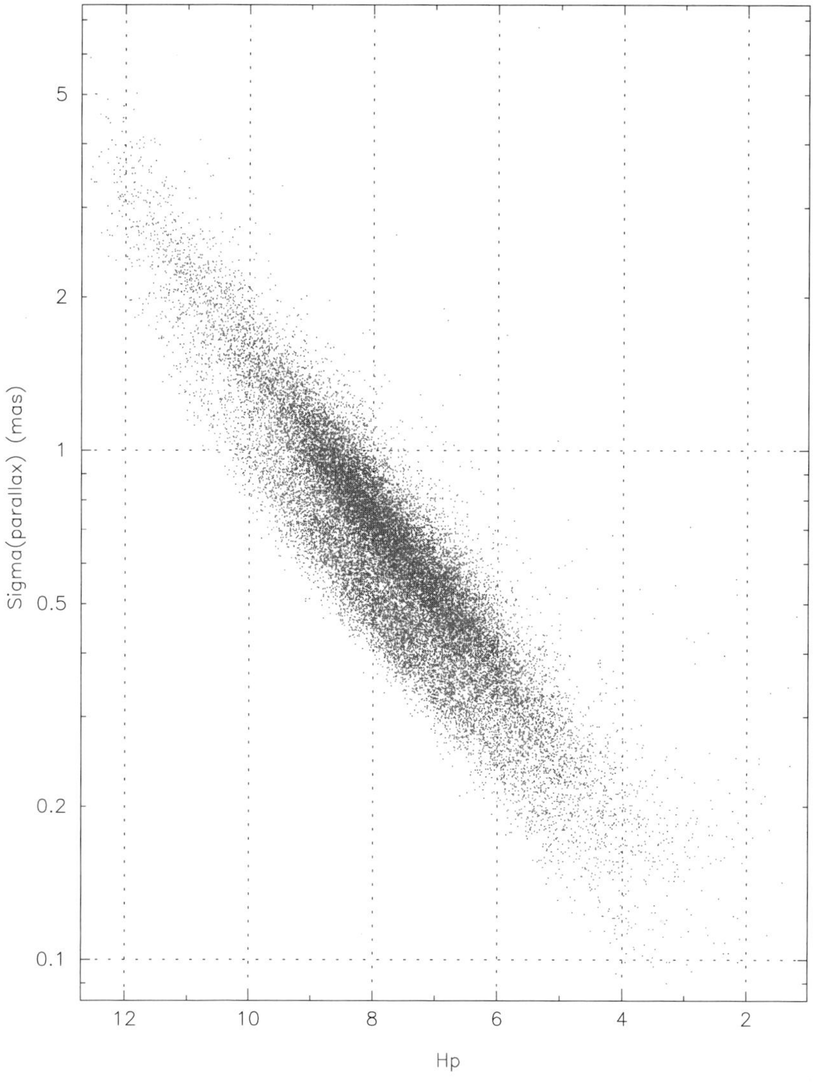

Figure 2.19. The parallax accuracies as a function of magnitude for 26 324 stars with parallaxes errors below 10 per cent. The bi-modal structure is a reflection of the scanning strategy, with stars near the ecliptic poles having much better coverage, and thus higher parallax accuracies, than stars near the ecliptic plane. The separation between the two groups is at ecliptic latitude of 47°

are dealt with in more detail in the next two chapters, here we just summarize the problems that may be encountered.

The most common problem is the presence of one or more companions, which may or may not be physically connected to the observed star. When visible, the companions will add their own modulated signal to that of the target star, and the combined signal has to be examined to reveal details on the relative positions and magnitudes of the stars involved and the astrometric parameters of the target star. This is described in Chapter 4. A companion may also cause disturbance when it is not visible, or only becomes noticeable for a variable star at the minimum of its light curve. The first case concerns orbital motion, described in Section 4.6, the second case is referred to as variability induced motion, described in Section 3.5, where the treatment of the remainder of unresolved cases is also presented. For long-term (longer than about 5 years) orbital period disturbances can be taken care of by introducing a first, and in a few cases a second, order derivative in the proper motion.

PART II

EXPLORING THE HIPPARCOS ASTROMETRIC DATA

Part II describes the use of the Hipparcos astrometric data for single stars and the analysis of non-single stars. The extraction of common parameters for groups of stars (parallaxes and proper motions for star clusters, luminosity calibrations for pulsating stars) is demonstrated through various examples. The extraction of kinematic information for the solar neighbourhood is also shown through examples.

Chapter 3

INDIVIDUAL, SINGLE STARS

3.1 Precisions and accuracies

It will become clear from the discussions that follow that availability of reliable estimates of the formal errors on the parallaxes is crucially important for their scientific exploitation. The problem is, that although precisions can be determined quite easily, accuracies are much harder to obtain. In general, errors can originate from two sources: internal (photon statistics, random noise introduced in instrument modelling), and external errors (systematic imperfections of the instrument modelling). We will first look at what the statistics of the abscissa residuals at different stages of the reductions tell us about the internal errors, and then at establishing the level of the external errors. Then we will look at how errors can affect the scientific interpretation of parallaxes. This is followed by an investigation concerning the accuracies of the proper motions and positions. The various types of disturbed solutions are described. Finally, it is shown that there is no significant transformation required to align the new catalogue with the optical realisation of the ICRS (the published *Hipparcos* catalogue). A comparison is made with a set of radio sources with high-accuracy astrometric data.

3.1.1 The Internal Errors

In the new reduction, the dominant error source by far for most of the data is the photon noise of the underlying measurements: the total photon-count as recorded for a star during its transit through the FOV. The new reduction of the *Hipparcos* data preserves this information with the reduced data from the first measurements to the final astrometric parameter determinations, and uses it to assess, verify and understand the characteristics of the formal errors and dispersions.

As described in Section 2.2.1, the one-dimensional positional measurements of *Hipparcos* are derived from the modulation phase β_3 (see Eq. 2.33) of light received from the stellar image as it transits the modulating grid in the focal plane of the instrument. This phase relates to a position relative to the grid pattern. The *a priori* (apparent) position of the star and the reconstruction of the satellite's attitude identify the actual grid line the star is transiting at any time of the observation. The projection of the grid on the sky for each FOV is described by a set of instrument parameters (Section 12.2).

The accuracy of the modulation-phase determination depends on the relative amplitude of the modulation ($M_1 \equiv \beta_2/\beta_1$) and the total photon count of the observation. As was described in Section 2.2.4, the average error on the modulation phase is determined as (Eq. 2.47, page 54):

$$\sigma_{\beta_3} \approx \frac{0.25}{M_1\sqrt{I_{\mathrm{tot}}}} \quad \mathrm{arcsec}, \tag{3.1}$$

where I_{tot} is the total photon count accumulated for the signal. The factor 0.25 in Eq. 3.1 varies by about 1 per cent over the mission and between the two fields of view. It represents the average over each field of view, and is affected by changes in focus. The average variation, as observed for one field of view over the mission, is shown in Fig. 3.1. Transits have been assigned formal errors

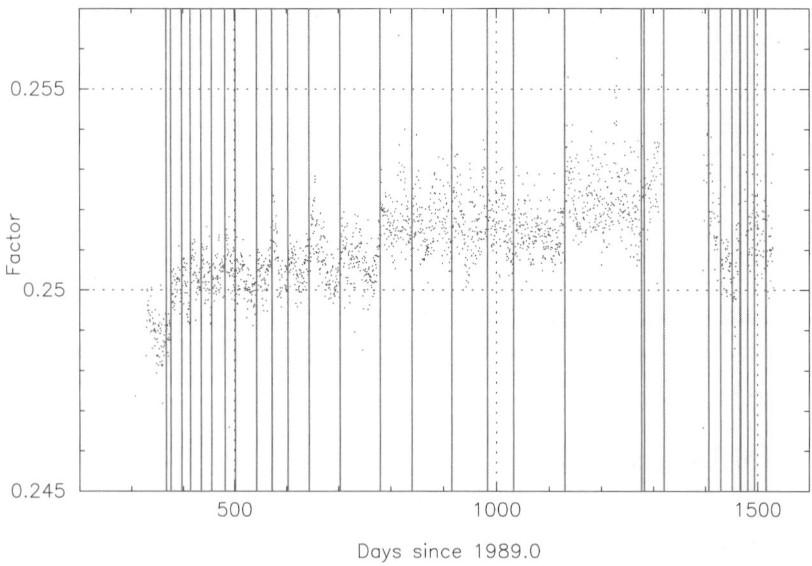

Figure 3.1. The evolution of the scale factor in Eq. 3.1 over the mission, and for one field of view. Re-focusing instances are shown by grey lines

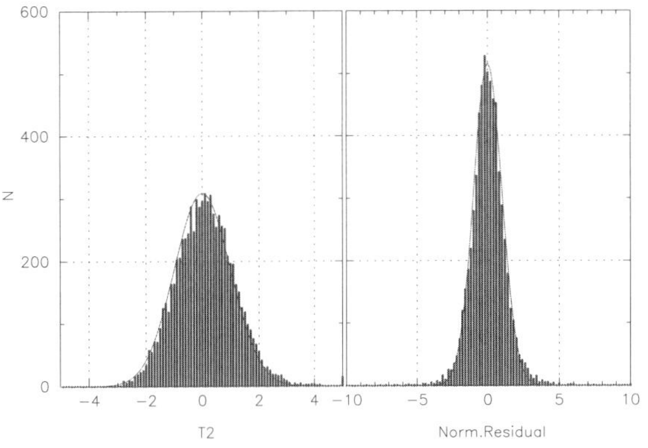

Figure 3.2. Statistics on field transits. Left: histogram of the F2 statistics for the formation of field transits from frame transits. Right: normalized residuals between field transit abscissae and the predicted positions, for the third iteration. The curves show the equivalent Gaussian distribution for the same number of observations. The data are for orbit 409

according to the observed total photon count and the nominal amplitude M_1 of the first harmonic, using Eq. 3.1, where the M_1 is slowly varying over the mission, and is a function of field of view and colour of the star. Its value is derived from the photometric calibrations as performed by NDAC (Evans et al. 1992). For the faintest stars a nominal background correction (see Chapter 7) is applied to the total photon count, depending on the total number of samplings.

Combining data from individual measurements to field transits is the first test on the reliability of the formal errors. An individual measurement spans a fixed amount of 32/15 s in time, while a complete transit of the field of view takes on average 9 of these units of time. There will generally be between 8 and 10 observations from which a field transit is constructed. The agreement between those observations, as based on their formal errors, is shown through the F2 statistics for the determination of the mean values, and through the distribution of unit-weight residuals relative to the predicted positions. These distributions have been collected for all orbits and formed part of the trend monitoring. An example is shown for orbit 409 in Fig. 3.2.

The final check on the internal errors is provided by the reconstruction of the astrometric parameters. As the majority of stars obey a simple 5-parameter astrometric model, the statistics of the residuals for all stars that don't show signs of duplicity (and would normally have disturbed modulation phases), can be used as a further check on the assigned formal errors. This final check showed originally the presence of a non-Gaussian noise component, which was identified as caused largely by the scan-phase discontinuities and unidentified hits described in Section 1.4.3 (see also van Leeuwen and Fantino 2005). The dispersions as a function of the total photon count for the contributing field transits are shown in Fig. 2.16, page 65. The flattening of the relation for the highest counts is due to limitations of the attitude modelling, which adds a variable noise to the reduced data. The dispersion calculations have not been "clipped"

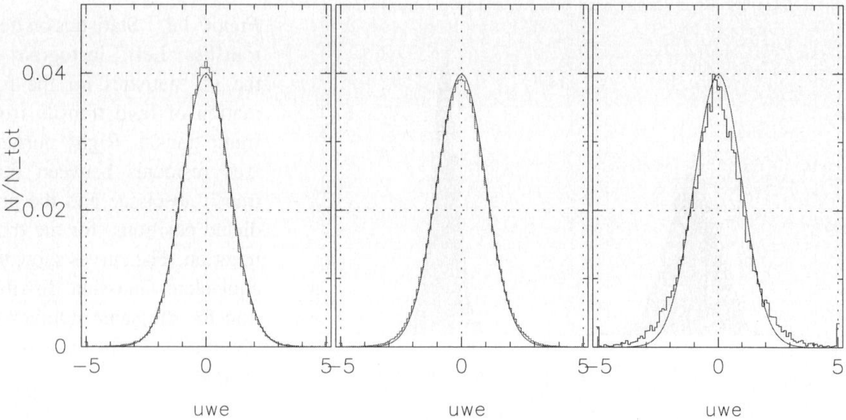

Figure 3.3. Three examples of the distribution of the unit-weight abscissa residuals, for, from left to right, faint (I=1096), medium (I=17 400) and bright (I=275 400) transits. In each case the histogram of residuals is compared with the expected Gaussian distribution

for outliers and contain everything accepted in the astrometric parameter determinations. The distributions of the residuals are shown for three different intensities in Fig. 3.3. It is clear from that figure that for the highest intensities the distribution is not ideal. The main reason for this is the difficulty to provide a sufficiently accurate reconstruction of the along-scan attitude while ensuring that data from both fields contributes significantly. This often implies that the more accurate bright transits can't be used at their full strength in the presence of only fainter stars in the other field of view. In such cases there will be a higher than average noise contribution from the along-scan attitude, creating the extended wings in the residuals distribution for the brightest transits. This also means that formal errors on the astrometric parameters for the brightest stars are generally less well determined than those on the fainter stars.

Overall the internal consistency of the formal errors indicates that the disturbing influence of the data modelling has been small in the new reduction. In particular, the noise contribution from the along-scan attitude is 4 to 5 times lower in the new reduction when compared with the published data.

3.1.2 The External Errors

The difficulties and limitations of testing the external errors in the *Hipparcos* parallaxes are described in detail by Lindegren (1995). Lindegren mentioned four methods for gaining information on the reliability of the formal errors quoted for the *Hipparcos* astrometric data:

1 Careful analysis and modelling of all error sources extraneous to the data reduction process;

2 Comparing with independent measurements of the same stars;

3 Introducing spectrophotometric, kinematic or dynamical information to constrain the ratio between observed and true parallaxes;

4 Through a statistical interpretation of the distribution of observed parallaxes.

The first of these has been covered in the preceding section. The second could still work for the published data, although only for a very small number of objects, using parallaxes measured with large radio telescopes (see further Arenou et al. 1995), but parallax accuracies as reached in the new reduction (down to 0.1 mas) are out of reach for this method. The third method effectively eliminates from the data the possibility to observe a non-compliance with assumed models. The fourth and final method relies almost entirely on the distribution of negative parallaxes, of which there are very few left for the highest parallax accuracies. This is the method used here to establish the level of the external errors on the new *Hipparcos* parallaxes, and is briefly explained here. For a more detailed derivation of the equations below see Lindegren (1995).

The Probability Density Function (PDF) f_P of the observed parallaxes p represents the convolution of two PDFs: the distributions of the actual errors f_X and of the true parallaxes ϖ: f_Π, where:

$$f_X(p - \varpi) \equiv f_{P|\Pi=\varpi}(p). \tag{3.2}$$

The convolution integral is obtained as the marginal PDF:

$$f_P(p) = \int_{-\infty}^{\infty} f_\Pi(\varpi) f_X(p - \varpi) \mathrm{d}\varpi. \tag{3.3}$$

For the true parallaxes we know that their value can never be less than zero, which is why the negative tail in the observed parallax distribution becomes a powerful tool in "measuring" the actual error distribution function f_X. Lindegren (1995) gives a comprehensive overview of how this criterion has been used and has evolved over the past 80 years to assess parallax errors. The distribution of negative parallaxes is under all conditions dominated by the distribution of the parallax errors.

Following Lindegren, we write the joint PDF for the observed parallaxes as:

$$f_{P,\Pi}(p, \varpi) = f_P(p) f_{\Pi|P=p}(\varpi), \tag{3.4}$$

where the second term on the right is the PDF of the true parallax ϖ, conditional on the observed value p. Now, using Eq. 3.3 we can write:

$$f_{\Pi|P=p}(\varpi) = \frac{f_\Pi(\varpi) f_X(p - \varpi)}{\int_0^\infty f_\Pi(\varpi') f_X(p - \varpi') \mathrm{d}\varpi'}, \tag{3.5}$$

which is Bayes's formula, where f_Π is called the prior density and $f_{\Pi|P=p}$ the *a posteriori* density, based on the observation p.

We now consider a sample of stars for which we assume that the true probability density functions f_Π and f_X apply. Each observed value p_i provides an *a posteriori* density for ϖ, using Eq. 3.5. Averaging this over all N stars in the sample gives the PDF:

$$h(\varpi) = \frac{1}{N}\sum_{i=1}^{N} \frac{f_\Pi(\varpi) f_X(p_i - \varpi)}{\int_0^\infty f_\Pi(\varpi') f_X(p_i - \varpi')\mathrm{d}\varpi'}. \tag{3.6}$$

As $N \to \infty$, h converges to f_Π. Under the assumption that we start off with a reasonable estimate of h, the following iterative formula is obtained:

$$h_{k+1}(\varpi) = h_k(\varpi) \times \frac{1}{N}\sum_{i=1}^{N} \frac{f_X(p_i - \varpi)}{\int_0^\infty f_\Pi(\varpi') f_X(p_i - \varpi')\mathrm{d}\varpi'}, \tag{3.7}$$

where k represents the iteration step. Again following Lindegren, we use as initial distribution function:

$$h_0(\varpi) = \begin{cases} 1 & \text{if } \varpi \geq 0 \\ 0 & \text{otherwise} \end{cases} \tag{3.8}$$

Eq. 3.7 is known as the Richardson-Lucy deconvolution algorithm (Richardson 1972; Lucy 1974).

In our application of the deconvolution we assume that the PDF f_X contains two contributions, which are both Gaussian distributions. The first is the formal error described above, and the second the external error or "background noise" in the parallax reference frame. The latter will normally only play a significant role for those stars with the smallest formal errors.

The parallaxes for all 98723 single stars (after iteration 12, not the final data) with a new astrometric solution were sorted on increasing formal error, and divided in ten deciles. The limits for each decile, the number of negative parallaxes and the average parallax error

$$\langle\sigma\rangle \equiv \sqrt{\frac{\sum_{i=0}^{N}\sigma_i^2}{N-1}} \tag{3.9}$$

are shown in Table 3.1. Also shown is the "apparent" unit weight standard deviation of the negative parallaxes. It is defined as

$$\sigma_{\mathrm{neg}} \equiv \sqrt{\frac{\sum_{i=0}^{n}(\varpi_i/\sigma_i)^2}{\sum_{i=0}^{n}(1/\sigma_i)^2}}, \tag{3.10}$$

Table 3.1. Distribution of formal errors over 10 deciles, giving the number of stars, number of negative parallaxes, lower and upper error limits, the mean error for the interval, the standard deviation for the negative parallaxes, and the number of iterations applied

N	neg	σ_l	σ_u	$\langle\sigma\rangle$	$\sigma_{\rm neg}$	itr
9873	32	0.09	0.47	0.36	0.79	3
9873	64	0.47	0.60	0.54	0.91	8
9873	72	0.60	0.72	0.66	0.75	11
9873	97	0.72	0.82	0.77	0.75	12
9873	153	0.82	0.92	0.87	0.80	15
9873	230	0.92	1.03	0.98	0.74	12
9873	335	1.03	1.16	1.10	0.74	11
9873	438	1.16	1.33	1.24	0.73	14
9873	480	1.33	1.61	1.45	0.77	13
9866	813	1.61	3.50	2.05	0.86	4

where the stars are sorted on parallax, starting at the lowest (negative) value, and a total of n negative parallaxes are found. In case the real parallaxes for these stars are all equal to zero, and the formal errors are correct, the expectation value for $\sigma_{\rm neg}$ would be one. All observed values are below one, as should be expected when the formal errors are correct and the objects concerned are not infinitely far away. It shows that these negative parallaxes are in fact measurement residuals relative to small but significant actual parallaxes: even the negative parallaxes still contain a little information. The final decile is troubled by a wide range of formal errors and was not fitted. The same was found in the application by Lindegren (1995).

The most interesting decile is the first, with the highest accuracy parallaxes. To what extent are these formal errors trustworthy? Fitting experiments for this decile were performed with a variable amount of external noise added. The best fits were obtained for an additional noise of around 0.15 mas, as shown in Fig. 3.4. Application of this noise to subsequent deciles has very little effect due to the higher formal errors.

It is possible that the additional background noise is non-Gaussian in character, as it is almost certainly affected by the local distribution of stars in number and brightness, but without an external data set of higher quality than the *Hipparcos* reduction this is not possible to assess.

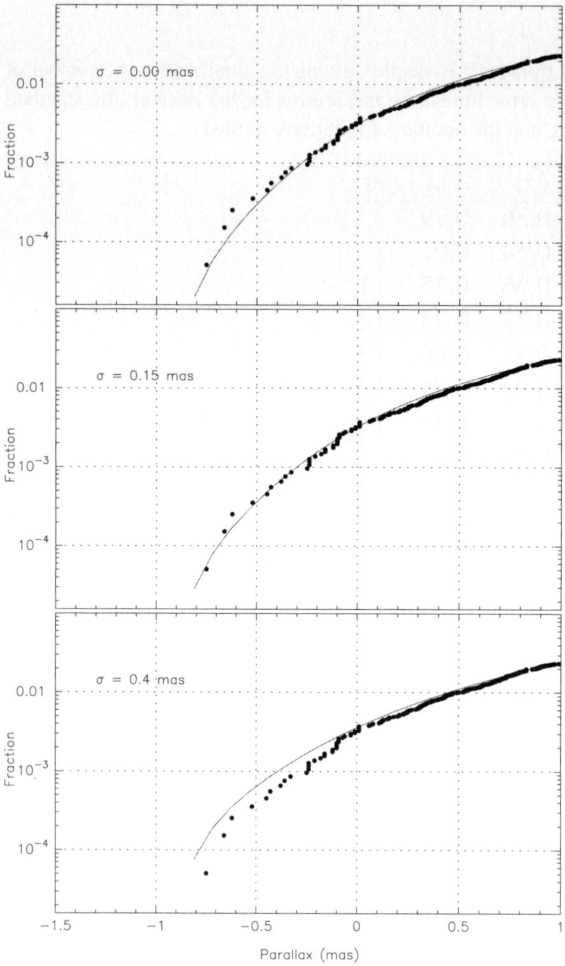

Figure 3.4. The cumulative distribution of the smallest parallaxes for the highest accuracy decile (see text). The three graphs show the fitted deconvolution model with no noise added (top), 0.15 mas noise added (middle) and 0.40 mas noise added

3.2 Correlations

3.2.1 Abscissa Error Correlations

Considering the very much reduced level of abscissa-error correlations in the new reduction (see Fig. 2.11, page 58), it is is interesting to see what for example the parallax differences between the old and the new solution show in terms of correlations as function of separation on the sky. Claims of large-scale correlations in the parallax errors for the published data (at a level of 5 to 10 degrees) have been made by Narayanan and Gould (1999), but such levels appeared to be unrealistic simply from the statistics of Great-Circle solutions in common between stars at various separations (van Leeuwen 1999b), to which we refer as the level of coincidence (Fig. 3.6). However, a map of the parallax differences between the old and new solution clearly does show small-scale

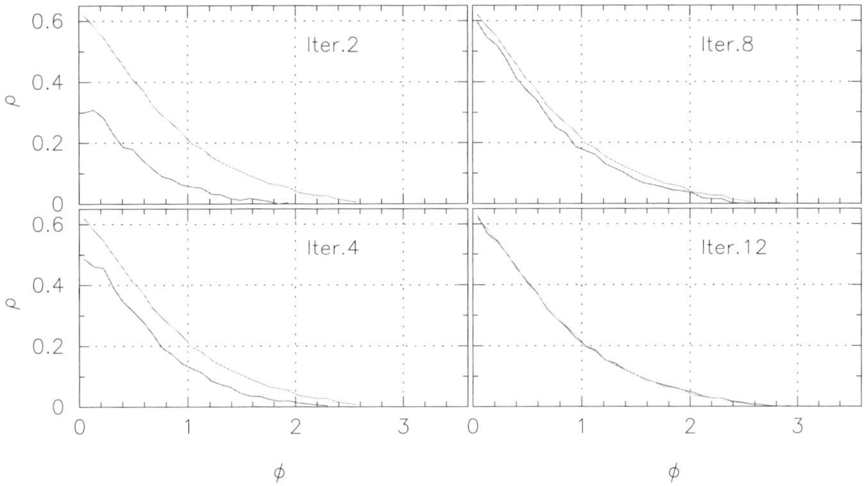

Figure 3.5. Correlations (ρ) in the differences between the parallax determinations in the old and new solutions as a function of stellar separation (ϕ). The figures show how the correlations developed over the early iterations and then stabilized. The upper curve in each figure is the situation for the final iteration. A comparison with the abscissa correlations (Fig. 2.11) shows clearly that the source of these correlations is entirely in the published data

structure. A correlation test over the entire catalogue (excluding large outliers) shows a very significant, and higher than expected, correlation level of around 0.6 for very small separations, completely disappearing at separations beyond 2.6 degrees. The typical Full Width Half Maximum (FWHM) of the correlation distribution is between 1.2 and 1.4 degrees. As the iterations of the new reduction progressed, these correlations became more apparent, as is shown for a few examples in Fig. 3.5. It should be noted here that these kind of corre-

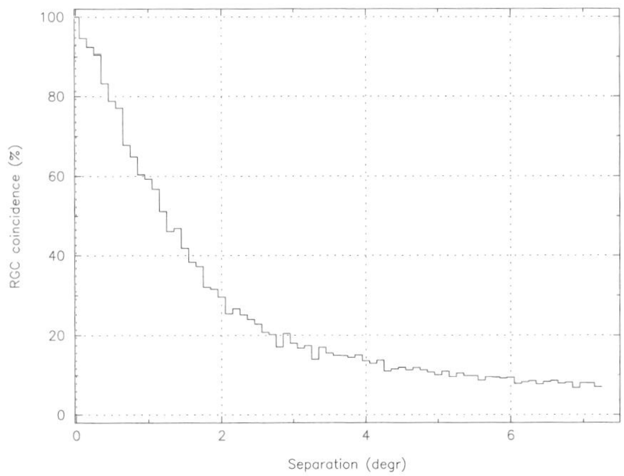

Figure 3.6. The coincidence of Reference Great Circles (RGCs) between a target star and all *Hipparcos* stars in a 7°.3 radius around it. The histogram shows the averaged results for 20 randomly chosen target stars as a function of distance from the target star (from van Leeuwen 1999b)

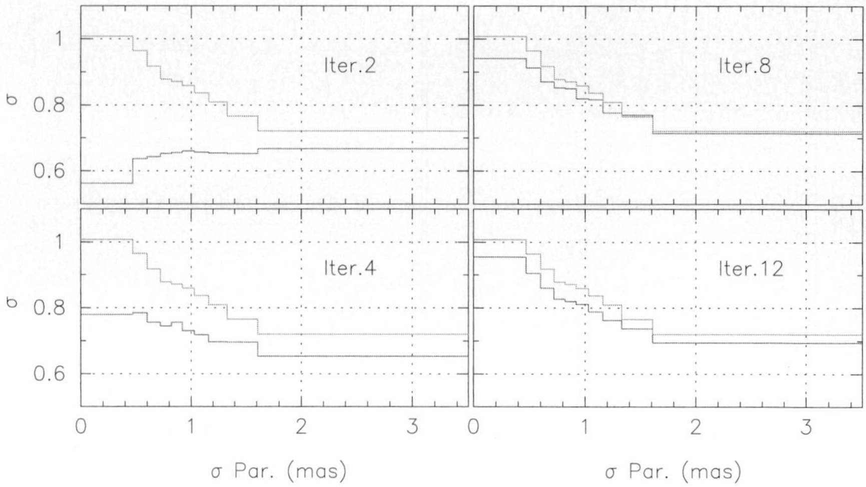

Figure 3.7. The dispersion of the unit weight differences between the parallax determinations in the old and new solution. The data have been divided in 10 deciles as described in the previous section. The applied weights are derived from the formal errors in the published data. As for the highest accuracy data the new errors are much smaller, the dispersions move to one for that group, and get close to one for the next few groups. The upper curve in each figure is the situation for the final iteration

lations can also be observed when examining the differences between the FAST and NDAC parallax determinations in the published data. Their scale length is, however, considerably smaller than was claimed to exist by Narayanan and Gould (1999), and is in agreement with what could be expected based on the average levels of coincidence of Great Circles for neighbouring stars (Fig. 3.6).

A related statistic is provided by the development over the iterations of the differences between the old and new parallax determinations. Considering that for the highest accuracy parallax determinations in the new solution the formal errors are about 3 to 5 times smaller than in the published data, these differences, when weighted by the formal errors of the published solutions, should develop towards a standard deviation close to one. This is shown to happen in Fig. 3.7 for the same iterations as shown in Fig. 3.5. These iterations are described in more detail by van Leeuwen and Fantino (2005). What they show is a gradual "loss-of-memory" of the starting conditions: the astrometric parameters of the published solution. How fast this process takes place depends much on the maximum weight ratio between the two fields of view as applied in the along-scan attitude solution (see further Section 10.5.3, page 272).

The nearly stable dispersion level of 0.7 for the lowest-accuracy group is probably a reflection of a phenomenon noticed before in the published data: in the comparisons between the NDAC and FAST data it was noted that the correlations between the results for the faintest stars did not get close to 100 per cent.

This would have been the case if the noise on the data for these stars was dominated by photon statistics. The photon-noise statistics are identical for any reduction of the *Hipparcos* data. Instead, there was an additional noise component for the faintest stars. This noise component was largely removed in the new reduction, as shown through the improved accuracies and the variances of the lowest intensity abscissa measurements (Fig. 2.16).

3.2.2 Proper Motion Error Correlations

The scan performed by *Hipparcos* inevitably leads to coverage variations depending on direction on the sky. These variations are simplest when expressed in ecliptic coordinates, reflecting the global symmetry of the scan with respect to the ecliptic plane. The error ellipses for the positional and proper motion coordinates tend to be elongated in the direction of ecliptic longitude. All calculations of the astrometric parameters, however, were carried out, and are presented, in equatorial coordinates. For error discussions a transformation of these parameters and their covariance matrix is often beneficial. In other situations, one may like to express the proper motions in the galactic reference frame, and still have available the proper covariances. In the following the transformation equations for proper motions between different reference systems are described.

The change with time in the position vector $\boldsymbol{R}$, expressed here in spherical coordinates:

$$\boldsymbol{R} = R \begin{bmatrix} \cos\alpha\cos\delta \\ \sin\alpha\cos\delta \\ \sin\delta \end{bmatrix}, \tag{3.11}$$

where R is the distance of the object, was derived in Section 2.1, Eq. 2.30 as:

$$\dot{\boldsymbol{R}} = R \begin{bmatrix} \cos\alpha\cos\delta & -\sin\alpha & -\cos\alpha\sin\delta \\ \sin\alpha\cos\delta & \cos\alpha & -\sin\alpha\sin\delta \\ \sin\delta & 0 & \cos\delta \end{bmatrix} \cdot \begin{bmatrix} \dot{R}/R \\ \mu_{\alpha*} \\ \mu_\delta \end{bmatrix}, \tag{3.12}$$

where $\mu_{\alpha*} \equiv \mu_\alpha \cos\delta$. The term $\dot{R}$ represents the radial velocity of the object. The transformation matrix is orthogonal and can also be written as:

$$\dot{\boldsymbol{R}} = R \begin{bmatrix} \boldsymbol{r} & \boldsymbol{p} & \boldsymbol{q} \end{bmatrix} \cdot \begin{bmatrix} \dot{R}/R \\ \mu_{\alpha*} \\ \mu_\delta \end{bmatrix}. \tag{3.13}$$

If $\boldsymbol{z} = (0, 0, 1)$ denotes the North pole of the coordinate system, then the following relations hold:

$$\begin{aligned} \boldsymbol{p} &= \langle \boldsymbol{z} \times \boldsymbol{r} \rangle \\ \boldsymbol{q} &= \boldsymbol{r} \times \boldsymbol{p}. \end{aligned} \tag{3.14}$$

These relations we use to derive the relations between proper motions in different coordinate systems.

Restricting the discussion to proper motions only (the radial direction will not change in the new coordinate system), we have in an arbitrary orthogonal right-handed coordinate system:

$$\dot{\boldsymbol{u}} = \dot{\boldsymbol{R}}/R = \begin{bmatrix} \boldsymbol{p}_c & \boldsymbol{q}_c \end{bmatrix} \cdot \begin{bmatrix} \mu_{l*} \\ \mu_b \end{bmatrix}, \tag{3.15}$$

where $\boldsymbol{p}_c$ and $\boldsymbol{q}_c$ can be derived using Eq. 3.14 and the position vector $\boldsymbol{z}_c$ of the pole of the new coordinate system within the equatorial reference system.

To relate the proper motions and their covariance matrix in the new coordinate system to those presented in the equatorial system, a simple set of equations has to be solved:

$$\begin{bmatrix} \boldsymbol{p}_c & \boldsymbol{q}_c \end{bmatrix} \cdot \begin{bmatrix} \mu_{l*} \\ \mu_b \end{bmatrix} = \begin{bmatrix} \boldsymbol{p} & \boldsymbol{q} \end{bmatrix} \cdot \begin{bmatrix} \mu_{\alpha*} \\ \mu_\delta \end{bmatrix}. \tag{3.16}$$

Multiplying left and right by $\begin{bmatrix} \boldsymbol{p}_c' \\ \boldsymbol{q}_c' \end{bmatrix}$ gives:

$$\begin{bmatrix} \mu_{l*} \\ \mu_b \end{bmatrix} = \begin{bmatrix} \boldsymbol{p}_c'\boldsymbol{p} & \boldsymbol{p}_c'\boldsymbol{q} \\ \boldsymbol{q}_c'\boldsymbol{p} & \boldsymbol{q}_c'\boldsymbol{q} \end{bmatrix} \cdot \begin{bmatrix} \mu_{\alpha*} \\ \mu_\delta \end{bmatrix}. \tag{3.17}$$

The transformation between proper motions in different reference systems amounts to an orthogonal rotation of the local coordinates. In the matrix of Eq. 3.17 the following relations apply:

$$\begin{aligned} \boldsymbol{p}_c'\boldsymbol{p} &= \boldsymbol{q}_c'\boldsymbol{q} &= \cos\phi \\ \boldsymbol{p}_c'\boldsymbol{q} &= -\boldsymbol{q}_c'\boldsymbol{p} &= \sin\phi, \end{aligned} \tag{3.18}$$

where ϕ is the local inclination of the coordinates in the new system relative to the equatorial system. Using the same formalism as in Section 2.2.3 (page 50), Eq. 2.42 and 2.43, the Jacobian matrix for the proper motion covariance matrix transformation is given by:

$$\mathsf{J} = \begin{bmatrix} \cos\phi & \sin\phi \\ -\sin\phi & \cos\phi \end{bmatrix}. \tag{3.19}$$

When applying it to the inverse of the covariance matrix, in order to retrieve formal errors, the inverse or transpose of J should be used in the Jacobian product.

Figure 3.8 shows some aspects of the proper motion correlation coefficients: close to the ecliptic plane larger values are found than near the ecliptic poles, and when expressed in ecliptic coordinates, the correlation coefficients tend to

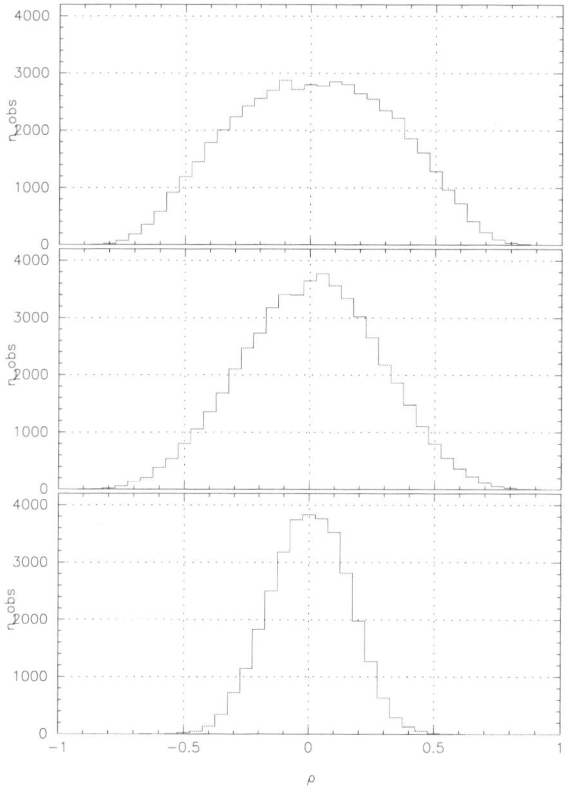

Figure 3.8. Histograms of the correlation coefficient between the two proper-motion components. The two top diagrams are for single stars in a zone around the ecliptic plane ($|\beta| < 40$ degrees). The top diagram for equatorial, the middle diagram for ecliptic coordinates. The bottom diagram represents the two ecliptic-pole regions, in ecliptic coordinates

be smaller than when expressed in equatorial coordinates. It is also clear from these histograms that the correlation coefficients are often quite significant and have to be accounted for when interpreting proper motions, and in particular when transforming data to a different reference system.

3.3 Parallaxes

In this section we look at the application of the astrometric parameters in scientific exploration: how this is affected by uncertainties in those parameters and how the level of these uncertainties can be measured or determined.

3.3.1 The Effects of Parallax Errors on Derived Parameters

When interpreting parallax information in the form of distances D or distance moduli $m - M$ (observed minus absolute magnitude), it is important to realize the effects of the non-linear relation between the measured and derived quantities:

$$\begin{aligned} D &= 1/\varpi, \\ m - M &= 5(\log D + 2) = 10 - 5\log\varpi, \end{aligned} \tag{3.20}$$

have on the error propagation (here D is measured in kpc, and ϖ in mas). When we look at the way the errors on ϖ propagate in first and second order into our determination of D and $m - M$, we find:

$$\begin{aligned} \delta D &\approx \frac{\delta\varpi}{\varpi^2}\left(-1 + \frac{\delta\varpi}{\varpi}\right), \\ \delta(m - M) &\approx \frac{-5}{\ln 10}\cdot\left[\frac{\delta\varpi}{\varpi} - 0.5\left(\frac{\delta\varpi}{\varpi}\right)^2\right]. \end{aligned} \tag{3.21}$$

When the ratio $\delta\varpi/\varpi$ gets above 0.1 (a 10 per cent formal error), the relative contribution of the quadratic terms becomes important, and interpretation of the derived quantities more complicated. The general rule is therefore not to use derived parameters when the relative error on a parallax is larger than 10 per cent.

3.3.2 The Lutz-Kelker Effect

The relative error on the parallax also affects what we determine to be the "most probable distance" for a star. Even for relative errors as small as 2.5 per cent there is a systematic offset of 0.01 mag. between the most likely and the directly derived distance modulus (Lutz and Kelker 1973). This effect is generally referred to as the "Lutz-Kelker" effect, after the two authors Thomas E. Lutz and Douglas H. Kelker who first presented a detailed mathematical description of its effects on estimates of absolute magnitudes. In the following we shall refer to it as the LKE.

The principle behind the LKE is simple: we examine a sample of stars for which the *observed* parallaxes are all identical and given by ϖ_0, and for which the formal errors are all the same, σ. The distribution of the formal errors is assumed Gaussian:

$$f(\varpi_0|\varpi) = \frac{1}{\sqrt{2\pi}\sigma}\exp\left(-\frac{(\varpi - \varpi_0)^2}{2\sigma^2}\right). \tag{3.22}$$

The question is, what is the probability distribution of parallaxes ϖ for the stars that give this observed value of ϖ_0. If space were one-dimensional, there would be no problem: the distribution would simply be given by the Gaussian distribution of the formal errors. But space is 3-dimensional, and as a result there are more stars for parallaxes smaller than ϖ_0 than there are for parallaxes larger than ϖ_0. Assuming a homogeneous stellar density, the number of stars

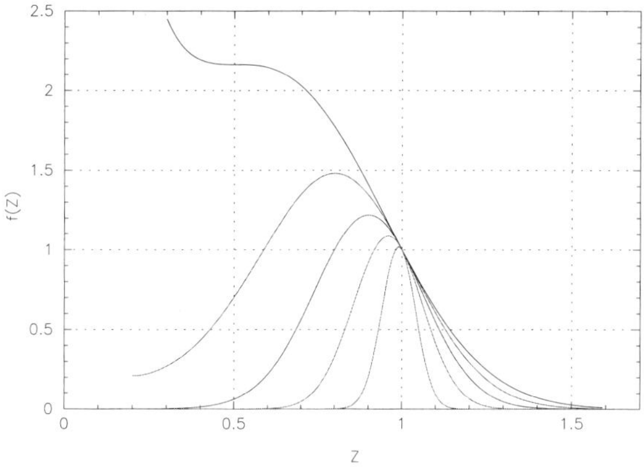

Figure 3.9. The probability distributions for the ratio $Z = \varpi/\varpi_0$ at different values of the relative error ϖ_0/σ. Starting from the innermost curve, the relative errors σ/ϖ_0 increase from 0.05 to 0.10, 0.15, 0.20 and ultimately 0.25. Beyond a value of 0.15 the observed parallaxes are clearly no longer useful in this interpretation

increases with decreasing parallax as:

$$N(\varpi) \propto \varpi^{-4}. \tag{3.23}$$

The probability function for the distribution of the actual parallaxes given an observed parallax ϖ_0 and formal errors σ is now given by:

$$f(\varpi_0|\varpi) \propto \left(\frac{\varpi_0}{\varpi}\right)^4 \exp\left(-\frac{(\varpi/\varpi_0 - 1)^2}{2(\sigma/\varpi_0)^2}\right), \tag{3.24}$$

where the scaling is chosen such that $f = 1$ for $\varpi = \varpi_0$. This function is shown for different relative noise values in Fig. 3.9. There are two important quantities in Eq. 3.24: ϖ/ϖ_0 and ϖ_0/σ. The latter of these we encountered already in the preceding section as the determining quantity in the propagation of parallax errors in derived parameters. Here we look at the probability distribution for ϖ/ϖ_0 as a function of ϖ_0/σ. Table 3.2 shows the peak values in the distribution curves and weighted mean values for the parallaxes and the offset in the absolute magnitude. It shows that even at a relative error of 10 per cent systematic effects at a level of 0.1 mag. occur.

Above all it should be realized that the LKE is a statistical effect, which may reasonably-well represent the cumulative effects on groups of stars of very clear and sharp criteria, but it should not be applied to the measurement of an individual star (see also Smith 2003).

3.3.3 Working in "Parallax Domain"

Can the data with larger relative errors be used still? The answer is affirmative, but those data can only be used in "parallax domain". This means that the predicted astrophysical quantities are translated to predicted parallaxes, which can subsequently be compared with the observed parallaxes and their formal errors. Say for example that a model predicts an absolute magnitude for a given

Table 3.2. Maximum and mean values of parallax and absolute magnitude corrections for the Lutz-Kelker effect at different values of relative noise σ/ϖ_0. These corrections are statistical, and should not be applied to individual stars. Column 1: Relative error on parallax; column 2: Most probable relative parallax; column 3: equivalent magnitude shift; column 4: mean parallax; column 5: equivalent magnitude shift

σ/ϖ_0	$\hat{\varpi}/\varpi_0$	$\Delta\hat{M}$	$\langle\varpi/\varpi_0\rangle$	$\langle\Delta M\rangle$
0.01	0.998	-0.004	1.000	-0.001
0.02	0.996	-0.009	0.998	-0.004
0.03	0.994	-0.013	0.996	-0.009
0.04	0.992	-0.017	0.994	-0.016
0.05	0.988	-0.026	0.990	-0.025
0.06	0.983	-0.037	0.985	-0.036
0.07	0.978	-0.048	0.980	-0.050
0.08	0.972	-0.062	0.974	-0.066
0.09	0.964	-0.080	0.966	-0.085
0.10	0.956	-0.098	0.958	-0.106
0.11	0.947	-0.118	0.948	-0.131
0.12	0.937	-0.141	0.937	-0.160
0.13	0.925	-0.169	0.925	-0.193
0.14	0.912	-0.200	0.911	-0.231
0.15	0.898	-0.234	0.896	-0.276

type of stars. Rather than deriving estimates for the absolute magnitudes using the parallax measurements for the observed stars, the differences between the predicted absolute and observed magnitudes (the predicted distance modulus) are converted for each star to predicted parallaxes by inverting the relation in Eq. 3.20. These predicted parallaxes are then compared with the observed parallaxes without the complication of formal error distortions. This way a larger sample of individually not extremely accurate parallax measurements can still be used. Examples of this kind of implementation are presented in Chapter 5.

However, even working in parallax domain doesn't entirely solve our problem, and the reason for this is the cosmic (intrinsic) variation of the physical parameters we are looking for. Cosmic variance will in a first instance occur as a result of unresolved properties of the stars we are investigating: most noticeably age and chemical composition, but also unresolved companions, disturbing chromospheric effects, rotation and the inclination of the rotation axis. This cosmic noise implies that our model-predicted parameters are not uniquely determined but have an intrinsic dispersion, and this dispersion will act in a similar way as the formal errors on the parallaxes, depending on the relative amplitude of the dispersion. However, in this case, which could be described as a "reverse LKE", there is no distance dependence.

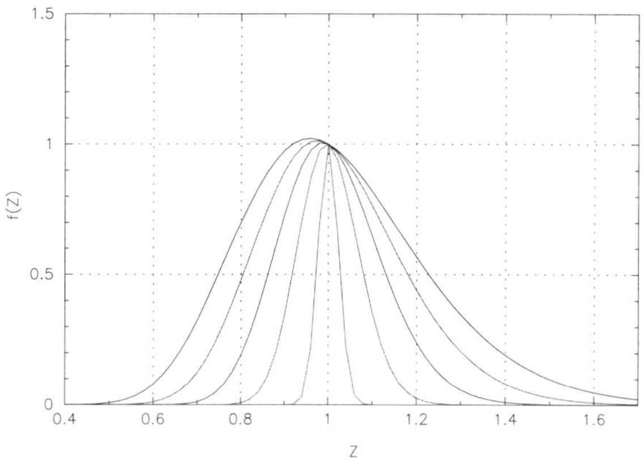

Figure 3.10. Reflection of a Gaussian cosmic dispersion on an absolute magnitude in the distribution of parallaxes, for dispersions (starting at the inner curve) of 0.05, 0.15, 0.25, 0.35 and 0.45 magnitudes

Rewrite Eq. 3.20 as:

$$\varpi/\varpi_0 = \exp(-0.4605(\mathrm{M} - \mathrm{M}_0)), \tag{3.25}$$

where ϖ_0 and M_0 are the reference absolute magnitude and the reference parallax given the apparent magnitude of the object. We further write $\Delta\varpi = \varpi - \varpi_0$ and $\Delta\mathrm{M} = \mathrm{M} - \mathrm{M}_0$, which gives:

$$\begin{aligned} \Delta\varpi/\varpi_0 &= \exp(-0.4605\Delta\mathrm{M}) - 1, \\ \Delta\mathrm{M} &= 2.1715\ln(1 + \Delta\varpi/\varpi_0). \end{aligned} \tag{3.26}$$

If we have a distribution function $f_{\mathrm{M}}(\Delta\mathrm{M})$ for the magnitudes, then we should expect the following distribution function for $\Delta\varpi$ (see for example Papoulis 1991):

$$f_\varpi(\Delta\varpi) = \frac{f_{\mathrm{M}}(2.1715\ln(1 + \Delta\varpi/\varpi_0))}{0.4605(1 + \Delta\varpi/\varpi_0)}, \tag{3.27}$$

or, using $Z = \varpi/\varpi_0$:

$$f_\varpi(\Delta\varpi) \propto \frac{f_{\mathrm{M}}(2.1715\ln(Z))}{Z}, \tag{3.28}$$

where the "normalization" has been chosen in the same way as before: such that all curves equal 1 for $\varpi = \varpi_0$. Examples of the parallax distribution when the underlying $\Delta\mathrm{M}$ distribution is Gaussian are shown in Fig. 3.10 for a few different values of the dispersion in the magnitudes. Similarly, Table 3.3 shows how the mean and dispersion of the parallax distribution gets affected by different levels of (Gaussian) noise on the absolute magnitudes. Clearly the effects are quite small.

Table 3.3. The reflection of a Gaussian distribution of ΔM values in the distribution parameters for the parallaxes: mean (column 2) and dispersion (column 3). The equivalent dispersion in the magnitudes assuming a simplified linear relation with the parallax dispersion is shown in column 4

$\sigma_{\Delta M_0}$	$\langle\varpi/\varpi_0\rangle$	$\sigma_{\Delta\varpi}/\varpi_0$	$\equiv\sigma_{\Delta M_0}$
0.030	1.000	0.014	0.030
0.060	1.000	0.028	0.060
0.090	1.001	0.042	0.091
0.120	1.002	0.056	0.121
0.150	1.002	0.070	0.151
0.180	1.003	0.084	0.181
0.210	1.005	0.098	0.212
0.240	1.006	0.112	0.243
0.270	1.008	0.126	0.274
0.300	1.010	0.140	0.305
0.330	1.012	0.155	0.336
0.360	1.014	0.169	0.368
0.390	1.016	0.184	0.400
0.420	1.019	0.199	0.431
0.450	1.021	0.213	0.463
0.480	1.024	0.227	0.494
0.510	1.026	0.241	0.523

We can also try to derive from the observed distribution of the parallax residuals the underlying cosmic noise on the magnitudes (using Eq. 3.25 and 3.26):

$$f_M(\Delta\mathrm{M}) \propto \frac{f_\varpi(E-1)}{(2.1715/\varpi)} \propto f_\varpi(E-1)E, \tag{3.29}$$

where $E \equiv \exp(-0.4605\Delta\mathrm{M})$. Application of this relation should be restricted to cases where the formal errors on the parallaxes are significantly smaller than the dispersion caused by the variation in absolute magnitude for a sample.

3.3.4 G8V and K0V Stars as an Example

The G8V and K0V stars form a reasonably homogeneous sample, without the difficulty of disentangling effects of post-main sequence evolution seen for earlier spectral types. A selection was made based on spectral classifications in the five Michigan Catalogs, the first four of which had been incorporated into the *Hipparcos* catalogue (ESA 1997; Houk and Swift 1999). Stars with spectral classes G8V, G8/K0V and K0V were selected. The parallax and photometric information for the 356 stars selected show that nearly 20 per cent are

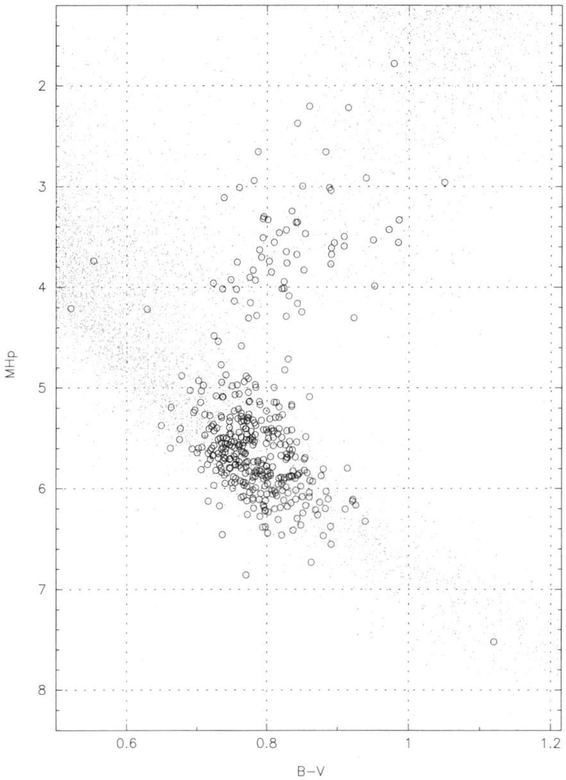

Figure 3.11. The HR diagram for G and K stars, showing the selection of G8V, G8/K0V and K0V stars as classified in the Michigan surveys

over luminous, and occupying the region between the main sequence and the red clump giants (Fig. 3.11). A few stars appear lost between earlier or later spectral types. Eliminating these stars leaves 286 candidates, of which 238 have relative parallax errors below 7 per cent. The distribution of these stars over the colour index B − V is shown in Table 3.4, and shows considerable overlap between the three groups.

It is clear from Fig. 3.11 that an absolute magnitude for this sample can only be calibrated as a function of the colour index $B - V$:

$$\mathrm{M_{Hp}} = t + s(\mathrm{B} - \mathrm{V} - 0.8), \tag{3.30}$$

where we assume initial values of $t = 6.0$ and $s = 6.5$. This relation, with the observed values of Hp, provides the predicted parallaxes ϖ_0, which combined with the observed parallaxes ϖ provides the distribution of Z values used in Eq. 3.28. The observed distribution for Z is affected by the actual cosmic dispersion in the magnitudes of the stars involved and by our choice of t and s in Eq. 3.30. The observed distribution has been fitted by the expected distribution of absolute magnitudes under different assumptions of the magnitude dispersion $\sigma\Delta\mathrm{M}$, offset t and slope s. These three parameters have been op-

Table 3.4. The distribution over the colour index B − V for G8V, G8/K0V, and K0V stars, showing considerable overlap between the different spectral classifications

B − V	G8V	G8/K0V	K0V	B − V	G8V	G8/K0V	K0V
0.66	1	0	0	0.78	4	6	7
0.67	3	0	0	0.79	1	7	13
0.68	1	0	0	0.80	2	5	11
0.69	3	0	0	0.81	3	4	7
0.70	4	1	1	0.82	0	3	13
0.71	5	2	1	0.83	3	1	12
0.72	10	2	0	0.84	0	0	6
0.73	16	5	0	0.85	0	1	5
0.74	14	1	5	0.86	0	0	5
0.75	12	6	6	0.87	0	0	3
0.76	14	8	5	0.88	0	0	4
0.77	16	4	6				

timized by examining the sum of squared differences between the observed and predicted distributions. The observed distribution has been oversampled by a factor two in order to reduce the effects of accidental noise on the results. The values thus obtained are: $t = 5.80 \pm 0.05$, $s = 6.60 \pm 0.05$ and $\sigma\Delta\mathrm{M} = 0.39 \pm 0.01$, where all errors are estimated. The data and fitted curve using those values are shown in Fig. 3.12.

There is still a small amount of bias possible in these data: due to the restriction on relative parallax error there will be more intrinsically bright stars than faint stars selected at the same colour index when errors are close to the cut-off

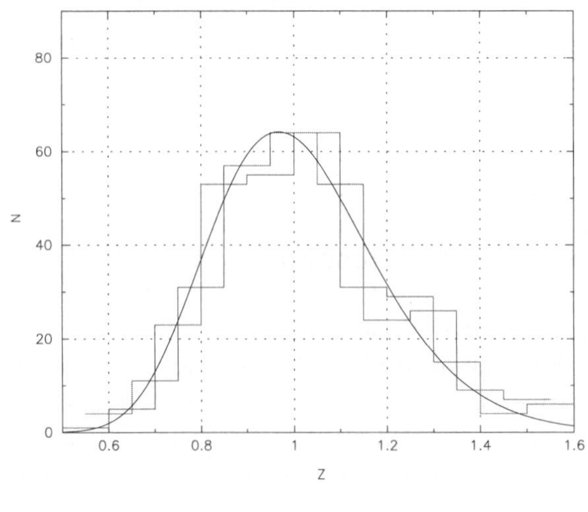

Figure 3.12. The distribution of $Z = \varpi/\varpi_0$ for G8V and K0V stars, compared with the best fit distribution expected from a Gaussian distribution of absolute magnitude residuals

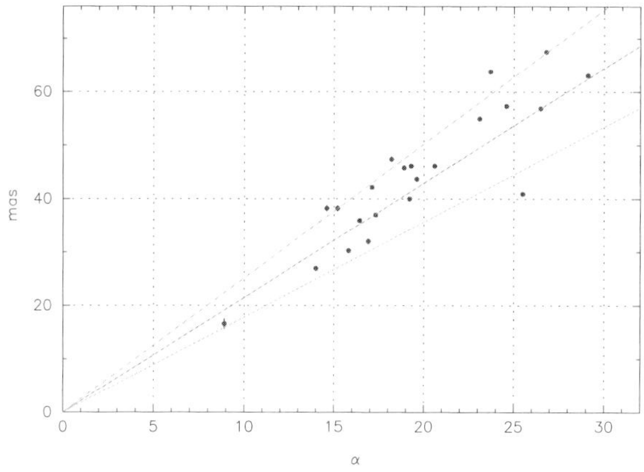

Figure 3.13. A repeat of the calibration of the parallax calibration of NGC 752 through the use of photometric boxes as was done by Nicolet (1981), but using *Hipparcos* instead of ground-based parallax determinations. The middle curve gives a fitted parallax of 2.15 ± 0.36 mas, the outer lines represent the one-sigma boundaries for this fit

value. Here the cut-off value is 15 per cent, with only 7 per cent of the stars having errors between 10 and 15 per cent. The errors on the parallaxes increase roughly with $I^{-0.5}$ (Eq. 3.1), where I is the intensity of the star. The actual parallax errors near the bottom envelope of the main sequence distribution for the late G and early K dwarfs could therefore be about a factor 1.5 higher than near the top envelope at the same colour index. However, part of this difference has been compensated for through the observing strategy of *Hipparcos*, which allocated systematically more observing time to fainter objects. It is further disturbed by the spread in parallax accuracies for stars with the same apparent magnitude: the scanning strategy used by *Hipparcos* caused variations by more than a factor two in the number of observations available per star (Fig. 2.19). Thus, a sharp cut off in the parallax accuracies will not reflect similarly in the brightness distribution. Together with the relatively small number of stars that could possibly be affected by it, it would seem that bias has no significant role in the experiment described above.

In comparison, a similar study carried out in magnitude space (Butkevich et al. 2005a,b) was dominated by the effects of Malmquist bias. In order to reduce and control it, only a small amount of data could in the end be used, and results still remained hard to interpret.

3.3.5 Photometric Distances

Photometric parameters have been, and are still, widely used to derive parallax estimates for stars. The assumption is that given an observed spectral type, an accurate estimate of the absolute magnitude can be obtained. This assumption is fundamental in the parallax zero-point calibration for differential trigonometric parallax studies (Gatewood et al. 1990; Gatewood and de Jonge 1994; Gatewood 1995; Gatewood et al. 2000; Soderblom et al. 2005). As is clear from the discussion and figures shown above, there is considerable

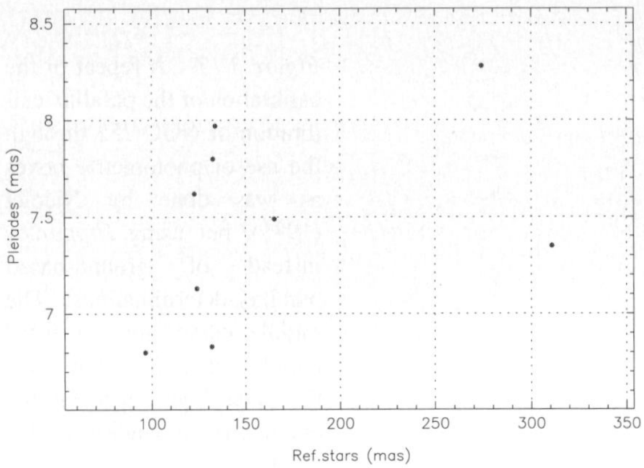

Figure 3.14. The parallax of the Pleiades as derived from comparisons between 11 nearby stars and the Pleiades main sequence for F and G type stars. The spread in derived parallaxes reflects the intrinsic dispersion in absolute magnitudes and the uncertainty related to this type of distance determination

uncertainty associated with the relation between colour and spectral type plus luminosity class, between colour and absolute magnitude, and between spectral type plus luminosity class and absolute magnitude. Some, but not all, of these variations may be understood from the tracks of theoretical isochrones derived for various chemical compositions.

A large intrinsic variation of absolute magnitudes is also observed for stars that are apparently very similar when observed in colour-colour diagrams of for example the Geneva photometry. A study by Nicolet (1981) on the distance calibrations for 43 open clusters based on such photometric boxes can be repeated using the *Hipparcos* rather than ground-based parallaxes for the comparison stars. Even though the stars selected for this study appear very similar in all colours, their absolute magnitudes still show a considerable spread, with a sigma of around 0.4 magnitudes (Fig. 3.13).

A very similar result is obtained when comparing the Pleiades main sequence for F and G stars with parallaxes of similar, nearby stars. This exercise was done using the Pleiades observations in Walraven photometry by van Leeuwen et al. (1986). The mean track of the cluster main sequence for F and G stars was determined, considering reddening and duplicity effects. This track was compared with nearby stars in the same spectral range and with parallax measurements from several observatories. These stars too were measured in the Walraven photometric system, the magnitude differences providing nominal distance ratios between the cluster and those stars. Substituting the ground-based parallaxes with the new *Hipparcos*-based determinations shows again the spread in absolute magnitudes and the uncertainty associated to a distance calibration based on such measurements. The mean cluster parallax derived from these data (Fig. 3.14) is of little or no value as long as the observed spread is considerably larger than expected from theoretical isochrones, while at the same time the main sequences for star clusters tend to be very narrow and well defined.

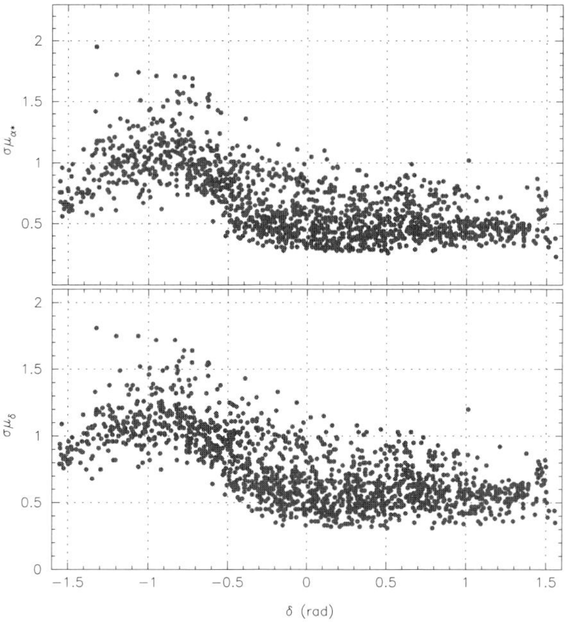

Figure 3.15. Formal errors in mas/yr of proper motions in the *FK5* catalogue, reflecting the difficulties encountered with the reconstruction of the Southern sky before the publication of the *Hipparcos* catalogue

3.4 Proper motions

The accuracy of the proper motions is even more difficult to estimate than that of the parallaxes. Proper motions are very sensitive to reference-frame definition and to the range in epoch over which they are obtained. There are proper motions available with compatible formal errors, but these tend to be obtained over long time intervals (50 to 150 years typically) and are often differential rather than absolute. A very long time interval tends to smooth out the temporal effects of orbital motions to which a short time interval is sensitive: those originating from orbital periods of around 5 to 10 years. In the following sections, the *Hipparcos* proper motions are compared with the all-sky absolute proper motions provided by the *FK5* catalogue and with differential proper motions obtained from open cluster studies. Subsequently we have a look at some applications of the proper motion data.

3.4.1 Comparison with the FK5

Before the publication of the *Hipparcos* catalogue, the *FK5* (Fricke et al. 1988, 1991) provided the definition of the optical reference frame. Its construction was very different from that of the *Hipparcos* catalogue: built from primarily meridian observations obtained over often well over 100 years, with a wide range of different telescopes, each covering zones in declination. Virtually all those observations were visual recordings of transits of reference stars, affected by a wide range of systematics, ranging from telescope- to season- and

Table 3.5. The spin components for the *FK5* catalogue as determined for the published (left) and new (right) reductions

Comp.	Spin (publ.) mas/yr	Spin (new) mas/yr
ω_x	-0.10 ± 0.10	-0.42 ± 0.07
ω_y	$+0.43 \pm 0.10$	$+0.56 \pm 0.06$
ω_z	$+0.88 \pm 0.10$	$+0.85 \pm 0.06$

observer-specific effects (see for example van Herk and van Woerkom 1961, for a summary of typical problems with classical meridian observations). In the construction of the *FK5* great care was taken to avoid such errors penetrating the catalogue, and the comparison with the *Hipparcos* data shows that for the Northern hemisphere this was indeed largely successful. For the Southern hemisphere the data coverage is of lower density, with the additional problem that the southern-most stars could only be observed at quite large zenith distances, for lack of telescopes at the southern-most latitudes. These problems are reflected in the formal errors for the proper motions in the *FK5* catalogue (Fig. 3.15).

All 1535 *FK5* and *FK5* extension stars are included in the *Hipparcos* catalogue, but 297 are (or are suspected of being) disturbed by companions and/or orbital motions, though some of these detections are marginal. This leaves 1238 stars for comparison tests on the proper motions in the two catalogues. For all tests the differences in proper motions were normalized by the formal error, consisting of the errors from the two catalogues added in quadrature. These should, if the formal errors are correct, be normalized errors, for which we expect under ideal conditions a Gaussian distribution with $\sigma = 1$.

Before any detailed comparison can be made, rotations between the two reference frames will have to be removed. In the proper motion comparison between *FK5* and *Hipparcos* this is equivalent to removing relative spin components, which are related to the different residual rotations of the two reference frames with respect to the inertial extragalactic frame. The *Hipparcos* frame has been fixed to the extragalactic frame to an accuracy of around 0.1 mas/yr (Hemenway et al. 1997; Kovalevsky et al. 1997), so in effect this transformation will describe the spin corrections required to adjust the *FK5* catalogue to the inertial reference system as defined by the International Celestial Reference System (ICRS), to the accuracy of the inertial adjustment of the *Hipparcos* reference frame. This rotation is described in Chapter 18 of Volume 3 of (ESA

Table 3.6. Variation of the ω_z component with declination, as observed over 9 declination zones. (ϵ gives the formal error on the mean, σ gives the unit-weight standard deviation for the solution

zone (deg.)	Stars	Rej.	Mean	ϵ	σ
			mas/yr		
-90 - -65	87	0	-14.42	1.64	4.32
-65 - -45	113	1	1.27	0.43	2.37
-45 - -25	162	2	2.72	0.21	2.50
-25 - -10	145	4	0.88	0.15	2.76
-10 - $+10$	212	4	1.25	0.10	2.63
$+10$ - $+25$	127	3	0.21	0.15	2.87
$+25$ - $+45$	178	4	0.55	0.15	2.73
$+45$ - $+65$	125	5	0.75	0.21	2.52
$+65$ - $+90$	89	3	-2.03	0.80	4.04

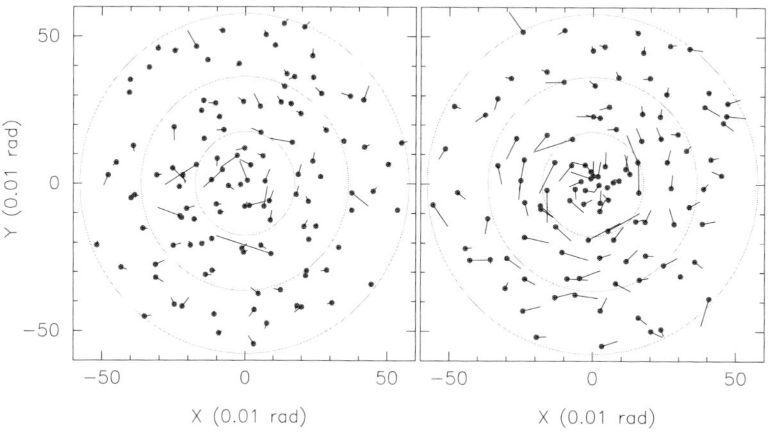

Figure 3.16. Proper motion residuals at the North (left) and South (right) polar caps. The circles indicate declinations of ±80, +70 and ±60 degrees. The proper motion differences are indicated by the lines, starting from the stellar position. The scale for these vectors is measured as 1 mas/yr per 0.01 rad on the map

1997) by:

$$\begin{bmatrix} -\sin\delta\cos\alpha & -\sin\delta\sin\alpha & \cos\delta \\ \sin\alpha & -\cos\alpha & 0 \end{bmatrix} \boldsymbol{\omega} = \begin{bmatrix} (\mu_{\alpha*})_{\rm F} - (\mu_{\alpha*})_{\rm H} \\ (\mu_\delta)_{\rm F} - (\mu_\delta)_{\rm H} \end{bmatrix}, \quad (3.31)$$

where the vector $\boldsymbol{\omega}$ represents the spin difference between the reference frames. The values found for the new reduction are given in Table 3.5, together with those for the published catalogue.

The main rotation component, ω_z, varies noticeably over sky. A solution for only this component, in 9 zones of declination, is presented in Table 3.6. The most striking discrepancy is at the South equatorial pole. Figure 3.16 shows that these rotations vary in addition with right ascension. Similar discrepancies were noted by Schwan (2001), who represented these by means of spherical harmonics, and for the declinations between ± 65 degrees by Mignard and Froeschle (2000), who provided a tabular representation of corrections.

Systematic errors up to 0.1 arcsec in the *FK5* were already detected before from comparisons with modern meridian observations using the Carlsberg Automatic Meridian Circle (Morrison et al. 1990), through CCD observations of positions relative to the extragalactic reference frame (Stone 1997), and through astrolabe observations (Noel 1997).

A comparison between the *Hipparcos* and *FK5* proper motions is complicated by the differences in epoch coverage: while the *Hipparcos* mission covered only 3 to 3.5 years, the data accumulated for the *FK5* covered 100 to 150 years. This creates a noticeable difference in sensitivity to orbital motions (Wielen et al. 1997). For orbital periods of order years the *FK5* data will relegate the orbital disturbances to the noise contributions, while the *Hipparcos* data will tend to fit it as part of the proper motions. Proper motion comparisons between those two catalogues can therefore be compromised. This has been quoted as the main reason for an unexpected level of discrepancies between the *FK5* and *Hipparcos* proper motions in the published catalogue. In some comparisons an additional cosmic noise of 2 mas/yr was added to the *Hipparcos* proper motion accuracies to account for these effects. The new reduction has further emphasized this discrepancy, in particular for some of the brighter stars. However, from the results described above it is also clear that there is a significant contribution to the noise level on these comparisons originating from systematic errors in the *FK5* catalogue as well.

To test the possible origin of these discrepancies further, the sample was divided in two according to parallax. For those stars nearest to us the proper motions can be expected to be most sensitive to short-period orbital disturbances. However, there are no significant differences noted between stars with parallaxes above or below 50 mas. It was noted though that the fraction with smaller parallaxes is much larger: many *FK5* stars are intrinsically bright stars at relatively large distances, and their proper motions should therefore be fairly insensitive to short-period orbital disturbances at the measuring accuracies of *Hipparcos* and the *FK5*. A set of histograms of the normalized residuals in different declination intervals gives the impression of difficulties with assigning errors to proper motions closer to the poles (Fig. 3.17), where errors in the *FK5* catalogue appear to have been underestimated. The equatorial poles have no meaning for the *Hipparcos* data, which, through the scanning strategy, was bound to the ecliptic reference system.

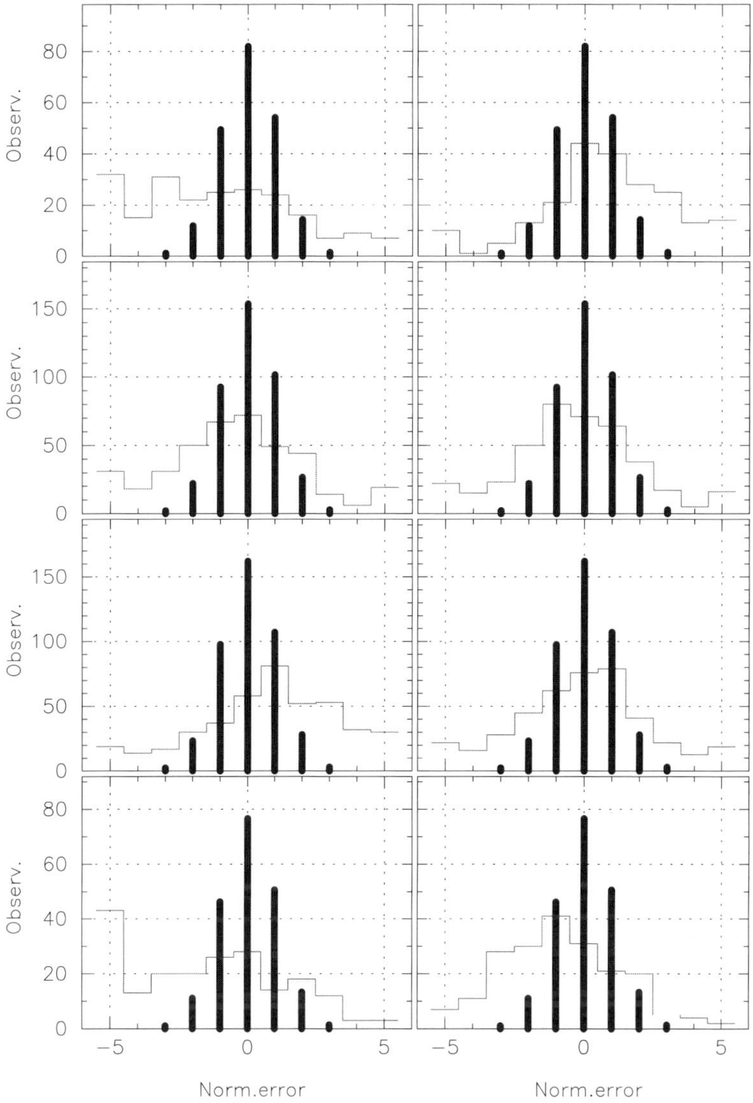

Figure 3.17. Histograms of normalized differences between the *FK5* and *Hipparcos* proper motions, in RA (left) and Declination (right), in bands of declination. From top to bottom: +90 to +45, +45 to 0, 0 to -45, and -45 to -90 degrees. The bars in each graph show the equivalent $\sigma = 1$ Gaussian distributions

3.4.2 Star Clusters and the Magellanic Clouds

The analysis of star-cluster data is presented in Chapter 5. However, clusters like the Pleiades and Praesepe, spanning between 5 and 10 degrees diameter fields on the sky, as well as the LMC and SMC, also offer some insight into the

Table 3.7. The astrometric parameters for the LMC and SMC as determined from the weighted and decorrelated means of the astrometric parameters of the individual member stars. Column 1: RA, Declination; Column 2: radius (degrees); Column 3: parallax and formal error (mas); Column 4: $\mu_{\alpha*}$ and formal error, mas yr^{-1}; Column 5: μ_δ and formal error, mas yr^{-1}; Column 6: standard deviation for the solution; Column 7: number of observation (3 times number of stars) and rejected observations; Columns 8 and 9: correlation coefficients (times 100) between parallax and proper motions, and between proper motions

Object	1	2	3	4	5	6	7	8	9
LMC	82.5	10.0	0.12	1.88	-0.17	1.09	87	5	0
	-69.5		0.26	0.24	0.29		2		1
SMC	13.5	7.0	-0.71	1.50	-1.47	0.64	24	5	0
	-72.5		0.33	0.39	0.35		0		1

reliability of the proper motions on a differential scale much larger than the typical intrinsic correlation scale of the data. Based on the assumption that the dispersion of the internal motions is either negligible or approximately known, and the radial velocity of the system is known, the projection effects of the space velocity can be calculated and incorporated to "correct" the observed proper motions (see Section 5.2.2). This provides effectively a check on the internal consistency of the *Hipparcos* proper motions over an area up to 10 degrees diameter. One of the complications is, though, that membership determination for open clusters is often based on proper motions, and open clusters tend to be surrounded by a halo of escaped members with generally higher velocity dispersion and wider space distribution (Terlevich 1987; Kroupa et al. 2001).

The solutions for the mean astrometric parameters of the LMC and SMC, (following the procedure presented in Section 5.2.2), are given in Table 3.7. The standard deviations of these solutions are a clear indication of internal agreement between the proper motions over at least the areas covered by these systems. This is also clear from the proper motion vector-point diagrams, shown in Fig. 3.18. This figure shows the improvement of the new solution, even for these faint stars.

3.5 Disturbed solutions

The basic 5-parameter solution for the positional variations can be disturbed by the close presence, as projected on the sky, of another star. In most cases this concerns a star with which the target star forms a binary or multiple system. When the companion star caused a disturbance on the signal modulation (stars classified as "C"), the data is treated separately (see Chapter 4). Orbital systems (classified as "O") are described in Section 4.6. All other systems (those classified as unperturbed or classified as "G", "X" or "V" in the published data) followed nearly the same reduction procedure:

1. If a star is variable, try the VIM solution, if significant (see for criteria below), accept;
2. Else, try the 5-parameter solution, and if the F2-statistic is less than 2.4, accept;
3. Else, if originally a stochastic solution was applied and F2 more than 5.0, apply stochastic;
4. Else, try the 7-parameter solution, if not significant, accept the 5-parameter solution;
5. Else, try the 9-parameter solution, if not significant, accept the 7-parameter solution; if significant, accept;

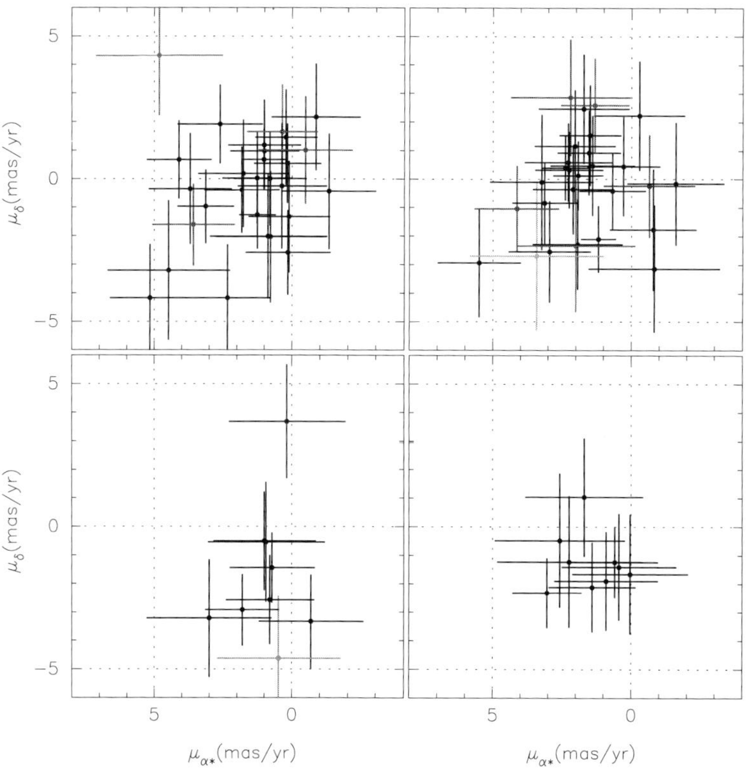

Figure 3.18. The proper motions for stars in the LMC (top) and SMC (bottom), for the published data (left) and the new solution (right)

Table 3.8. A comparison of the single star solutions between the published and new reductions. The columns represent the solutions in the published data, the rows solutions in the new reduction. For example, in the new reduction only 588 of the original 1561 stochastic solutions as presented in the published data were confirmed as stochastic, while 962 have now been solved as simple 5-parameter solutions. Similar effects can be seen for the number of 7 and 9 parameter solutions. The figures on the Variability Induced Movers (VIMs) are affected by the low detection cut-off in the published data, and the increased cut-off in the new reduction (see below), allowing a relatively high number of spurious cases among the detections in the published catalogue. New detections of stochastic solutions were expected as a result of the higher measurement accuracy, exposing more disturbed cases

Publ.Solution type:	5-par	7-par	9-par	Stoch.	VIM	All types
Number of entries	100038	2163	459	1561	288	104509
New solution 5-par	99131	1310	129	962	252	102072
New solution 7-par	609	568	23	4	9	1213
New solution 9-par	38	28	32	3	1	102
New solution Stoch	242	257	275	588	5	1367
New solution VIM	18	0	0	4	21	43

6 Else, if F2 for the 5-parameter solution is larger than 5.0, apply the stochastic solution.

The stochastic solution was the final resort, applied when all other attempts to obtain an acceptable solution had failed. It most likely represents unresolved orbital motion on a time scale much less than the mission length of about 3 years. Table 3.8 summarises the overall statistics on the different solutions, showing that the new reduction succeeded in resolving a significant number of previously troublesome solutions, while at the same time detecting disturbance for systems with improved measurement accuracies. The different types of solutions are detailed further in the sections below.

3.5.1 The Effect of Radial Velocity on the Proper Motion

In a small number of cases the radial velocity of a star can have a significant effect on the proper motion. A recent review on this subject of perspective acceleration can be found in (Dravins et al. 1999). It can in principle be accounted for by the seven-parameter solutions, but it is better to include where relevant the measured radial velocity as a (small) correction to the astrometric solution. The relative change in the proper motion, assuming a constant velocity of the star, is determined by the relative change in distance R of the star. The relation between the tangential motion and the observed proper motion of

Table 3.9. The stars with the largest acceleration components in their proper motions due to their radial velocity

HIP	V_R	$\dot{\mu}$	ϖ σ_ϖ	$\mu_{\alpha*}$ $\sigma_{\mu_{\alpha*}}$	μ_δ σ_{μ_δ}	Hp B − V	Name
	km s^{-1}	mas yr^{-2}	mas	mas yr^{-1}			
3829	263.0	-0.18	249.03	1242.60	-2706.20	12.559	Van Maanen 2
			5.90	5.26	3.19	0.554	
24186	245.2	-0.56	255.66	6505.08	-5730.84	8.932	Kapteyn's star
			0.91	0.98	0.96	1.543	
54035	-85.6	0.16	392.64	-580.27	-4765.85	7.506	
			0.67	0.62	0.64	1.502	
87937	-109.7	0.64	548.31	-798.58	10328.12	9.490	Barnards's star
			1.51	1.72	1.22	1.570	
104217	-65.3	0.10	285.88	4106.90	3144.68	6.147	NSV 13546, 61 Cyg B
			0.54	0.32	0.44	1.309	

a star is given by (Eq. 2.32):

$$\mu = V_T \frac{\varpi}{\kappa}, \tag{3.32}$$

where V_T is the transverse space velocity, and $\kappa = 4.74047$ is the constant relating the measurements of the proper motion in mas s^{-1}, the parallax measured in mas, and the velocities measured in km s^{-1}. The change in proper motion due to the change in parallax is thus given by:

$$\frac{\mathrm{d}\mu}{\mathrm{d}t} = \frac{V_T}{\kappa}\frac{\mathrm{d}\varpi}{\mathrm{d}t} = -\frac{V_T\varpi}{\kappa R}\frac{\mathrm{d}R}{\mathrm{d}t} = -\frac{\mu}{R}\frac{\mathrm{d}R}{\mathrm{d}t}, \tag{3.33}$$

with R measured in kpc. Introducing the radial velocity V_R, measured in km s^{-1}, gives:

$$\dot{\mu} = -\mu\varpi V_R \times 1.0227 \cdot 10^{-9} \mathrm{mas\ yr}^{-2}, \tag{3.34}$$

where the numerical factor takes care of the transformation from km s^{-1} to kpc yr^{-1}. The effect will be very small for all but a few of the most nearby stars, and this correction therefore plays only a minor role in the *Hipparcos* data. As it depends itself on the astrometric parameters, the measured parallax and proper motion, determining the astrometric parameters for such stars is in principle an iterative procedure. However, the adjustments to the measured parallax will have only third order, negligible effects on the solution, while the proper motion correction can be either directly incorporated in the solution as a time-dependent factor, or simply added to the predicted abscissae.

Only for six stars (two of which are part of a double system) we find $|\dot{\mu}| \geq 0.1$ mas yr^{-2}, and does this correction play any role. Even for those stars, shown in Table 3.9, this role is small.

3.5.2 Seven and Nine-parameter Solutions

In the published catalogue are included 2623 astrometric solutions for which the proper motion was solved as a first or even second order function in time.

These solutions were mainly introduced to cover realistic effects such as disturbance by a companion. They can also occasionally allow for the accidental modelling of other disturbances, such as caused by inaccuracies in the reconstruction of the along-scan attitude.

Following the procedures for the published data, the reference times for the second and third order time dependencies of the proper motions were chosen such that the actual proper motion at the reference epoch of the mission, 1991.25, is largely unaffected (the integral of these coefficients over time for the mission will be close to zero for most stars):

$$p(t) = p_0 + \mu_p \cdot t + \dot{\mu}_p(t^2 - 0.81)/2 + \ddot{\mu}_p(t^2 - 1.69)t/6, \tag{3.35}$$

where t is measured in years since 1991.25, and p can be either of the two coordinate directions on the sky, for example (α_*, δ). The significance of the acceleration coefficients was assessed by means of the statistic $F_{\dot{\mu}}$:

$$F_{\dot{\mu}}^2 \equiv \begin{bmatrix} \dot{\mu}_{\alpha*} & \dot{\mu}_\delta \end{bmatrix} \begin{bmatrix} \sigma^2_{\dot{\mu}_{\alpha*}} & \rho_\mu \sigma_{\dot{\mu}_{\alpha*}} \sigma_{\dot{\mu}_\delta} \\ \rho_\mu \sigma_{\dot{\mu}_{\alpha*}} \sigma_{\dot{\mu}_\delta} & \sigma^2_{\dot{\mu}_\delta} \end{bmatrix}^{-1} \begin{bmatrix} \dot{\mu}_{\alpha*} \\ \dot{\mu}_\delta \end{bmatrix}. \tag{3.36}$$

A very similar relation defines $F_{\ddot{\mu}}$. The same criterion as applied to the published data was used also in the new reductions: a detection was considered significant when $F_{\dot{\mu}} > 3.44$, accepting 0.27 per cent of false detections in case the acceleration terms are random variables. On a total of 100 000 stars this would still cause 270 false detections.

A comparison between detections in the published and new solutions shows that the majority of the 650 confirmed detections are probably real (Fig. 3.19).

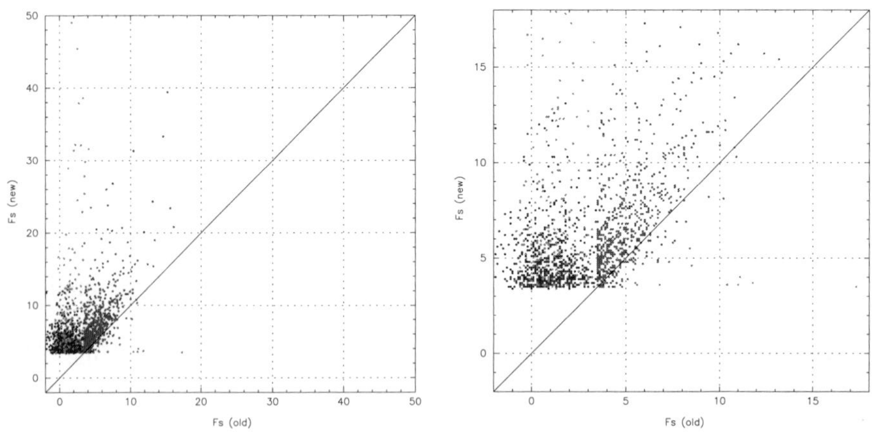

Figure 3.19. A comparison of the detection statistic for 7-parameter solutions in the published data (old) and the new reduction (new). On left, the large-scale overview, on the right focused in on the main distribution of data. The group of data points towards the left of the diagram constitute new detections, many of which are for double stars

The new reduction gives in general a higher probability for these detection, which, considering the higher accuracy should also be expected. Similarly, an increase in detections should be expected and appears to be present in the data. In the new reduction also double stars have been included for the 7-parameter solution, though these are not included in the Table 3.8. It is here that all of the strongest detection are made, but this may also be a reflection of the general level of disturbance these solutions suffer from.

The situation for the 9-parameter solutions is similar, in that for the small number (7.6 per cent) of confirmed solutions a reasonable correlation is observed between the detection statistics of the published and the new solutions, with the new reduction showing the higher values for the detection statistic. No 9-parameter solutions were applied to double stars.

3.5.3 Variability-induced Movers

The variability-induced movers, or VIM solutions, were introduced in the *Hipparcos* data reductions to account for double stars in which the secondary component only became noticeable at minimum light of the primary component (or maximum light of the secondary component). These systems generally contained insufficient information for a full double-star solution as described in Chapter 4. Instead, the effects on the shift in the photo-centre of the system as caused by the variability of one of its components is modelled as part of the astrometric-parameter solution. It is possible to model these effects without knowledge of the actual separation and magnitude difference of the stars involved.

The position of the photo-centre of a double system with respect to the position $\boldsymbol{p}_a$ of the primary can be represented by:

$$\boldsymbol{p}_{pc} = \boldsymbol{p}_a + \boldsymbol{\rho}\frac{r}{1+r} = \boldsymbol{p}_a + \boldsymbol{\rho}\frac{I_b}{I_{a+b}}, \tag{3.37}$$

(see also Section 4.2, Eq. 4.30, page 127), where ρ is the vector describing the position of the secondary relative to the primary star, $r \equiv I_b/I_a$, and I_a and I_b are the pseudo-intensities of the two components of the system. Similarly, I_{a+b} is the intensity of the combined system. The difference in the position of the photo-centre caused by a change in the intensity of the primary to I'_a is then given by:

$$\Delta\boldsymbol{p} = \frac{\boldsymbol{\rho} I_b}{I_{a+b}}(1 - I_{a+b}/I'_{a+b}). \tag{3.38}$$

The pseudo-intensities are related to the magnitudes through:

$$I_{a+b} = 10^{-0.4 m_{a+b}}. \tag{3.39}$$

Substituting in Eq. 3.38 gives:

$$\Delta p = \frac{\rho I_b}{I_{a+b}}(1 - 10^{-0.4(m_{a+b} - m'_{a+b})}), \qquad (3.40)$$

where m_{a+b} is an arbitrary reference magnitude for the system, and m'_{a+b} the observed magnitude. Introducing the coefficient $(1 - 10^{-0.4(m_{a+b} - m'_{a+b})})$ as a degree of freedom in the astrometric solutions for variable stars will therefore provide an estimate of the vector $\rho 10^{-0.4(m_b - m_{a+b})}$, defining a relation between the separation between the two components and the magnitude of the secondary. This relation is limited by the range in possible brightness of the secondary, which can never be brighter than the system at minimum light. Small photometric variations are a problem for these corrections, as the coefficients remain close to one and the resulting parameters become both poorly determined and strongly correlated with the positional correction in the astrometry. The application of this correction has therefore been limited to variable stars with relatively large amplitudes (above 0.4 mag.).

To solve for the possible presence of a disturbing component for a variable star, two additional parameters were added to Eq. 2.52 in the astrometric solutions for all variable stars that are not already part of the double star selection:

$$\begin{aligned} \mathrm{d}a &= \cdots + v_{\alpha *} \cos\psi(1 - 10^{-0.4(m_{a+b} - m'_{a+b})}), \qquad (3.41) \\ &\quad + v_\delta \sin\psi(1 - 10^{-0.4(m_{a+b} - m'_{a+b})}), \end{aligned}$$

where ψ is, as before, the local inclination of the scan direction. The significance of these parameters was assessed through the statistic F_D, defined in the same way as for the acceleration coefficients (Eq. 3.36). However, a less stringent criterion for selection was adopted, $F_D > 2.15$, in the published data, while in the new reductions the criterion was again set at $\mathrm{F_D} > 3.44$. The reason for this was the very low level of confirmation of cases selected by the old and the new solution. The possibility that many of the VIM detections in the published data are in fact spurious was also raised by Pourbaix et al. (2003). Most of these VIM "detections" should preferably be used only as an indication of possible duplicity, certainly not as proof. The most significant in the new reduction are shown in Table 3.10, and an example of the light curve for one of these is shown in Fig. 3.20.

3.5.4 Stochastic Solutions

Solutions were classified as stochastic in cases where the dispersion in the abscissa residuals was significantly higher than expected, but did not allow modelling with the seven or nine parameter model. Of the number of such solutions in the published data, 1561, only 573 have been confirmed in the

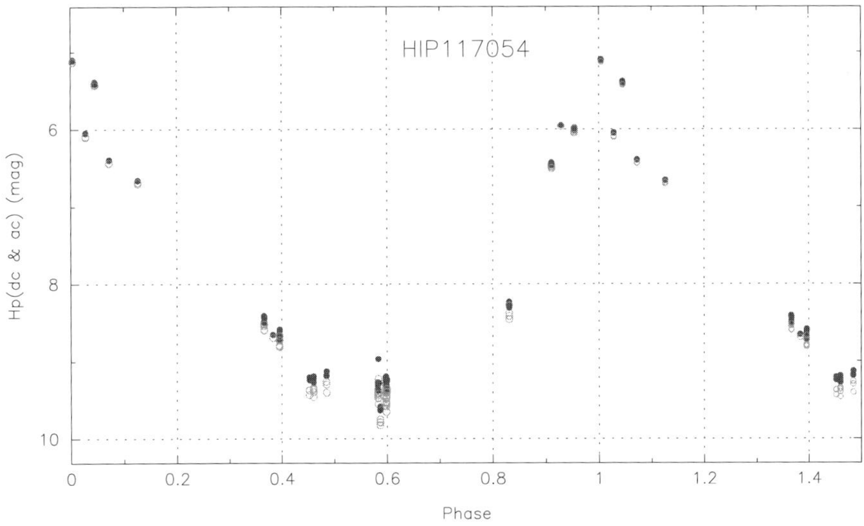

Figure 3.20. The light curve of the Mira star R Aqr, HIP 117054. The filled dots show the H_{DC} (or Hp), the open circles the H_{AC} photometry. Towards the minimum of the light curve, the effects of a companion become visible through the increase in difference between the DC and AC magnitudes. The presence of the companion decreases the amplitude of the first harmonic, and therefore make the system look fainter when examined in AC magnitudes. The presence of the companion is noted in the astrometry, and resolved through a so-called VIM solution

Table 3.10. Some of the significant detections of variability-induced movers in both the published and new reduction. The amplitude of the variations of the star is given by ΔHp

HIP	F_D(old)	F_D(new)	Hp	ΔHp	Name	Type
6759	5.5	5.2	6.41	1.13	R Scl	SRB
67419	2.6	7.2	6.10	2.02	W Hya	SRA
95032	2.5	4.0	11.44	2.42	HO Lyr	M
105485	2.5	8.0	11.99	1.46	MZ Cyg	CWA
109089	8.7	6.6	8.33	3.10	RZ Peg	M
109340	6.4	5.0	9.19	0.73	Y Lac	DCEP
117054	7.1	12.6	8.71	4.80	R Aqr	M

new reduction, while 947 of these cases were successfully solved as unperturbed five-parameter solutions. This is a reflection of the significant reduction in modelling noise in the construction of the new catalogue. For the confirmed stars the most probable explanation is the presence of unresolved orbital motions. For each such solution, the distribution of those residuals will need further analyses.

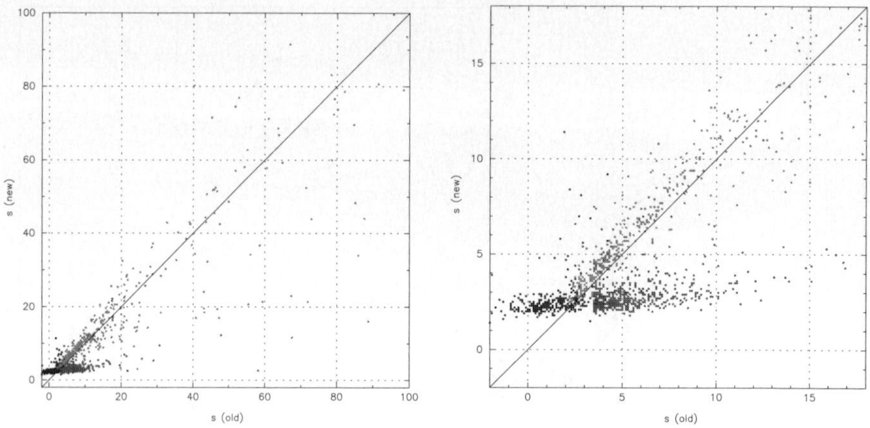

Figure 3.21. A comparison of the cosmic dispersion added for stars solved with the stochastic solution in the old and the new solution. For new detections the F2 statistic is given for the old solution. These can be recognized as the horizontal distributions in the lower part of the graph. To the left, these are mainly normal 5-parameters solutions, to the right they are mainly 7- or 9-parameter solutions in the published data

With the reduction in abscissa errors for the brighter stars, a number of new potentially disturbed solution are exposed. Many of these had been solved in the published data as 7- or 9-parameter solutions (Fig. 3.21).

3.5.5 Conclusions on Single-star Solutions

The single-star solutions have shown in a variety of ways both the quality of the new solution and where the problems are in the published data: in the correlated abscissa errors, most of which were caused by the effects of scan-phase discontinuities, which, combined with the great-circle reduction process, replicated themselves up to 9 times over the great circle. These correlated errors could, and did, accidentally locally collect to create local systematic errors in the astrometric parameters. The scale length over which this took place is predictable from the level of coincidence of great circles as a function of separation, and is as such observed in the correlations for the difference between the parallaxes in the published and the new reduction. As abscissa-error correlations in the new reduction are observed to be a factor 40 smaller than those in the published data, it seems more likely that these differences represent the correlated errors in the published data. The typical scale length of these correlations is much smaller than what some authors have claimed is present in the published *Hipparcos* data.

3.6 Comparison with the ICRS

3.6.1 The Transformation from the New to the Old Catalogue

In order to preserve the definition of the ICRS, the catalogue as produced by the new reduction is transformed, through an orthogonal rotation, to the old catalogue. This involves two transformations, one for the positions at the reference epoch, and one for the proper motions.

The transformation for the reference positions can be described as three successive rotations, which are applied to the new positions, expressed as direction cosines, to obtain the equivalent in the reference frame for the published catalogue. As the rotation angles involved are likely to be very small (at or below mas level), the transformation can be accurately approximated by:

$$\begin{bmatrix} \cos\alpha\cos\delta \\ \sin\alpha\cos\delta \\ \sin\delta \end{bmatrix} \approx \mathsf{C}\cdot\begin{bmatrix} \cos\alpha'\cos\delta' \\ \sin\alpha'\cos\delta' \\ \sin\delta' \end{bmatrix}, \tag{3.42}$$

where

$$\mathsf{C} = \begin{bmatrix} 1 & -\epsilon_z & \epsilon_y \\ \epsilon_z & 1 & -\epsilon_x \\ -\epsilon_y & \epsilon_x & 1 \end{bmatrix}, \tag{3.43}$$

and the accented values apply to the catalogue created through the new reduction. Subtracting on both sides the direction cosine vector for the new catalogue gives:

$$\begin{bmatrix} \cos\alpha\cos\delta - \cos\alpha'\cos\delta' \\ \sin\alpha\cos\delta - \sin\alpha'\cos\delta' \\ \sin\delta - \sin\delta' \end{bmatrix} \approx \mathsf{C}'\cdot\begin{bmatrix} \cos\alpha'\cos\delta' \\ \sin\alpha'\cos\delta' \\ \sin\delta' \end{bmatrix}, \tag{3.44}$$

with

$$\mathsf{C}' \equiv \mathsf{C} - \mathsf{I} = \begin{bmatrix} 0 & -\epsilon_z & \epsilon_y \\ \epsilon_z & 0 & -\epsilon_x \\ -\epsilon_y & \epsilon_x & 0 \end{bmatrix}. \tag{3.45}$$

Given the very small displacements, Eq. 3.44 can be further approximated by:

$$\begin{bmatrix} -\sin\alpha & -\cos\alpha\sin\delta \\ \cos\alpha & -\sin\alpha\sin\delta \\ 0 & \cos\delta \end{bmatrix} \cdot \begin{bmatrix} \mathrm{d}\alpha_* \\ \mathrm{d}\delta \end{bmatrix} \approx \mathsf{C}' \cdot \begin{bmatrix} \cos\alpha'\cos\delta' \\ \sin\alpha'\cos\delta' \\ \sin\delta' \end{bmatrix}, \tag{3.46}$$

where $\mathrm{d}\alpha_* \equiv (\alpha - \alpha')\cos\delta$ and $\mathrm{d}\delta \equiv \delta - \delta'$. Multiplying left and right in Eq. 3.46 by the matrix:

$$\begin{bmatrix} -\sin\alpha & \cos\alpha & 0 \\ -\cos\alpha\sin\delta & -\sin\alpha\sin\delta & \cos\delta \end{bmatrix}$$

and equating $\alpha' \approx \alpha$, $\delta' \approx \delta$ when multiplied by ϵ, gives:

$$\begin{bmatrix} \mathrm{d}\alpha_* \\ \mathrm{d}\delta \end{bmatrix} \approx \begin{bmatrix} -\cos\alpha\sin\delta & -\sin\alpha\sin\delta & \cos\delta \\ \sin\alpha & -\cos\alpha & 0 \end{bmatrix} \cdot \begin{bmatrix} \epsilon_x \\ \epsilon_y \\ \epsilon_z \end{bmatrix}. \tag{3.47}$$

The matrix in Eq. 3.47 is the one encountered earlier in the proper motion transformation as given by Eq. 3.31. Assigning weights to the observations is ambiguous, as the noise on the input data consists of the correlated photon noise and the largely uncorrelated attitude noise, entering at different ratios depending on the brightness and number of observations for each star. The observed rotations are, however, very small, and depend little on what weights are assigned to the data. The following values are observed, using 98 446 single stars with 5-parameter solutions in both catalogues:

$$\begin{aligned} \epsilon_x &= 0.030 \pm 0.004 \quad \text{mas} \\ \epsilon_y &= -0.018 \pm 0.004 \quad \text{mas} \\ \epsilon_z &= 0.028 \pm 0.005 \quad \text{mas}, \end{aligned} \tag{3.48}$$

and a unit-weight standard deviation of 1.4. These rotations are much smaller than the accuracy of about 0.5 mas assigned to the definition of the optical reference frame through the published *Hipparcos* catalogue, which makes their application rather meaningless. It shows, however, that the new reduction, though using a very different approach from that used by FAST and NDAC for the published catalogue, did not cause any significant distortion or rotation in the new catalogue.

Solving the same rotations for the proper motion differences between the two solutions provides the following values:

$$\begin{aligned} \omega_x &= 0.008 \pm 0.004 \quad \text{mas/yr} \\ \omega_y &= 0.012 \pm 0.004 \quad \text{mas/yr} \\ \omega_z &= 0.015 \pm 0.005 \quad \text{mas/yr}, \end{aligned} \tag{3.49}$$

again much smaller than the definition of the reference frame, which is accurate to about 0.1 mas s^{-1}. As a reference frame, the new catalogue is therefore considered as equivalent to, though more precise than, the published catalogue.

3.6.2 The VLBI Radio Stars

A small number of stars also active at radio wavelengths was especially measured around the same epoch as the *Hipparcos* mission took place. The 12 stars for which accurate measurements were obtained are also included in the catalogue, and played a major role in linking the *Hipparcos* catalogue to the ICRS as defined through the radio-telescope network. The data for 11 of these stars have been published by Lestrade et al. (1999), while the method for linking those data with the *Hipparcos* catalogue is described by Lestrade et al. (1995). Dr. Lestrade kindly provided the data on the remaining object (for which the publication of the full details is in preparation), and the relevant data for all 12 stars are presented in Table 3.11. A comparison with the parallax measurements in the new reduction shows an overall good agreement.

Table 3.11. VLBI-based astrometric data for 12 radio stars, measured especially for the linking of the *Hipparcos* catalogue to the ICRS. All data has been transformed to the *Hipparcos* reference epoch, 1991.25. Parallax and proper motion data are given in mas or mas yr^{-1}. The last column gives the parallax in the final reduction.

HIP Ident	α(rad) $\sigma\alpha_*$ (mas)	δ(rad) $\sigma\delta$ (mas)	ϖ $\sigma\varpi$	$\mu_{\alpha,*}$ $\sigma\mu_{\alpha,*}$	μ_δ $\sigma\mu_\delta$	ϖ_H $\sigma_{\varpi H}$
12469	0.7004344004	1.0686534506	0.26	0.97	−1.21	-0.29
LS161303	3.08	1.94	0.61	0.26	0.32	2.99
14576	0.8210414073	0.7148109586	33.32	2.79	−0.64	36.25
Algol	0.57	0.71	0.73	0.14	0.18	1.40
16042	0.9014161172	0.5011772353	19.37	41.23	−104.01	19.37
UX Ari	1.84	2.13	0.39	0.18	0.20	0.70
16846	0.9459181887	0.0102652416	33.88	−31.59	−161.69	32.59
HR1099	0.54	0.53	0.47	0.33	0.31	0.64
19762	1.1092236018	0.4922434532	6.47	0.42	−23.25	10.46
HD283447	2.76	5.17	0.25	0.29	0.28	2.68
23106	1.3015753830	−1.3138351727	4.02	−4.16	−4.15	3.47
HD32918	0.93	3.15	0.80	0.30	1.00	0.52
66257	3.5552172283	0.6489559693	22.21	85.50	−9.22	21.90
HR5110	0.54	0.73	0.45	0.13	0.16	0.23
79607	4.2528612978	0.5909479582	43.93	−267.05	−86.66	47.35
σ2CrB	0.28	0.30	0.10	0.04	0.05	1.20
98298	5.2288376221	0.6143842215	0.73	−3.79	−6.25	1.69
Cyg X1	1.24	1.75	0.30	0.17	0.21	0.91
103144	5.4711440032	0.7746890644	8.59	26.60	−1.24	9.87
HD199178	2.21	2.24	0.33	0.41	0.43	0.46
109303	5.7974646723	0.7983509095	23.97	−52.08	47.03	23.39
AR Lac	0.76	0.97	0.37	0.13	0.19	0.35
112997	5.9910083647	0.2939354481	10.28	−20.59	−27.53	11.18
IM Peg	1.42	1.57	0.62	0.46	0.40	0.34

Chapter 4

THE ASTROMETRIC DATA FOR COMPOSITE IMAGES AND ORBITAL BINARIES

In this chapter we examine the *Hipparcos* data for double and multiple star systems, some of which exhibit detectable orbital motions. Many of these systems had been included in the *Hipparcos* Input Catalogue for their astrophysical importance: a resolved double star system with well determined orbital parameters and an accurate parallax provides direct measurements of stellar masses, and in case the system also shows eclipses, stellar radii. The processing of *Hipparcos* data for double and multiple systems was, however, complex, and for multiple systems it required *a priori* information. The further complexities of a combination of orbital motion and duplicity stretched the interpretation possibilities of the data to the limit. Despite these complications, a large number of new binary systems, and even a number of orbital systems, were discovered and analysed using the mission data.

To start, we examine the effects a double star has on the observed signal, and how these effects are used to analyse the signal to provide information on separation, orientation and magnitude differences for relatively small (< 0.3 arcsec), intermediate (the remaining combined measurements) and large (individual components both measured) separations. Next we look at deriving astrometric parameters for double stars, and how this is affected by separation and magnitude difference. Further complications, such as caused by one of the stars being variable, or systems with more than two components, are also described. Finally, the effects of orbital motions, from wide binaries to planetary companions, are described.

4.1 The modulated signal for small-separation double stars

A transit of the image of a point source over the grid in the focal plane of the *Hipparcos* telescope produced a very regularly modulated signal, fully described by the mean intensity and the first and second harmonics of the grid period, with a well-defined relation between the phases and amplitudes of the two harmonics (described by the OTF, see Section 12.1). However, often more than one image was visible within the Instantaneous Field of View (IFOV), Fig. 2.5, page 48), the small area of the grid seen by the Image Dissector Tube or IDT. The grid had been optimized especially to enable the disentangling of these double-star images: the ratio between the widths of the transparent and non-transparent lines on the grid was chosen such that a sufficiently significant second harmonic would be produced, while retaining enough light to fulfil the overall astrometric goals of the mission. In the chosen configuration the average amplitude of the first harmonic is about 0.7, and the second harmonic about 0.2, times the mean response level. The presence of a significant second harmonic in the modulated signal meant that the combined signal for two images would be distinctly different from that of a single image. The difference between the combined signal and individual signal contained information on the projected separation and magnitude difference of the system components. For a single transit of a double star there remained, however, an ambiguity in the separation, which can only be determined modulus the grid period of 1.2074 arcsec. This ambiguity is resolved, together with the orientation of the system and the magnitude difference, through combining observations obtained in different scan directions over the mission.

4.1.1 The Combined Signal

The signal of a single star as it transits the modulating grid in the focal plane is accurately represented by a simple five-parameter model, representing the mean signal strength and the amplitudes and phases of the first and second harmonics in the modulation:

$$A_k = B + I \cdot \big(1 + M_1 \cos(p_k + \phi_1) + M_2 \cos(2p_k + \phi_2)\big), \tag{4.1}$$

where B is the local background, I the signal intensity, M_1 and M_2 the relative amplitudes of the first and second harmonic, ϕ_1 and ϕ_2 the phases of the first and second harmonic, and p_k the relative phase of sample k, producing the response A_k. This is the same signal as represented by the parameters β_1 to β_5 in Eq. 2.33, page 50:

$$I_k = \beta_1 + \beta_2\big[\cos(p_k + \beta_3) + \beta_4 \cos 2(p_k + \beta_3) + \beta_5 \sin 2(p_k + \beta_3)\big], \tag{4.2}$$

where the parameters β_4 and β_5, together, represent the relative phase and amplitude of the second harmonic. For a single star these two parameters are

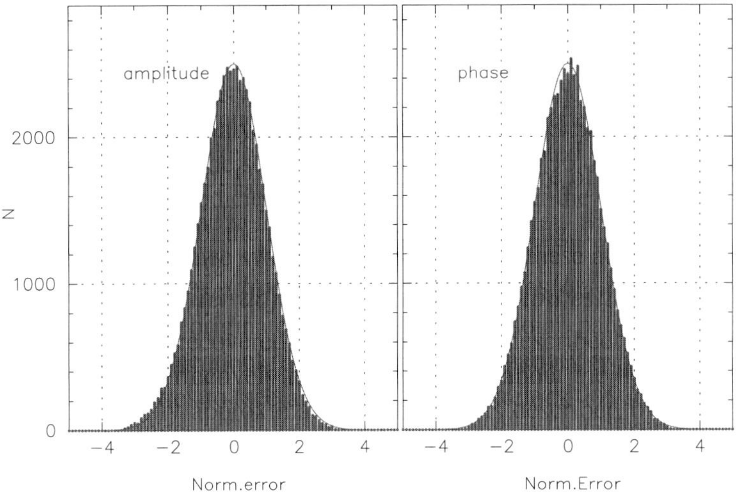

Figure 4.1. Histograms of normalized formal errors on $\hat{\beta}_4 - 1$ (left) and $\hat{\beta}_5$ (right). The curves show the Gaussian distributions for the same number of observations, and $\sigma = 1$. The data are from orbit 401

instrument parameters which are calibrated as a function of position on the grid and colour of the star, the so-called optical transfer function or OTF calibration (see Section 12.1, page 301) . Application of this calibration allows for a normalization of the observed β_4 and β_5 as is described in more detail in Section 12.1 and by Eq. 12.5, repeated here:

$$\begin{aligned} \tilde{\phi}_2 &\equiv \arctan(\tilde{\beta}_5/\tilde{\beta}_4), \\ \tilde{R} &\equiv \sqrt{\tilde{\beta}_4^{\,2} + \tilde{\beta}_5^{\,2}}, \\ \hat{\beta}_4 &\equiv \left[\ \ \beta_4 \cos(2\tilde{\phi}_2) + \beta_5 \sin(2\tilde{\phi}_2)\right]/\tilde{R}, \\ \hat{\beta}_5 &\equiv \left[-\beta_4 \sin(2\tilde{\phi}_2) + \beta_5 \cos(2\tilde{\phi}_2)\right]/\tilde{R}. \end{aligned} \tag{4.3}$$

where $\tilde{\beta}_4$ and $\tilde{\beta}_5$ are the values predicted by the OTF model for the second-harmonic. The transformation described above produces the normalized parameters $\hat{\beta}_4$, centred on 1.0, and $\hat{\beta}_5$, centred on zero. The errors on β_4 and β_5 are transformed using similar relations. An example of the observed distributions for these two parameters in observations of single stars only is shown in Fig. 4.1. The good agreement with the expected Gaussian distributions shows that the OTF calibration model has performed well, which is important when using these data for analysing the combined signals of double stars.

The effect of applying the transformations given by Eq. 4.3, to Eq. 4.1 is to set $\phi_2 = 2\phi_1$, which simplifies the further calculations. Furthermore, we know from the OTF calibrations that the main contribution to the variations in

β_4 and β_5 originates from the measurement-position on the grid, with only a small contribution related to the colour index of the star (see Section 12.1.2, Fig. 12.3, page 303). The variations in these parameters due to a colour index difference of 1.0 mag. amounts to 2 to 4 per cent, generally well below their formal errors. In the processing by NDAC and FAST for the published data colour indices of individual components of double and multiple systems were poorly known, and the ratio $R(= M_2/M_1)$ was assumed the same for the two components. Better information on colours is available now (Fabricius and Makarov 2000), but the complications of incorporating this information in the reductions are considerable and the benefit is low. Therefore, the same assumption on R has been made for the new reduction. This also applies to the modulation factor M_1. The background for both signals is naturally the same.

We can now add together the signals from two point sources (a and b) at close separation (close in this case means no more than about 30 arcsec., beyond which the IFOV response is too low and too poorly known):

$$\begin{aligned} A_{k,a} &= B + I_a \cdot \left[1 + M_1\left(\cos(p_k + \phi_a) + R\cos(2p_k + 2\phi_a)\right)\right], \quad (4.4) \\ A_{k,b} &= B + I_b \cdot \left[1 + M_1\left(\cos(p_k + \phi_b) + R\cos(2p_k + 2\phi_b)\right)\right], \end{aligned}$$

and the observed combined signal:

$$A_{k,ab} = B + I_{ab} \cdot \left(1 + M_{1,ab}\cos(p_k + \phi_{ab}) + M_{2,ab}\cos(2p_k + 2\phi_{ab})\right). \quad (4.5)$$

The combined signal will have to include also the influence of the IFOV, of which the main effect is to reduce the light received from a companion when centred on one of the components of a double system. There is, however, also an indication that the modulation parameters and the OTF calibration are affected by the position in the IFOV (see for example Fig. 11.6, page 296), but this was not possible to calibrate to a useful level of accuracy. The IFOV corrections are in particular relevant for systems with separate entries, those with separations above 12 arcsec (see further Section 4.3). For systems with shorter separations the IFOV was generally pointed at a position somewhere between the components. The effect of the IFOV will enter the equations below simply as a reduction of the intensity I_b (when centred on component a), or I_a when centred on component b.

Expressed in the observed β parameters for the double system, we find for the combined signal the following relations:

$$\begin{aligned} \beta_1 &= B + I_{ab}, \quad (4.6) \\ \beta_2 &= I_{ab} \cdot M_{1,ab}, \\ \beta_3 &= \phi_{ab}, \\ \beta_4 &= \left(M_{2,ab}/M_{1,ab}\right)\cos(2\phi_1 - \phi_2), \\ \beta_5 &= \left(M_{2,ab}/M_{1,ab}\right)\sin(2\phi_1 - \phi_2). \end{aligned}$$

4.1.2 Recovering the Differential Parameters

For the recovery of the differential parameters of a double star we introduce the magnitude difference Δm or intensity ratio r and the apparent phase difference $\Delta\phi$:

$$r \equiv I_b/I_a = 10^{-0.4\Delta m}, \tag{4.7}$$

$$\Delta\phi \equiv \phi_b - \phi_a. \tag{4.8}$$

The phase difference $\Delta\phi$ depends on the double star's differential parameters: separation ρ and orientation θ, on the orientation ψ of the scan on the sky, and on the period of the modulating grid as projected on the sky, $s = 1.2074$ arcsec. These quantities are shown in Fig. 4.2.

The local orientation of the scan circle ψ is given through Eq. 2.50 by the instantaneous position of the satellite's spin axis at the time of observation (α_s, δ_s) and the position of the star (α_*, δ_*). The value for ψ for each field transit is contained in the intermediate astrometry or abscissae files for the new reduction. The actual separation as projected along the scan direction is then:

$$\Delta G = \rho \sin(\theta + \psi), \tag{4.9}$$

and the phase difference $\Delta\phi = 2\pi \bmod(\Delta G/s, 1)$.

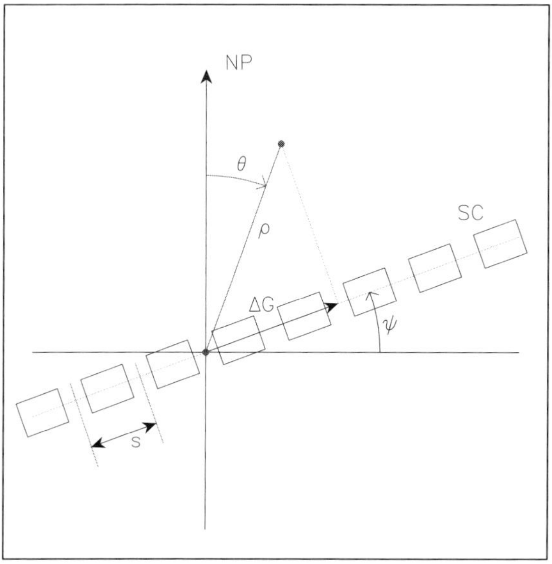

Figure 4.2. Definition of the angles involved in describing the geometry of a double-star transit over the modulating grid as defined by the orientation ψ of the scan circle (Scan Circle (SC))

Applying Eq. 4.8 to Eq. 4.4 gives:

$$\begin{aligned} A_{k,a} &= B + I_a \cdot \big[1 + M_1\big(\cos(p_k + \phi_a) + R\cos(2p_k + 2\phi_a)\big)\big] \quad (4.10)\\ A_{k,b} &= B + r \cdot I_a \cdot \big[1 + M_1\big(\cos(p_k + \phi_a + \Delta\phi) + \\ &\quad R\cos(2p_k + 2\phi_a + 2\Delta\phi)\big)\big]. \end{aligned}$$

To derive the combined signal, we introduce:

$$\begin{aligned} \cos\Delta\phi_{ab} &\equiv \frac{r\cos\Delta\phi + 1}{\sqrt{1 + 2r\cos\Delta\phi + r^2}}, \\ \sin\Delta\phi_{ab} &\equiv \frac{r\sin\Delta\phi}{\sqrt{1 + 2r\cos\Delta\phi + r^2}}, \end{aligned} \tag{4.11}$$

and therefore:

$$\begin{aligned} \cos 2\Delta\phi_{ab} &= \frac{r^2\cos 2\Delta\phi + 2r\cos\Delta\phi + 1}{1 + 2r\cos\Delta\phi + r^2}, \\ \sin 2\Delta\phi_{ab} &= \frac{2r\sin\Delta\phi + r^2\sin 2\Delta\phi}{1 + 2r\cos\Delta\phi + r^2}, \end{aligned} \tag{4.12}$$

where $\Delta\phi_{ab}$ represents the phase offset for the first harmonic in the combined signal, relative to the expected phase for the first harmonic of the signal of the primary star. The combined signal can then be written as:

$$\begin{aligned} A_{k,ab} &= B + I_a \cdot \big[(1 + r) + \\ &\quad M_1\big(\sqrt{1 + 2r\cos\Delta\phi + r^2}\cos(p_k + \phi_a + \Delta\phi_{ab}) + \\ &\quad R(1 + r\cos 2\Delta\phi)\cos(2p_k + 2\phi_a) - Rr\sin 2\Delta\phi\sin(2p_k + 2\phi_a)\big)\big]. \end{aligned} \tag{4.13}$$

To make this relation compatible with Eq. 4.2, we shift the phase of the second harmonic to agree with that of the first harmonic, and write the combined signal as:

$$\begin{aligned} A_{k,ab} &= B + I_a \cdot \Bigg[(1 + r) + \\ &\quad M_1\Bigg(\sqrt{1 + 2r\cos\Delta\phi + r^2}\cos(p_k + \phi_a + \Delta\phi_{ab}) + \\ &\quad R(1 + r\cos 2\Delta\phi)\big(\cos(2p_k + 2\phi_a + 2\Delta\phi_{ab})\cos 2\Delta\phi_{ab} + \\ &\quad \sin(2p_k + 2\phi_a + 2\Delta\phi_{ab})\sin 2\Delta\phi_{ab}\big) \\ &\quad -Rr\sin 2\Delta\phi\big(\sin(2p_k + 2\phi_a + 2\Delta\phi_{ab})\cos 2\Delta\phi_{ab} - \\ &\quad \cos(2p_k + 2\phi_a + 2\Delta\phi_{ab})\sin 2\Delta\phi_{ab}\big)\Bigg)\Bigg]. \end{aligned} \tag{4.14}$$

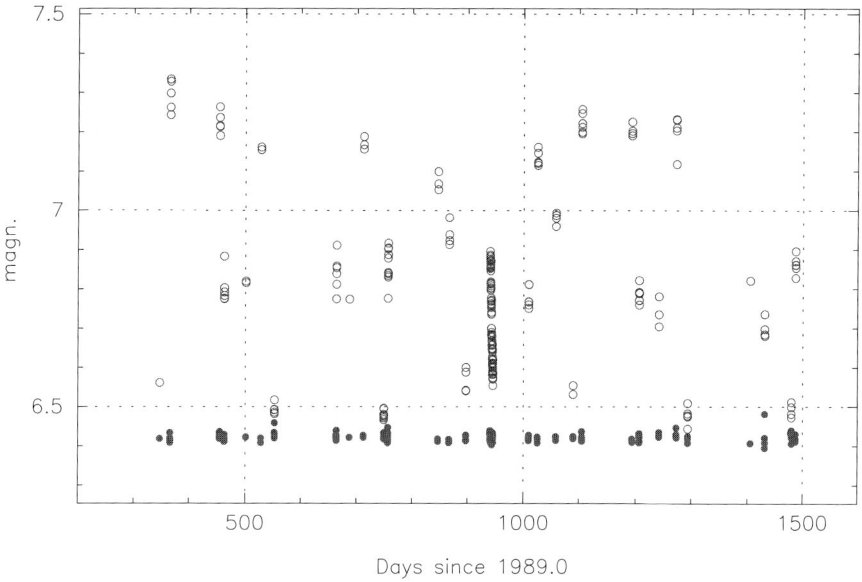

Figure 4.3. The H_{AC} (open circles) and H_{DC} (filled circles) magnitude for the double star HIP 25, as measured over the mission. The fainter AC-magnitudes are the result of the dampening of the first harmonic in the modulated signal due to the superposition of the signals from the two double star components

Combining the relevant terms for the second harmonic gives:

$$
\begin{aligned}
A_{k,ab} = {} & B + I_a \cdot \Bigg[(1+r) + \\
& M_1 \Bigg(\sqrt{1 + 2r\cos\Delta\phi + r^2} \cos(p_k + \phi_a + \Delta\phi_{ab}) + \\
& R \, \big(\cos(2p_k + 2\phi_a + 2\Delta\phi_{ab}) \\
& ((1 + r\cos 2\Delta\phi)\cos 2\Delta\phi_{ab} + r\sin 2\Delta\phi \sin 2\Delta\phi_{ab}) + \\
& \sin(2p_k + 2\phi_a + 2\Delta\phi_{ab}) \\
& ((1 + r\cos 2\Delta\phi)\sin 2\Delta\phi_{ab} - r\sin 2\Delta\phi\cos 2\Delta\phi_{ab})\big)\Bigg)\Bigg].
\end{aligned} \tag{4.15}
$$

Substituting in these relations those of Eq. 4.11 and 4.12 gives:

$$
\begin{aligned}
A_{k,ab} = {} & B + I_a \cdot \big[(1+r) + \\
& M_1 \sqrt{1 + 2r\cos\Delta\phi + r^2} \big(\cos(p_k + \phi_a + \Delta\phi_{ab}) + \\
& R F_c(r, \Delta\phi) \cos(2p_k + 2\phi_a + 2\Delta\phi_{ab}) + \\
& R F_s(r, \Delta\phi) \sin(2p_k + 2\phi_a + 2\Delta\phi_{ab}) \big) \big],
\end{aligned} \tag{4.16}
$$

which now has the same form as Eq. 4.2. The coefficients $F_c(r, \Delta\phi)$ and $F_s(r, \Delta\phi)$ are given by:

$$\begin{aligned} F_c(r, \Delta\phi) &= \frac{1 + (r + r^2)(2\cos\Delta\phi + \cos 2\Delta\phi) + r^3}{(1 + 2r\cos\Delta\phi + r^2)^{3/2}}, \\ F_s(r, \Delta\phi) &= \frac{(r - r^2)(2\sin\Delta\phi - \sin 2\Delta\phi)}{(1 + 2r\cos\Delta\phi + r^2)^{3/2}}. \end{aligned} \tag{4.17}$$

This leads to the following relations for the modulated signal of a double star:

$$\begin{aligned} \beta_1 &= B + I_a(1 + r), \\ \beta_2 &= I_a \cdot M_1 \sqrt{1 + 2r\cos\Delta\phi + r^2}, \\ \beta_3 &= \phi_a + \Delta\phi_{ab}, \\ \beta_4 &= R \cdot F_c(r, \Delta\phi), \\ \beta_5 &= R \cdot F_s(r, \Delta\phi). \end{aligned} \tag{4.18}$$

The functions $F_c(r, \Delta\phi)$ and $F_s(r, \Delta\phi)$ have already been derived for the signal in the form of $\hat{\beta}_4$ and $\hat{\beta}_5$ respectively (Eq. 4.3). We thus have two equations to resolve the phase difference $\Delta\phi$ and intensity ratio r of a double-star transit. For a single star ($r = 0$) we find $F_c(r, \Delta\phi) = 1$ and $F_s(r, \Delta\phi) = 0$, which are the expectation values for $\hat{\beta}_4$ and $\hat{\beta}_5$.

A third equation can be obtained from the photometric calibrations as based on β_1 (the DC magnitudes) and β_2 (the AC magnitudes), $\mathrm{H_{DC}}$ and $\mathrm{H_{AC}}$ respectively:

$$\mathrm{H_{AC}} - \mathrm{H_{DC}} = -2.5\log\left(\frac{\sqrt{1 + 2r\cos\Delta\phi + r^2}}{(1 + r)}\right). \tag{4.19}$$

Figure 4.3 shows an example of the effect a double system has on the determination of the magnitudes. The problem with these three relations is, however, that the phase difference relates to the positions on the modulating grid, and therefore has an ambiguity equal to the grid period, 1.2074 arcsec as projected on the sky. Also, when the separation as projected along-scan, and therefore perpendicular to the grid lines, becomes small, or magnitude differences large, determination of parameters becomes very uncertain, and "accidental solutions" can no longer be distinguished easily from proper solutions. The functions $F_c(r, \Delta\phi)$ and $F_s(r, \Delta\phi)$, for small $\Delta\phi$ values, differ from their single-star values only through a dependence on $\Delta\phi^2$ and $\Delta\phi^3$ respectively, as described in Section 4.1.3.

4.1.3 Re-adjusting Double-star Parameters

If we assume that the published data provide a very good estimate of the differential double star parameters (ρ, θ and Δm), then we can use the relations presented above to obtain differential corrections and assess the accuracies and correlations of those parameters. The procedure is simple: using the double star parameters as published, one can derive for each field transit the predicted values for $\hat{\beta}_4$, $\hat{\beta}_5$ and $H_{AC} - H_{DC}$. One can also derive the changes to those predicted values due to small changes in the double-star parameters. The differences between the observed and predicted values are then fitted by least squares as a function of those differential corrections, providing a new solution for the differential parameters. An example of this procedure is presented here for the double star HIP 25 (CCDM 00003−4417). The three functions identified above are shown for this system in Fig. 4.4.

It is also possible to fit a time-dependent correction to the separation and orientation angle. This is appropriate for representing long-period orbital binaries and accidental superpositions. Such linear dependencies were also fitted for many well-resolved binaries in the published data. Table 4.1 shows an example of this type of solution, which will be referred to as *L-type* solutions.

For double stars with separations less than 0.3 arcsec the orientation of the the system follows directly from the distribution of $H_{AC} - H_{DC}$ magnitude differences. This can be seen from an evaluation of Eq. 4.19 for small values

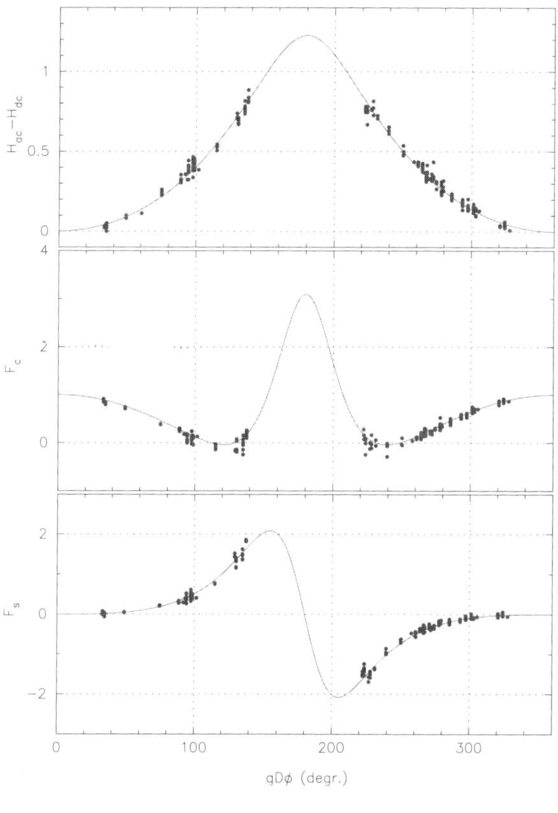

Figure 4.4. Fitting the double star parameters for HIP 25 as a function of the projected separation of the two components. The dots show the actual observations, the curves show their expected values based on the derived double-star parameters

Table 4.1. Published and new determinations for the differential parameters of HIP 25

Parameter	Old	New	σ	units
θ	315.80	315.79	0.05	degrees
$\dot{\theta}$	0.80	0.75	0.06	degree/yr
ρ	0.463	0.464	0.002	arcsec
$\dot{\rho}$	−0.009	−0.013	0.003	arcs/yr
dM	0.65	0.721	0.009	magnitudes

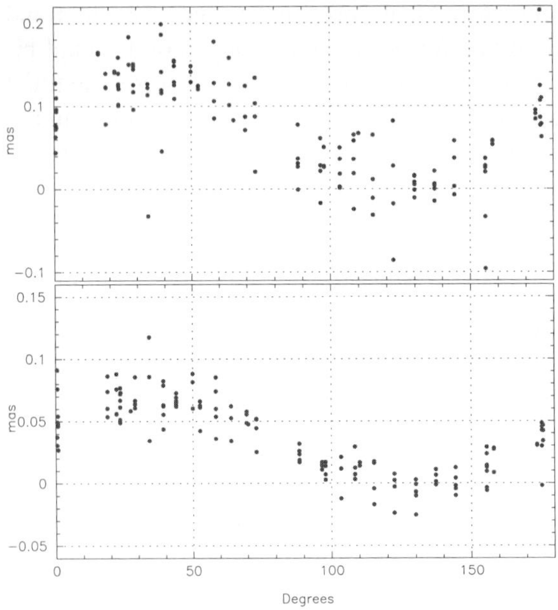

Figure 4.5. Variations in $1 - \hat{\beta}_4$ (top) and $\mathrm{H_{AC}} - \mathrm{H_{DC}}$ (bottom) as a function of the scan-orientation angle ψ for a double star with a separation of 0.192 arcsec (HIP 274). At a small separation these parameters become proportional to $\sin(\theta + \psi)^2$, as shows clearly in the graphs. The maximum in the distributions defines the orientation angle θ, for this star $\theta = 37°$

of $\Delta\phi$:

$$\begin{aligned} \mathrm{H_{AC}} - \mathrm{H_{DC}} &\approx -2.5\log\left(1 - \frac{r}{(1+r)^2}\frac{\Delta\phi^2}{2}\right), \\ &= -1.086\ln\left(1 - \frac{r}{(1+r)^2}\frac{\Delta\phi^2}{2}\right), \\ &\approx 0.543\,\frac{r}{(1+r)^2}\Delta\phi^2. \end{aligned} \tag{4.20}$$

The phase difference $\Delta\phi$ is for these small separations directly related to the projection of the separation ρ and the orientation of the binary with respect to the scan (see Fig. 4.2):

$$\Delta\phi = 2\pi\frac{\rho}{s}\sin(\theta + \psi), \tag{4.21}$$

Table 4.2. Published and new determinations for the differential parameters of HIP 274

Parameter	Old	New	σ	units
θ	38	36.9	1.1	degrees
ρ	0.180	0.194	0.011	arcsec
r	0.192	0.165	0.012	
dM	1.79	1.95	0.08	magnitudes

where as before s is the grid period, measured in the same units as ρ. Combined with Eq. 4.20 we find:

$$\mathrm{H_{AC}} - \mathrm{H_{DC}} \approx 0.543 \frac{r}{(1+r)^2} \left(\frac{2\pi\rho}{s} \sin(\theta + \psi) \right)^2. \tag{4.22}$$

An example of this relation is shown in Fig. 4.5. The angle θ (modulus π) can be derived directly from this relation, being equal to $\pi/2 - \psi$ at the value of ψ for which the above function has its maximum.

In a similar approximation for $\hat{\beta}_4$ we find:

$$1 - \hat{\beta}_4 \approx 1.5 \frac{r}{(1+r)^2} \Delta\phi^2. \tag{4.23}$$

A comparison with Eq. 4.20 shows that for those close binaries the information from the magnitude differences and $\hat{\beta}_4$ is identical and can't be used to resolve the ambiguity between r and ρ. For that we need $\hat{\beta}_5$, for which the information is even less sensitive at small values of $\Delta\phi$:

$$\hat{\beta}_5 \approx \frac{r - r^2}{(1+r)^3} \Delta\phi^3 = (1 - \hat{\beta}_4) \frac{2}{3} \cdot \frac{1-r}{1+r} \Delta\phi. \tag{4.24}$$

The procedure for finding the characteristic parameters for small separation binaries is now as follows, illustrated by the application for HIP 274:

1 Use the $\mathrm{H_{AC}} - \mathrm{H_{DC}}$ magnitude differences to derive the orientation θ. The amplitude of the variation gives the first relation between ρ and r. The values found for HIP 274 are:

$$\begin{aligned} \theta &= 36.^\circ 9 \pm 1.^\circ 1, \\ \frac{\sqrt{r}}{1+r} \frac{2\pi\rho}{s} &= 0.3520 \pm 0.0034. \end{aligned} \tag{4.25}$$

2 Next, keep the orientation θ fixed and fit the amplitude of the modulation of $\hat{\beta}_5/(1-\hat{\beta}_4)$, which gives the second relation between ρ and r. The value found for HIP 274 is:

$$\frac{1-r}{1+r} \frac{2\pi\rho}{s} = 0.723 \pm 0.036. \tag{4.26}$$

3 These two relations determine r and ρ. For HIP 274 we find:

$$\frac{\sqrt{r}}{1-r} = 0.487 \pm 0.025, \tag{4.27}$$

from which $r = 0.165 \pm 0.012$ and, through back substitution, $\rho = 0.194 \pm 0.011$ arcsec.

The values as presented in (ESA 1997) are shown with the new determinations in Table 4.2. This type of solution will be referred to as *S-type* solutions.

4.1.4 Comparisons with the Published Data

As part of the new reductions a new determination of the differential parameters for all simple binary stars was made. Stars with separations below 0.28 arcsec were treated with the *S-type* solution, those with larger separations with the *L-type* solution. The limit of 0.28 arcsec represents the separation beyond which the simple linear approximations used above to derive the orientation of the system are no longer sufficient to describe the observations. The *S-type* solutions were only accepted if the two relations between r and ρ shown above both had a significance of 6 σ or more. For lower significance the relative errors will cause a significant bias in the results, which makes the solution unreliable. Solutions for only 592 of the 2109 binaries with separation less than 0.28 arcsec could be accepted, all others were more or less marginal in as far as the *Hipparcos* data are concerned. The remaining 1517 stars will have, without the use of external data, rather unreliable solutions, where the *Hipparcos* data alone cannot give unambiguous or even significant information on the parameters of the system: separation, orientation and magnitude difference. The *L-type* solution has been successfully applied to 6882 binaries. Here the limit was set by the separation: for large separations the differential corrections are no longer reliable, so only separations up to 6 arcsec have been solved again.

The most striking difference with the published data is a systematic offset for the magnitude differences. A mean difference of 0.1 magnitude is observed for both L and S solutions, as can be seen in Fig. 4.6, top graph. There are also systematic differences in the separations, possibly related to the grid period of 1.2074 arcsec. The relation between the grid period and the separation of the stars in a binary system creates a specific pattern in the distribution of measurements, to which different solution methods show different sensitivities. These differences can be seen in the bottom graph of Fig. 4.6. There appears to be progression of these differences independent of the solution method. For the *S-type* solutions the corrections to the separation and magnitude difference are clearly correlated, as can be seen in Fig. 4.7.

The formal errors on the determinations of the double star parameters show their general reliability as a function of separation and magnitude difference.

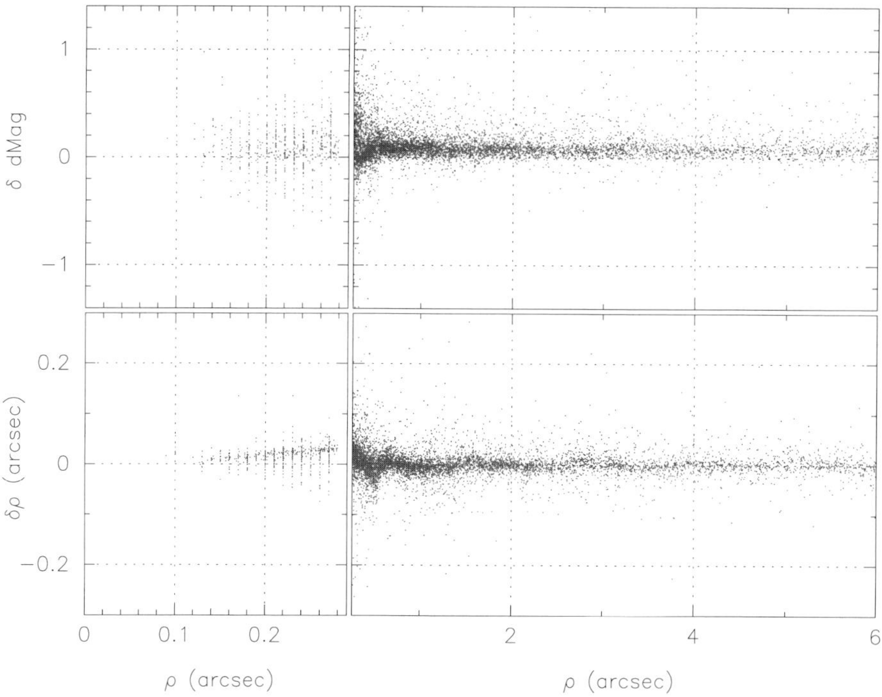

Figure 4.6. Differences between the published and the new solutions for the differential parameters ρ (separation, bottom) and $\mathrm{d}Mag$ (magnitude difference, top). The magnitude differences in the new solutions are systematically 0.1 magnitude larger than in the published data. The differences in separation show systematics that are at least partly related to the grid period of 1.2074 arcsec

When plotted as a function of separation (Fig. 4.8) they show systematics similar to those observed for the corrections, again related to the geometry of the double-star system with respect to the periodicity of the modulating grid. This graph also shows the main reason for the switch in solution types: formal errors for *L-type* solutions increase dramatically towards separations below 0.3 arcsec, and a more explicit control on the solutions, as provided by the *S-type* solutions, is required. When formal errors are plotted as a function of magnitude difference, the different sensitivities of the two types of solution show up (Fig. 4.9).

In the left-hand graphs of both Fig. 4.6 and Fig. 4.8 one can recognize two populations of stars: one with separations as presented in the published catalogue are all rounded off to the nearest 10 mas in separation, and one with the separation resolved down to 1 mas. This reflects the accuracies of the published data, which can also be seen from the dispersions of the differences between

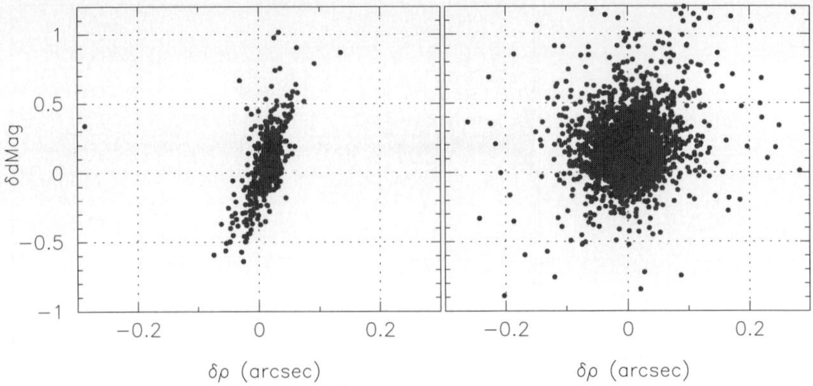

Figure 4.7. The relation between $\delta\rho$ and δdMag for the *S-type* solutions (left) and *L-type* solutions (right). A strong correlation between the corrections for these two parameters for the *S-type* solutions was expected, given the way these two parameters are derived together from two complex relations. The offset of the magnitude difference is clear in both diagrams

the old and new solutions, which are clearly smaller for systems given with a higher precision separation in the published data. Most of the lower accuracy systems in this group refer to systems with large magnitude differences, as could be expected.

4.2 Astrometric parameters for double stars

For the astrometric-parameter determination of single-entry double stars we distinguish two different cases: a determination for the photo centre or a determination for the primary star. The first type of solution is applied generally to systems with separations smaller than about 0.3 arcsec, the second type of solution to systems with larger separations.

In the first type of solution the observed phases of the first and second harmonics in the combined signal have to be related to the expected phases for these harmonics for the primary star alone. This was done already in Section 4.1.2, where the difference is given by $\Delta\phi_{ab}$. Using the relations given in Eq. 4.11 we find the following expression:

$$\Delta\phi_{ab} = \arctan(r \sin\Delta\phi/(r\cos\Delta\phi + 1)). \tag{4.28}$$

A further complication is the reduction in amplitude of the first harmonic, as this amplitude is used in assigning formal errors to the measurements. This amplitude can be derived from Eq. 4.16:

$$M_1{}' = M_1 \frac{\sqrt{1 + 2r\cos\Delta\phi + r^2}}{1 + r}. \tag{4.29}$$

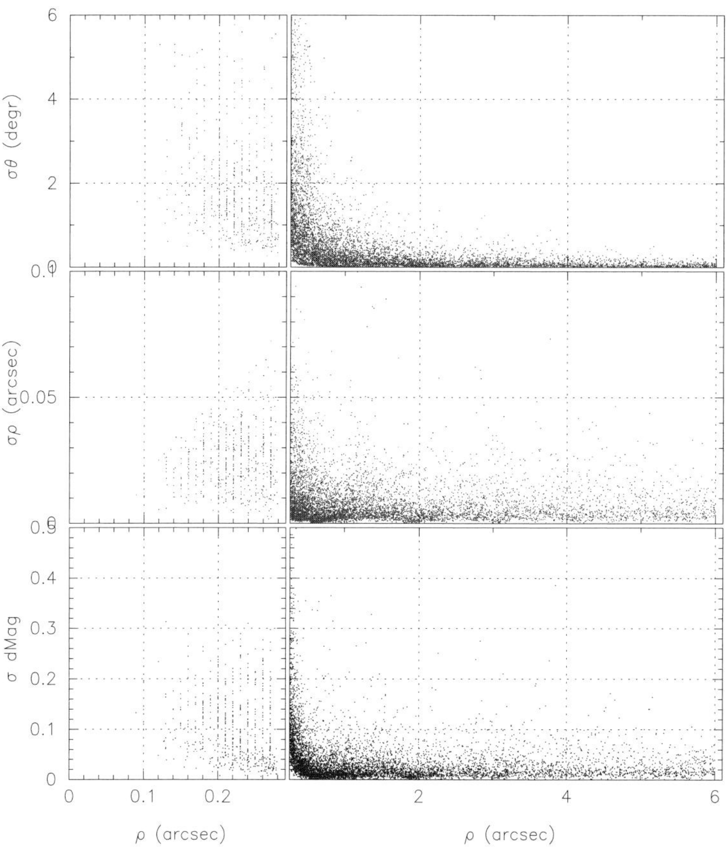

Figure 4.8. Formal errors on the double star parameters as determined with the *S-type* solutions (left) and *L-type* solutions (right). systematic variations can clearly be seen for the magnitude difference and separation determinations, depending on the relation between the separation and the grid period. The separations are as given in the published data

Thus, $M_1{}'$ varies between M_1 and $M_1(1-r)/(1+r)$, and can effectively vanish when the intensity ratio of the two components equals one. An example of this type of solution is shown in Fig. 4.10 and Table 4.3.

The second type of solution, which refers the astrometric parameters to the photo centre of the system, can only be applied to systems with small (less than 0.3 arcsec) separation. The photo centre is defined in the usual way as the mean position of the two components, weighted by their intensities:

$$\boldsymbol{p}_{pc} = (I_a \boldsymbol{p}_a + I_b \boldsymbol{p}_b)/(I_a + I_b) = \boldsymbol{p}_a + \boldsymbol{\rho}\frac{r}{1+r} = \boldsymbol{p}_b - \boldsymbol{\rho}\frac{1}{1+r}, \quad (4.30)$$

where the vector $\boldsymbol{\rho}$ represents the scalar separation ρ and orientation θ. For these systems, the projected separation ΔG (see Eq. 4.9) simply relates to the

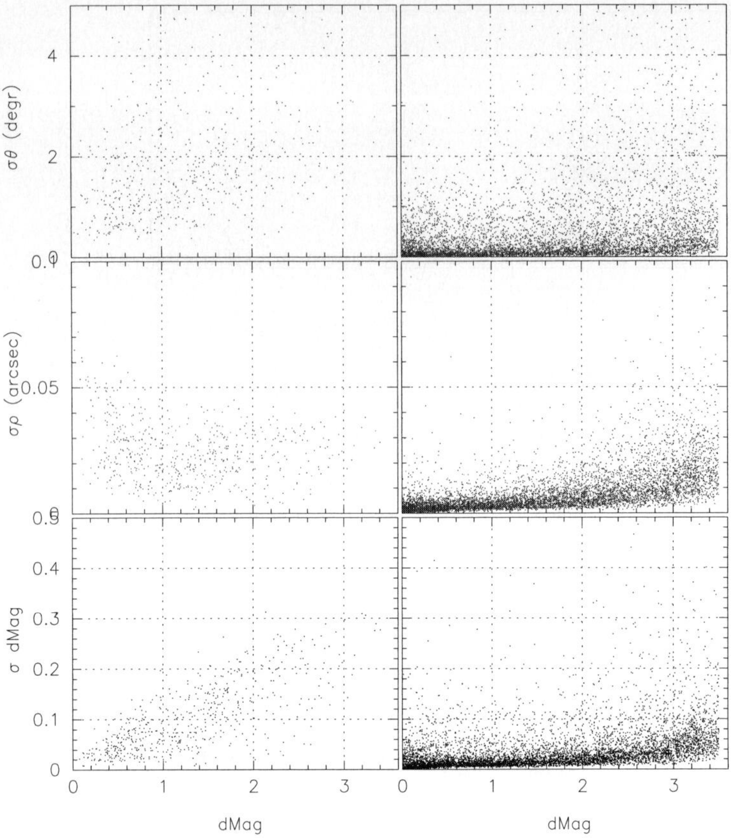

Figure 4.9. Formal errors on the double star parameters as determined with the *S-type* solutions (left) and *L-type* solutions (right)

Table 4.3. Published and new determinations for the differential parameters and astrometry of the primary component of HIP 55

Parameter	Old	σ	New	σ	units
θ	273.60		273.66	0.05	degrees
ρ	3.810		3.816	0.004	arcsec
dM	1.80		1.85	0.01	magnitudes
ϖ	14.66	0.98	15.06	0.56	mas
$\mu_{\alpha*}$	162.88	0.82	162.83	0.45	mas/yr
μ_δ	−28.82	0.82	−29.25	0.51	mas/yr

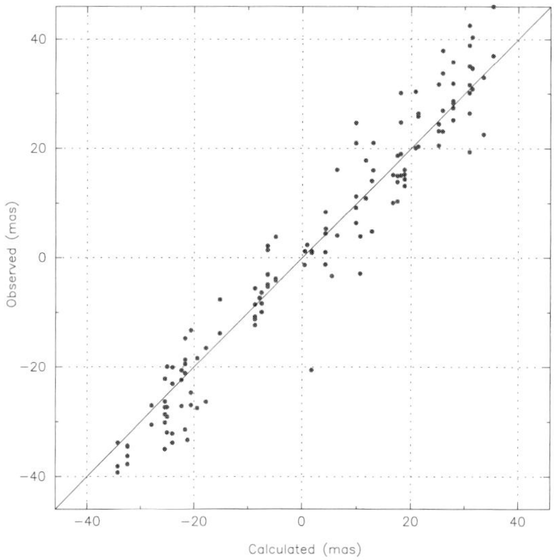

Figure 4.10. A comparison between predicted and observed values for the phase corrections $\Delta\phi_{ab}$ for the double star HIP 55

phase difference between the first and second harmonics as:

$$\Delta\phi = 2\pi\Delta G/s, \tag{4.31}$$

where s is the grid period of 1.2074 arcsec. When we relate the measurements to the photo centre, we have to divide the phase shift $\Delta\phi$ between the two contributions according to their intensities, to give, as the equivalent of Eq. 4.14 above:

$$\begin{aligned} A_{k,a} &= B + I_a \cdot \big[1 + M_1\big(\cos(p_k + \phi_p - \Delta\phi\frac{r}{1+r}) + \\ &\quad R\cos(2p_k + 2\phi_p - 2\Delta\phi\frac{r}{1+r})\big)\big], \\ A_{k,b} &= B + r \cdot I_a \cdot \big[1 + M_1\big(\cos(p_k + \phi_p + \Delta\phi\frac{1}{1+r}) + \\ &\quad R\cos(2p_k + 2\phi_p + 2\Delta\phi\frac{1}{1+r})\big)\big]. \end{aligned} \tag{4.32}$$

The equivalent of Eq. 4.11 now becomes:

$$\begin{aligned} \cos\Delta\phi_p &\equiv \frac{r\cos(\Delta\phi\frac{1}{1+r}) + \cos(\Delta\phi\frac{r}{1+r})}{\sqrt{1 + 2r\cos\Delta\phi + r^2}}, \\ \sin\Delta\phi_p &\equiv \frac{\sin(\Delta\phi\frac{r}{1+r}) - r\sin(\Delta\phi\frac{1}{1+r})}{\sqrt{1 + 2r\cos\Delta\phi + r^2}}, \end{aligned} \tag{4.33}$$

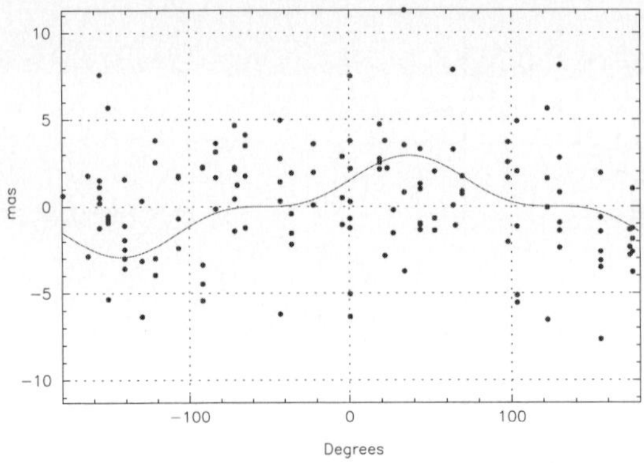

Figure 4.11. A comparison between predicted and observed values for the phase corrections $\Delta\phi_p$ of the photocentre for the double star HIP 274. The curve shows the expected variation

leading to the following combined signal:

$$\begin{aligned} A_{k,p} &= B + I_a \cdot \Big[(1+r) + \\ &\quad M_1\big(\sqrt{1+2r\cos\Delta\phi + r^2}\cos(p_k+\phi_p+\Delta\phi_p) + \\ &\quad R(\cos(2\Delta\phi\frac{r}{1+r}) + r\cos(2\Delta\phi\frac{1}{1+r}))\cos(2p_k+2\phi_p) + \\ &\quad R(\sin(2\Delta\phi\frac{r}{1+r}) - r\sin(2\Delta\phi\frac{1}{1+r}))\sin(2p_k+2\phi_p)\big)\Big], \end{aligned} \tag{4.34}$$

which can be worked out further to give:

$$\begin{aligned} A_{k,p} &= B + I_a \cdot \Big[(1+r) + \\ &\quad M_1\sqrt{1+2r\cos\Delta\phi + r^2}\big(\cos(p_k+\phi_p+\Delta\phi_p) + \\ &\quad RF_c(r,\Delta\phi)\cos(2p_k+2\phi_p+2\Delta\phi_p) + \\ &\quad RF_s(r,\Delta\phi)\sin(2p_k+2\phi_p+2\Delta\phi_p)\big)\Big], \end{aligned} \tag{4.35}$$

very similar to Eq. 4.16. The factors F_c and F_s don't change, as they describe the properties of the second harmonic relative to the first harmonic.

For small values of $\Delta\phi$, $\Delta\phi_p$ goes to zero. This condition is always fulfilled for stars with separations much smaller than the grid period. For these stars the phase of the photo centre differs only from the phase of the combined signal in higher orders of $\Delta\phi$, and can be used to derive the astrometric parameters for the system. It is further interesting to see what the effect is of an uncertainty in the identification of the primary star in the system. A change in identification between the primary and secondary stars is equivalent to substituting r by $1/r$ in Eq. 4.33. It is easy to see that this changes the sign in the lower equation. The residual effects of the combined modulation at the photo centre can therefore indicate the position of the primary star. The same result is obtained when applying a 180 degrees rotation to the orientation θ of the system, which results in changing the sign of $\Delta\phi$ (Eq. 4.9). An example of residuals in the phase of

Table 4.4. Published and new determinations for the astrometry of the primary and secondary components of HIP 70&71

Star	Param.	Old	σ	New	σ	units
70	ϖ	5.25	13.87	5.95	3.08	mas
71	ϖ	9.13	1.84	7.33	1.36	mas
70	$\mu_{\alpha *}$	−46.74	16.42	−23.25	3.27	mas/yr
71	$\mu_{\alpha *}$	−24.50	2.05	−23.61	1.39	mas/yr
70	μ_δ	−0.88	11.24	−18.58	1.92	mas/yr
71	μ_δ	−19.44	1.40	−20.55	0.85	mas/yr

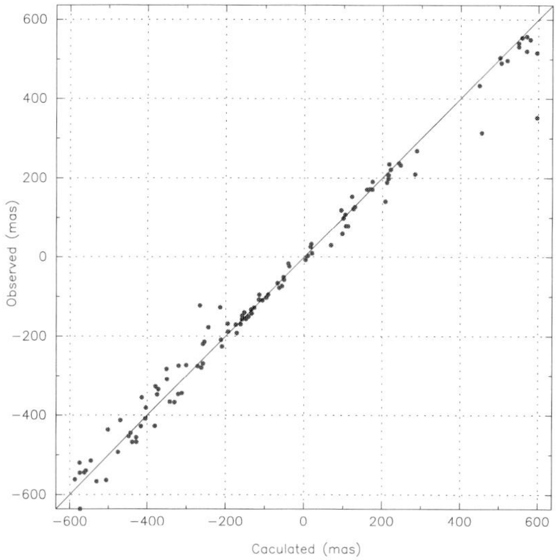

Figure 4.12. A comparison between predicted and observed values for the phase corrections $\Delta\phi_p$ of the first harmonic for the secondary star HIP 70. Note the much larger range of values than seen in Fig. 4.10 for a primary component

the first harmonic as observed and expected for binary HIP 274 is shown in Fig. 4.11. Clearly, the non-linear effects that could indicate the orientation of the system are small and in many cases they are not significant, leaving orientations often undetermined. All photo-centre systems were solved in the new reduction for both options, in the vast majority of cases the identification of the primary as given in the published data was confirmed, though in many cases the difference with the alternative solution was small, and the alternative also provided an acceptable solution.

4.3 Double stars with two catalogue entries

The *Hipparcos* catalogue contains 957 double-star systems with two entries, representing systems with component separations generally above 12 arcmin. The processing of these systems is very sensitive to assigning the correct scan direction to each observation, in this case an individual field transit. Measurements for the secondary are much more sensitive in this respect than those of the primary star. This can be seen from Eq. 4.11, where r is generally much less than one for the primary, while being larger than one for the secondary. Data averaged over a 10 hour interval and projected on a great circle, as used in the published data, cannot provide sufficiently accurate and detailed information on the instantaneous inclination of the scan for these larger separations. even relatively small variations of the orientation reflect very significantly in the modulation phase of the component. This resulted in rather large errors on the astrometry of the fainter components of these systems.

As an example, we look at the system HIP 70+HIP 71. This system has a separation of 15.35 arcsec, and has $r = 4.636$ for the secondary. The phase corrections for the secondary star cover a wide range, much wider than for the primary. In our example the phase corrections for the primary (HIP 71) had a range of ± 18 mas, and for the secondary (HIP 70) ± 600 mas (Fig 4.12). Inaccuracies in the assumed scan direction have considerable effect on these phase corrections. Over a single great circle interval, the local tilt of the scan can change by as much as 1.5 degrees, but even changes by only 10 to 15 arcmin already seriously affect the signal of the secondary star. If approximated by the mean scan direction for the reference great circle, residuals of the order of several tens of mas remain. The level of errors on the astrometric parameters as seen in the published data for these secondary components strongly gives the impression that this was the case in the published data. The new reduction presents results fully resolved in field transits with detailed information on the instantaneous inclination of the scan for each measurement. Implementing this information results in a dramatic reduction of the errors on the astrometric data as can be seen from the data in Table 4.4. It is clear from these data that the much improved accuracies of the astrometric parameters, that result from the proper resolution of the measurements, lead to a very significant improvement in the understanding of the system: what appeared to be a possible accidental superposition of two stars now turns out to be most likely an actual physical double star, considering the similarity in proper motion and parallax.

4.4 Variable-brightness of one component

Variability of one of the components of the double system means that the intensity ratio r has to be adjusted according to the light curve as determined by the total light. For this, there are naturally two possible situations: either the primary is variable, or the secondary. When both are variable there is little if any hope for reconstruction without further external information.

The intensity ratio r between the two components varies with the epoch of observation: $r = r(t)$. To evaluate r for different epochs, we need to define a reference ratio for a reference combined magnitude. In the published data this reference ratio is loosely defined as referring to the mean brightness. Given these assumed values, the magnitudes of the primary and secondary star are defined as a function of the variable total magnitude, provided we can identify which component is variable.

We introduce the pseudo-intensities I_a and I_b for the two components at the reference epoch:

$$\begin{aligned} m_{a+b} &= -2.5\log(I_a + I_b) \\ m_b - m_a &= -2.5\log(I_b/I_a) = -2.5\log(r), \end{aligned} \tag{4.36}$$

where m_{a+b} is the magnitude of the combined system, m_a and m_b the magnitudes of the primary and secondary star in the system. From this we derive:

$$\begin{aligned} m_a &= m_{a+b} + 2.5\log(1 + r) \\ m_b &= m_{a+b} + 2.5\log(1 + 1/r), \end{aligned} \tag{4.37}$$

where variability enters through m_{a+b} and r. As neither m_a nor m_b can be brighter than the combined system at any time, there is a constraint on the possible values r can take:

$$\begin{aligned} 2.5\log(1 + r) &> \Delta m_{a+b}, \\ 2.5\log(1 + 1/r) &> \Delta m_{a+b}, \end{aligned} \tag{4.38}$$

where Δm_{a+b} is the difference between the magnitude at minimum light and the reference magnitude m_{a+b}. The first relation applies when the secondary, the second when the primary is variable. As by definition the ratio r (at the reference point) is less than one, the second relation only becomes relevant when $\Delta m_{a+b} > 0.75$, a condition which is reached for only five variable double stars in the *Hipparcos* catalogue: three Miras and two Algol-type eclipsing binaries.

We define $(r_0, m_{a+b,0})$ for the arbitrary reference point, and derive from this the values of r at any other epoch of observation, using the observed value of m_{a+b} and the knowledge (or assumption) that either m_a or m_b is constant:

$$\begin{aligned} m_{a+b} + 2.5\log(1 + 1/r) &= m_{a+b,0} + 2.5\log(1 + 1/r_0) \\ m_{a+b} + 2.5\log(1 + r) &= m_{a+b,0} + 2.5\log(1 + r_0), \end{aligned} \tag{4.39}$$

where the first equation applies when the primary, and the second when the secondary is variable. These equations can also be written as:

$$\begin{aligned} \frac{1 + r}{r} &= \frac{1 + r_0}{r_0} 10^{-0.4(m_{a+b} - m_{a+b,0})} \\ 1 + r &= (1 + r_0)10^{-0.4(m_{a+b} - m_{a+b,0})}, \end{aligned} \tag{4.40}$$

where r is a function of the epoch of observation.

To obtain estimate-adjustments for r_0, we need to know how changes in r_0 affect the estimates of r. The magnitudes of the two components vary with changes to r_0 as:

$$\begin{aligned} dm_{a,0} &= 1.086\frac{1}{1+r_0}dr_0 \\ dm_{b,0} &= -1.086\frac{r_0}{1+r_0}dr_0. \end{aligned} \tag{4.41}$$

Together with Eq. 4.37 (or directly from Eq. 4.40) we find:

$$\begin{aligned} dr &= \frac{r_0}{r}\frac{1+r}{1+r_0}dr_0 \\ dr &= \frac{1+r}{1+r_0}dr_0, \end{aligned} \tag{4.42}$$

where, as before, the first relation is used when the primary star, and the second when instead the secondary is variable. With these relations in place, we can process a double star with one variable component using what was described above as the *L-type* solution.

For the *S-type* solution the situation is more complex, as the difference $\mathrm{H_{AC}} - \mathrm{H_{DC}}$, used to determine the orientation angle θ, is now dependent on a variable intensity ratio. Given the determination as published, we can instead use to estimate the orientation angle θ:

$$(\mathrm{H_{AC}} - \mathrm{H_{DC}})\frac{(1+r)^2}{r} \approx 0.543\left(\frac{2\pi\rho}{s}\sin(\theta-\psi)\right)^2, \tag{4.43}$$

where r now depends on the epoch of observation and our initial estimate r_0. The modulation amplitudes as defined by Eq. 4.22 and Eq. 4.24 similarly become affected by the light variations, and are best approached as Maximum Likelihood (ML) corrections to the assumed values r_0 and ρ, starting again with the value obtained in the earlier reductions. To reduce the noise level on the estimates of r, a light curve can be fitted for in particular periodic variables, which then provides a best estimate of m_{a+b} at any given phase.

The *Hipparcos* catalogue contains 284 binary systems of which one component is identified as a variable star. More than half of these, 162, concern eclipsing binaries of some sort (for example 68 Algol types, 60 β Lyrae types), which probably in most of these cases implies triple, or even more complex systems (the astrometrically unresolved stars in the eclipsing binary, and the resolved objects in the visual binary). An extreme example is the system τ CMa, where the eclipsing binary is part of a close, just resolved, binary system, which in turn is part of a wide binary system, with all stars being of type late O or early B (van Leeuwen and van Genderen 1997). There are also

11 Cepheids and 3 RR Lyrae stars, as well as 3 Miras among the resolved double stars in the *Hipparcos* catalogue. The variability amplitudes span a wide range, from a few hundredths of a magnitude for the 23 Alpha-2 CVn (ACV) and 16 Slowly-pulsating B stars (SPB) stars, to over 2 magnitudes for two of the Miras. The methods outlined above will clearly have little meaning when variability amplitudes get below a certain value. This value is determined by the variations introduced on the magnitude difference, and how this compares with the expected accuracy for the determination of this parameter in the absence of variability.

The effect on r of small variations in intensity of one of the components of the double star can be derived from Eq. 4.40:

$$\mathrm{d}r/r = \pm 0.92\mathrm{d}m_{a+b}$$

where $+$ applies when the primary, and $-$ when the secondary is variable. Naturally, when $\mathrm{d}m_{a+b}$ is negative due to the primary getting brighter, $\mathrm{d}r/r$ will get smaller.

The "light curves" for most of the small-amplitude variables in the *Hipparcos* data can be represented by a simple sinusoidal function. For these variables the dispersion in magnitudes is close to 0.35 times the total amplitude of the variations (Chapter 7).

4.5 Multiple systems

The *Hipparcos* catalogue contains 182 triple and 8 quadruple systems. Of the triple systems, 129 have one, 50 two, and three have three entries in the catalogue, reflecting the separations between the components. Of the quadruple systems, 6 have single entries, one has two entries of each two components, and one has three entries. The analysis of some of these systems is further complicated by variability: 23 components in 22 triple systems are classified as variable in the *Hipparcos* catalogue. In not all cases does this concern actual variability. For example, the one system with three entries (HIP 106884, 106886, 106890) has two of its components classified as variable, but this variability appears to be more likely due to the separation of 12 arcsec between 106886 (Hp= 5.3) and 106890 (Hp= 8.1). Measurements of 106890 are therefore very sensitive to the positioning of the IFOV and the position of 106886 on the slope of the IFOV. A similar, but smaller effect is seen for measurements of 106886 itself. It is therefore not surprising that 106886 shows a variability of a few hundredth, and 106890 of a few tenths of a magnitude in the *Hipparcos* observations. A further examination of components and separations for the 22 variable systems shows that there are two systems for which the variability is beyond any doubt (HIP 5348 = ζ Phe and 103542 = KZ Pav), both Algol-type eclipsing binaries. HIP 24019 (V1156 Tau) possibly also contains an eclipsing binary with a period twice that given in (ESA 1997). The distribution of magnitudes for HIP 56769 also strongly indicates the presence of an eclipsing binary in the system. A further examination shows a possible period of about 2.734 d. In 11 systems the variability appears to be most likely the result of IFOV positioning. As was shown above, the main problem with variability in multiple systems is the need to identify the actual variable component, which affects the phase corrections.

The astrometric-parameter analysis of these systems all follow the same procedure. Using the astrometric and differential parameters of the multiple system as published, the geometric configuration of the system is reconstructed, relative to the first component given in the catalogue. The reconstruction can contain time-dependent elements if separations and orientation angles included first derivatives in time. This geometric system is then shifted to the target component, which can be the photo-centre of one or more components. The actual measurement target for the satellite could be a component, the geometric centre or the photo-centre of the (sub)system (this information is available through the *Hipparcos* Catalogue (ESA 1997)), and it is relative to this position that the separations of other catalogue entries for the system are to be calculated. Knowing the separation, the IFOV profile is used to get an attenuation factor or correction to the magnitude, required for the further calculations. Figure 4.13 shows an example of three different targets in a well-known quadruple system (the Trapezium in Orion), and how this affected, through the IFOV attenuation, the appearance of the system for the measurements.

The instantaneous scan direction ψ is used to rotate the system coordinates to align with the scan. The projections of the separations between the components and the astrometric target are the ΔG values encountered before

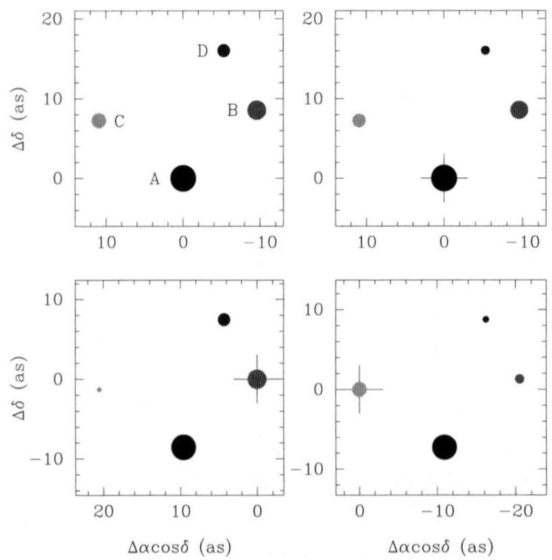

Figure 4.13. The quadruple system θ^1 Ori or Trapezium. The stellar magnitudes are indicated by the sizes of the circles. To left: as the system appears on the sky; top right: as the system appears for observations centred on component A (HIP 26221); bottom left: similarly, centred on component B for measurements of components B and D (HIP 26220); bottom right: as centred (HIP 26224) on component C. Note that there are two different systems for identifying the components (Evans 2000)

(Eq. 4.9). As before, the ΔG values are transformed to $\Delta\phi$ values, and the signals of all components are evaluated relative to the reference position for the astrometry:

$$\begin{aligned} A_{k,i} = B &+ I_i \cdot \big(1 \\ &+ M_1 \cos(p_k + \phi_c) \cos\Delta\phi_i - \sin(p_k + \phi_c) \sin\Delta\phi_i) \\ &+ M_2 (\cos(2p_k + 2\phi_c) \cos 2\Delta\phi_i - \sin(2p_k + 2\phi_c) \sin 2\Delta\phi_i)\big), \end{aligned} \tag{4.44}$$

where ϕ is the correction for the reference position of the system. The intensities I_i have been, where required, corrected for the IFOV response function.

The combined signal from all components is obtained by taking the sum over the components:

$$\begin{aligned} A_{k,c} &= B + \sum_i I_i \\ &+ M_1 \left[\cos(p_k + \phi_c) \sum_i I_i \cos\Delta\phi_i - \sin(p_k + \phi_c) \sum_i I_i \sin\Delta\phi_i\right] \\ &+ M_2 \left[\cos 2(p_k + \phi_c) \sum_i I_i \cos 2\Delta\phi_i - \sin 2(p_k + \phi_c) \sum_i I_i \sin 2\Delta\phi_i\right], \end{aligned} \tag{4.45}$$

from which the effective phase shift $\Delta\phi_c$ is derived:

$$\Delta\phi_c = \arctan\left(\frac{\sum_i I_i \sin\Delta\phi_i}{\sum_i I_i \cos\Delta\phi_i}\right). \tag{4.46}$$

The modulation amplitude of the first harmonic is given by:

$$\begin{aligned} M_{1,c} &= M_1 \cdot \frac{\sqrt{(\sum_i I_i \sin\Delta\phi_i)^2 + (\sum_i I_i \cos\Delta\phi_i)^2}}{\sum_i I_i} \\ &= M_1 \cdot \frac{\sqrt{\sum_i \sum_j I_i I_j \cos(\Delta\phi_i - \Delta\phi_j)}}{\sum_i I_i}. \end{aligned} \tag{4.47}$$

Very similar relations can be derived for the second harmonic phase and amplitude. As I_i is always positive, $M_{1,c}$ will always be smaller than M_1, which reflects in the precision of the abscissa measurements and the astrometric parameters as derived for multiple systems.

4.6 Orbital motions

The orbital solutions cover a range of different situations, from testing for the presence of possible orbital disturbances due to planetary companions (Mazeh et al. 1999), through the detection of low-mass stellar companions (Bernstein 1997, 1999; Reffert and Quirrenbach 2006) and new orbital systems (Mason et al. 1999; Jorissen et al. 2004; Balega et al. 2005, 2006), systematic examination of spectroscopic binaries (Pourbaix and Jorissen 2000; Pourbaix and Boffin 2003; Pourbaix 2004), to providing a group of measurements for well established, mainly long-period orbitals (Lindegren et al. 1997). In most cases orbital information can only be incorporated in the *Hipparcos* analysis when external information is present. This external information ranges from individual measurements of separation and/or magnitude differences, through partially determined orbits from radial velocity measurements to fully determined orbits from ground-based astrometric data. The available volume of such external data has steadily increased since the analysis for the published data, making it possible to build a new table of reference data for the analysis of the *Hipparcos* orbital binaries.

4.6.1 The Reference Data

A new list of orbital binary solutions as available for the current study has been compiled from the published data (ESA 1997), two studies on short-period binaries (Martin et al. 1998; Soederhjelm 1999), papers by Balega et al. (2005) and Balega et al. (2006) presenting orbital parameters for 12 systems, results from the Center for High Angular Resolution Astronomy (CHARA) speckle program (Mason et al. 1999; Hartkopf et al. 2000), and supplementary data on magnitude differences (ten Brummelaar et al. 2000; Horch et al. 2004). The new list contains 541 entries for 491 systems, of which 235 were presented as such in the published data. Of the remaining 256 systems, in the published data 42 received a standard 5-parameter solution, 17 an accelerated solution, 12 were noted as stochastic, and 185 were only solved as double stars (i.e. with fixed relative positions and magnitude differences). For 5 entries two solution options are given. The accumulated reference data are presented in Appendix F with some further details.

4.6.2 The Reflection of Orbital Motion in Hipparcos Astrometry

The complications of the description of an orbital motion by means of the one-dimensional *Hipparcos* data are severe. What can be observed by *Hipparcos* is the projection of the separation of the two components, or, in case the images are too close together as projected on the scan circle, only a slightly distorted modulation phase. When the separation is less than about 15 arcsec and with magnitude differences between the components of less than 3.5 mag., the signals have to be analysed as a double star with a variable separation and orientation that satisfy the orbital solution for the system (see Appendix E). A detailed description for solving these complications is presented by Martin et al. (1997). Where the secondary component is too faint, being for example a brown dwarf or planetary companion, the problem is slightly more straightforward, with the abscissa residuals describing effectively the actual orbit of the observed star (rather than that of the photo centre).

Astrometric solutions involving orbit parameters come in three types, as specified in Appendix E. The data for the 235 systems presented in the published catalogue refer to the orbital description of the photo centre of the system (Fig. 4.14). Data published since enabled resolving 14 of these systems.

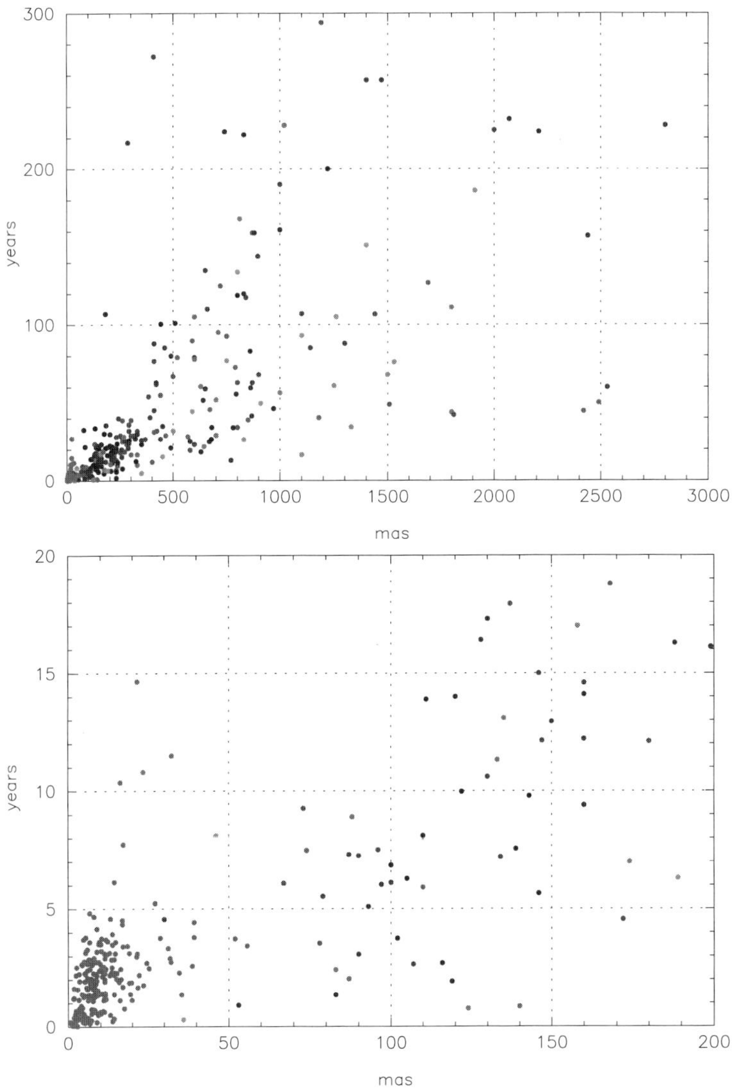

Figure 4.14. The distribution over period and semi-major axis for orbital binaries in the *Hipparcos* catalogue. The top figure shows the data for 99 per cent of the binaries, the lower figure shows the distribution for small separations and short periods. The concentration on the left in this diagram is entirely due to orbits determined for photo centres

For newly discovered orbits usually only a circular orbit can be assumed as a solution. More detailed orbital information requires additional information: ground-based observations, both astrometric and radial velocity data, and speckle interferometry for small separations (primarily from the CHARA and US Naval Observatory program, Hartkopf et al. 1989, 1996). Without these additional data, orbital parameters are often poorly determined.

4.6.3 Solving Orbital Binaries

The reduction procedure for the recovery of astrometric parameters for orbital binaries is as follows:

- For the epoch of each observation, calculate using the procedures described in Appendix E, the displacements of the primary star in the binary system as projected on the sky, $(\Delta\xi, \Delta\nu)$;
- Using the local instantaneous scan direction ψ (see Fig. 4.2), determine the along-scan projection of the separation;
- Determine the scan phase of the primary component or for the photo-centre (depending on the semi-major axis a), either through a resolution of the combined modulated signal, using the magnitude difference and projected separation as derived above, or directly from the observed abscissae for systems where the secondary is much fainter than the primary;
- For long-period binaries, correct the observed abscissae for the predicted offset based on the orbital parameters, as projected along the scan direction; for short-period binaries enter the pre-calculated offsets in the astrometric parameters solution with a single scaling factor;
- Solve the remaining residuals for the correction to be applied to the assumed astrometric parameters and the scaling factor for short-period binaries.

In these calculations the final phase residuals are obtained modulus the grid period of 1.2074 arcsec, which can cause a problem for binaries with separations larger than the grid period.

4.6.4 Restricting Planetary and Brown Dwarf Companion Solutions

Orbit determinations based on radial velocities alone can only establish the value of $a \sin i$ for the orbit, leaving the masses involved undetermined (see Appendix E). In some cases the *Hipparcos* data can be used to put limits on the possible values of i. For different test values of i the reflection of the orbit in the astrometric data is calculated and compared with the observations. This needs to be done for different values of Ω, as that angle also remains undetermined when using radial velocity data only. Each complete set of orbital parameters provides predicted offsets for the observed abscissae. After applying these orbital corrections, the remaining residuals can be solved for the five-parameter astrometric model. The minimum of the χ^2 values obtained for different values of Ω at a given value of i is derived, giving the optimal Ω value. These χ^2 values are transformed into a pseudo-Gaussian variable σ, using the degrees of freedom from the solutions. The minimum σ values for all attempted values of i indicates the most likely value for the inclination. This method was first applied by Mazeh et al. (1999) on the data for υ And (HIP 7513) and subsequently to HIP 8159 (Zucker and Mazeh 2000) and a list of 47 possible planetary and brown-dwarf companions (Zucker and Mazeh 2001). The radial velocity measurements for υ And established $a \sin i = 0.56$ mas, and combined with the *Hipparcos* data the inclination was found to be restricted in the range 131.4 to 163.9 degrees, giving $a = 1.4 \pm 0.6$ mas (Fig. 4.15).

The newly reduced intermediate astrometric data have a noise level about four times lower than the published data, and are in addition resolved at field transit rather than orbit level, allowing for a much improved detection of possible outliers. Improvements have also been made on the orbital parameters for many potential planetary companions (Butler et al. 2006). Using these new

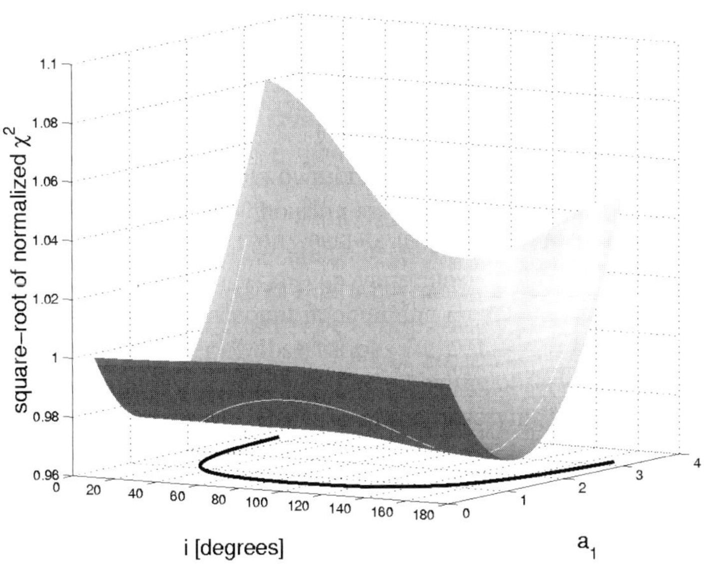

Figure 4.15. χ^2 values for the astrometric solution of HIP 7513, as a function of values for the inclination and the semi-major axis. The line shows the constraint imposed by the radial-velocity measurements (from Mazeh et al. 1999)

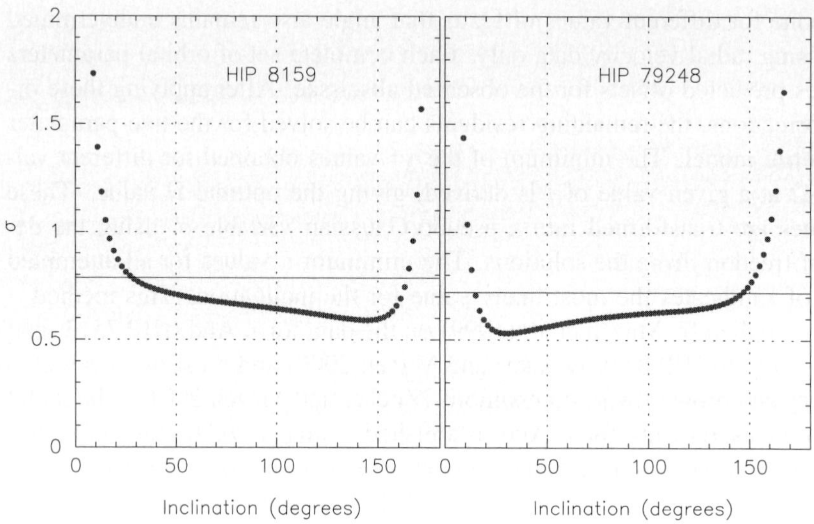

Figure 4.16. σ values for the astrometric solutions of HIP 8159 (left) and HIP 79248 (right), as a function of values for the inclination at fixed values of $a \sin i$. The horizontal lines show the equivalent σ values for solutions without any corrections applied

orbital parameters and the new *Hipparcos* reduction, applications of the detection method were applied to HIP 8159 and HIP 79248, the results of which are shown in Fig. 4.16. The inclination values found for these systems are 145 and 31 degrees respectively, with Ω values of 5 and 35 degrees. The inclination values put the mass at an upper limit of approximately 10–11 M_J for both companions.

Chapter 5

GROUPS OF SINGLE STARS

Groups of single stars offer calibration possibilities that enable measurement of astrophysical parameters well beyond what is exposed by the measurements of individual stars alone. This is made possible primarily by exploiting parameters common to groups of stars, such as parallax and proper motion for members of a star cluster, or period-luminosity relations for groups of pulsating stars. In this chapter techniques used to extract such parameters, and thereby exploring the *Hipparcos* data well beyond the accuracy levels for individual stars, are described, extensively illustrated by means of actual applications.

5.1 Solving for common parameters

We would like to obtain solutions for parameters common to specific groups of stars, such as members of an open cluster (parallax and proper motion) or pulsating stars (luminosities). Solving for common parameters has been considerably simplified with the new reduction results: the abscissa correlations have been reduced to a level where they no longer play a significant role (Fig. 2.11), and the data are now derived from a single analysis. In the following sections it will be shown how, using the new reduction results directly or as intermediate data, we can combine the astrometric and luminosity data from individual stars to solve for common astrometric or astrophysical parameters. This is illustrated by a few actual implementation examples. These examples are solely intended as an illustration of how to use the *Hipparcos* astrometric data in a range of typical astronomical investigations. They are in all cases based on the results from iterations 12, 13 or 14, and not the final catalogue. First, however, the principles of the two ways in which data can be combined are described: either through using the astrometric results directly, or through using the underlying abscissa measurements.

5.1.1 Combining Astrometric Data for Individual Stars

When using the astrometric data directly (as parallaxes and proper motions), they become observations with the covariance matrix of their solution as an indication of the noise and correlation levels. In the simplest case, when we only consider parallax information, the observations describe the relation between the observed parallaxes (weighted according to their formal errors and noise intrinsic to the model) and the modelled quantities. The intrinsic noise represents, for example, the dispersion of parallaxes along the line of sight for members of an open cluster resulting from the space-density distribution of the cluster members.

In a more complex situation, when considering parallaxes as well as proper motions, the correlations between the measurements of these parameters need to be taken into account using the upper-triangular matrix U as provided with the data (see Appendix C). Take a matrix A relating the observed quantities $\boldsymbol{z} = (\varpi, \mu_{\alpha*}, \mu_\delta)$ to a set of model parameters $\boldsymbol{x}$:

$$\mathsf{A}\boldsymbol{x} = \boldsymbol{z}. \tag{5.1}$$

Multiplying left and right by U provides a set of uncorrelated observations with unit weight variances for the estimated formal errors, which can be combined with other similar observations (see Appendix C for further details). Unlike the published data (van Leeuwen and Evans 1998), error correlations between neighbouring stars for the underlying abscissa data are, at a maximum level of just over 0.01, and therefore no longer significant (Fig. 2.11, page 58) (see also van Leeuwen and Fantino 2005).

5.1.2 Combining Abscissa Data

The parallaxes and proper motions for members of an open cluster vary little, in particular when the cluster is observed at a distance of more than about 200 pc. In that case one may assume those data to be identical, within the observational errors, for all cluster members, and obtain a “cluster solution”. The “cluster solution” solves for a single cluster parallax and proper motion using the abscissa residuals of the individual cluster members. The main advantage of this approach is the improved chance for the detection of faulty measurements, which for an individual star may more easily be able to hide behind a slightly distorted solution. The “cluster solution” was first proposed by van Leeuwen and Evans (1998), and used to produce the cluster parallax estimates by van Leeuwen (1999a) and Robichon et al. (1999). It had been designed primarily to allow correction for the abscissa error correlations. However, as has become clear since (van Leeuwen 2005a,b), these correlations are more complicated than was assumed, in particular in the way they can affect the measurements of the brighter stars.

The abscissa residuals $\mathrm{d}a_j$ are given (in the old as well as the new reductions) relative to the adopted astrometric solution. To do a "cluster solution" we first need to refer all abscissa residuals to the same reference cluster proper motion and parallax. The choice of these reference values is non-critical, as the adjustment process is linear. With every abscissa are given the scan direction, ordinate and epoch of observation, from which the so-called "derivatives" can be derived. These derivatives describe the relation between changes to the abscissae and changes to the astrometric parameters (see Section 2.5 and in particular Eq. 2.52, page 68).

If we denote by $\boldsymbol{z}_i$ the astrometric parameters of cluster star i, and by $\boldsymbol{z}_c$ the assumed astrometric parameters of the cluster (covering, as above, the parallax and proper motion only), then the corrected abscissa residual j of star i is given by:

$$\mathrm{d}a'_{i,j} = \mathrm{d}a_{i,j} + (\boldsymbol{z}_i - \boldsymbol{z}_c) \cdot \frac{\partial a_{i,j}}{\partial \boldsymbol{z}}. \tag{5.2}$$

These corrected abscissa residuals can now be used to fit a correction to the assumed cluster proper motion and parallax. In this solution, the positional corrections for each star are fitted too, giving a total of $2n + 3$ unknowns for n cluster members. The decrease in degrees of freedom for the combined abscissa-residuals solution should potentially provide a more stable solution and better outlier detection. It is, however, more difficult to incorporate in this solution the effects of intrinsic dispersions in the parallaxes and proper motions. The cluster solution is therefore now only used for clusters beyond 200 pc, where such effects can be ignored.

5.2 Application to star clusters

For the application of the methods presented above we have to distinguish between three situations:

- the relatively nearby Hyades cluster, covering a 30 degrees diameter field;
- the intermediate distance (50-200 pc) clusters (Pleiades, Praesepe etc), covering up to 10 degrees on the sky;
- the more distant (200-550 pc) clusters.

These different groups require each a different approach: for the Hyades we need to fully resolve the space velocity and three-dimensional structure of the cluster; for the Pleiades group it is sufficient to include first order corrections for parallax dispersion and proper motion projections, allowing for a determination based on averaging results for individual members; for the more distant clusters the field covered is small enough to consider proper motions and parallaxes all the same for the cluster members, and the determination is based on the cluster solution applied to the abscissae of the cluster members.

There are two further complications: the internal velocity dispersions and the presence of a halo of (semi-) detached (former) cluster members. The presence of such halo is very clear from the Hyades studies of Perryman et al. (1998) and Madsen et al. (2001). There are also indications in studies of the Pleiades cluster of the presence of escaping members, and the phenomenon is clearly reproduced in numerical simulations of open clusters (Terlevich 1987; Kroupa et al. 2001). The typical internal velocity dispersion for a young open cluster is of the order of 0.8 km s^{-1} (van Leeuwen 1994). For an older cluster it tends to drop to around 0.6 km s^{-1}. The velocity dispersion of the escaped halo stars is, however, generally significantly higher. These internal dispersions, which are most relevant for the clusters in the Pleiades group, can be taken into account when determining mean parallax and proper motion data from the individual member results.

5.2.1 The Hyades: Kinematic Parallaxes

The Hyades cluster offers a unique possibility for a detailed study based on the *Hipparcos* data: its relatively short distance and therefore large spread on the sky, combined with a large radial velocity and proper motion, make it possible to resolve the cluster in three dimensions. The same characteristics made this cluster in the past a suitable target for the so-called "convergent point" distance determination. This effectively constituted a comparison between the radial velocity of the cluster and its apparent contraction as measured from the absolute proper motions of the cluster members.

The stars in a cluster share the same space velocity, apart from a small amount of motion within the cluster. This space velocity can be derived from the observed proper motions, parallaxes and radial velocities. To do so, we have to invert Eq. 2.32 to obtain:

$$\begin{bmatrix} \kappa\mu_{\alpha*}/\varpi \\ \kappa\mu_\delta/\varpi \\ V_R \end{bmatrix} = \begin{bmatrix} -\sin\alpha & \cos\alpha & 0 \\ -\cos\alpha\sin\delta & -\sin\alpha\sin\delta & \cos\delta \\ \cos\alpha\cos\delta & \sin\alpha\cos\delta & \sin\delta \end{bmatrix} \cdot \begin{bmatrix} \dot{X} \\ \dot{Y} \\ \dot{Z} \end{bmatrix}, \qquad (5.3)$$

where, as before, $\kappa = 4.74047$. Here, instead of ecliptic coordinates, equatorial coordinates are used.

Applying Eq. 5.3 to the selection of Hyades members as presented by Perryman et al. (1998), but using the new reduction data, gives the solution for the space velocity of the cluster shown in the first column of Table 5.1, using 85 cluster members with both radial velocity and proper motions and the proper motions for 27 spectroscopic binaries. The study by Lindegren et al. (2000) showed systematic discrepancies, as a function of spectral type, between the measured radial velocities and those derived from the astrometric data under the assumption of a shared space velocity. This discrepancy is also observed in the decrease of the unit-weight standard deviation of the solution

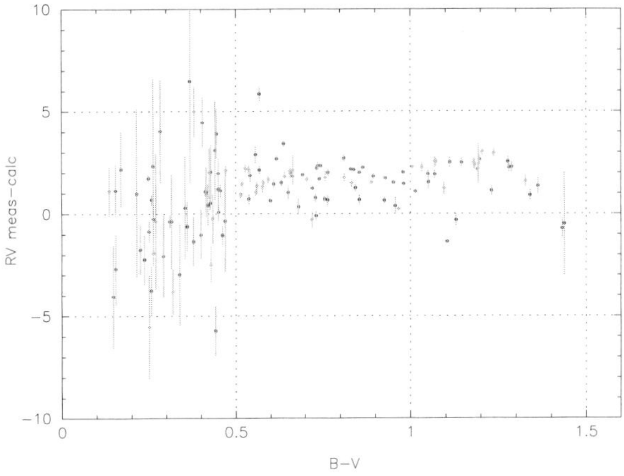

Figure 5.1. The differences between the observed and predicted radial velocities for members of the Hyades cluster, as a function of the photometric index B − V. The differences may be attributed to a combination of effects, ranging from convective motions to gravitational red shift. The zero point of the offsets shows a variation of about 0.5 km s^{-1} depending on the assumed space velocity of the cluster

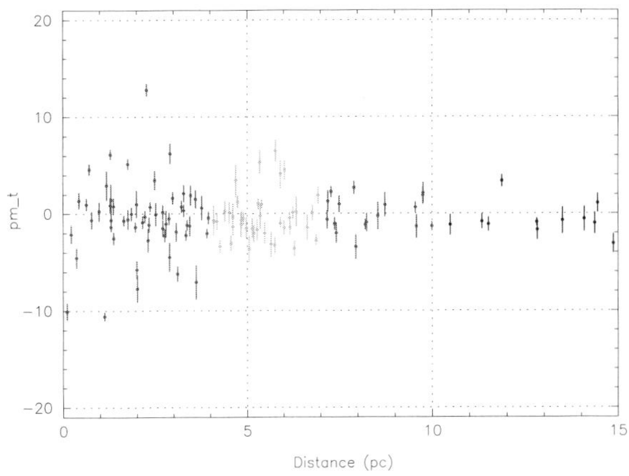

Figure 5.2. The distribution of the proper motions perpendicular to the direction of the space motion of the cluster, shown as a function of distance from the cluster centre. The dispersion towards the centre of the cluster is between 4.5 and 5 mas s^{-1}, dropping rapidly away from the centre

for the space velocity: when including the radial velocities it is around 2.2, without radial velocities it is down to 1.3. The radial velocity differences observed for a solution based only on proper motions are shown as a function of colour in Fig. 5.1. Part of this discrepancy can possibly be attributed to stellar gravitational redshift (see for example Hentschel 2005).

In the application of Eq. 5.3 formal errors need to be assigned to the observed proper motions in relation to the model. These errors are affected by the measurement accuracy (proper motion and parallax) and by the internal velocity dispersion in the cluster. Lindegren et al. (2000) estimate the internal velocity dispersion at 0.31 ± 0.02 km s^{-1}, while Perryman et al. (1998) consider it possibly higher, but not possible to determine. Here we determine the velocity dispersion from the variances in the proper motion component perpendicular to the direction of the projected space velocity vector at the position on the sky for each star. This direction is independent of the distance of the star, and well determined by the accuracy of the space velocity vector (see below).

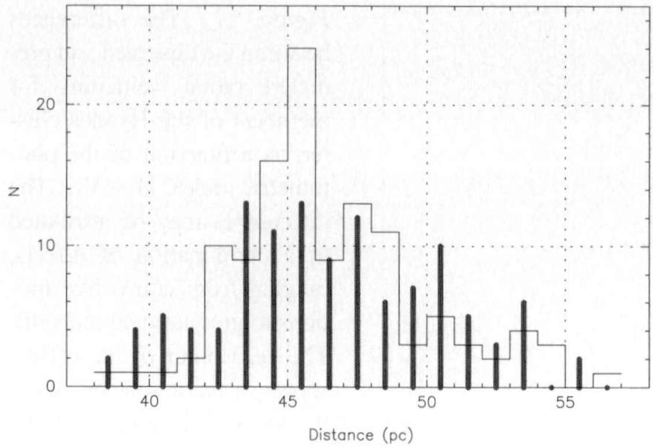

Figure 5.3. The distribution of distances of the Hyades stars as projected on the line of sight for the centre of the cluster. The bars show the distribution based on the observed parallaxes, the histogram contour shows the distribution for the same stars, using the proper-motion-based distance estimates

Table 5.1. Values for the Hyades space velocity components. Column 1: New data, including radial velocities, but using for spectroscopic binaries only the proper motions, restricted to a 20 degree (15.5 pc) diameter field; Column 2: As Column 1, but without radial velocities; Column 3: Perryman et al. (1998), inner 20 pc; Column 4: Perryman et al. (1998), inner 10 pc; Column 5: Lindegren et al. (2000), inner 15 pc

Component	1	2	3	4	5
$\dot{X}$	-5.54 ± 0.11	-6.05 ± 0.49	-6.28	-6.32	-5.90 ± 0.13
$\dot{Y}$	46.42 ± 0.08	44.57 ± 1.15	45.19	45.24	45.65 ± 0.34
$\dot{Z}$	5.87 ± 0.19	5.21 ± 0.40	5.31	5.30	5.56 ± 0.10
stars	112	107	180	134	168

A similar approach was used by Makarov et al. (2000), who derived an internal velocity dispersion of 0.3 km s^{-1}, averaged over the whole cluster. Here we observe the dispersion as a function of the projected distance to the cluster centre (Fig. 5.2), revealing the internal proper motion dispersion to be significantly higher in the core of the cluster than towards the halo (see also Madsen et al. 2001). The dispersions observed projected on the core are just below 1 km s^{-1}, a value that may be affected by the presence of halo stars that are no longer bound to the cluster. Away from the cluster centre the dispersion drops to around 0.35 km s^{-1}.

The observed dispersions have been added in quadrature to the formal errors of the proper motions in the solution of Eq. 5.3. Whether the central proper motion dispersion is representative for the actual velocity dispersion in the cluster centre is irrelevant, as the proper motion components along the space velocity direction are likely to be similarly affected. As a result of the higher dispersions assigned to the central proper motions, the formal error on the derived space velocity components has increased relative to previous studies. The results are presented in Table 5.1. As a comparison, solutions based on the published data derived by Perryman et al. (1998) and Lindegren et al. (2000) are also shown.

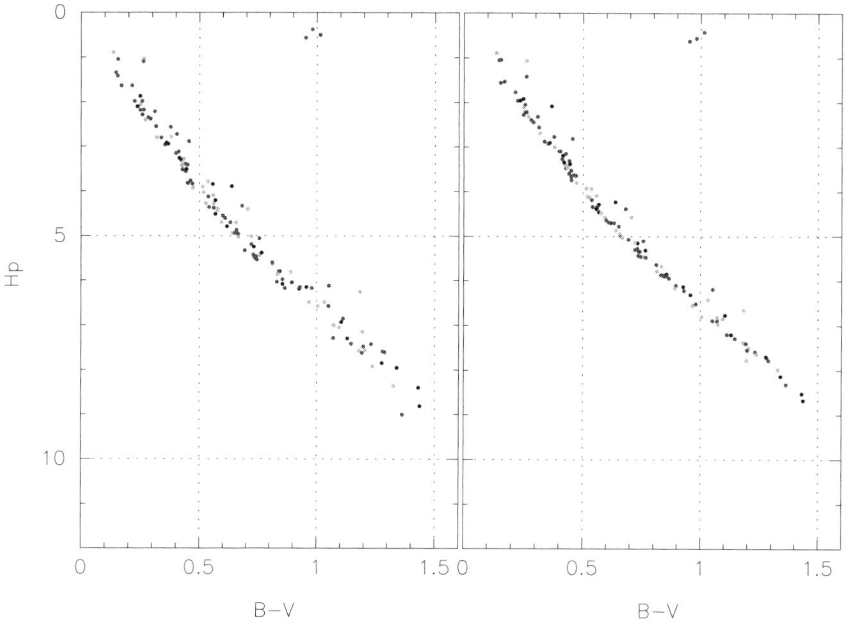

Figure 5.4. The HR diagram for the Hyades cluster. On the left, the directly measured parallaxes are used to calculate the distance moduli of the individual stars. On the right, the distance moduli have been derived using distance estimates based on the relation between the space motion of the cluster, and the distance, position and observed proper motion of each star

With the space velocity determined, distances to individual stars can be estimated. As was stated before, the space velocity reflects in a predicted proper motion vector. The direction of this vector depends only on the position of the star on the sky, the length depends on the position of the star and its distance. The procedure is therefore to transform for each star the observed proper motion and associated errors to this direction. It is important to incorporate in this calculation the covariance between the proper motion components to obtain the correct error estimates: the error distributions for *Hipparcos* astrometry tend to be significantly different in ecliptic longitude and latitude (see Section 3.2.2, page 83). The perpendicular component has been used above to estimate the internal proper motion dispersions. The comparison between the component along the space velocity direction and the projected space velocity provides the distance estimate for an individual star. The distribution of these distances is shown in Fig. 5.3, together with a similar distribution derived from distances based on the measured *Hipparcos* parallaxes. The new distribution clearly shows a reduced width, and corresponds to a mean distance, for the sample of stars used here, of 46.5 ± 0.3 pc. The dispersion of distances is 3.3 pc. The improvements are also clearly seen in the HR diagram of the cluster

(Fig. 5.4), which exposes also some of the spectroscopic binaries, positioned up to 0.75 mag above the cluster main sequence.

5.2.2 Other Nearby Clusters

Table 5.2. Reference data for the nearby clusters. Columns 1 to 3: parallax estimates in mas; Column 1: Kharchenko et al. (2005); Column 2: Pinsonneault et al. (1998); Column 3: The original *Hipparcos* solution by van Leeuwen (1999a); Column 4: R_V (km s^{-1}); Column 5: E(B – V); Column 6: log(age)

Cluster	1	2	3	4	5	6
Praesepe	5.34	5.86	5.32 ± 0.37	33.6	0.01	8.90
Coma Ber	10.49	12.36	11.11 ± 0.23	-1.2	0.00	8.80
Pleiades	7.69	7.59	8.45 ± 0.25	5.4	0.03	8.08
IC 2391	5.68		6.84 ± 0.23	16.0	0.01	7.88
IC 2602	6.25		6.88 ± 0.20	19.0	0.03	7.83
NGC 2451	5.32		5.30 ± 0.20	21.7	0.00	7.76
α Per	5.26	5.68	5.46 ± 0.20	-2.6	0.09	7.55

We now consider the parallax determinations for the nearby clusters (within 200 pc from the Sun), for which the formal error on the new *Hipparcos* parallax is less than 5 per cent, giving a formal error on the distance modulus of less than 0.1 magnitude. Next to the Hyades, seven clusters fit this criterion. A summary of reference data and earlier parallax estimates for the selected clusters is presented in Table 5.2. The differences for some of these clusters between the *Hipparcos* results and those expected mostly as based on main-sequence fitting procedures have led to controversy concerning the *Hipparcos* astrometric data since its publication in 1997, leading to many claims and counter claims on its reliability. With problem areas in the *Hipparcos* astrometric data now identified and cleared by the new reduction, the mean astrometric parameters for the nearby clusters need to be re-determined.

Each of these clusters extends over several degrees over the sky, which implies that projection effects have to be considered for both proper motions and parallaxes. A first approximation for projection corrections can be derived from Eq. 5.3. We rewrite this equation as:

$$\boldsymbol{V} = \mathsf{A}\boldsymbol{S}, \tag{5.4}$$

where:

$$\boldsymbol{V} \equiv \begin{bmatrix} \kappa\mu_{\alpha*}/\varpi \\ \kappa\mu_\delta/\varpi \\ V_R \end{bmatrix}, \quad \boldsymbol{S} \equiv \begin{bmatrix} \dot{X} \\ \dot{Y} \\ \dot{Z} \end{bmatrix}, \tag{5.5}$$

and A the matrix as defined in Eq. 5.3. We denote by $\boldsymbol{V}_0$ the value of $\boldsymbol{V}$ at a fiducial reference point at the projected centre of the cluster, $\boldsymbol{V}_0 = \mathsf{A}_0\boldsymbol{S}$, and we want to find the difference $\Delta\boldsymbol{V}_i$ for a star i at position $\Delta\alpha_i = \alpha_i - \alpha_0$ and $\Delta\delta_i = \delta_i - \delta_0$. This is given by:

$$\Delta\boldsymbol{V}_i \equiv \boldsymbol{V}_i - \boldsymbol{V}_0 = \left(\frac{\partial\mathsf{A}_0}{\partial\alpha}\Delta\alpha_i + \frac{\partial\mathsf{A}_0}{\partial\delta}\Delta\delta_i\right)\boldsymbol{S}. \tag{5.6}$$

Substituting $\boldsymbol{S} = \mathsf{A}_0^{-1}\boldsymbol{V}_0$, we get:

$$\Delta\boldsymbol{V}_i = \left(\frac{\partial\mathsf{A}_0}{\partial\alpha}\Delta\alpha_i + \frac{\partial\mathsf{A}_0}{\partial\delta}\Delta\delta_i\right)\mathsf{A}_0^{-1}\boldsymbol{V}_0. \tag{5.7}$$

The matrix products can be worked out to give:

$$\frac{\partial\mathsf{A}_0}{\partial\alpha}\mathsf{A}_0^{-1} = \begin{bmatrix} 0 & \sin\delta_0 & -\cos\delta_0 \\ -\sin\delta_0 & 0 & 0 \\ \cos\delta_0 & 0 & 0 \end{bmatrix}, \tag{5.8}$$

and

$$\frac{\partial\mathsf{A}_0}{\partial\delta}\mathsf{A}_0^{-1} = \begin{bmatrix} 0 & 0 & 0 \\ 0 & 0 & -1 \\ 0 & 1 & 0 \end{bmatrix}. \tag{5.9}$$

Thus, the proper motion and radial velocity corrections are in first approximation given by:

$$\begin{aligned} \Delta\mu_{\alpha*} &= \Delta\alpha\left(\mu_\delta\sin\delta_0 - \frac{V_R\varpi}{\kappa}\cos\delta_0\right) \\ \Delta\mu_\delta &= -\Delta\alpha\mu_{\alpha*}\sin\delta_0 - \Delta\delta\frac{V_R\varpi}{\kappa} \\ \Delta V_R &= \Delta\alpha\frac{\kappa\mu_{\alpha*}}{\varpi}\cos\delta_0 + \Delta\delta\frac{\kappa\mu_\delta}{\varpi}. \end{aligned} \tag{5.10}$$

For example, in the cluster Coma Ber (with a near zero radial velocity) the maximum corrections are no more than about ±0.3 mas yr^{-1} in proper motion and ±0.2 km s^{-1} in radial velocity, but for the Pleiades cluster corrections can get ten times higher. The small size of these corrections for Coma Ber makes this cluster unsuitable for the methods described above for the Hyades cluster, despite its relative proximity. The same conclusion was reached by Lindegren et al. (2000).

The astrometric-parameter solutions for these clusters were done through weighted averaging of the parallax and proper motion data for the individual members. The internal velocity dispersion as well as the spread in the parallaxes, due to the depth of the cluster along the line of sight, contribute to the dispersions in proper motions and parallaxes, and these dispersions have to be added to the noise matrices when combining the measurements of the individual stars. The noise matrix for the astrometric parameters is given by (following Section 5.1):

$$\mathsf{N}_i = \mathsf{U}_i^{-1} \cdot (\mathsf{U}_i^{-1})^T \tag{5.11}$$

To this is added the cluster cosmic noise matrix:

$$\mathsf{N}_c = \begin{bmatrix} 0 & 0 & 0 & 0 & 0 \\ 0 & 0 & 0 & 0 & 0 \\ 0 & 0 & \sigma_p^2 & 0 & 0 \\ 0 & 0 & 0 & \sigma_\mu^2 & 0 \\ 0 & 0 & 0 & 0 & \sigma_\mu^2 \end{bmatrix}, \tag{5.12}$$

where σ_p is the dispersion in parallaxes, approximated by $(\sigma_r \varpi)$, in which ϖ is the cluster parallax measured here in arcsec, and σ_r is measured in parsec. For most of the clusters considered here $\sigma_r \approx 3$ pc. Similarly, σ_μ is the internal proper motion dispersion, and is approximated as:

$$\sigma_\mu = \sigma_V \cdot \varpi/\kappa, \tag{5.13}$$

and $\sigma_V \approx 0.8$ km s^{-1}.

The noise matrix for the contribution of the astrometric parameters for a cluster member to the mean cluster astrometric parameters is given by $\mathsf{N}'_i = \mathsf{N}_i + \mathsf{N}_c$. From this noise matrix we derive the upper triangular matrix U'_i, defined in relation to N' in the same way as U_i relates to N_i in Eq. 5.11.

The astrometric parameter estimates $\boldsymbol{p}_i \equiv (\varpi_i, \mu_{\alpha*,i}, \mu_{\delta,i})$ for each cluster member i provide three equations for the estimates of the mean astrometric parameters of the cluster, $\tilde{\boldsymbol{p}}_c$:

$$\mathsf{U}_i \cdot \tilde{\boldsymbol{p}}_c = \mathsf{U}_i \boldsymbol{p}_i + \boldsymbol{\epsilon}_i, \tag{5.14}$$

where only the lower-right three-by-three part of the upper-triangular matrix U is used. The expectation value of the noise vector $\boldsymbol{\epsilon}$ is zero, and the standard deviation equals 1 for all elements, as the multiplication by U provides a set of decorrelated and normalized observations.

This method has been applied to seven nearby clusters. Cluster members were selected on basis of proper motion limits, and occasionally a cut-off on parallax to eliminate a clear background star. The solutions were iterated, but very few observations were rejected. The results are presented in Table 5.3.

The improvement in accuracy of the new solutions is clear to see from a comparison with the earlier results presented in Table 5.2, and follows directly from the improved accuracies of the astrometric parameters of bright stars in the new reduction. The combined HR diagram (Fig. 5.5) shows good agreement between the observational isochrones of the older clusters (the Hyades

Table 5.3. The astrometric parameters for the nearby clusters as determined from the weighted and decorrelated means of the astrometric parameters of the individual member stars. Column 1: RA, Declination; Column 2: radius (degrees); Column 3: parallax and formal error (mas); Column 4: $\mu_{\alpha *}$ and formal error, mas yr^{-1}; Column 5: μ_δ and formal error, mas yr^{-1}; Column 6: standard deviation for the solution; Column 7: number of observation (3 times number of stars) and rejected observations; Column 8: Distance modulus and formal error; Columns 9 and 10: correlation coefficients (times 100) between parallax and proper motions, and between proper motions

Cluster	1	2	3	4	5	6	7	8	9	10
Praesepe	130.0	2.7	5.49	-35.68	-12.72	1.28	69	6.30	-13	-10
	19.7		0.19	0.30	0.25		0	0.07		-3
Coma Ber	186.0	6.0	11.53	-11.72	-8.78	0.82	87	4.69	-1	1
	26.0		0.12	0.25	0.23		2	0.02		-1
Pleiades	56.5	5.0	8.18	20.00	-45.34	1.37	159	5.44	-1	-1
	24.1		0.13	0.27	0.25		1	0.03		11
IC 2391	130.0	2.5	6.78	-24.76	22.98	1.28	33	5.85	0	2
	-53.1		0.13	0.37	0.36		0	0.04		1
IC 2602	161.0	2.5	6.64	-17.18	11.06	1.02	45	5.89	1	0
	-64.4		0.09	0.24	0.23		0	0.03		1
NGC 2451	115.3	2.0	5.39	-21.38	15.54	1.35	42	6.34	0	-1
	-38.5		0.11	0.26	0.27		0	0.04		-1
Alpha Per	52.5	5.0	5.63	22.57	-26.15	1.25	147	6.25	5	-2
	49.0		0.09	0.17	0.17		1	0.04		3

and Praesepe sequences effectively coincide) and those of the younger clusters (the Pleiades, IC 2391. IC 2602, α Per all coincide). Reddening corrections were applied, but no corrections for possible differences in chemical composition.

Despite the apparent internal consistency of these results, there are some significant differences with earlier estimates, in particular for Coma Ber and the Pleiades. The result for Praesepe agrees with those of Gatewood and de Jonge (1994) and Pinsonneault et al. (1998) within the error margins given. The same applies to NGC 2451 (Platais et al. 2001) and α Per (Pinsonneault et al. 1998). Distance determinations for IC 2391 and IC 2602 preceding *Hipparcos* are quite uncertain due to the limited material available for these clusters. Both appear closer than stated by Kharchenko et al. (2005), and agree within the error margins with Nicolet (1981). For the Pleiades and Coma Ber, however, the differences are often beyond what could be expected from the formal errors quoted for different determinations. The parallax definition for these two clusters is demonstrated in Fig. 5.6. A comparison with the results for two member stars of Coma Ber by Gatewood (1995) is shown in Table 5.4. Again the differences are much larger than the formal errors would suggest possible. The determination by Makarov (2003) for Coma Ber, based on a re-interpretation of the published *Hipparcos* data, is superseded by the current new reduction of the raw *Hipparcos* data, which takes care much more thoroughly of the problems that were identified by Makarov, and eliminates them at a level where this can be done reliably.

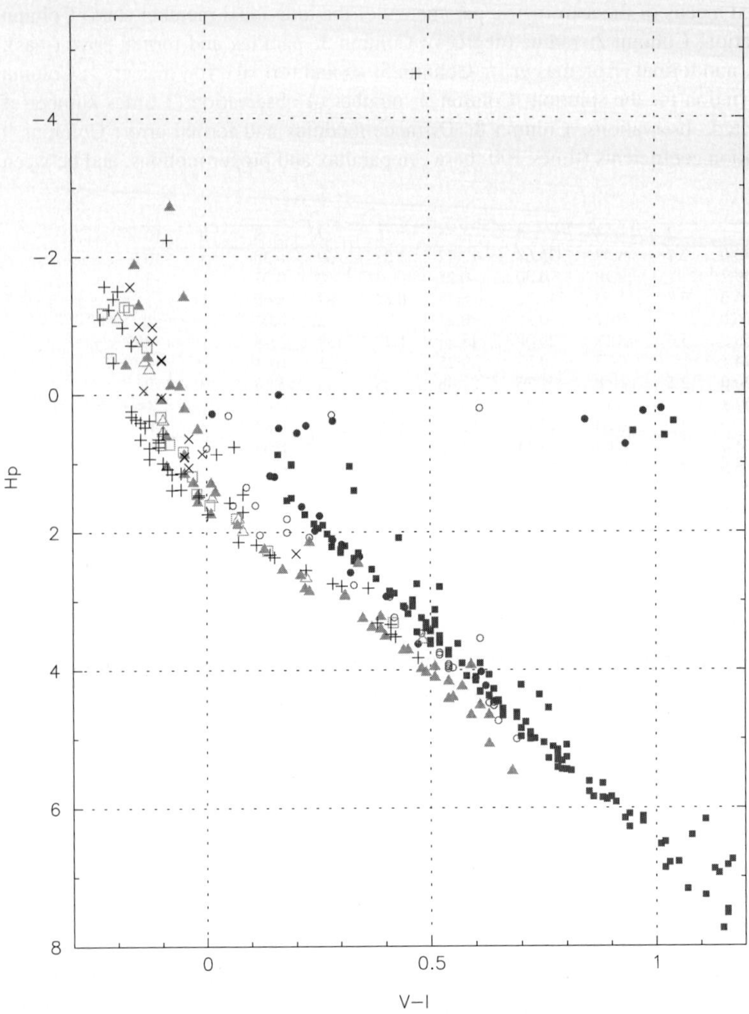

Figure 5.5. The combined HR diagram for the eight nearest open clusters: Hyades (■); Praesepe (●); Coma Ber (○); Pleiades (▲); IC 2391(△); IC 2602 (□); NGC 2451 (×); α Per (+)

Table 5.4. Comparison between parallaxes in the Coma Ber field determined by Gatewood (1995) and as determined in the new reduction (final catalogue)

HIP	AO	V	Gatewood	*Hipparcos* old	*Hipparcos* new	Note
60233	1143	8.60	2.3 ± 0.6	4.27 ± 0.92	3.46 ± 0.71	Comp.star
60351	1270	4.83	13.2 ± 0.5	11.93 ± 0.72	11.07 ± 0.24	Clust.memb.
60406	1155	9.10	14.5 ± 0.7	11.86 ± 1.22	11.56 ± 1.12	Clust.memb.

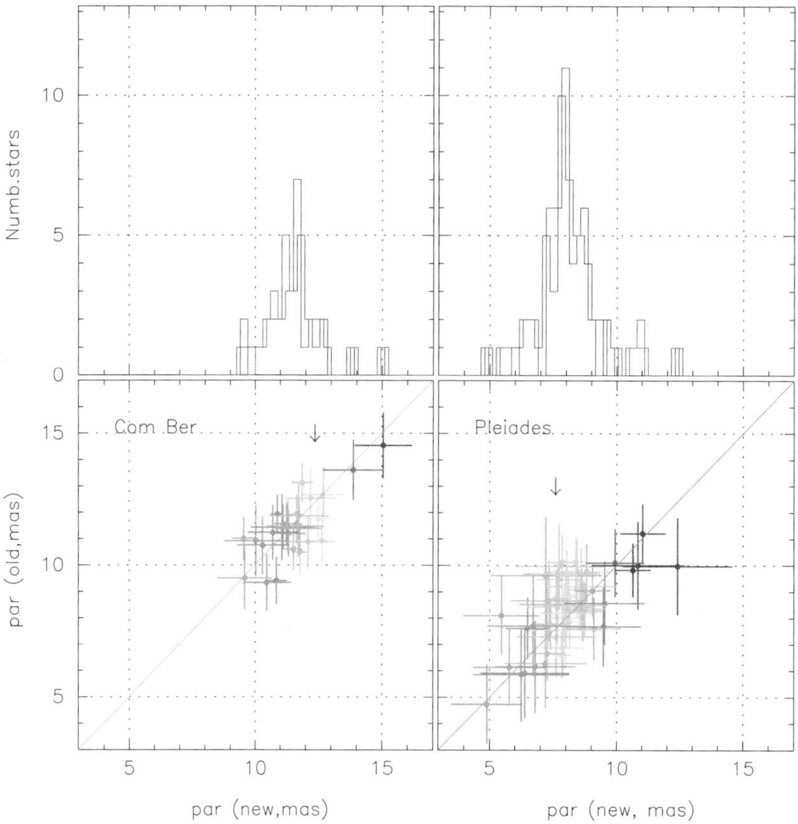

Figure 5.6. A comparison between the parallaxes for Coma Ber (left) and Pleiades (right) members as determined in the new and published reductions. The arrows shows the expected parallax according to Pinsonneault et al. (1998), which is not too far off for the Pleiades, but disagrees significantly for Coma Ber

For the Pleiades there are a few compatible determinations, involving two double stars (Pan et al. 2004; Munari et al. 2004; Zwahlen et al. 2004) and HST observations of three cluster members (Soderblom et al. 2005). The last paper boldly claimed to have delivered the "confirmation of errors in the *Hipparcos* parallaxes", but completely fails to provide the necessary background information (measurement accuracies, reduction methods, rejection criteria etc.) to verify the accuracy assessment of their own study. It provides the parallaxes for three fainter members in the inner halo of the cluster, and simply equates the mean parallax of these three stars to the cluster parallax. If we were to select randomly three stars of these magnitudes in the cluster and compare their mean position on the sky with the approximate centre of the cluster, one might get a better idea of the uncertainty associated with this determination. There is the additional complication that low-mass cluster members are very scarce in the cluster centre due to the combined effects of mass segregation (van Leeuwen 1980) and tidal friction (Terlevich 1987). Most surprisingly under these circumstances, the distance derived is exactly the one they expected to find. Although correlated errors exist in the published data (see Section 3.2), their role is on a much smaller scale than the extent of the Pleiades cluster, and the relative parallax determination for three faint cluster members in the Pleiades is by no means the conclusive evidence for widespread errors in the *Hipparcos* parallaxes. In the new reduction, these correlated errors no longer exist, but a significant difference in the parallax determination remains.

The parallaxes of the two binary stars are within 1 and 2 sigma from the individual results of the new reduction for these stars. The larger difference is for Atlas (HIP 17704), for which the *Hipparcos* astrometry is likely to be disturbed by an orbital motion which has not been taken into account in the *Hipparcos* reductions. Again, one cannot equate the parallax of a single star to that of the cluster without adding the additional uncertainty of the star's position in the cluster.

The new *Hipparcos* reduction results largely confirm the earlier results, including what has been referred to as errors in the published data: the parallaxes of the Pleiades and Coma Ber. This is not entirely unexpected: although the published data had some problems with error correlations, the typical correlation scale length (as set by the coincidence of great circles used for determining the astrometric parameters) falls far short of the areas of sky covered by these clusters. The new reduction leaves little, if any, room for an explanation of these differences as due to errors in the *Hipparcos* data.

5.2.3 The More Distant Clusters

Here we consider the determinations and results for a group of clusters with estimated distances between 200 and 550 pc. In most cases the derived parallaxes have relative errors around or below 10 per cent. However, for most of the youngest clusters it remains unclear as to whether the group of stars is a dynamically bound open cluster or an accidental density fluctuation in an OB association. This is indicated by both the distribution of "cluster" stars on the sky and by the proper motion distribution. Table 5.5 summarises the data obtained for these clusters. The actual errors on the parallaxes may well be larger due to the membership uncertainties and low number of stars for some of these clusters.

A most interesting case is Cep OB6, first identified as a possible sub-group by de Zeeuw et al. (1999) (see also Feast 1999), which counts among its possible members the star δ Cep (HIP 110991). The new reduction has confirmed and strengthened the identification of this group, even suggesting it might be a star cluster (Figure 5.7).

Table 5.5. The astrometric parameters for 12 clusters with distances between 200 and 550 pc, as determined through "cluster solutions" applied to the abscissa data of the individual members. Column 1: RA, Declination; Column 2: $\log(age)$ and $\mathrm{E(B-V)}$; Column 3: parallax and formal error (mas); Column 4: $\mu_{\alpha*}$ and formal error, mas yr^{-1}; Column 5: μ_δ and formal error, mas yr^{-1}; Column 6: standard deviation for the solution; Column 7: number of abscissae and rejected observations; Column 8: Distance modulus and formal error; Columns 9 and 10: correlation coefficients (times 100) between parallax and proper motions, and between proper motions

Cluster	1	2	3	4	5	6	7	8	9	10
Blanco 1	0.8	8.32	4.14	18.44	1.27	1.03	1993	6.91	14	0
	-30.1	0.01	0.17	0.17	0.09		2	0.16		-16
NGC 6475	268.5	8.22	3.31	1.92	-4.73	1.06	1786	7.40	-8	9
	-34.8	0.10	0.13	0.16	0.09		4	0.12		- 9
NGC 7092	322.9	8.44	3.36	-7.71	-20.04	0.94	799	7.37	0	-11
	48.5	0.01	0.20	0.20	0.20		0	0.10		-14
NGC 2232	97.0	7.73	2.83	-5.62	-2.89	0.95	632	7.74	-7	4
	-4.8	0.03	0.17	0.15	0.13		8	0.17		4
IC 4756	279.0	8.70	2.27	-0.43	-6.62	0.99	900	8.22	6	9
	5.4	0.19	0.34	0.29	0.26		2	0.62		11
NGC 2516	119.5	8.10	2.90	-4.29	11.47	1.00	1545	7.69	0	0
	-60.8	0.07	0.08	0.09	0.09		3	0.07		-19
Trump 10	131.9	7.54	2.52	-12.72	6.81	1.01	896	8.00	10	10
	-42.4	0.03	0.19	0.17	0.16		2	0.24		-8
NGC 3532	166.5	8.49	2.16	-9.87	4.48	0.91	976	8.33	1	2
	-58.7	0.04	0.19	0.19	0.17		1	0.25		29
Coll 140	110.8	7.55	2.67	-7.39	3.99	1.00	1350	7.87	11	-9
	-32.1	0.03	0.13	0.09	0.12		7	0.12		-5
NGC 2547	122.5	7.56	2.48	-9.30	4.18	0.97	2211	8.03	6	11
	-49.2	0.04	0.12	0.12	0.11		8	0.17		-6
NGC 2422	114.2	7.86	2.19	-6.68	0.82	1.01	1974	8.30	-23	-4
	-14.5	0.07	0.16	0.12	0.11		7	0.22		3
Cep OB 6	337.2		3.68	16.35	4.37	1.04	805	7.17	1	3
	58.4		0.13	0.14	0.12		4	0.09		2

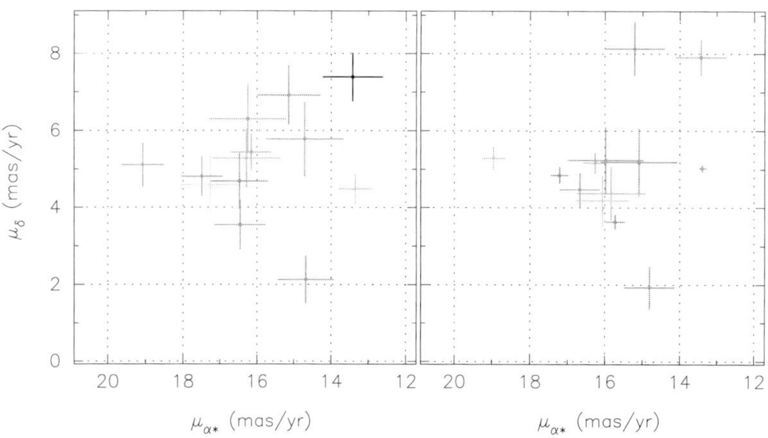

Figure 5.7. Proper motions in the region of δ Cep, left for the published data, right for the new reduction. The new reduction increases the probability that δ Cep is a member of an open cluster, to which the identification Cep OB6 has been assigned

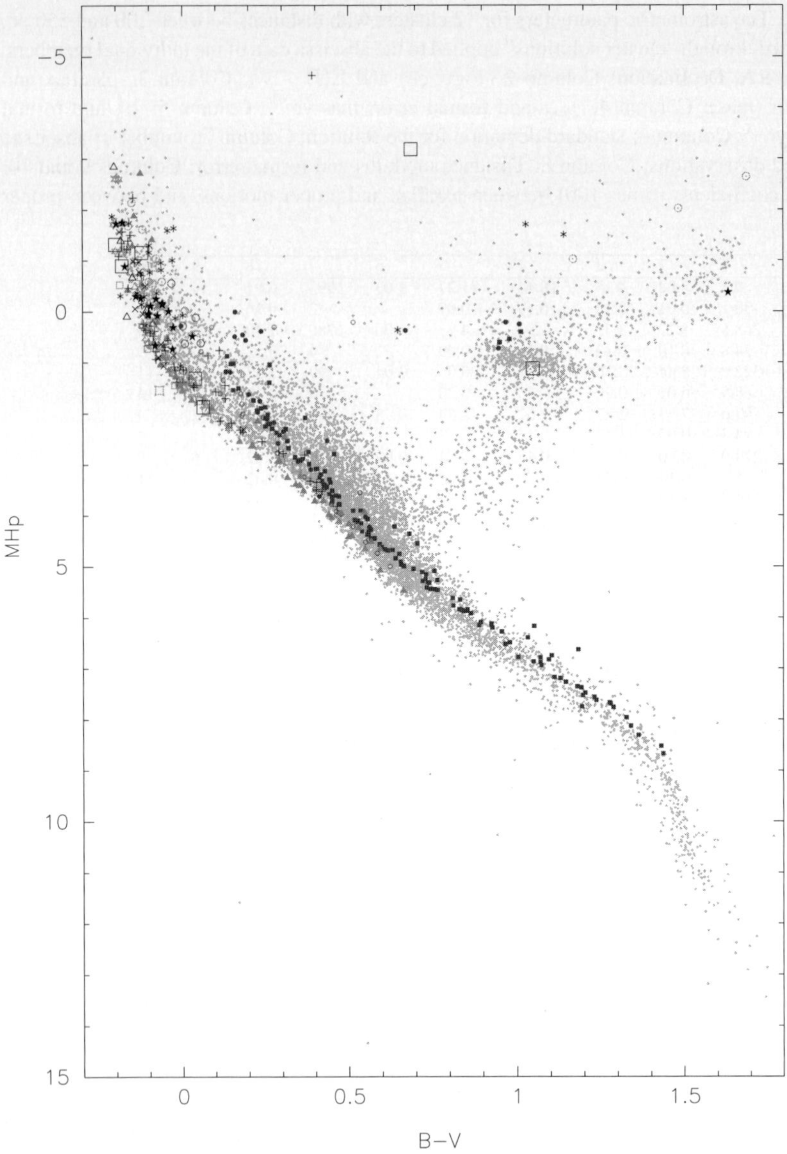

Figure 5.8. The HR diagrams for 20 open clusters compared with the same data for 9344 single stars with parallax accuracies better than 5 per cent

A comparison in the colour-magnitude diagram of the open-cluster data with 9344 (apparently) single stars (for the *Hipparcos* observations, i.e. including astrometrically unresolved binaries) with parallax accuracies better than 5 per cent is shown in Fig. 5.8. Contrary to claims made earlier by for example

(Pinsonneault et al. 1998; Soderblom et al. 1998), there is no discrepancy between the loci of a young open cluster like the Pleiades and the lower bound of the HR diagram as defined by the nearby stars. On the other hand, the impression of an as yet unresolved age-related effect behind the difference between Hyades and Pleiades is confirmed by the close coincidence of the Hyades and Praesepe loci and those of the Pleiades and clusters like IC 2602, IC 2391 and α Per. The cluster Coma Ber, with an age between the Hyades and Praesepe on one side and the Pleiades on the other, is clearly situated in the HR diagram between those clusters.

5.2.4 The Orion OB1 Association

The much improved parallaxes and proper motions for the brighter stars in the new reduction allow for a more detailed investigation of the OB associations. Here, as an example, the data in the Orion region are briefly investigated. The Orion OB1 association has, due to its galactic position with respect to the Sun, a rather small proper motion, which makes it unsuitable for the application of the methods described by de Zeeuw et al. (1999): variations in the projection of the space velocity on the sky are small.

The Orion OB1 association is a complex structure. Part of the reason may be that, as seen from the Sun, the line of sight goes through a spiral arm rather than being nearly perpendicular to it as in some easier-to-study associations. The result is a relatively wide range of distance moduli being observed for stars in this region, and, at first sight, no clear clustering being present. On a relatively small scale, fields of 0.5 to 1.5 degrees diameter, this was noted in high-accuracy differential proper motion studies (van Altena et al. 1988; Jones and Walker 1988; McNamara 1976; McNamara et al. 1989; Tian et al. 1996), where a clustering in neither space nor proper motions was recognized, though areas with different densities of stars are clearly visible. This is also the case on a larger scale: the Orion OB1 association can be traced over an area of at least 20 degrees diameter. Blaauw (1964) identified 3 sub groups, a concept further evolved by Warren and Hesser (1977b,a, 1978). A complicating issue is the differential reddening in this region (see for example Lee 1968). Some of these problems were tackled with an extensive study using VBLUW photometry (Lub and Pel 1977; Brown et al. 1994, 1995), although possible variations in the extinction law remain difficult to incorporate.

The Orion OB1 association, as covered by the *Hipparcos* catalogue (Fig. 5.9), contains a large number of bright to relatively bright stars, for which the new reduction provides much improved parallax and proper motion determinations. As abscissa-error correlations have effectively been eliminated, the new reduction will allow for a more detailed study of the dense Orion region than was possible before. This applies in particular to details in the proper motion and parallax distributions over the area covered by Orion OB1. The

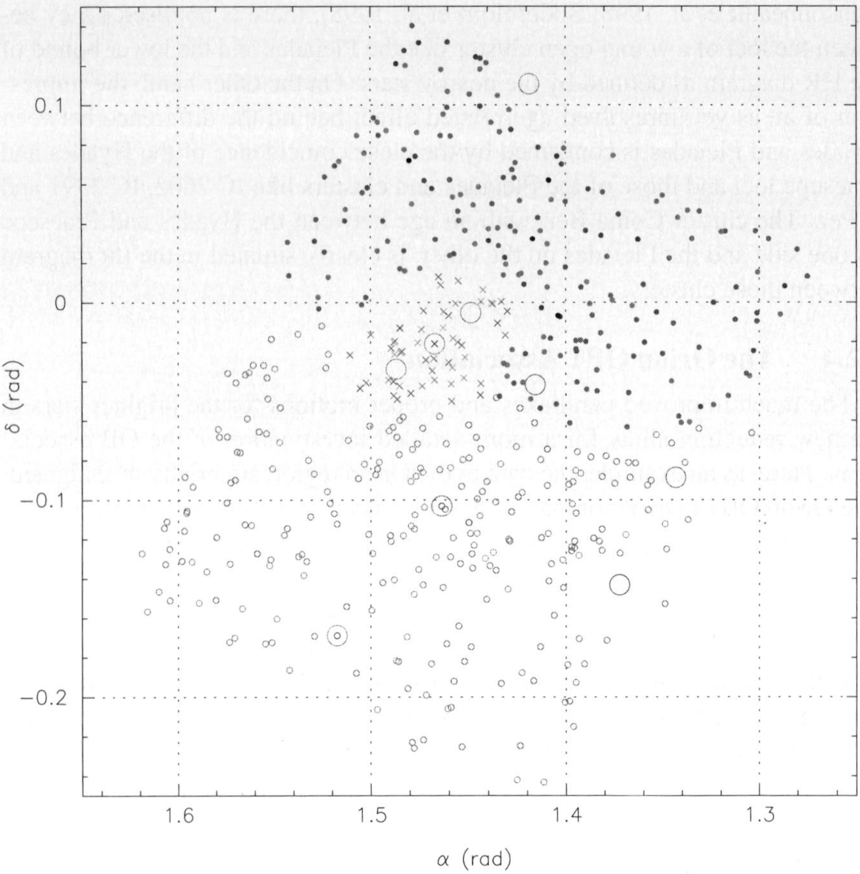

Figure 5.9. Distribution of targets in the *Hipparcos* catalogue for the Orion OB1 association. Three sub-groups as distinguished by (Blaauw 1964) are indicated by filled circles (group a), crosses (group b) and small open circles (group c). The brightest stars in this field are shown as large circles. The group assignments follow Brown et al. (1994)

selection of stars to be included in the *Hipparcos* survey had been largely determined by a proposal for stars to be included in the *Hipparcos* survey, of which also the studies by Brown et al. (1994, 1995) were a part. Those studies are used here as the starting point for some further investigations. As they were finished before the publication of the *Hipparcos* catalogue, they do not include the updated parallax and proper motion information presented in ESA (1997).

Brown determined membership of Orion OB1 based on distance modulus estimates derived from physical parameters (surface gravity and effective temperature). The present reduction allows for a comparison between those distance moduli and the observed parallaxes (Fig. 5.10). Compared to the expected relation, also shown in the figure, there appears to be some tendency

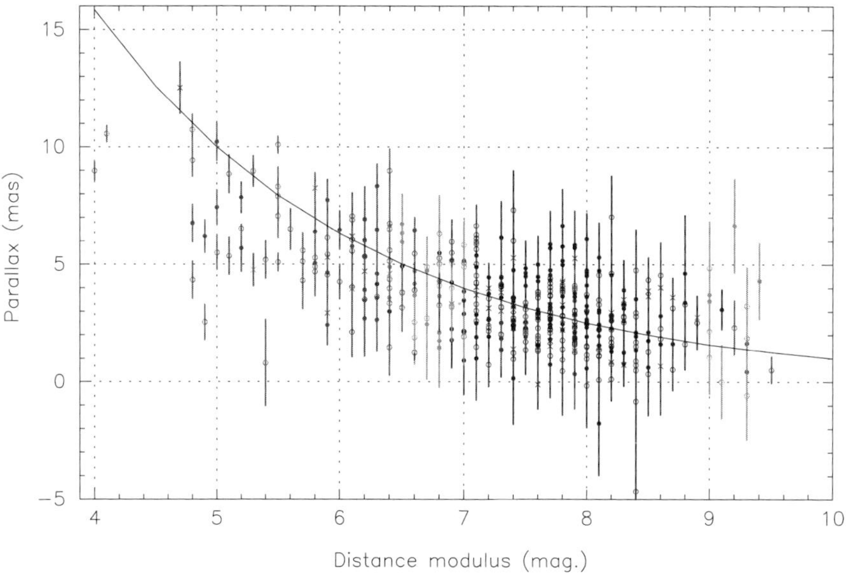

Figure 5.10. Distance moduli in the Orion OB1 region as determined by Brown et al. (1994), compared with the parallaxes for the same stars as obtained in the new *Hipparcos* reduction. The solid line shows the theoretical relation between these parameters. Although there is general overall agreement, there is also a clear indication of underestimated distance moduli for 10 to 20 stars, originating in all three subgroups of the association

to underestimate distance moduli as based on physical parameters. In fact, most of the stars assigned as possible members and some stars assigned as non-members by Brown now appear to be at least good candidate members as based on the parallax data.

A further selection of members was made by Brown based on the proper motions, which were obtained from the PPM catalogue (Roeser and Bastian 1989). The errors on the PPM proper motions are, at 3 mas yr^{-1}, about five times larger than in the *Hipparcos* data for the same stars. This increase in accuracy for the *Hipparcos* data exposes new details in the distribution of the proper motions. Figure 5.11 shows the distribution of the proper motions for the sub-groups as well as for all groups together, covering largely those stars selected as members by Brown. Two main features are visible: feature A is a dense concentration centred on (+1.0,-0.05) mas yr^{-1} approximately, with a typical dispersion of the order of 1 to 1.5 mas yr^{-1}, and feature B is a much lower density distribution with its centre at (-1.0,-3.0) mas yr^{-1} and a dispersion of the order of 4 to 5 mas yr^{-1}. Some structures may be visible in both distributions. Feature A coincides with the "Orion cluster" as studied in the differential proper motion studies mentioned earlier, and contains most of the brightest Orion stars within a very small range (of order 1 mas yr^{-1}) of proper motions.

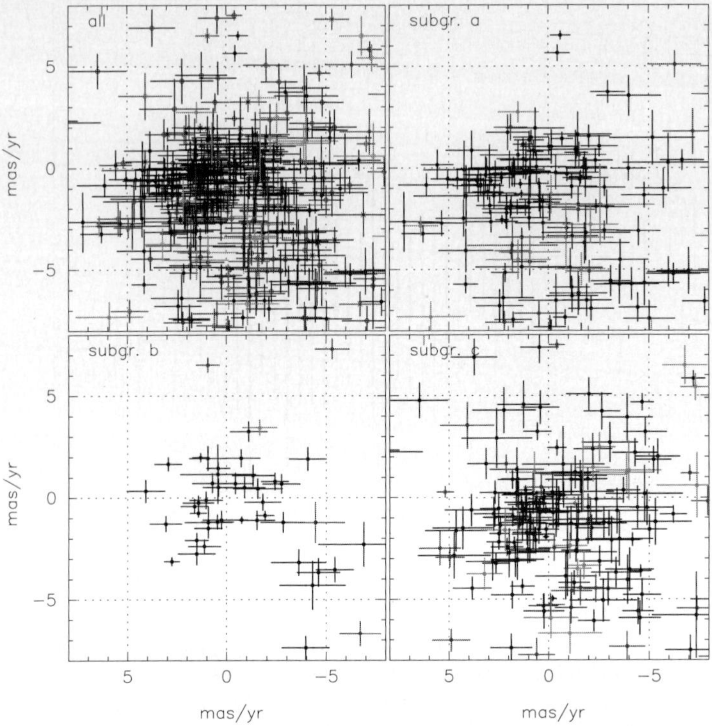

Figure 5.11. Proper motion vector-point diagrams for the Orion OB1 association, for the entire field (top left) and the three sub-groups individually (as indicated). The dense concentration close to the centre of the plot is referred to here as feature A, the wider distribution as feature B

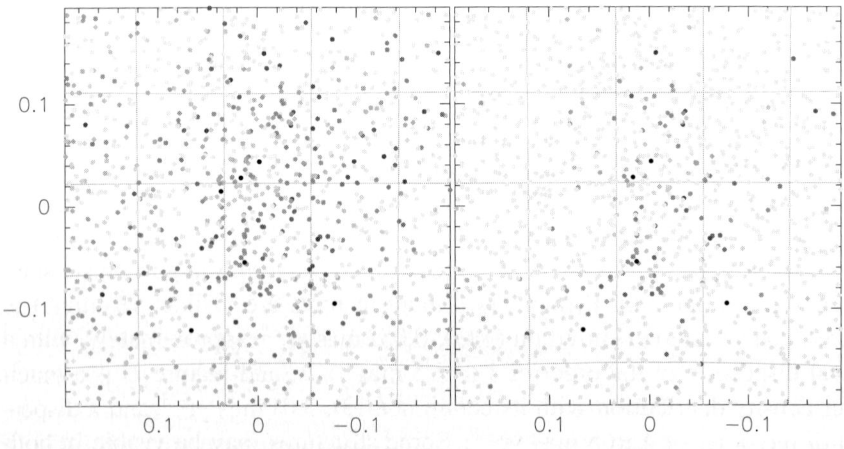

Figure 5.12. Maps of the stars in the Orion region. The centre of each map is at $\alpha = 83.1$, $\delta = -2.8$ degrees, the scales are in radians, the field size is 22 degrees square. The map on the left shows all stars for proper-motion features A and B together, the map on the right shows all stars for proper-motion feature A. Not-selected stars are shown light grey

A crescent like structure in Orion was identified by Jones and Walker (1988) as the main feature of the Orion cluster, as shown for example by the distributions of T Tauri and in particular by the H α emission-line stars in the region. This is confirmed by the *Hipparcos* data: feature A in the proper motion diagram primarily consists of stars in a wider extension of this same region, while feature B applies to stars outside this region (Fig. 5.12), and seems to extend well beyond the regions assigned as subgroups a and c, which each partly cover the crescent. For a new investigation of Orion OB1, stars from feature A are probably the more suitable targets.

5.3 Calibrating luminosities

5.3.1 Calibration Principles

In its most general form, a luminosity calibration based on parallax measurements is best described as a differential correction to an assumed model, through the relation between the parallax, the apparent magnitude (reddening corrected) and the predicted absolute magnitude. Say the calibration model for the absolute magnitude $\hat{\mathrm{M}}$ is described as a function of a set of parameters a_i:

$$\hat{\mathrm{M}} = \mathrm{M}(a_i) \tag{5.15}$$

The relation between the observed magnitude m (corrected for interstellar reddening and measured in the same photometric passband as M), the calibrated magnitude $\hat{\mathrm{M}}$ and the parallax ϖ (here expressed in mas) can be derived from Eq. 3.20 (page 86) as:

$$\hat{\varpi} = 100. \times \mathrm{e}^{-0.4605(\mathrm{m}-\mathrm{M}(a_i))}. \tag{5.16}$$

Thus, a correction to the estimated parallax $\hat{\varpi}$ is in first approximation given by:

$$\Delta\varpi = 0.4605\hat{\varpi} \sum_{i=1}^{n} \frac{\partial \mathrm{M}}{\partial a_i} \Delta a_i. \tag{5.17}$$

What is immediately clear from Eq. 5.17, and also quite obvious, is that input to these determinations decreases with decreasing predicted parallax: they are in most cases dominated by the few nearest objects. But this can be partly compensated by the increase in number for the more distant objects.

In the specific case where the only model parameter to be fitted is a constant, a direct solution between the observed magnitudes and parallaxes can be obtained. This was implemented by Feast and Catchpole (1997) in their calibration of the zero point in the Cepheids PL relation. Otherwise, Eq. 5.17 will require iterations.

Table 5.6. Reference information for 10 large-amplitude δ Sct and SX Phe stars

HIP	HD	Name	V_J	Π_0	ϖ_p	ϖ_H	$\sigma\varpi_H$	[Fe/H]
5321	6870	BS Tuc	7.48	0.0650	10.01	9.97	0.42	-0.8
11390	15165	VW Ari	6.69	0.1606	7.80	7.37	0.60	-0.5
28321	40535	V474 Mon	6.15	0.1361	11.19	10.91	0.51	0.0
38473	64191	AD CMi	9.31	0.1230	2.80	4.80	1.60	0.0
40330	69213	AI Vel	6.56	0.1117	10.60	10.23	0.34	-0.2
42594	73857	VZ Cnc	7.73	0.2263	3.83	6.13	0.71	0.2
56327	100363	SU Crt	8.28	0.1490	3.95	6.08	1.15	-0.3
86650	160589	V703 Sco	7.85	0.1500	4.79	5.58	0.67	0.0
107231	206379	RS Gru	8.38	0.1469	3.81	4.56	0.87	-0.5
117254	223065	SX Phe	7.33	0.0550	12.01	12.46	0.53	-1.3

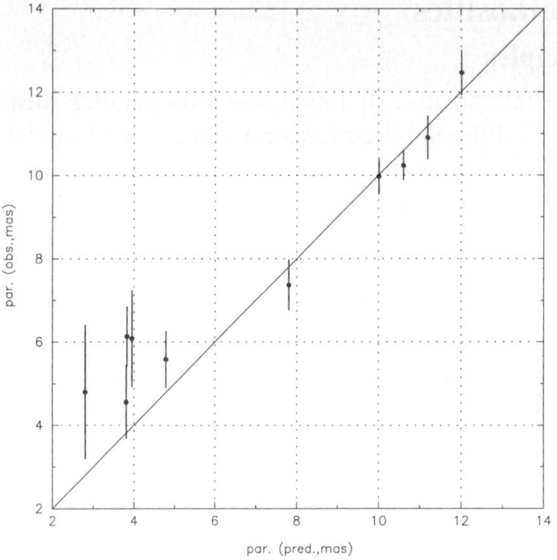

Figure 5.13. A comparison between the model-predicted and observed parallaxes for 10 δ Sct stars with large amplitudes or classified as probable SX Phe stars

5.3.2 δ Scuti and SX Phe Stars

Petersen and Hoeg (1998) identified 7 large-amplitude δ Scuti stars with significant to good parallax measurements (including SX Phe) in the *Hipparcos* catalogue. A further three SX Phe stars were selected by Antonello and Mantegazza (1997). Information on the selected stars as derived from those papers and from Rodriguez and Breger (2001) is presented in Table 5.6. A problematic case in the studies by Petersen and Hoeg (see also Hoeg and Petersen 1997) was the *Hipparcos* parallax for AD CMi. In the new reduction, the parallax of this star is much smaller, and the discrepancy effectively removed. The PL relation as based on these 10 stars, and derived using Eq. 5.17, is given by:

$$M_V = (-3.39 \pm 0.39)\log(\Pi_0/0.1) + (1.848 \pm 0.066), \qquad (5.18)$$

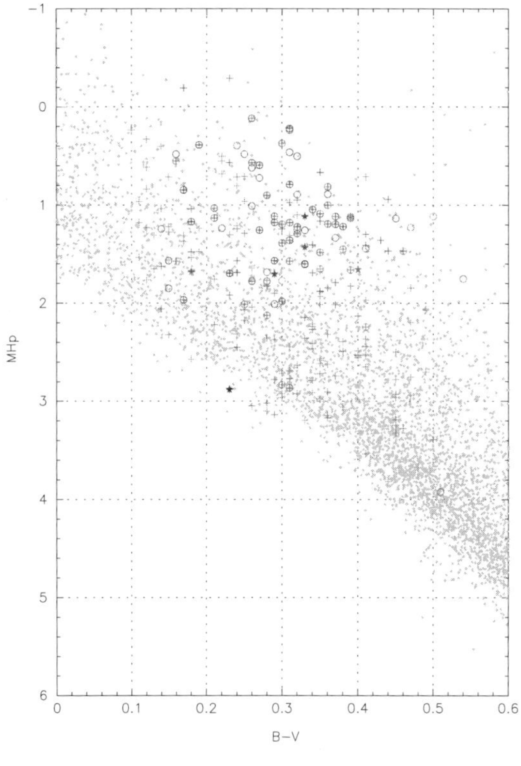

Figure 5.14. The HR diagram in the δ Sct region, showing as a background all stars with parallax accuracies better than 5 per cent. The known or suspected δ Sct stars are indicated by circles, the SX Phe-like stars by asterisks. All periodic variables with $B - V$ between 0.1 and 0.5 are indicate with + signs

with a standard deviation of 1.6. The comparison between predicted and observed parallaxes as based on this relation is shown in Fig. 5.13. There is a possible selection bias for the more distant stars, with a preference for larger observed parallaxes. For the five to six largest parallaxes the above relation appears to fit well. When based on these six stars the above relation becomes:

$$M_V = (-3.67 \pm 0.20)\log(\Pi_0/0.1) + (1.816 \pm 0.034), \qquad (5.19)$$

with a standard deviation of 0.81. The slope as determined by these relations is reasonably close to the theoretically predicted value of -3.13 to -3.18 (Templeton et al. 2002), and the zero-point is close to the value expected for fundamental-mode pulsators. Including the metallicity in the calibration did not produce any significant result.

The HR diagram for all stars with $B - V$ between 0.0 and 0.6 is shown in Fig. 5.14, with known or suspected δ Sct stars indicated, as well as all periodic variables (but excluding eclipsing binaries) with $B - V$ between 0.1 and 0.5 and relative errors on the parallax of less than 12 per cent. The δ Sct stars are predominantly found in the less populated and relatively more luminous part of the diagram, with very few nearer the blue boundary. Figure 5.15 shows the same stars in a period-luminosity diagram. Two groups can be distinguished,

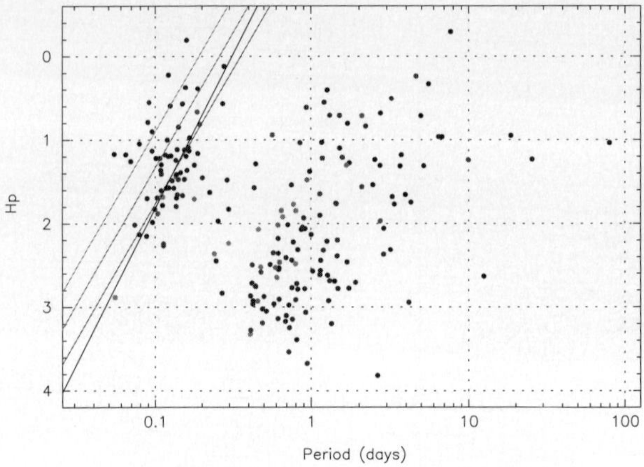

Figure 5.15. The relations between periods and luminosities for variable stars with B – V between 0.1 and 0.5. The group on the left are the δ Sct stars, for which the grey lines indicate the PL relations for fundamental, first and second overtone modes as derived from stellar models by Templeton et al. (2002), and as a black line the same relation as given by Eq. 5.19. The group on the right are mostly γ Dor stars (Kaye et al. 1999)

the δ Sct stars with short periods and high luminosities, and a larger group of generally longer periods and reaching to much lower luminosities, consisting mainly of γ Doradus stars (Kaye et al. 1999; Warner et al. 2003; de Cat et al. 2006), a relatively new class of pulsating variable stars. Many of these stars have periods around one day and were discovered from the *Hipparcos* photometric data. It is interesting to note that in Fig. 5.15 some form of PL relation also appears to exist for these stars.

The lack of δ Sct stars among the lower luminosities may be a selection effect, caused by the difficulties associated with detecting small, often multi-amplitude, variations with periods less than one hour. As a comparison, the diagram also shows the PL relations for fundamental, and first and second overtone pulsations as derived from stellar models by Templeton et al. (2002), and the relation derived above for the SX Phe stars. No attempt has been made to apply the corrections between Hp and V, which are small in this spectral region. Once these relations are put in place, probable fundamental-mode pulsators can be recognized as the concentration of stars close to the lowest of the three lines. One of those stars is classified as RRc (HIP 12113, DX Cet), which seems to be confirmed by its light curve and high space motion.

5.3.3 RR Lyraes

The parallaxes for all but two of the RR Lyrae stars as measured by *Hipparcos* have relative errors that make them useless on an individual basis. Only for RR Lyrae itself (HIP 95497, RRab) and DX Cet (HIP 12113, RRc) we have significant parallax measurements. The latter star has also been referred to as an SX Phe type variable. The new distance estimate for RR Lyrae is 258^{+25}_{-29} pc, giving an absolute magnitude of $0.65^{+0.20}_{-0.26}$, well within 1 σ from the absolute magnitude derived from kinematic studies (Fernley et al. 1998). However, using the combined information of RRab stars in the catalogue it is possible to obtain a still better overall absolute magnitude calibration. For this we use a selection of 121 RRab stars with $\mathrm{V_J}$ magnitudes and reddenings as given in Table 1 of Fernley et al. (1998). Following Eq. 5.17 the relation between the observed parallax and reddening corrected magnitude is given by:

$$\varpi + \epsilon = Q \cdot 100\mathrm{e}^{-0.4605\mathrm{V_J}}, \tag{5.20}$$

where $Q \equiv \mathrm{e}^{0.4605\mathrm{M_V}}$. Weighting each observation by its formal error ϵ, a least squares solution for Q is obtained:

$$Q = 1.32 \pm 0.12, \tag{5.21}$$

which translates into an absolute magnitude of:

$$\mathrm{M_V} = 0.61 \pm 0.09. \tag{5.22}$$

The data and the fitted relation are shown in Fig. 5.16, and show that the solution is much dominated by the parallax measurement for the star RR Lyrae.

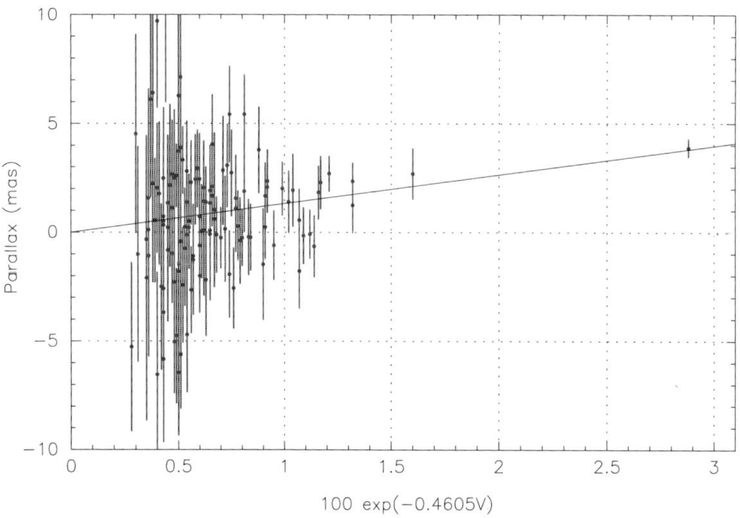

Figure 5.16. The input data for the calibration of the absolute magnitudes of RR Lyrae stars. The diagonal line shows the least-squares fit to the data, from which an absolute magnitude of 0.61 ± 0.09 is derived. The relation is much dominated by the parallax value obtained for the star RR Lyrae, the right-most data point in the diagram

The new determination is in good agreement with the kinematic study by Hawley et al. (1986) and its further evaluation using *Hipparcos* proper motions by Fernley et al. (1998), which gives $M_V = 0.77 \pm 0.15$ at $[Fe/H] = -1.53$. It is also in excellent agreement with Luri et al. (1998) ($M_V = 0.65 \pm 0.23$), using the LM method (Luri et al. 1996).

5.3.4 Sub Dwarfs

Next to RR Lyraes, also metal-poor sub dwarfs are used to calibrate distances of globular clusters. It is therefore interesting to see what the new data tell about the position of the main sequence for very low metallicity stars. Gratton et al. (1997) (G97 from here on) presented a study which claimed globular-cluster distances significantly longer than generally accepted. The selection of low to very-low metallicity dwarfs as used by G97, but with the new parallax data, is presented in Table 5.7.

When emphasizing the low metallicity dwarfs in the HR diagram for G and early K sub dwarfs (Fig. 5.17), the effects of extreme metallicities on the luminosities of those stars are clearly visible. The linear relation between colour and magnitude for 15 stars with $[Fe/H] < -1.2$ shows a slope:

$$\Delta(B-V)/\Delta M = 0.196 \pm 0.010, \tag{5.23}$$

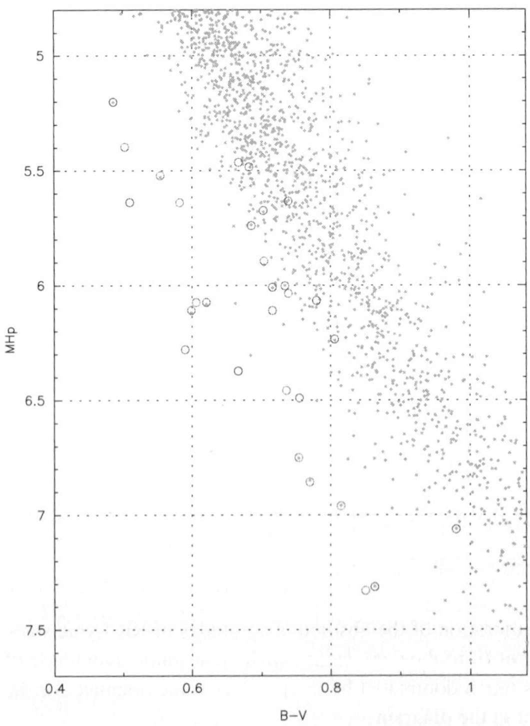

Figure 5.17. The positions of 32 sub dwarfs in the HR diagram. The sub dwarfs are shown as circles against the background of the general HR diagram for stars with parallax accuracies better than 5 per cent. The lower sequence consists almost entirely of dwarfs with metallicity index less than -1

Table 5.7. Summary of data on Sub Dwarfs. Column 2: solution type (see Section sec:solutions); column 3 (notes from G97): SB: spectroscopic binary; IR: Suspected binary based on Infra Red observations; S?: possible binary; SO: spectroscopic binary with orbital solution; AB: suspected astrometric binary (*Hipparcos* data)

HIP	isol	Note	Parallax mas	Hp	V (G97)	B − V Hip	[Fe/H]
999	5	SO	25.76 ± 1.26	8.580 ± 0.031	8.515 ± 0.075	0.739	-0.56 ± 0.13
5336	5	SB	132.10 ± 1.08	5.290 ± 0.006	5.170 ± 0.031	0.704	-0.87 ± 0.07
14594	5		25.72 ± 0.89	8.150 ± 0.008	8.050 ± 0.010	0.486	-1.91 ± 0.07
15797	5		38.92 ± 1.11	9.113 ± 0.017	8.971 ± 0.009	0.980	-0.41 ± 0.07
16404	5	AB	18.46 ± 1.67	10.042 ± 0.031	9.910 ± 0.000	0.667	-1.92 ± 0.07
18915	5		54.29 ± 0.91	8.639 ± 0.012	8.506 ± 0.001	0.863	-1.69 ± 0.07
24316	5		14.87 ± 0.80	9.535 ± 0.017	9.470 ± 0.040	0.503	-1.44 ± 0.07
38541	5	SB	34.19 ± 0.86	8.404 ± 0.012	8.280 ± 0.010	0.621	-1.48 ± 0.07
38625	15	SB	49.55 ± 2.54	7.560 ± 0.008	7.430 ± 0.000	0.739	-0.93 ± 0.07
39157	5	SB	60.14 ± 0.58	7.113 ± 0.008	6.990 ± 0.010	0.716	-0.50 ± 0.13
57450	5		13.30 ± 1.41	10.021 ± 0.036	9.920 ± 0.030	0.582	-1.26 ± 0.07
57939	5		108.96 ± 0.46	6.564 ± 0.010	6.425 ± 0.005	0.754	-1.24 ± 0.07
60956	5	SB	20.33 ± 1.01	9.134 ± 0.017	9.009 ± 0.019	0.703	-0.58 ± 0.07
62607	5	S?	31.27 ± 0.79	8.264 ± 0.011	8.139 ± 0.008	0.686	-0.52 ± 0.07
66509	5		20.10 ± 1.06	8.948 ± 0.015	8.820 ± 0.010	0.668	-0.55 ± 0.07
70681	5		21.48 ± 1.19	9.413 ± 0.014	9.310 ± 0.030	0.606	-1.09 ± 0.07
72998	5		19.67 ± 1.16	9.640 ± 0.017	9.512 ± 0.002	0.716	-0.63 ± 0.07
74234	5		35.66 ± 1.53	9.568 ± 0.024	9.441 ± 0.001	0.850	-1.28 ± 0.07
74235	5		34.00 ± 1.32	9.200 ± 0.016	9.073 ± 0.002	0.770	-1.30 ± 0.07
78775	5		69.27 ± 0.35	6.799 ± 0.009	6.660 ± 0.000	0.734	-0.52 ± 0.13
79537	5		72.49 ± 0.63	7.660 ± 0.010	7.531 ± 0.001	0.815	-1.15 ± 0.13
81170	5	SB	22.09 ± 1.49	9.737 ± 0.018	9.612 ± 0.012	0.736	-1.14 ± 0.07
89215	5	IR	15.79 ± 1.80	10.499 ± 0.033	10.430 ± 0.030	0.755	-1.51 ± 0.16
94931	5	IR	27.89 ± 0.84	9.008 ± 0.018	8.865 ± 0.005	0.806	-0.46 ± 0.09
95727	5		24.03 ± 1.42	9.162 ± 0.023	9.004 ± 0.003	0.780	-0.44 ± 0.07
98020	5	SB	27.04 ± 1.07	8.948 ± 0.011	8.830 ± 0.003	0.599	-1.37 ± 0.07
99267	5		12.14 ± 1.09	10.098 ± 0.028	9.990 ± 0.060	0.470	-1.81 ± 0.16
100568	5		22.48 ± 0.99	8.762 ± 0.012	8.652 ± 0.002	0.554	-1.00 ± 0.07
103269	5		14.92 ± 1.39	10.321 ± 0.036	10.180 ± 0.060	0.560	-1.48 ± 0.16
106924	5		15.49 ± 1.31	10.331 ± 0.035	10.190 ± 0.060	0.501	-1.60 ± 0.16
112811	5		16.02 ± 1.26	9.460 ± 0.019	9.333 ± 0.003	0.683	-0.66 ± 0.07

less steep than the value of 0.152 determined for more average abundance stars of similar spectral types in Section 3.3.4 (page 90). Using this slope, the data can be folded to either show the relative distribution of the colours or the relative distribution of the absolute magnitudes. Here the *Hipparcos* Hp rather than the V magnitudes are used, as these provide the most homogeneous and accurate magnitudes over the whole sky. The relation between Hp and V_J in this spectral range $(0.5 < (B - V) < 0.9)$ is given by:

$$\mathrm{Hp} \approx \mathrm{V_J} + 0.1(\mathrm{B} - \mathrm{V}) - 0.05. \tag{5.24}$$

Thus, the slope used in folding the data is not much different in Hp or $\mathrm{V_J}$, with the empirical slope of 0.196 for Hp equivalent to a slope of 0.20 for $\mathrm{V_J}$. As comparison, G97 determined a slope of 0.168 ± 0.004 for $\mathrm{M_V} < 6$ and 0.216 ± 0.007 for $\mathrm{M_V} > 6$, folding the data with $\mathrm{M_V} = 6$ as reference point. These slopes had been derived from mean globular cluster loci.

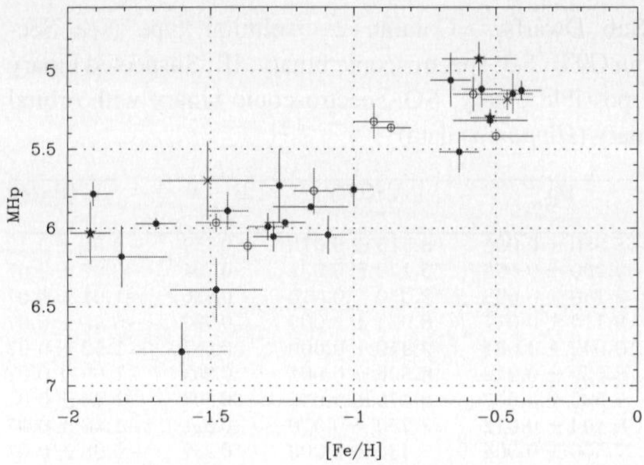

Figure 5.18. The relation between [Fe/H] and absolute magnitude in the *Hipparcos* broad band (MHp). Single stars are shown by filled dots, known spectroscopic binaries by open circles, IR binaries by crosses, and other possible binaries by asterisks. The projection reference point is at B − V = 0.6

The data are shown in Fig. 5.18 as a magnitude-metallicity diagram, to show directly the relevance to absolute magnitude calibrations for very metal-poor stars. The diagram shows the offsets in Hp of the individual stars from the calibration relation Eq. 5.23. The relation between the absolute magnitude offset and metallicity has been fitted with a second-order polynomial in [Fe/H], in the same way this was done by G97, using only probably-single stars. The least-squares fit with its covariance matrix allows for an estimate with formal error for the expected absolute magnitudes offsets at different [Fe/H] values.

The least-squares solution can be done either using weights for individual observations, assuming the main noise contribution comes from the uncertainties in the absolute magnitudes, or using equal-weights for all observations, assuming that the dominating noise contribution is cosmic and similar for all stars. The weighted solution is dominated by a small number of stars with very accurate parallaxes, while the equal-weight distribution ignores the relative quality of the absolute magnitudes. The covariance matrix and standard deviation of the least-squares solution are used to derive estimates and formal errors for specific [Fe/H] values. Using the inverse of the slope as defined in Eq. 5.23, relations for the absolute magnitude for different metallicities are obtained. For example, in the weighted solution at $[\mathrm{Fe/H}] = -1.5$, the relation is found to be:

$$\mathrm{MHp} = (5.84 \pm 0.10) + (5.10 \pm 0.27)(\mathrm{B} - \mathrm{V} - 0.6), \tag{5.25}$$

putting these low-metallicity stars close to 1 magnitude below the main sequence as defined by stars with more average metallicities. At $[\mathrm{Fe/H}] = -2.0$, the constant in Eq. 5.25 equals 6.01 ± 0.24. In the unit-weight solution the zero points at these two metallicity values are 6.09 ± 0.07 and 6.07 ± 0.21 respectively, showing both the sensitivity of these values to the assumptions made for the solution and the uncertainty in their determination. Finding nearly the same absolute magnitudes for the selected sub dwarfs with $[\mathrm{Fe/H}] < -1.5$ could be expected from an inspection of the data in Fig. 5.18. When expressed in V instead of Hp, the constants are increased by 0.05 and the slope is decreased by 0.1.

Equation 4 in G97 gives $(B-V)=0.554$ at $M_V=6$ for $[Fe/H]=-2.0$. The equivalent values presented here are $(B-V)=0.59\pm0.05$ or 0.58 ± 0.04 for the weighted or unit-weight solutions respectively, equivalent to the absolute magnitudes of these stars being 0.17 ± 0.24 or 0.12 ± 0.21 magnitudes fainter. This small difference may be partly due to inclusion or exclusion of some of the brightest sub dwarfs, though both G97 and the current study suggest that this has little effect on the results.

5.3.5 Cepheids

The *Hipparcos* data provided the first opportunity to calibrate independently the critical parameters in the PLC relation for classical Cepheids, even though in the published data only very few of these stars had marginally significant individual parallax measurements, restricting the calibration of the PL to the zero point only. Using the method described in Section 5.3.1, and describing the calibration entirely in parallax space, Feast and Catchpole (1997) (FC97 from here on) provided the first such calibration. Although this method has been questioned by some authors (see for example Oudmaijer et al. 1998), subsequent studies using Monte-Carlo simulations have fully endorsed it (Luri et al. 1998). A comprehensive review of the various attempts made to analyse the *Hipparcos* data on the Cepheids was presented by Pont (1999), who also did various test to show the problems associated with some of the determinations, in particular those that are magnitude-based. That study too confirmed the validity of the calibration method used by FC97 and further described and used here.

The distribution in apparent brightness of the Cepheids in the *Hipparcos* catalogue means that these stars have been very positively affected by the improved accuracies in the new reduction, creating for the first time the possibility to attempt a full parallax-based PL calibration.

Reddening corrections used here were obtained from Tammann et al. (2003). As was explained by Feast and Whitelock (1997), small errors in reddening will have only minor effects on the PL calibration, which is given by:

$$\langle M_V \rangle = \delta \log P_0 + \rho, \tag{5.26}$$

where P_0 is the fundamental-mode period in days. For a small number of stars, including α UMi, the fundamental mode period has been derived from the actual first-overtone period, by means of (Alcock et al. 1995):

$$P_1/P_0 \approx 0.716 - 0.027 \log P_1. \tag{5.27}$$

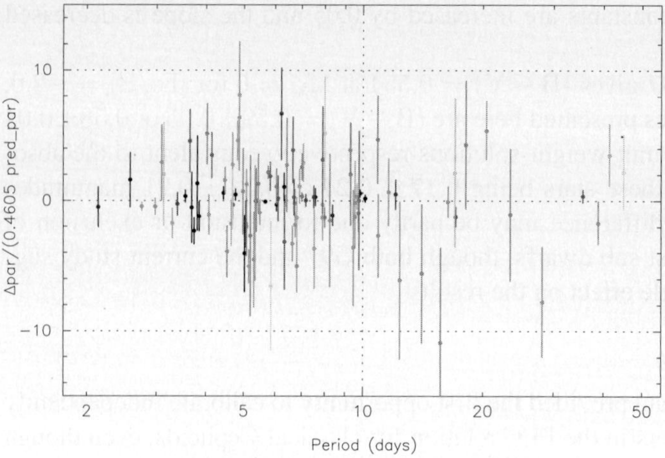

Figure 5.19. A graphical representation of the parallax-based PL calibration for 100 Cepheids. Formal errors on the parallaxes of the selected stars are 1 mas or less. The observations are given by $(\varpi - \hat{\varpi})/0.4605\hat{\varpi}$, the formal errors by $\sigma_\varpi/(0.4605\hat{\varpi})$ (see text). Residuals shown in this graph are used to determine corrections to the PL relation

Then, following Eq. 5.17, the calibration is carried out as a differential correction to the assumed relation:

$$\Delta\varpi = \varpi - \hat{\varpi} = 0.4605\hat{\varpi}\big(\Delta\delta \log P_0 + \Delta\rho\big). \tag{5.28}$$

To reduce the correlation between $\Delta\rho$ and $\Delta\delta$, Eq. 5.28 is reorganized as follows:

$$\Delta\varpi' = \frac{\varpi - \hat{\varpi}}{0.4605\hat{\varpi}} = \Delta\delta(\log P_0 - 0.78) + \Delta\rho_6, \tag{5.29}$$

where $\Delta\rho_6$ is the correction to the zero point of the PL relation as evaluated at $\log P_0 = 0.78$, or $P_0 = 6.0$ days. The advantage of taking the zero point at that position, rather than at $\log P_0 = 0.0$ as is traditionally done, is its reduced dependence on the determination or assumption of the slope δ.

Table 5.8. Parameters used and derived for selected Cepheids

HIP	Name	logP	⟨B⟩	⟨V⟩	EBV	BV_0	M_V	$\hat{\varpi}$	ϖ mas	$\sigma\varpi$
2085	TU Cas	0.330	8.369	7.753	0.109	0.507	-2.32	1.14	0.88	0.77
2347	DL Cas	0.903	10.121	8.968	0.479	0.674	-3.93	0.54	1.84	1.04
7192	V636 Cas	1.083	8.555	7.173	0.666	0.716	-4.43	1.31	1.61	0.64
11767	α UMi	0.754	2.326	1.978	-0.007	0.355	-3.51	7.92	7.67	0.12
13367	SU Cas	0.440	6.671	5.970	0.273	0.428	-2.62	2.86	2.31	0.32
19057	RX Cam	0.898	8.879	7.683	0.536	0.660	-3.91	1.08	0.94	0.75
21517	SZ Tau	0.651	7.377	6.530	0.280	0.567	-3.22	1.71	2.94	0.74
23360	RX Aur	1.065	8.632	7.674	0.276	0.682	-4.38	0.59	0.76	0.65
23768	CK Cam	0.517	8.521	7.545	0.426	0.550	-2.84	1.58	0.32	0.79
26069	β Dor	0.993	4.557	3.757	0.089	0.711	-4.18	2.96	3.74	0.28

Continued on next page

Table 5.8 – continued from previous page

HIP	Name	logP	⟨B⟩	⟨V⟩	EBV	BV_0	M_V	$\hat{\varpi}$	ϖ mas	$\sigma\varpi$
27119	ST Tau	0.605	9.070	8.221	0.339	0.510	-3.09	0.91	3.04	0.93
30827	RT Aur	0.572	6.039	5.448	0.049	0.542	-3.00	2.20	-0.17	0.99
31404	W Gem	0.898	7.868	6.955	0.266	0.647	-3.91	1.00	0.41	0.78
33874	V526 Mon	0.580	9.206	8.619	0.088	0.499	-3.02	0.54	0.86	0.86
34088	ζ Gem	1.006	4.712	3.915	0.033	0.764	-4.22	2.49	2.69	0.18
35212	RY CMa	0.670	8.952	8.106	0.223	0.623	-3.27	0.74	1.68	0.80
38907	AP Pup	0.706	8.211	7.380	0.241	0.590	-3.37	1.01	1.61	0.45
40155	AH Vel	0.782	6.264	5.688	0.070	0.506	-3.59	1.55	1.33	0.24
40178	AT Pup	0.823	8.769	7.982	0.167	0.620	-3.70	0.59	1.65	0.64
40233	RS Pup	1.617	8.446	7.009	0.453	0.984	-5.93	0.52	1.36	0.51
41588	V Car	0.826	8.243	7.371	0.157	0.715	-3.71	0.77	0.86	0.42
42257	RZ Vel	1.310	8.201	7.080	0.293	0.828	-5.07	0.58	2.02	0.54
42321	T Vel	0.667	8.958	8.030	0.271	0.657	-3.26	0.83	0.16	0.59
42492	AP Vel	0.495	11.109	9.999	0.490	0.620	-2.78	0.58	1.12	1.03
42831	SW Vel	1.370	9.271	8.120	0.337	0.814	-5.24	0.36	0.18	0.79
42926	SX Vel	0.980	9.173	8.285	0.250	0.638	-4.14	0.48	0.44	0.65
42929	ST Vel	0.768	10.907	9.697	0.496	0.714	-3.55	0.48	-0.96	1.05
44847	BG Vel	0.840	8.850	7.662	0.439	0.749	-3.75	1.02	1.54	0.54
47854	l Car	1.551	4.960	3.698	0.160	1.102	-5.75	1.66	1.92	0.26
50655	RY Vel	1.449	9.752	8.373	0.554	0.825	-5.46	0.40	-1.03	0.73
50722	AQ Car	0.990	9.785	8.855	0.158	0.772	-4.17	0.32	0.32	0.73
51142	UW Car	0.728	10.441	9.430	0.439	0.572	-3.43	0.52	0.11	1.02
51262	YZ Car	1.259	9.829	8.709	0.372	0.748	-4.93	0.33	0.98	0.70
51338	UX Car	0.566	8.947	8.302	0.091	0.554	-2.98	0.63	0.48	0.75
51653	Y Car	0.561	8.754	8.139	0.169	0.446	-2.96	0.77	2.40	0.82
51909	UZ Car	0.716	10.212	9.331	0.184	0.697	-3.40	0.38	-0.09	1.03
52570	SV Vel	1.149	9.696	8.588	0.365	0.743	-4.62	0.40	-0.63	0.94
52661	SX Car	0.687	10.013	9.086	0.310	0.617	-3.32	0.53	2.07	0.95
53536	XX Car	1.196	10.422	9.353	0.343	0.726	-4.75	0.25	-1.02	0.81
53589	U Car	1.589	7.435	6.253	0.287	0.895	-5.85	0.59	0.13	0.37
53945	XY Car	1.094	10.510	9.294	0.408	0.808	-4.46	0.33	-0.60	0.88
54621	GH Car	0.916	10.121	9.159	0.394	0.568	-3.96	0.43	0.13	0.97
54659	V898 Cen	0.701	8.535	7.963	-0.046	0.618	-3.36	0.51	-0.44	0.64
54715	IT Car	0.877	9.092	8.109	0.209	0.774	-3.85	0.56	-0.39	0.73
54862	GI Car	0.802	9.040	8.326	0.166	0.548	-3.64	0.52	-0.24	0.82
55726	AY Cen	0.725	9.797	8.811	0.295	0.691	-3.43	0.56	-0.61	0.90
56176	V419 Cen	0.898	8.954	8.189	0.167	0.598	-3.91	0.49	1.19	0.88
57260	RT Mus	0.490	9.830	8.990	0.292	0.548	-2.77	0.69	1.62	0.87
59551	S Mus	0.985	6.955	6.125	0.220	0.610	-4.16	1.22	1.42	0.48
60259	T Cru	0.828	7.494	6.564	0.178	0.752	-3.71	1.15	1.06	0.50
60455	R Cru	0.766	7.553	6.761	0.150	0.642	-3.54	1.09	2.03	0.49
61136	BG Cru	0.678	6.077	5.459	0.050	0.568	-3.29	1.91	2.32	0.29
61981	R Mus	0.876	7.057	6.305	0.134	0.618	-3.85	1.14	1.29	0.36
61981	R Mus	0.876	7.057	6.305	0.134	0.618	-3.85	1.14	1.29	0.36
62986	S Cru	0.671	7.360	6.597	0.162	0.601	-3.27	1.35	1.15	0.65
64969	V378 Cen	0.969	9.515	8.481	0.376	0.658	-4.11	0.53	1.03	0.93
66696	XX Cen	1.039	8.814	7.820	0.258	0.736	-4.31	0.55	-0.13	0.77
67566	V381 Cen	0.706	8.467	7.673	0.195	0.599	-3.37	0.83	0.06	0.80
70203	V339 Cen	0.976	9.908	8.695	0.412	0.801	-4.13	0.51	-0.34	1.01
71116	V Cen	0.740	7.702	6.821	0.264	0.617	-3.47	1.30	0.76	0.52
71492	V737 Cen	0.849	7.695	6.727	0.228	0.740	-3.77	1.12	2.62	0.69
75018	R TrA	0.530	7.370	6.653	0.134	0.583	-2.88	1.52	1.34	0.56
75430	GH Lup	0.968	8.847	7.631	0.346	0.870	-4.11	0.76	2.13	0.69
78476	S TrA	0.801	7.124	6.380	0.082	0.662	-3.64	1.12	1.67	0.49
78978	U TrA	0.410	8.524	7.947	0.084	0.493	-2.54	0.90	1.57	0.83
79932	S Nor	0.989	7.370	6.422	0.178	0.770	-4.17	1.00	0.56	0.60
83059	RV Sco	0.782	7.995	7.032	0.338	0.625	-3.59	1.25	1.88	0.74
83674	BF Oph	0.610	8.220	7.340	0.247	0.633	-3.10	1.18	0.59	0.79
85035	V636 Sco	0.833	7.587	6.654	0.212	0.721	-3.73	1.15	0.45	0.81
85701	V482 Sco	0.656	8.956	7.964	0.340	0.652	-3.23	0.96	-0.77	0.83
86269	V950 Sco	0.683	8.088	7.305	0.254	0.529	-3.31	1.10	3.44	0.72
87072	X Sgr	0.846	5.318	4.564	0.201	0.553	-3.77	2.91	3.38	0.22
87495	Y Oph	1.234	7.521	6.169	0.623	0.729	-4.86	1.61	0.70	0.50
88567	W Sgr	0.880	5.415	4.670	0.112	0.633	-3.86	2.33	2.63	0.66

Continued on next page

Table 5.8 – continued from previous page

HIP	Name	logP	$\langle B \rangle$	$\langle V \rangle$	EBV	BV_0	M_V	$\hat{\varpi}$	ϖ mas	$\sigma\varpi$
89276	AP Sgr	0.704	7.754	6.950	0.174	0.630	-3.37	1.12	0.20	0.89
89968	Y Sgr	0.761	6.602	5.745	0.188	0.669	-3.53	1.86	3.71	0.30
90836	U Sgr	0.829	7.794	6.694	0.403	0.697	-3.72	1.52	0.22	0.60
92013	V350 Sgr	0.712	8.372	7.466	0.295	0.611	-3.39	1.05	-1.39	0.57
92370	YZ Sgr	0.980	8.363	7.328	0.285	0.750	-4.14	0.78	0.33	0.79
92491	BB Sgr	0.822	7.917	6.947	0.276	0.694	-3.70	1.13	-0.59	0.64
93124	FF Aql	0.806	6.128	5.373	0.213	0.542	-3.65	2.15	2.18	0.35
94004	V496 Aql	0.833	8.903	7.746	0.393	0.764	-3.73	0.92	-1.20	0.80
94094	FM Aql	0.786	9.545	8.274	0.617	0.654	-3.60	1.07	1.16	0.82
95820	U Aql	0.846	7.474	6.448	0.371	0.655	-3.77	1.58	3.47	0.69
96458	U Vul	0.903	8.404	7.128	0.593	0.683	-3.93	1.51	0.62	0.48
97150	SU Cyg	0.585	7.433	6.862	0.088	0.483	-3.03	1.20	0.53	0.57
97717	SV Vul	1.653	8.675	7.226	0.518	0.931	-6.03	0.50	1.00	0.60
97794	V1162 Aql	0.731	8.671	7.790	0.187	0.694	-3.44	0.75	0.76	0.80
97804	η Aql	0.856	4.690	3.901	0.133	0.656	-3.79	3.53	3.58	0.71
98085	S Sge	0.923	6.416	5.614	0.112	0.690	-3.98	1.43	0.69	0.46
98852	CD Cyg	1.232	10.221	8.953	0.486	0.782	-4.85	0.36	0.72	0.92
102276	X Cyg	1.215	7.530	6.393	0.261	0.876	-4.80	0.86	0.83	0.38
102949	T Vul	0.647	6.389	5.753	0.067	0.569	-3.21	1.78	2.27	0.31
104185	DT Cyg	0.550	6.315	5.775	0.037	0.503	-2.93	1.91	2.22	0.33
105369	V532 Cyg	0.670	10.121	9.087	0.508	0.526	-3.27	0.72	0.75	0.71
107899	VZ Cyg	0.687	9.836	8.959	0.274	0.603	-3.32	0.53	2.49	0.97
108426	IR Cep	0.474	8.711	7.795	0.413	0.503	-2.72	1.46	1.15	0.35
110968	V411 Lac	0.617	8.691	7.950	0.154	0.587	-3.12	0.77	2.00	0.56
110991	δ Cep	0.730	4.614	3.953	0.068	0.593	-3.44	3.68	3.63	0.17
111972	Z Lac	1.037	9.514	8.417	0.378	0.719	-4.30	0.51	1.60	0.74
112026	RR Lac	0.808	9.731	8.847	0.296	0.588	-3.66	0.49	1.11	0.77
112430	CR Cep	0.794	11.048	9.654	0.697	0.697	-3.62	0.64	2.59	1.00
112626	V Lac	0.697	9.813	8.940	0.315	0.558	-3.35	0.56	0.74	0.81
112675	X Lac	0.736	9.306	8.406	0.339	0.561	-3.46	0.70	1.13	0.68

In the study of FC97 the calibration was based on 25 to 26 Cepheids with parallax accuracies better than 1.0 mas, and only the zero point ρ was determined, fixing the slope to the value observed for the LMC. With the new reduction we have available just over 100 Cepheids with formal errors on their parallaxes of 1.0 mas or less (Table 5.8), which allows for the first time a calibration of the slope and the zero point based on *Hipparcos* parallaxes:

$$\begin{aligned} \delta &= -2.22 \pm 0.45 \\ \rho_6 &= -3.58 \pm 0.07. \end{aligned} \tag{5.30}$$

The fitting of this relation is shown in Fig. 5.19. Using the slopes as given by Laney and Stobie (1994): $\delta = -2.87 \pm 0.07$, and Caldwell and Laney (1991): $\delta = -2.81 \pm 0.06$ (as used by FC97), the zero point is found to be:

$$\begin{aligned} \rho_1 &= -1.345 \pm 0.089, \quad \text{or} \\ \rho_1 &= -1.392 \pm 0.084 \end{aligned} \tag{5.31}$$

respectively, confirming the earlier result by FC97. The difference between the value for the slope as found here and that determined by Caldwell and Laney (1991) is no more than 1.3σ, and therefore not significant.

Table 5.9. Summary of data used in the PL calibration of the Mira stars (van Leeuwen et al. 1997c) (last two columns: see text)

HIP	Name	$\log P$	A_V	K_0	$m_{\rm bol}$	ϖ	$\sigma\varpi$	f	σf
10826	o Cet	2.521	0.01	-2.50	0.70	9.65	1.35	1.71	0.24
12193	R Tri	2.426	0.14	0.93	4.04	1.90	1.09	1.41	0.81
13502	R Hor	2.611	0.02	-0.94	2.22	4.47	0.67	1.88	0.28
23203	R Lep	2.630	0.08	-0.01	3.45	2.86	0.64	1.90	0.43
28041	U Ori	2.566	0.23	-0.64	2.54	2.29	1.06	1.03	0.48
46806	R Car	2.490	0.13	-1.35	1.74	6.53	0.70	1.87	0.20
48036	R Leo	2.491	0.02	-2.55	0.69	10.60	1.26	1.75	0.21
49751	S Car	2.175	0.35	1.84	4.65	1.73	0.41	1.30	0.31
65835	R Hya	2.590	0.03	-2.48	0.66	7.23	0.97	1.45	0.19
69754	R Cen	2.737	0.21	-0.72	2.38	1.55	0.46	0.88	0.26
82912	RR Sco	2.449	0.20	-0.25	2.88	3.17	0.96	1.41	0.43
93820	R Aql	2.454	0.23	-0.78	2.34	2.30	0.82	0.81	0.29
97629	χ Cyg	2.611	0.14	-1.93	1.39	5.42	0.88	1.45	0.23
104451	T Cep	2.589	0.11	-1.71	1.50	4.57	0.69	1.30	0.20
117054	R Aqr	2.588	0.01	-1.02	2.26	0.70	1.60	0.27	0.63
118188	R Cas	2.633	0.12	-1.80	1.40	6.17	0.84	1.81	0.25

5.3.6 Miras

A study on the *Hipparcos* data for 11 Mira variables was presented by van Leeuwen et al. (1997c). The Mira stars are affected in two ways by the new reductions: the general improvements of the solution and the much improved handling of the colour variations for these stars thanks to the introduction of epoch-resolved colour information, as derived by Platais et al. (2003) and Pourbaix et al. (2003), in the data reductions.

The PL relation for the same 11 Miras (Table 5.9) is examined in the way described by van Leeuwen et al. (1997c): the slope is assumed known and fixed at the value observed for the LMC (Feast et al. 1989), where, in IR pass bands, these stars show a well-determined PL relation with small scatter. Here, only the zero point is calibrated for galactic Miras. The relations for the K and $M_{\rm bol}$ magnitudes are then given by:

$$\begin{aligned} \mathrm{K} &= -3.47\log P + \beta_1 \\ \mathrm{M_{bol}} &= -3.00\log P + \beta_2. \end{aligned} \tag{5.32}$$

The observed quantities are the apparent magnitudes (reddening corrected K_0 and $m_{\rm bol}$) and the parallaxes ϖ (see also Eq. 3.20):

$$\begin{aligned} 0.2\,(\mathrm{K_0}-\mathrm{K}) &= -\log(0.01\varpi), \\ 0.2\,(\mathrm{m_{bol}}-\mathrm{M_{bol}}) &= -\log(0.01\varpi), \end{aligned} \tag{5.33}$$

with the main uncertainty coming from the parallaxes. Additional noise will enter the solution originating from the cosmic spread of the absolute magnitudes with respect to the PL relation. This is similar to what is described in Section 3.3.3. This intrinsic scatter, as estimated from the LMC measurements, is about 0.14 mag (Feast et al. 1989), which adds approximately 0.06ϖ noise to the data (see Table 3.3).

Substituting Eq. 5.33 into Eq. 5.32 we can obtain the expressions for the observed parallaxes, as this was also used by van Leeuwen et al. (1997c):

$$\begin{aligned} q1 &\equiv 10^{0.2\beta_1} = 0.010\,\varpi\,\mathrm{e}^{0.4605(3.47\log P+\mathrm{K_0})}, \\ q2 &\equiv 10^{0.2\beta_2} = 0.010\,\varpi\,\mathrm{e}^{0.4605(3.00\log P+\mathrm{m_{bol}})}. \end{aligned} \tag{5.34}$$

The first of these relations is shown in Table 5.9 under column 9, with the formal error in column 10.

The values derived for $q1$ and $q2$ are:

$$\begin{aligned} q1 &= 1.49 \pm 0.09, \\ q2 &= 3.71 \pm 0.24, \end{aligned} \tag{5.35}$$

from which the following values for β_1 and β_2 are derived:

$$\begin{aligned} \beta_1 &= 0.87 \pm 0.13, \\ \beta_2 &= 2.85 \pm 0.14. \end{aligned} \tag{5.36}$$

Both values confirm the earlier results of van Leeuwen et al. (1997c), but as for the Cepheids, formal errors have been significantly reduced.

5.4 Conclusions

The increased accuracies as obtained in the new *Hipparcos* reduction have benefited groups of stars and the associated calibrations of luminosities and empirical isochrones. A significant improvement has been obtained in the detail visible in these calibrations, also by virtual elimination of the abscissa error correlations which so much troubled the analysis of the open cluster data in the published catalogue.

Chapter 6

KINEMATICS OF THE SOLAR NEIGHBOURHOOD

The statistical properties of distributions in position and proper motion of the stars over the sky reflect the kinematics of the solar neighbourhood, the local distribution of mass, and the peculiar motion of the Sun with respect to the local standard of rest. This chapter presents a brief overview of how various aspects of the local kinematics are derived from the *Hipparcos* proper motion and parallax data.

6.1 Systematic motions

In this section we investigate systematic motions (apparent and real) as observed within the solar neighbourhood: the solar motion and galactic rotation.

6.1.1 The Local Standard of Rest and Solar Motion

The Local Standard of Rest (LSR) is defined as the velocity $V_0 = R_0\Omega_0$ in a circular orbit around the centre of the galaxy at the Galactocentric distance of the Sun, R_0. The solar motion describes the velocity vector of the Sun with respect to the LSR. The average motion of a group of stars with respect to the LSR is zero radially inwards and vertically upwards, and differs in the direction of galactic rotation by the asymmetric-drift velocity, which depends linearly on the velocity dispersion of the selected stars (Dehnen and Binney 1998).

The reflection of the solar motion in the proper motions can be directly derived from inverting Eq. 2.32 (page 46):

$$\begin{aligned} \kappa\mu_{l*,s}r &= -\sin(l)u_0 + \cos(l)v_0, \\ \kappa\mu_{b,s}r &= -\cos(l)\sin(b)u_0 - \sin(l)\sin(b)v_0 + \cos(b)w_0. \end{aligned} \tag{6.1}$$

Here we expressed the relations in galactic coordinates (l, b) and similarly the velocity vector as u_0 (radially inwards), v_0 (in the direction of galactic rotation) and w_0 (vertically upwards) as this was done by Dehnen and Binney (1998). Instead of the parallax, we use the distance r. With r expressed in kpc, and the proper motions in mas s^{-1}, the conversion factor $\kappa = 4.74047$. The dependence on the distance means that the reflection of the solar motion is strongest for nearby stars.

6.1.2 Galactic Rotation

Also reflected in the proper motions are the differential effects of galactic rotation. The exact relation for the projection of galactic rotation in observed radial velocities and proper motions for stars in the galactic plane can be derived from Fig. 6.1:

$$\begin{aligned} v_{\text{rad}} &= (\Omega - \Omega_0) R_0 \sin(l), \\ v_t = \kappa \mu_{l*,g} r &= (\Omega - \Omega_0) R_0 \cos(l) - \Omega r, \end{aligned} \tag{6.2}$$

where v_{rad} is part of the observed radial velocity. From Fig. 6.1 follows:

$$R = \sqrt{R_0^2 - 2rR_0\cos(l) + r^2}, \tag{6.3}$$

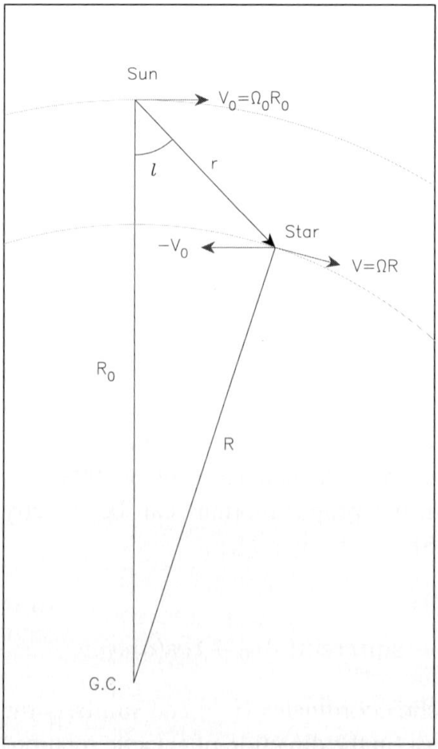

Figure 6.1. Identification of coordinates in the projection of galactic rotations

which for small values of r/R_0 is approximated by:

$$R \approx R_0\left[1 - \frac{r}{R_0}\cos(l) + \left(\frac{1}{2}\frac{r^2}{R_0^2}\sin^2(l)\right)\right]. \tag{6.4}$$

The local change of Ω with distance to the centre of the galaxy is represented by a simple gradient:

$$\Omega = \Omega_0 + \frac{d\Omega}{dR}(R - R_0) \approx \Omega_0 - \frac{d\Omega}{dR}r\left[\cos(l) - \frac{r}{2R_0}\sin^2(l)\right]. \tag{6.5}$$

With these substitutions, and restricting to a first-order approximation in r/R_0, Eq. 6.2 simplifies to (see also Feast and Whitelock 1997):

$$\begin{aligned} v_{\rm rad} &= A\sin(2l)r, \\ \kappa\mu_{l*,g} &= A\cos(2l) + B = 2A\cos^2(l) - \Omega_0, \end{aligned} \tag{6.6}$$

where A and B are Oort's constants (Oort 1927b,a; Binney and Tremaine 1987):

$$\begin{aligned} A &= -\frac{1}{2}R_0(d\Omega/dR)_0, \\ B &= -\Omega_0 - \frac{1}{2}R_0(d\Omega/dR)_0. \end{aligned} \tag{6.7}$$

The constant A represents the local shear and B the vorticity of the rotation in our galaxy. While the first is an actual physical effect, the latter is a combination of physical and projection effects: no real vorticity should normally exist. If the galaxy were to rotate as a solid body, the rotation speed Ω would be constant and $A = 0$. An effect would still be measured in the proper motions, but none in the radial velocities. Radial velocity observations were used by Oort (1927b) to show for the first time the differential rotation of our galaxy. A further evaluation involving also local expansion and non-tangential shear is described by Binney (1999).

A comparison between Eqs. 6.1 and 6.6 shows that while the effects of solar motion decrease with distance, the effects of galactic rotation remain constant. The more distant stars are therefore better tracers of galactic rotation, while the nearer stars will reflect more clearly the solar motion. However, when examining the proper motion data for more distant stars, it is important to realize the approximations made to derive the Oort constants. Already at distances of 2 kpc the first-order approximation of Eq. 6.6 is not very accurate, and the effect of second-order terms need to be considered. We re-write Eq. 6.5 as follows:

$$\Omega \approx \Omega_0 + 2A\frac{r}{R_0}\left[\cos(l) - \frac{r}{2R_0}\sin^2(l)\right], \tag{6.8}$$

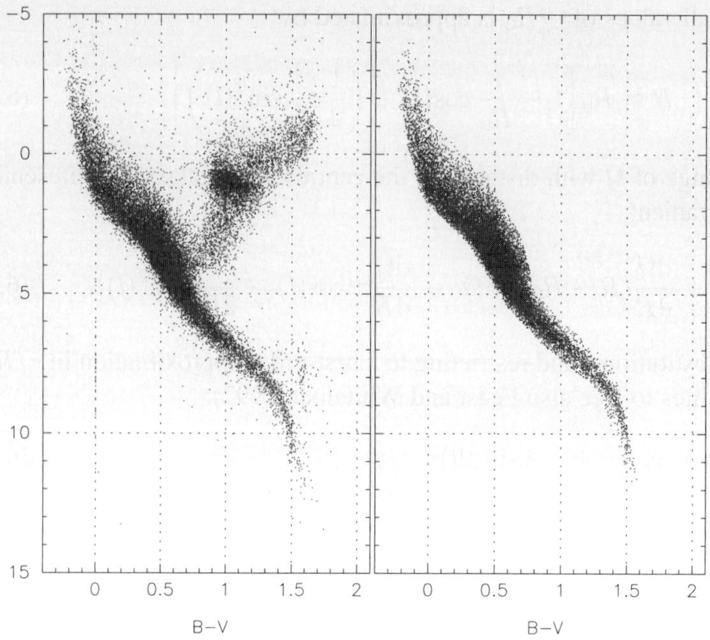

Figure 6.2. Left: the HR diagram for 25 995 single stars with parallax accuracies better than 10 per cent in the new catalogue. Right: the selection of 19 842 main sequence stars used in the solar-motion determinations

which can be substituted in Eq. 6.2 to give:

$$\begin{aligned} v_{\mathrm{rad}} &= 2A\sin(l)r\big[\cos(l) - \frac{r}{2R_0}\sin^2(l)\big], \\ \kappa\mu_{l*,g} &= 2A\cos(l)\big[\cos(l) - \frac{r}{2R_0}(\sin^2(l))\big] - \Omega_0. \end{aligned} \tag{6.9}$$

Uncertainties in R_0 introduce only a third-order correction, while third-order corrections in r/R_0 are still no more than 5 per cent at $r = 7$ kpc. To account for the galactic latitude of the stars, an additional coefficient $\cos(b)$ is introduced:

$$\begin{aligned} v_{\mathrm{rad}} &= 2A\sin(l)r\big[\cos(l) - \frac{r}{2R_0}\sin^2(l)\big], \\ \kappa\mu_{l*,g} &= 2A\cos(l)\big[\cos(l) - \frac{r}{2R_0}(\sin^2(l))\big] - \Omega_0. \end{aligned} \tag{6.10}$$

The second of these equations will be used later to derive A and Ω_0 from the proper motion data of the Cepheids.

Table 6.1. Solar motion determinations using single stars in different colour intervals on the main sequence and. The standard deviation σ is an indication of the velocity dispersion of the sample

BV_-	BV_+	obs	u_0	v_0	w_0	σ
-0.25	0.00	1243	12.36 ± 0.30	14.11 ± 0.35	7.43 ± 0.27	8.67
0.00	0.10	1087	9.25 ± 0.41	10.18 ± 0.45	6.86 ± 0.38	11.03
0.10	0.20	1074	9.66 ± 0.47	8.88 ± 0.51	6.28 ± 0.46	12.75
0.20	0.25	580	11.76 ± 0.67	8.72 ± 0.74	6.68 ± 0.67	13.53
0.25	0.30	655	11.76 ± 0.68	11.33 ± 0.74	6.40 ± 0.66	14.43
0.30	0.35	801	10.04 ± 0.65	9.09 ± 0.71	6.20 ± 0.67	15.57
0.35	0.40	962	12.31 ± 0.64	10.60 ± 0.67	6.68 ± 0.65	16.45
0.40	0.45	1572	9.72 ± 0.54	10.91 ± 0.58	6.59 ± 0.56	18.06
0.45	0.50	1872	9.14 ± 0.59	13.69 ± 0.61	7.13 ± 0.61	21.23
0.50	0.55	1897	10.67 ± 0.63	16.61 ± 0.67	7.53 ± 0.67	23.27
0.55	0.60	1688	8.20 ± 0.77	19.57 ± 0.81	7.20 ± 0.81	26.54
0.60	0.65	1523	11.29 ± 0.90	24.94 ± 0.93	8.16 ± 0.94	29.27
0.65	0.70	1187	9.79 ± 1.06	25.38 ± 1.07	8.58 ± 1.09	30.03
0.70	0.80	1325	12.24 ± 1.00	26.52 ± 1.03	7.06 ± 1.06	30.49
0.80	0.90	678	9.24 ± 1.46	27.93 ± 1.53	9.16 ± 1.53	31.77
0.90	1.05	656	15.30 ± 1.46	28.73 ± 1.50	7.73 ± 1.53	31.21
1.05	1.25	506	14.46 ± 1.60	25.13 ± 1.60	7.57 ± 1.70	29.87
1.25	1.60	536	8.69 ± 1.59	24.82 ± 1.60	7.86 ± 1.73	30.73

6.1.3 Application to Main Sequence Stars

To study the solar motions and the peculiar mean motions of groups of Main-Sequence stars, stars with relative errors on the parallax less than 10 per cent, and absolute error less than 2.5 mas, were extracted from the catalogue. A total of 25 995 single stars was found this way. The distribution of these stars in the HR diagram is shown in Fig. 6.2. From this figure, the main sequence stars were selected, as shown on the right-hand side. This selection also eliminated a small number of stars without photometric data. The total number of selected stars, 19842, is nearly twice that used by Dehnen and Binney (1998), who also applied some further selection criteria to achieve a more homogeneous sample. There is, however, an inevitable element of inhomogeneity in the *Hipparcos* sample: the combination of the limiting magnitude of the *Hipparcos* sample together with the relation between space density and absolute magnitudes for main-sequence stars implies that samples of blue stars have average distances much larger than samples of red stars.

The sample has been sub-divided into colour bins. These bins will cover a (small) range of masses, and a large range of ages. In particular the range of ages will vary considerably for $(B - V) < 0.6$, close to the turn-off point for the oldest stars in our galaxy. The age gradient towards the earlier spectral types is reflected in a gradual decrease in the velocity dispersions for the different samples, and the associated decrease in v_0. This can be seen very clearly in Table 6.1 and Fig. 6.3. The change to v_0 is referred to as the asymmetric drift and its relation to the velocity dispersion is shown clearly in Fig. 6.4. Ignoring the A and B stars, and using $S^2 \equiv 2\sigma^2$ as defined by Dehnen and Binney

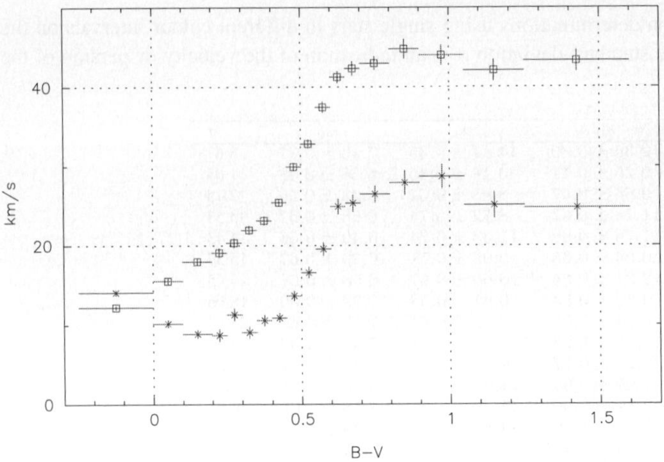

Figure 6.3. The relative motion of the Sun, in the direction of galactic rotation (v_0), with respect to main sequence stars in different bins of the colour index $(\mathrm{B}-\mathrm{V})$ (crosses), and the velocity dispersions for the same groups of stars (squares). The observed velocity dispersions were multiplied by $\sqrt{2}$ to be compatible with the quantity S as defined by Dehnen and Binney (1998)

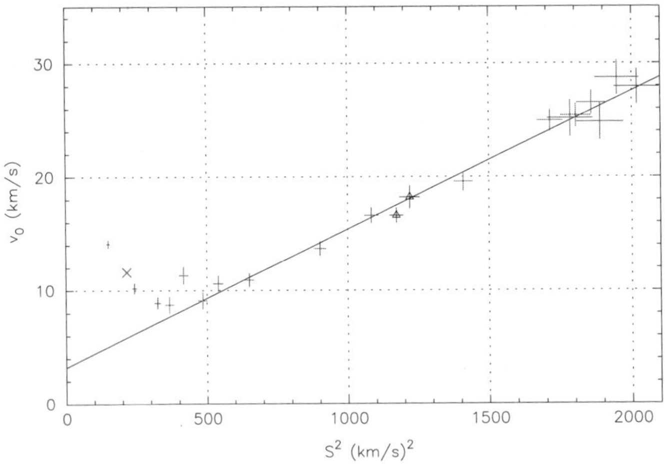

Figure 6.4. The asymmetric drift in the galactic rotation shown as the relation between the observed values of v_0 and $S^2 \equiv 2\sigma^2$ for main sequence stars and Red Clump and RGB stars (indicated by triangles). The value observed for the Cepheids is shown by $\times$

(1998), the observed relation is:

$$v_0 = (3.21 \pm 0.52) + (0.01217 \pm 0.00051)S^2. \tag{6.11}$$

The data for the A and B stars are much affected by large-scale systematic motions, such as imposed by the Sco-Cen OB association.

The weighted-mean values for u_0 and w_0 show that there are also some systematic variations for the observed values of u_0, while w_0 appears to be unaffected. This is shown through the unit-weight standard deviations of the values presented in Table 6.1: 2.44 and 0.95 respectively. This may possibly be due to the sample selection. The mean values and formal errors on the mean are found to be:

$$\begin{aligned} u_0 &= 10.77 \pm 0.36 \\ w_0 &= 7.04 \pm 0.14. \end{aligned} \tag{6.12}$$

The equivalent values presented by Dehnen and Binney (1998) are $u_0 = 10.00 \pm 0.36$, $v_0 = 5.25 \pm 0.62$ and $w_0 = 7.17 \pm 0.38$. A comparison with Eq. 6.11 and 6.12 shows again the somewhat larger variation associated with determining u_0 and v_0. We will come back to this in Section 6.2.

6.1.4 Post-main-sequence Stars

Table 6.2. Solar motion determinations using single stars in the Red Clump and RGB. The standard deviation σ is an indication of the velocity dispersion of the sample

Group	perc.	obs	u_0	v_0	w_0	σ
RC	10	2458	8.43 ± 0.58	16.63 ± 0.64	7.55 ± 0.60	24.22
RC	7	1391	8.62 ± 0.79	16.04 ± 0.86	7.24 ± 0.81	24.85
RGB	10	1054	10.24 ± 0.90	18.17 ± 0.99	7.47 ± 0.92	24.70
RGB	7	615	11.70 ± 1.15	17.10 ± 1.24	6.47 ± 1.17	23.87

Selections of stars in the Red Clump and the RGB have also been investigated for the way they reflect the solar motion. Values were calculated for different relative parallax error limits: 7 and 10 per cent, as shown in Table 6.2. The insignificant differences between the two solutions effectively shows that the sample of stars is sufficiently homogeneous. The values observed for v_0 and σ are in very good agreement with those observed for main sequence stars (Fig. 6.4), and correspond best with main sequence stars of $(\mathrm{B}-\mathrm{V}) = 0.55$. The value for w_0 is again well determined and essentially the same as observed for the main-sequence stars, while u_0 shows the somewhat larger deviations. The internal velocity dispersion, combined with the distribution in distances, for the selected stars is still much too high to use them for a meaningful determination of the Oort's constants. For that determination more distant stars are required.

6.1.5 Application to the Cepheids

To examine both the galactic rotation and solar motion as reflected in the proper motions we use, as was done before by Feast and Whitelock (1997),

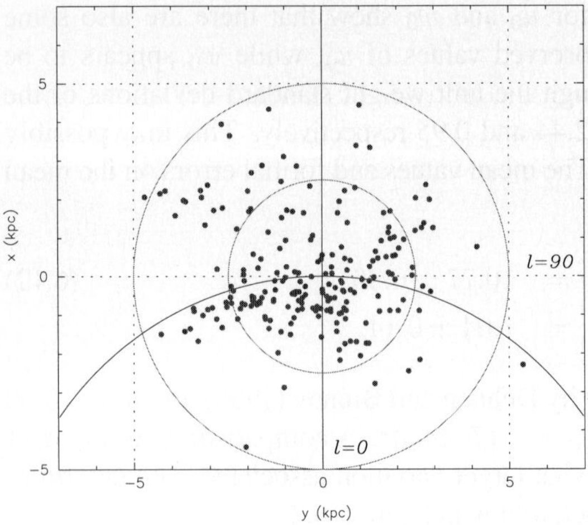

Figure 6.5. A map showing the galactic distribution of 213 Cepheids used in the determination of the Oort constants. The black curved line shows the track of a circular orbit for the galactocentric distance of the Sun, if $R_0 = 8.5$ kpc. The two concentric grey circles show projected distances of 2.5 and 5 kpc from the Sun

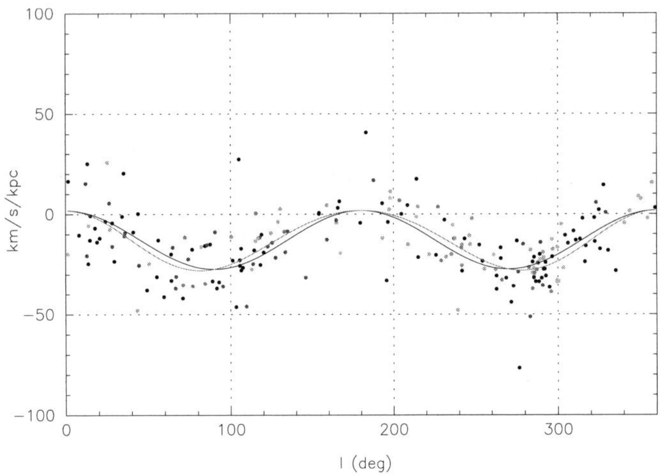

Figure 6.6. The quantity $\kappa\mu_l$, corrected for solar motion, as a function of galactic longitude, showing the effect of galactic rotation. The solid curve shows the fitting of the data for stars within 1 kpc from the Sun, the grey curve shows the same for stars with a projected distance of 5 kpc. Only stars with projected distances beyond 1.2 kpc are shown and used in the solution for the curves shown here

the Cepheid stars and their absolute magnitude calibration as described in Section 5.3.5:

$$M_V = -2.81(\log P - 0.78) - 3.58, \tag{6.13}$$

which, together with the observed mean V magnitudes and reddenings as determined by Tammann et al. (2003) provides distance moduli and distance estimates. A total of 213 Cepheids have been selected, with distances up to 7 kpc (Fig. 6.5). Figure 6.6 shows the variation of the proper motion in longitude for Cepheids, reflecting the galactic rotation.

Two solutions are made: the first using all stars, and solving for solar motion and galactic rotation, and the second using 152 stars with projected distances beyond 1.2 kpc, and solving only for galactic rotation. In the first solution defining the formal "errors" on the measurements is crucially important: these errors reflect both the measurement errors and the intrinsic velocity dispersion of the sample. In fact, the solution allows a determination of this intrinsic dispersion through an examination of the standard deviation as a function of the assumed intrinsic velocity dispersion. The velocity dispersion thus found is 10.4 km s^{-1}.

Both solutions used Eq. 6.10 for representing the galactic rotation, taking into account the second-order projection effects. The value of R_0 was taken from Pont et al. (1994) as 8.09 kpc. The results when using $R_0 = 8.5$ kpc are only marginally different. The first solution, using 213 stars and proper motions in galactic longitude and latitude, provided:

$$\begin{aligned} u_0 &= 7.9 \pm 1.3 \quad , \quad A = 15.27 \pm 0.85 \\ v_0 &= 11.4 \pm 1.5 \quad , \quad \Omega = 27.59 \pm 0.80 \\ w_0 &= 6.5 \pm 1.0 \end{aligned} \tag{6.14}$$

while the 152 stars in the second solution gave the following result:

$$A = 15.00 \pm 1.06 \quad , \quad \Omega = 28.02 \pm 0.85. \tag{6.15}$$

The two solutions are clearly in very good agreement. This measurement is shown by a cross in Fig. 6.4. The radial-velocity-only based study of Cepheids by Pont et al. (1994) gave the following values:

$$\begin{aligned} u_0 &= 9.3 \pm 0.8 \quad , \quad A = 15.92 \pm 0.34, \\ v_0 &= 11.2 \pm 0.7, \end{aligned} \tag{6.16}$$

values that are fully consistent with the astrometric study presented here. There is, naturally, also good agreement with the values obtained by Feast and Whitelock (1997), i.e. there are no systematic difference between the new and old reduction results.

The values of v_0 and σ found for the Cepheids produce an other data point in Fig. 6.4, where it is positioned between the B and early A type stars, indicating that the position in this graph of the B stars is a systematic rather than an accidental deviation from the otherwise linear relationship between v_0 and S^2. The discrepancy is about 10 km s^{-1}, very similar to what was observed for example by Humphreys (1970) for Supergiants based on radial velocity measurements, and which she attributed to spiral arms in the galaxy being the result of density waves (Lin et al. 1969; Yuan 1969; Roberts 1972; Bertin and Mark 1978).

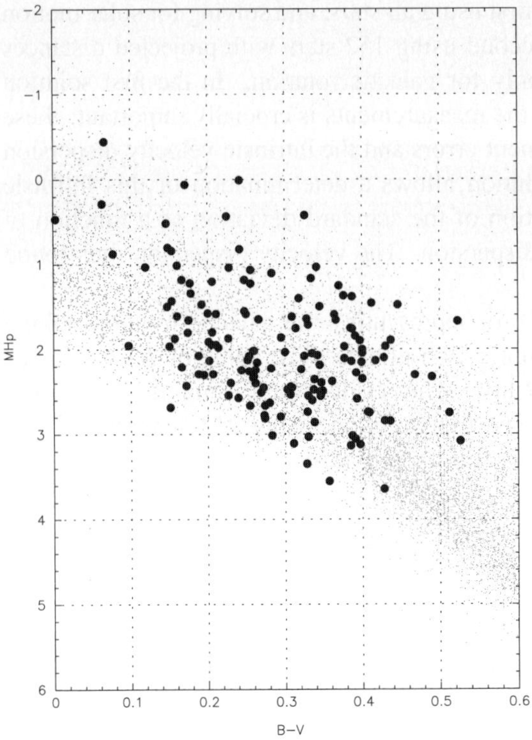

Figure 6.7. The HR diagram for A and F stars, emphasizing those stars classified as luminosity types earlier than IV. It appears that late A and early F type main sequence stars can be problematic in determining luminosity types

6.2 The distribution of nearby stars

The distribution of the nearby stars, and in particular those brighter than magnitude $V = 8$ and of spectral type A or F, can provide insight in the local distribution of stars, and especially on their distribution perpendicular to the galactic plane. Combined with their velocity distribution, an estimate can be obtained of the local mass-density of the galaxy and "K_z", the force law perpendicular to the galactic plane. This was shown for the published catalogue by Creze et al. (1998) (CCBP from hereon). Here we show, by following the methods presented by these authors, how this derivation is done, and what the results are when applied to the new reduction.

6.2.1 The Sample

The tracer sample consists of all apparent main-sequence stars with *Hipparcos* magnitude brighter than 8 and the colour index B − V between −0.1 and 0.6. As the catalogue is virtually complete for these criteria, the tracer sample can be used for the selection of subsets for mapping the local systematic density variations.

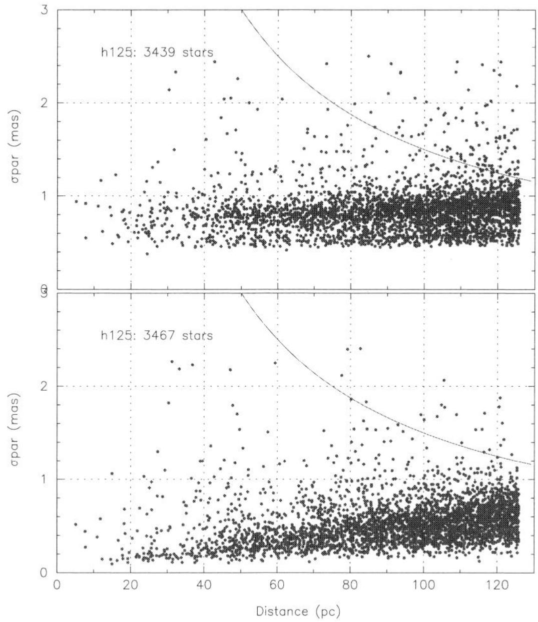

Figure 6.8. Parallax accuracies as a function of distance from the Sun for selection h125 as based on the published catalogue (top) and the new reduction (bottom). The curved line shows the limit above which relative parallax errors are worse than 15 per cent

In their selection, CCBP excluded stars based on their luminosity class, eliminating classes earlier than IV. In this particular area of the HR diagram, however, luminosity class determination appears problematic. Many stars have not been assigned luminosity classes, and among the stars that appear to make up the main sequence, there is a relatively low number of stars classified as luminosity class IV or V, while at the same time there is a scattering of about 150 stars with luminosity classes from Ib to III. Those stars are mainly concentrated in the interval $0.15 < \mathrm{B} - \mathrm{V} < 0.45$ (Fig. 6.7).

A total of 7508 stars were selected, compared to 7099 stars by CCBP (their Fig. 1), though there are some uncertainties about the selection by CCBP, as the numbers quoted in their Fig. 2 add up to 7666 stars, and the densities quoted in their Table 1 also indicate higher numbers of stars used.

A nearly complete and homogeneous sample is extracted from the tracer sample by selecting all stars with absolute magnitude less than 2.5 and distance less than 125.89 pc (distance modulus less than 5.5, ensuring that the sample remains within the completeness limit of $\mathrm{Hp} = 8$ for the catalogue). Using the new reduction, this sample contains 3464 stars, compared to the 2977 stars contained in the sample used by CCBP, although this figure too is uncertain due to internal inconsistencies in the data presented in their Tables and Figures. The main contribution of the new reduction for this sample is the improvement in the parallax accuracies, which above all affect the distance estimates of stars close to the distance limit of the selection. This sample is, as in CCBP, referred to as h125. Figure 6.8 compares the formal errors for the sample as based on the published data and the new reduction.

6.2.2 Measuring the Stellar Densities

The stellar densities are evaluated through measuring for each star the "empty" volume around it, the radius of which is given by the distance to the nearest neighbour within the sample. The average volume per star in the sample is given by the total volume of $8.36 \cdot 10^6$ cubic parsec and the total number of stars:

$$v_0 = 2410\text{pc}^3, \tag{6.17}$$

equivalent to $41.5 \cdot 10^{-5}$ stars per cubic parsec.

We assume that the number of stars counted in any sub volume within the sample is a Poisson variate with its expectation value equal to the mean density times the volume. This also applies to the volume around a single star. When we normalize the density with the average density, and introduce the variable $x = v/v_0$, then the expectation value for the empty volume around a star equals one, and its PDF is given by:

$$\mathrm{d}P(x) = \mathrm{e}^{-x}\mathrm{d}x, \tag{6.18}$$

or, expressed in the volume v:

$$\mathrm{d}P(v) = \frac{\mathrm{e}^{-v/v_0}}{v_0}\mathrm{d}v. \tag{6.19}$$

As was shown by CCBP, in the presence of a constant density, the average over the single-star volumes equals the average density:

$$v_0 = \langle v \rangle, \tag{6.20}$$

which makes an average over single star volumes a suitable estimate for local star densities.

6.2.3 The Observed Density Distributions

Nearest distances were determined for all members of sample h125, incorporating the corrections for edge effect as described by CCBP. The observed distribution is best displayed as function of $y = \ln(x)$, in which case the PDF is given by:

$$\mathrm{d}P(y) = \mathrm{e}^{y-\mathrm{e}^y}\mathrm{d}y. \tag{6.21}$$

Figure 6.9 shows the good agreement between the sample and the expected probability distribution. The scaling factor used for the latter is 1156, compared to 1020 as given for the smaller sample by CCBP.

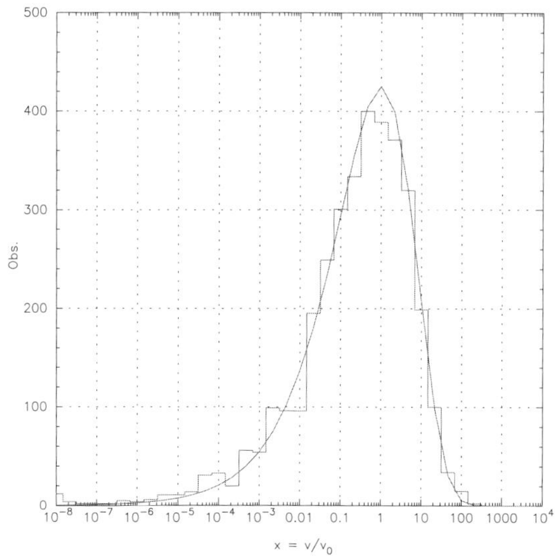

Figure 6.9. The distribution of single-star volumes in sample h125 and the scaled curve of the expected probability distribution

Table 6.3. Data for the fitted parabolic curves as defined by Eq. 6.22

Axis	Obs.	x_0	x_1	x_2	stdv
X	3416	1.009 ± 0.027	0.013 ± 0.036	-0.045 ± 0.063	0.16
Y	3416	1.010 ± 0.028	0.014 ± 0.037	-0.047 ± 0.064	0.17
Z	3416	0.845 ± 0.027	-0.080 ± 0.040	0.571 ± 0.072	0.16

The density profile was probed along three orthogonal axes: X along the direction of the galactic centre, Y along the direction of galactic rotation, and Z perpendicular to the galactic plane. Experiments with averaging over different numbers of stars were made, and a running mean over 51 stars was finally selected as a compromise between reach and noise level (a smaller number will get closer to the edges of the sample but will also be noisier). Figure 6.10 shows the resulting profiles. The data for the fitted parabolic curves (with h expressed in units of 0.1 kpc):

$$x = x_0 + x_1 h + x_2 h^2, \tag{6.22}$$

are presented in Table 6.3. For the Z coordinate, the quantity h represents the height above (or below) the galactic plane. The errors on the parameters as given in Table 6.3 take into account the oversampling inherent to a running mean as used in these determinations. The data for the Z axis provide an estimate of the height of the Sun above the galactic plane: $h_0 = 7.0 \pm 3.6$ pc. This compares with a recent determination of 15.5 ± 3 pc by Hammersley et al. (1995), based on large-scale surveys.

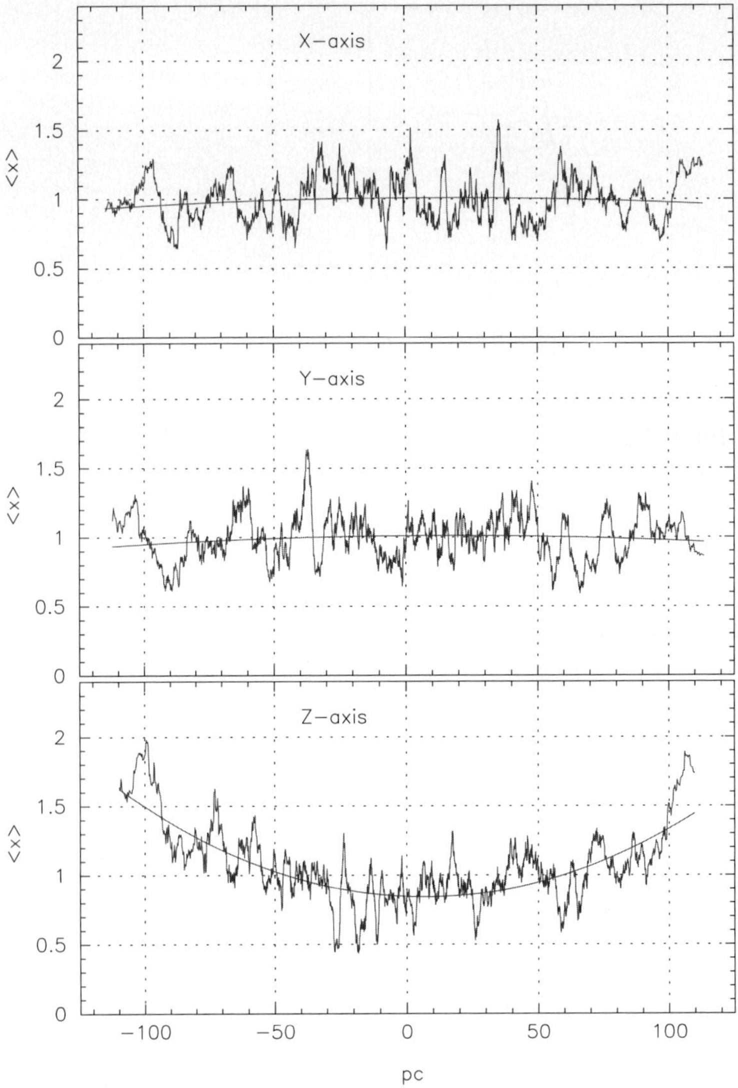

Figure 6.10. Running averages of the normalized single-star volumes along the three axes as indicated and as defined in the text

The curvature in the density profile perpendicular to the galactic plane is clearly demonstrated by the data for the Z-axis as shown in Fig. 6.10 and Table 6.3. However, as the total volume examined is spherical, some of this curvature is expected to reflect, but with opposite sign, in the observations for the X and Y axis. The question to ask is: how does a density profile with a quadratic dependence on the coordinate perpendicular to the galactic plane only, reflect in mean density observations along the three axes defined above, when the volume examined is a sphere around the Sun? Instead of densities we use here the inverse densities, or average "empty" volume around a star, normalized as described above, $\langle x \rangle \equiv \langle v \rangle / v_0$. The inverse-density profile is then described by

$$x(h) = x_0 \left(1 + x_2(h - h_0)^2\right), \tag{6.23}$$

where h is again the coordinate perpendicular to the galactic plane.

The mean values of x for each of the three axes are obtained by averaging data within a thin approximately cylindrical slice through the sphere. With the slice at distance t from the Sun, the radius of this cylinder equals $R' = \sqrt{R^2 - t^2}$, where $R = 125.89$ is the outer radius of the sample sphere. Applied to the X and Y axes, the quantity x as measured in the cylinder varies as described by Eq. 6.23, while applied to the Z axis it is constant at $x_0(1 + x_2(t - h_0)^2)$. For the X and Y axes the average of x over the cylinder is described by:

$$\langle x \rangle (t) = \frac{\int_0^{2\pi} \int_0^{R'} x(h) r \mathrm{d}r \mathrm{d}\theta}{\int_0^{2\pi} \int_0^{R'} r \mathrm{d}r \mathrm{d}\theta}, \tag{6.24}$$

where $h = r \sin\theta$. Substituting Eq. 6.23 and evaluating the integrals gives:

$$\langle x \rangle (t) = x_0 (1 + h_0^2 x_2 + \frac{1}{4} x_2 (R^2 - t^2)). \tag{6.25}$$

This relation applies to the data for the X and Y axes. The equivalent relation for the Z axis is:

$$\langle x \rangle (t) = x_0 (1 + x_2 (t - h_0)^2). \tag{6.26}$$

Thus, we expect to see one quarter of the curvature on the Z axis reflected with opposite sign on the X and Y axis. Although the observed ratio is more like 12 than 4, the formal errors on the curvatures as measured on X and Y are such that a ratio of 4 is only 1.5σ from the observed values.

We can re-organize the above equations to allow a simultaneous solution on the three axes for the parameters of Eq. 6.23:

$$\begin{aligned} \langle x \rangle (t)_X &= x_0' + x_2'(R^2 - t^2)/4 \\ \langle x \rangle (t)_Y &= x_0' + x_2'(R^2 - t^2)/4 \\ \langle x \rangle (t)_Z &= x_0' + x_2' t^2 + h_0' t, \end{aligned} \tag{6.27}$$

where we solve for three variables: $x_0' \equiv x_0 + h_0^2 x_2$, x_2' and $h_0' \equiv -2 x_2 h_0$. The values found in the combined solution are (with t and R measured in units of 0.1 kpc):

$$\begin{aligned} x_0' &= 0.844 \pm 0.023 \\ h_0' &= -0.077 \pm 0.040 \\ x_2' &= 0.503 \pm 0.066, \end{aligned} \tag{6.28}$$

or, expressed in the parameters of Eq. 6.23:

$$\begin{aligned} x_0 &= 0.841 \pm 0.026, \\ h_0 &= 7.6 \pm 4.0 \text{ pc}, \\ x_2 &= (0.598 \pm 0.081) \cdot 10^{-4} \text{ pc}^{-2}, \end{aligned} \tag{6.29}$$

very similar to the results obtained when solving for individual axes, showing that the data are internally consistent with the simple model and within the margins of the measurement errors.

We can do a similar solution for the density estimates as derived from the single-star volumes:

$$\langle \nu \rangle(t) = \nu_0 \left(1 + \nu_2 (t - h_0)^2\right), \tag{6.30}$$

where the values found are:

$$\begin{aligned} \nu_0 &= 1.198 \pm 0.040 \\ h_0 &= 5.2 \pm 4.7 \text{ pc} \\ \nu_2 &= (-0.416 \pm 0.062) \cdot 10^{-4} \text{ pc}^{-2}. \end{aligned} \tag{6.31}$$

The value for ν_2 will be used in deriving the local dynamic density, but to make that possible we first need to derive the velocity distribution.

6.2.4 The Velocity Distribution

To obtain an estimate of the mass density in the solar neighbourhood we need to combine the data on the density profile with the velocity distribution for the same sample of stars. The proper motions provide some information, but for extracting the velocity dispersion perpendicular to the galactic plane, the sample would need to be restricted to low galactic latitude. Combining the proper motion data with radial velocity measurements provides a more complete picture. Of the sample of 3467 stars, 2298 stars have radial velocities with formal errors below about 5 km s^{-1}. Given the quality of the ground-based measurements, most of the stars above this limit are likely components in binary or multiple systems. Above this limit, in addition, the errors on the radial velocities become significant with respect to the internal velocity dispersions of the sample. The current situation is a significant improvement with respect to CCBP, primarily due to the study by Nordstroem et al. (2004) on radial velocities of F and G stars. Other data was extracted from the SIMBAD data base.

Figure 6.11 shows the distributions of the velocities in two directions: along the direction of the galactic centre and perpendicular to the galactic plane. The distribution perpendicular to the galactic plane is well represented by the Gaussian curve with $\sigma = 8.90 \pm 0.13$ km s^{-1}, also shown in the graph.

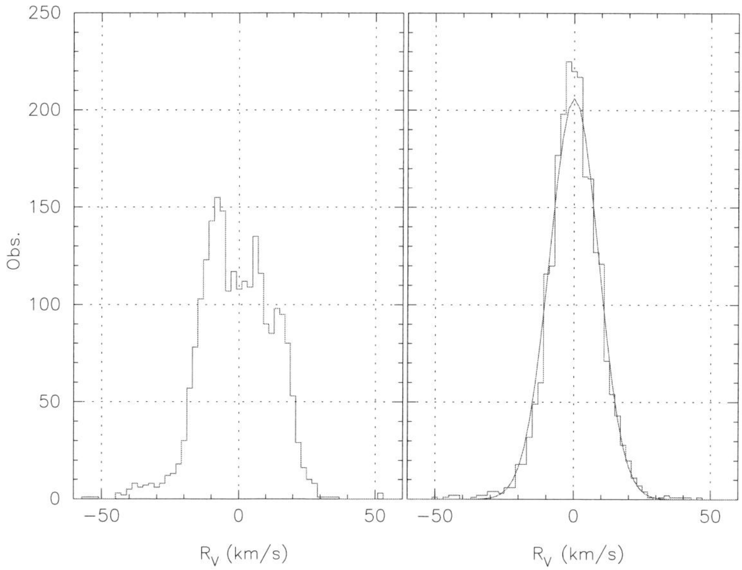

Figure 6.11. Velocity histograms for A & F stars along the direction of the galactic centre (left) and perpendicular to the galactic plane (right). The curve in the right-hand graph represents a single Gaussian distribution with $\sigma = 8.9$ km s^{-1}

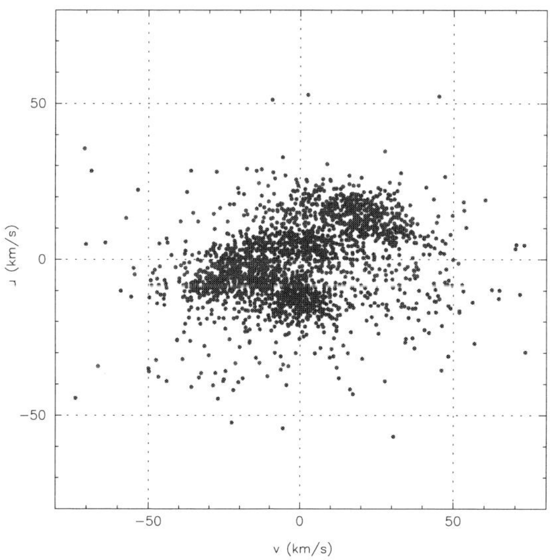

Figure 6.12. The vector-point diagram of the velocities of A & F stars as projected on the galactic plane, showing several clear concentrations

Before going on to deriving the characteristics of the potential well perpendicular to the galactic plane, it is worth noting the peculiarities of the velocity distribution along the direction to the galactic centre, which suggest at least three to four kinematically different groups. This may well be the reason for the relatively large variations in the equivalent component u_0 of the solar motion as noted in Section 6.1.3. This is shown even more clearly when we draw a diagram of the velocities of these stars as projected on the galactic plane (Fig. 6.12). An analysis of the substructure in the three-dimensional velocity diagram has been presented by Chereul et al. (1998). A diagram with similar features was produced by Myllari et al. (2001), based on the published catalogue and stars within 75 pc, and similar features were shown to exist for K and M giants by Famaey et al. (2005).

If the features in Fig. 6.12 have a long-term physical significance, and relate to groups of stars originating for example from an OB association or open cluster, then we expect to see strongly correlated features for stars in each group when displayed in the HR diagram. This is found to apply to only one feature, at $v = -30$ and $u = -7$ km s^{-1} (relative to the LSR): the Hyades cluster. All other features show fully random distributions in the HR diagram, with the only difference being the higher or lower abundance of the earlier type stars. These findings confirm the results of a study by De Simone et al. (2004), who showed that features as observed here form, and disappear again, naturally through interactions with spiral waves during the stellar orbits.

6.2.5 The Local Dynamical Density

For a stellar population at equilibrium in its potential well, the relation between the density distribution $\nu(h)$, the potential $\phi(h)$ and the velocity distribution $f(w)$ (all perpendicular to the galactic plane) was derived by Fuchs and Wielen (1993) and Flynn and Fuchs (1994) as:

$$\nu(\phi) = 2\int_{\sqrt{2\phi}}^{\infty} \frac{f(|w|)w}{\sqrt{w^2 - 2\phi}} \mathrm{d}w. \tag{6.32}$$

The velocity distribution was described in the preceding section (relative to the mean velocity) by means of a single Gaussian profile:

$$f(w) = \frac{1}{\sigma_w\sqrt{2\pi}} \mathrm{e}^{-w^2/2\sigma_w^2}, \tag{6.33}$$

with $\sigma_w = 8.90 \pm 0.13$ km s^{-1}. Inserting this in Eq. 6.32, we obtain a simpler expression for the density profile ν:

$$\nu(h) = \nu_0 \cdot \mathrm{e}^{-\phi(h)/\sigma_w^2}. \tag{6.34}$$

We now express the potential ϕ as observed near the galactic plane by means of a simple quadratic dependence on h:

$$\phi(h) \approx \alpha h^2. \tag{6.35}$$

The local density ρ_0 for a sample in equilibrium is given by the Poisson equation:

$$\rho_0 = \frac{1}{4\pi G}\frac{\mathrm{d}^2\phi}{\mathrm{d}h^2} \approx \frac{\alpha}{2\pi G}. \tag{6.36}$$

With ρ_0 measured in $(\mathrm{M}_\odot\ \mathrm{pc}^{-3})$, and α in $(\mathrm{km\ s}^{-1}\ \mathrm{pc}^{-1})^2$, the constant $1/2\pi$ $G = 37.004$ (see Creze et al. 1998).

Combining these equations with Eq. 6.30 gives to first-order approximation:

$$\alpha \approx -\nu_2\sigma_w^2 = 0.0033 \pm 0.0005, \tag{6.37}$$

from which is derived $\rho_0 = 0.122 \pm 0.019\ \mathrm{M}_\odot\ \mathrm{pc}^{-3}$. The difference with the value presented by CCBP (0.076 ± 0.015) originates in the density profiles, for which their relatively high densities away from the galactic plane could not be reproduced by the present experiment, neither with the published nor with new catalogue.

PART III

A DESCRIPTION OF THE CONTENTS AND PECULIARITIES OF THE HIPPARCOS PHOTOMETRIC DATA

Although not specifically designed to be a photometric survey, the Hipparcos data do provide a homogeneous all-sky photometric reference system. The scanning law, however, caused certain peculiarities in the distribution of data, which affected the periodicity analysis of variable stars. Methods used to cope with the resulting often poor window functions, are reviewed. The passband reconstruction and the way changes in the passband affected the astrometric and photometric data are also described.

Chapter 7

THE PHOTOMETRIC DATA

7.1 The Hipparcos photometric pass bands

The *Hipparcos* mission produced photometric data derived from transits over the main grid as well as for transits over the star mapper slits. For the main grid a wide pass band "Hp" was used, effectively determined by the transmission characteristics of the optics (the telescope mirrors and the lenses in the Image Dissector Tube assembly, IDT) and the S20 photocathode responses (Varma and Ghosh 1973). To be optimized for astrometric measurements, it was required to collect as much light as possible in the Hp band. For the star mapper slits the light was split into a "B$_T$" and a "V$_T$" pass band by means of a dichroic prism. The subscript T refers to the use of those transits to create the *Tycho* catalogue, a largely complete astrometric and photometric survey of about 2 million stars down to magnitude 11 (Hoeg et al. 2000b,a), which is based on the complete star mapper photon count records. All three pass bands evolved, and in particular the responses for the main grid photometry (Hp) deteriorated significantly over the mission (see below).

The purpose of the photometry for the main field was to act in the first instance as a detector of duplicity (see Chapter 4) through the relation between the mean signal and the amplitude of the first harmonic in the modulation, referred to as M_1 (Eq.2.33 and Section 2.2.4, page 54). The Hp photometry also served very well as a detector of variability (van Leeuwen et al. 1997a,b) and as a global uniform reference system for photometric calibration (ESA 1997). The photometry obtained from the star mapper transits provided colour information which was used in a range of instrument parameter calibrations. Unfortunately, it was limited to stars brighter than 10th magnitude only. For fainter stars, measured or derived ground-based data was used in those calibrations.

The aim of this chapter is to provide the necessary background information for the use of the *Hipparcos* photometry, which has been included in ESA (1997) as both mean values plus dispersions, and as calibrated measurements per field of view transit, the so-called epoch photometry. The new reduction does not include a new analysis of the photometric data, as there are no reasons to suspect the possibility of significant improvements: the main noise contribution on the reduced data appears to be the photon noise of the underlying measurements. This reflects in the high correlation between the FAST and NDAC reduction results. A comparison between the epoch-photometry residuals for constant stars in the FAST and NDAC reductions shows that these accidental errors were dominated by the identical photon noise in the data, and therefore highly correlated (Fig. 7.1).

In the following sections, the pass bands and their variation over the mission are described. This is followed by a summary of the photometric calibrations (a detailed description of the photometric calibration parameters is presented in Chapter 12) and a discussion of formal errors and variability detection. Finally, the distribution of epochs and how this affected the choice of variability analysis methods and tools concludes this chapter.

7.1.1 Response Evolution

The lenses and prisms used in the IDT and the Star Mapper channels were subjected to radiation, as a result of which their transmission decreased over the mission, most drastically for the IDT (Fig. 7.2). This change in transmission was wavelength dependent, leading to time-variable pass bands for all *Hipparcos* photometry. Most of these variations could be calibrated out by means of first and higher-order colour corrections. For these corrections a $V_J - I_C$ index was used which in most cases was reconstructed from information on a measured $B - V$ and luminosity class, but such information was not always reliable. In extreme cases the implementation of a faulty colour index could be

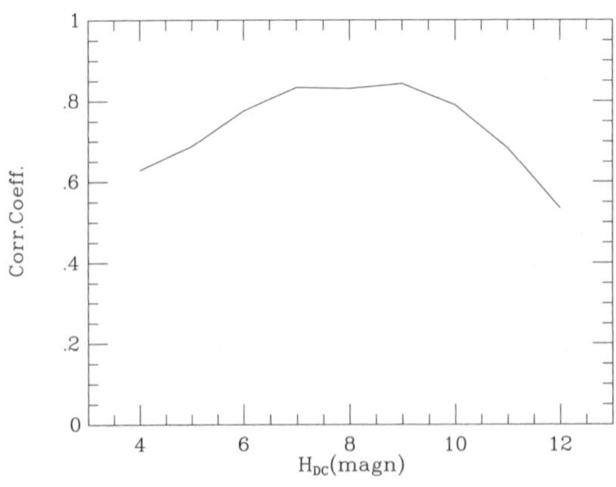

Figure 7.1. The correlation coefficient, as a function of magnitude, between the residuals in the FAST and NDAC epoch photometry. Only constant stars were used

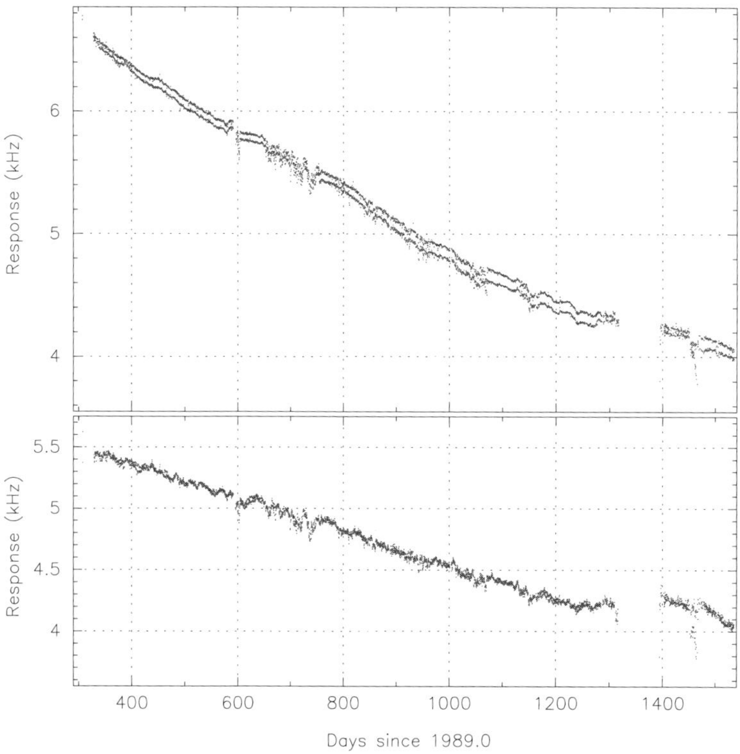

Figure 7.2. The change in response in kHz for an 8th magnitude star over the mission. Top graph: $(V - I)_c = 0$, bottom graph: $(V - I)_c = 2$. Data for both fields of view are shown, which in the top graph are separated, and in the bottom graph largely overlap

recognized from a residual, nearly linear, drift in the brightness of a star over the mission. Though the printed version of the published catalogue provides the best estimate available at that time for $V_J - I_C$, this value is not always the one used for the data processing. That value is available in the machine-readable version of the catalogue only.

All photometric data were ultimately reduced to a set of photometric standard stars (selected on the basis of their stability during the mission). The responses of these standards were adjusted to a reference pass band corresponding closely (but not identical) to the actual pass band in the preceding FOV for orbit 1780, on 1st of January 1992. The reference pass bands are shown in Fig. 7.3 (B_T and V_T), Fig. 7.4 (Hp) and Table 7.1. A different evaluation of the *Hipparcos* pass bands has been provided by (Bessell 2000), with the main difference being a lower response towards the blue edge. Section 12.3 describes in more detail the various parameters in the photometric calibrations that effectively represented corrections for the variations in the pass band. What is

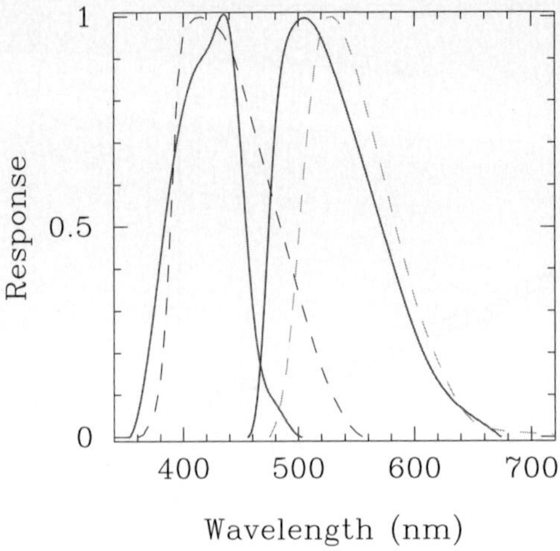

Figure 7.3. The references pass bands for the *Tycho* B_T and V_T photometry, shown by the solid lines. The dashed lines show as reference the B_J and V_J pass bands

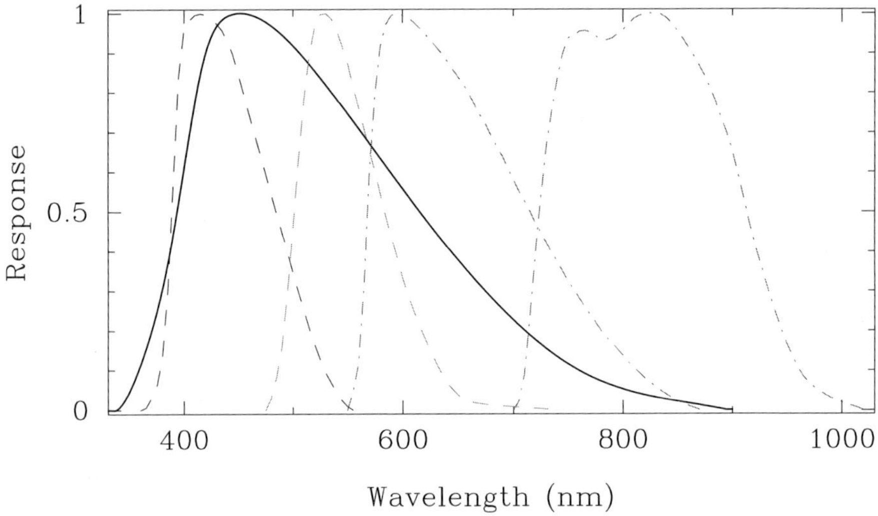

Figure 7.4. The reference pass band for the Hp photometry, shown by the solid line. The dashed lines show as reference the B_J and V_J pass bands, the dash-dot lines the I_C and R_C pass bands

of importance here is how the IDT photometric data can be corrected in case a faulty colour index has been used in the analysis.

7.1.2 A Posteriori Corrections for Colour-index Adjustments

After the photometric reductions by FAST and NDAC, the two calibrations were further examined by Michel Grenon, to ensure consistent application also

Table 7.1. The responses for the reference pass bands. Wavelength λ in nm

λ	B_T	V_T	Hp	λ	B_T	V_T	Hp	λ	V_T	Hp	λ	Hp
330	0.000	0.000	0.000	475	0.101	0.530	0.979	620	0.135	0.483	765	0.092
335	0.000	0.000	0.000	480	0.080	0.737	0.969	625	0.114	0.465	770	0.085
340	0.000	0.000	0.006	485	0.059	0.870	0.958	630	0.097	0.447	775	0.079
345	0.000	0.000	0.023	490	0.036	0.940	0.946	635	0.082	0.429	780	0.073
350	0.000	0.000	0.047	495	0.016	0.973	0.933	640	0.069	0.412	785	0.067
355	0.014	0.000	0.078	500	0.003	0.990	0.919	645	0.058	0.395	790	0.062
360	0.058	0.000	0.114	505	0.000	0.996	0.903	650	0.047	0.378	795	0.057
365	0.123	0.000	0.154	510	0.000	0.991	0.888	655	0.038	0.361	800	0.053
370	0.206	0.000	0.198	515	0.000	0.975	0.871	660	0.028	0.345	805	0.049
375	0.305	0.000	0.248	520	0.000	0.949	0.855	665	0.018	0.329	810	0.045
380	0.416	0.000	0.305	525	0.000	0.916	0.838	670	0.008	0.314	815	0.041
385	0.530	0.000	0.369	530	0.000	0.878	0.820	675	0.000	0.298	820	0.038
390	0.636	0.000	0.442	535	0.000	0.837	0.803	680	0.000	0.283	825	0.035
395	0.724	0.000	0.523	540	0.000	0.794	0.785	685	0.000	0.269	830	0.032
400	0.787	0.000	0.608	545	0.000	0.749	0.766	690	0.000	0.254	835	0.029
405	0.830	0.000	0.694	550	0.000	0.704	0.748	695	0.000	0.241	840	0.026
410	0.861	0.000	0.774	555	0.000	0.658	0.729	700	0.000	0.227	845	0.024
415	0.889	0.000	0.845	560	0.000	0.612	0.710	705	0.000	0.214	850	0.022
420	0.920	0.000	0.901	565	0.000	0.565	0.691	710	0.000	0.201	855	0.019
425	0.953	0.000	0.941	570	0.000	0.518	0.672	715	0.000	0.189	860	0.017
430	0.982	0.000	0.967	575	0.000	0.471	0.653	720	0.000	0.177	865	0.015
435	1.002	0.000	0.984	580	0.000	0.424	0.634	725	0.000	0.166	870	0.012
440	0.976	0.000	0.993	585	0.000	0.379	0.615	730	0.000	0.155	875	0.010
445	0.861	0.000	0.998	590	0.000	0.335	0.596	735	0.000	0.144	880	0.007
450	0.685	0.000	1.000	595	0.000	0.293	0.577	740	0.000	0.134	885	0.005
455	0.489	0.000	1.000	600	0.000	0.254	0.558	745	0.000	0.125	890	0.002
460	0.317	0.022	0.998	605	0.000	0.218	0.539	750	0.000	0.116	895	0.000
465	0.202	0.115	0.993	610	0.000	0.186	0.520	755	0.000	0.108	900	0.000
470	0.136	0.301	0.987	615	0.000	0.159	0.502	760	0.000	0.100		

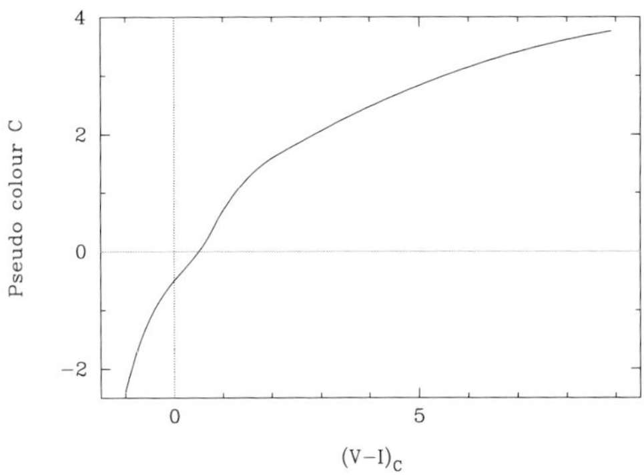

Figure 7.5. The pseudo-colour index C as a function of $(V-I)_C$

Table 7.2. Definition of the pseudo-colour index $C = a + b(\mathrm{V}-\mathrm{I}) + c(\mathrm{V}-\mathrm{I})^2 + d(\mathrm{V}-\mathrm{I})^3$ for different intervals of V − I

Interval	a	b	c	d
$\mathrm{V}-\mathrm{I} \leq 0.85$	-0.48729	0.98554	-0.31968	0.592
$0.85 < \mathrm{V}-\mathrm{I} \leq 2.00$	-1.03936	2.14720	-0.416	0.0
$\mathrm{V}-\mathrm{I} > 2.0$	0.5152	0.592	-0.0256	0.0

Table 7.3. Definitions of the colour correction factors $F(t)$ expressed as a polynomial in $t - t_0$, for DC and AC magnitudes. t and t_0 are measured in units of 1000 JD, $t_0 = 2448.6225 \equiv$ 1 January 1992

	DC	DC		AC	
	approx	preceding	following	preceding	following
Stand.dev.	0.0031	0.0012	0.0012	0.0017	0.0017
const	0.0084	0.00303	0.00873	0.0416	0.0478
$(t - t_0)$	−0.0537	−0.04022	−0.04098	−0.0431	−0.0389
$(t - t_0)^2$		0.09000	0.07720	0.1317	0.0338
$(t - t_0)^3$		−0.18863	−0.13772	−0.2384	−0.8277
$(t - t_0)^4$		−0.6746	−0.6735	−1.9387	−2.9383
$(t - t_0)^5$		0.598	0.1322	−2.9443	−3.6662
$(t - t_0)^6$		2.018	2.052	−1.3327	−1.5601
$(t - t_0)^7$		0.340	1.989		
$(t - t_0)^8$		−0.742	0.553		

for very red stars. For this purpose large numbers of red to very red stars in the *Hipparcos* catalogue were systematically observed from the ground during the mission. As nearly all of these stars are variable, this was the only way in which they could still be used for calibration purposes. With this information, and the FAST and NDAC colour coefficients, the entire colour correction of the photometric reductions was redone: subtracting the applied corrections, and adding the new corrections.

The new and final colour corrections use a pseudo-colour index C. Expressed in C, the colour corrections are effectively linear, contrary to the third-order polynomial correction in (V − I) used in the reductions, and can be extrapolated safely for application to very red stars. This colour index is defined in Table 7.2 and Fig. 7.5. As the final colour correction was done using this index, it is also possible to correct the published photometry in case a faulty colour index was used in the data reductions. Corrections to the published H-magnitudes can be determined as:

$$\mathrm{Hp}_{\mathrm{new}} = \mathrm{Hp}_{\mathrm{old}} + F(t) \times (C_{\mathrm{old}} - C_{\mathrm{new}}), \tag{7.1}$$

where the assumed colour index is C_{old} and the newly determined index C_{new}. The function $F(t)$ is approximated as a polynomial in time, the details of which are presented in Table 7.3.

7.2 Formal errors and variability indicators

The observing strategy used for *Hipparcos* implied that the amount of time spent on observing a star during its transit through one of the fields of view

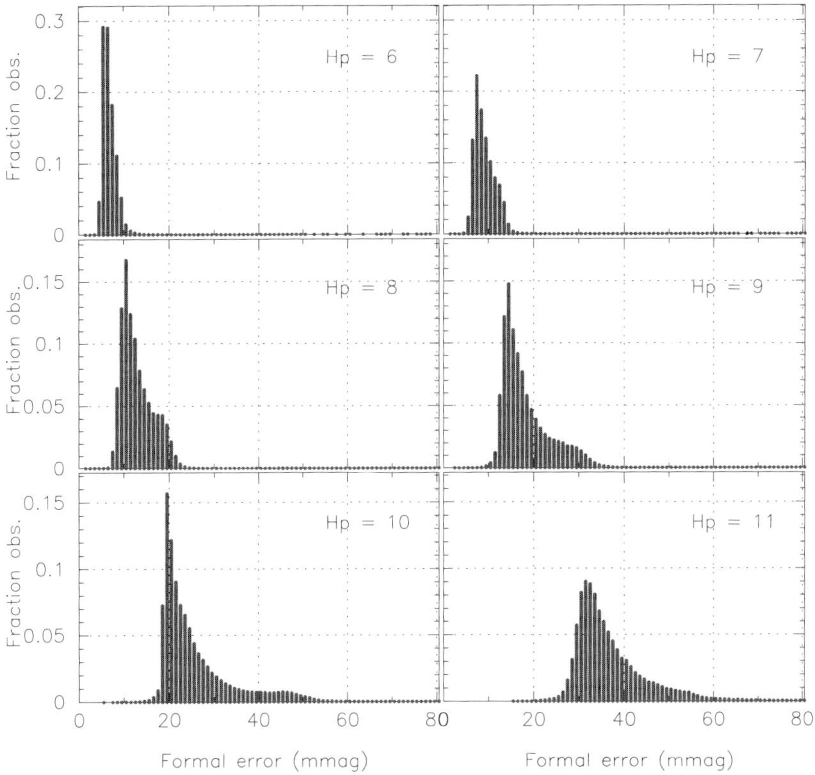

Figure 7.6. Histograms of the formal errors on the *Hipparcos* DC epoch photometric data. Each diagram represents an interval of 0.2 magnitudes around the indicated reference magnitude

depended not only on its own brightness, but also on the competition for observing time from neighbouring stars, those simultaneously observed through the same or the other field of view. Thus, individual errors for different measurements of the same star can vary considerably (Fig. 7.6), and have to be taken properly into account in obtaining means when studying variability.

The total number of observations for a star, and therefore the errors on the mean magnitudes, are further affected by the observing strategy and the differences in sky coverage. Because of the observing strategy, three zones can be distinguished:

- The ecliptic plane, between $\beta = \pm 47°$, with the lowest numbers of observations and where gaps of up to 130 days in the data coverage are not uncommon;
- The zones around $\beta = \pm 47°$, with the largest numbers of observations, sometimes in stretches of longer than one day; gaps in the data coverage are almost always less than 50 days;

- The ecliptic poles, with good and relatively homogeneous coverage, but never long uninterrupted data stretches; gaps in the data coverage are on average around 30 days.

These features can be recognized in Fig. 7.7. The observed distributions of formal errors on the mean photometry reflect all these effects, as well as stellar variability and for the AC component stellar duplicity (see Chapter 4). It is therefore not surprising that we find a relatively large range of formal errors, from below 1 mmag for some stars brighter than Hp= 6.5 to more than 0.1 mag for some of the faintest stars in the catalogue (Fig. 7.8). The errors on the magnitudes for the epoch photometry are sufficiently small to allow for weighted means of the magnitudes rather than the more proper, but also more complex, weighted means of pseudo-intensities: the bias caused by the nonlinear transformations is sufficiently small to be ignored.

With these differences in coverage and formal errors, one has to be careful in assessing variability. To detect variability, a general statistic was derived for each star, based on an analysis of the histogram of unit-weight residuals relative to the mean magnitude. The residuals were binned over a range of ± 5, and the number of bins was adjusted to the number of observations available: from 9 bins for less than 93 observations to 33 bins for more than 200. For every binned distribution there was an equally binned Gaussian distribution with $\sigma = 1$. The differences between the observed (h_o) and expected, equivalent Gaussian distributions (h_c), for bins with at least 3 observations, were accumulated in a χ^2 value (Pearson's test Papoulis 1991, page 273)):

$$\chi^2 = \sum_{i=-m}^{m} \frac{(h_o(i) - h_c(i))^2}{h_c(i)}. \tag{7.2}$$

The χ^2 value was transformed into a probability using the *NAG*-library function *G01ECF* (returns the lower or upper tail probability for a χ^2 distribution

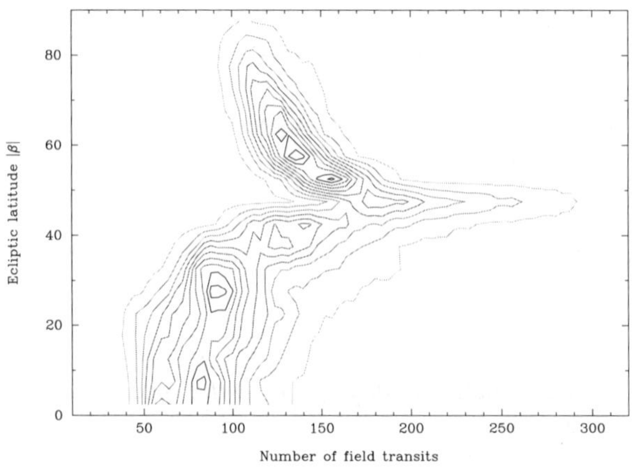

Figure 7.7. The total number of field transits per star as a function of ecliptic latitude. The bin size is 5° in latitude and 1 in number of field transits. The contour interval is 20 and the highest contour represents 200 stars per bin

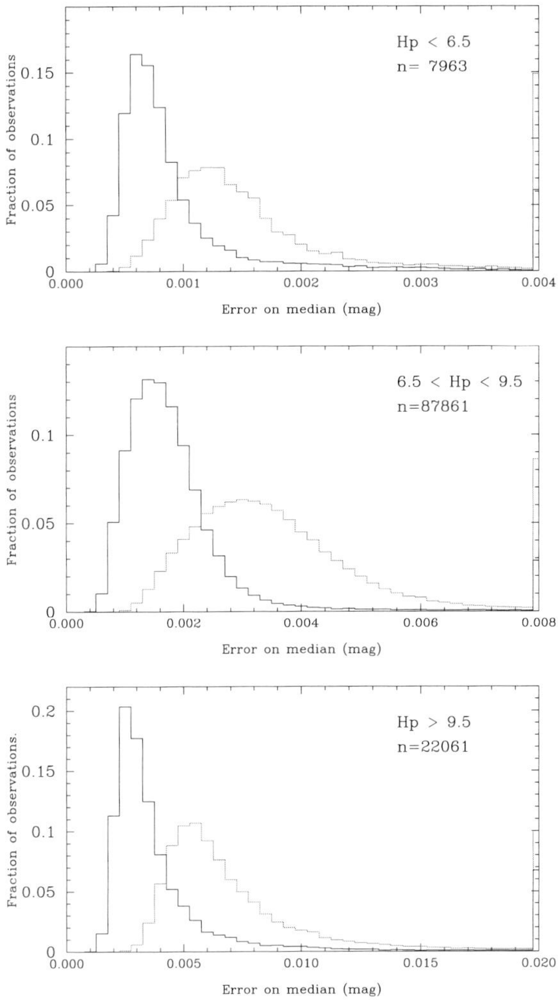

Figure 7.8. The distribution of errors on the medians in three magnitude intervals. The distribution for the dc magnitudes is shown by solid lines, for the ac magnitudes by dotted lines. The accumulation to the right of each histogram represents the variable stars, and for ac magnitudes also the double stars

with real degrees of freedom). Figure 7.9 shows the observed distribution of probabilities compared with the distribution that would have been obtained if no *Hipparcos* stars were variable. The difference between these two distributions indicate the presence of some 20 000 variables among the *Hipparcos* stars, but for many of those the variability was too small to allow further investigations. The probabilities are reflected in the main catalogue through a 'C' in field `H52` for stars found to have a more than 50 per cent probability of being constant (but note that this probability is very much magnitude dependent due to the increasing noise levels towards fainter stars, see Table 7.4), and by '␣' for stars with a probability of being constant between 0.5 and 10^{-4}.

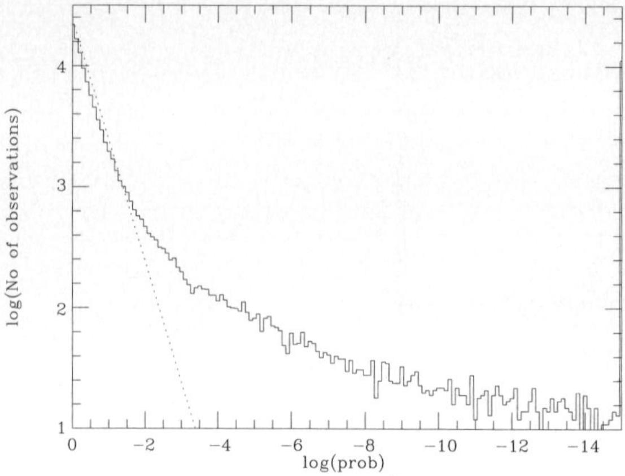

Figure 7.9. The observed distribution of χ^2 values (full line) compared with the distribution expected under the assumption that all stars are constant (dotted line). The lowest probability bin accumulated all observations with probabilities below 10^{-15}

Table 7.4. The minimum peak-to-peak amplitude for stars to be confirmed as variables. Stars having variability at these levels have an average probability of being constant of less than 0.001. The reference points are given for the average of 110 observations.

Hp	T_{110}	Hp	T_{110}	Hp	T_{110}
4.5	0.010	7.5	0.022	10.5	0.063
5.0	0.011	8.0	0.026	11.0	0.076
5.5	0.012	8.5	0.031	11.5	0.098
6.0	0.013	9.0	0.037	12.0	0.124
6.5	0.015	9.5	0.044	12.5	0.18
7.0	0.018	10.0	0.052	13.0	0.36

7.3 Variability analysis

The rather poor epoch coverage of the *Hipparcos* photometric data made the application of some standard period-searching algorithms somewhat unreliable, and a search for relatively stable methods that could be applied to these data was initiated within the context of the NDAC activities. A parallel study took place at Geneva Observatory, and the published results are based on combined information from both studies.

7.3.1 Periodicity Recovery

Two periodicity-search methods were selected, based on their capabilities to handle unevenly spaced data and their very different approach: the analysis of variance (Schwarzenberg-Czerny 1989, 1997) and Scargle's periodogram (Scargle 1982, 1989). The analysis of variance is more powerful to deal with relatively sharp features in light curves, provided the bin size used is small enough. Because of their different approaches to the analysis of the data, it was possible to use these two methods in a complementary way, where the discrepancy between results often indicated the uncertainty associated with period determinations based on the *Hipparcos* photometric data. In general, the temporal distribution of the *Hipparcos* data is such that periods between a few hours and 2 days can be recovered well as long as the amplitudes are not too small compared to the errors on the underlying data. The difficulty for periods from a few days to a few tens of days depends on the ecliptic latitude of the object, while for the very long (more than half a year) periods most objects show irregular light curves which require supplementary ground-based data to interpret.

7.3.1.1 The analysis of variance

The version and method as used for the analysis of the *Hipparcos* photometric data were derived from the paper by Schwarzenberg-Czerny (1989). This method compares the overall variance of a signal with that obtained for means and formal errors in a small number of phase bins, when the sample is folded with a given period. The reduction of the variance as observed for the folded and binned data is used as a measure for the likelihood that an actual period has been detected.

Let $X(t)$ denote a time series, usually assumed to be of the form $X(t) = S(t) + R(t)$, where $S(t)$ is the signal and $R(t)$ the noise. Let $X(t)$ consist of n elements $x(t)$ with individual measurement errors $\sigma(t)$, such as the magnitudes and their formal errors as deduced from the *Hipparcos* measurements. For a periodic signal, a given period relates the time t to a phase ϕ. The data are collected in k phase intervals, such that $x_{i,j}$ refers to measurement j in interval i. Then the overall mean $\tilde{x}$ is defined as:

$$\tilde{x} = \frac{\sum_{i=1}^{k} \sum_{j=1}^{n_i} x_{i,j} \cdot w_{i,j}}{\sum_{i,j} w_{i,j}}, \tag{7.3}$$

where n_i is the number of measurements in interval i, and $w_{i,j} \equiv 1/\sigma_{i,j}^2$. The number of intervals chosen is 4 to 5 for intrinsic variables, and around 8 for possible eclipsing binaries and RR Lyrae stars. These can be partly recognized from the skewness in their distribution of magnitudes. The actual number of intervals ultimately used also depends on the total number of data points available.

The data collected for each phase interval i defines a mean $\tilde{x}_i$ and formal error σ_i:

$$\begin{aligned} \tilde{x}_i &= \frac{\sum_j x_{i,j} \cdot w_{i,j}}{\sum_j w_{i,j}}, \\ \sigma_i &= \Big(\sum_j w_{i,j}\Big)^{-1/2}, \end{aligned} \tag{7.4}$$

from which the following statistics can be constructed:

$$(k-1)s_1^2 = \sum_i (\tilde{x}_i - \tilde{x})^2 \cdot w_i, \tag{7.5}$$

$$(n-k)s_2^2 = \sum_i \sum_j (x_{i,j} - \tilde{x}_i)^2 \cdot w_{i,j},$$

where $w_i \equiv 1/\sigma_i^2$. In case all weights $w_{i,j}$ are equal to one, $w_i = n_i$ (the number of observations per phase interval), as in Eq. 1 of Schwarzenberg-Czerny (1989).

In the absence of a signal S and in the presence of white Gaussian noise, the Analysis of Variance (AOV) statistic is defined as:

$$\Theta_{\rm AOV} = s_1^2/s_2^2, \tag{7.6}$$

and is the ratio of two independent χ^2 random variables. As such it will have an F-distribution, also known as a Fisher-Snedecor distribution, of $k-1$ and $n-k$ degrees of freedom with the following expectation value and variance (Schwarzenberg-Czerny 1989):

$$E[\Theta_{\rm AOV}] = \frac{n-k}{n-k-2}, \tag{7.7}$$

$$\mathrm{Var}[\Theta_{\rm AOV}] = \frac{2(n-k)^2(n-3)}{(k-1)(n-k-2)^2(n-k-4)}.$$

In its practical implementation for the *Hipparcos* photometric data, provisions had to be made for empty bins, while it was also clear that pseudo-periods could be detected in some cases because of empty bins.

7.3.1.2 Modified discrete Fourier transform

This method for the analysis of poorly distributed data was presented by Scargle (1982), with further analysis in (Scargle 1989). It is a modification of the Discrete Fourier Analysis (DFA), which corrects for non-orthogonality caused by uneven distribution of the data. The periodogram is defined as:

$$P_x(\omega) = \frac{1}{2}\left\{ \frac{\left[\sum_j x_j \cos\omega(t_j-\tau)\right]^2}{\sum_j \cos^2\omega(t_j-\tau)} + \frac{\left[\sum_j x_j \sin\omega(t_j-\tau)\right]^2}{\sum_j \sin^2\omega(t_j-\tau)} \right\}, \tag{7.8}$$

where τ is defined as:

$$\tan(2\omega\tau) = \frac{\sum_j \sin 2\omega t_j}{\sum_j \cos 2\omega t_j}, \tag{7.9}$$

and ω is the trial frequency. This modification maintains the exponential distribution of the power P for unevenly spaced data if x is pure Gaussian noise (as in the classical periodogram for evenly-spaced data).

7.3.1.3 *An application example*

In its application the AOV method is generally noisier than the DFA, as measurements switch between phase intervals when the test period is changed. As an example, the analysis of the RRc variable RS Scl (HIP 8163) is shown. This 11th magnitude star, with a period of 0.37736 days, was a relatively easy target for the periodicity searches. The results of the AOV and DFA analysis are shown in Fig. 7.10, the resulting light curve in Fig. 7.11.

7.3.1.4 *Further adjustments to the period estimates*

The periods recovered with the methods described above are not optimal: both methods effectively make assumptions about the light curve which are in most cases only approximately fulfilled. As described by van Leeuwen et al. (1997b), this situation can be much improved through actual light-curve fitting.

The procedure applied is as follows. The observations O_i, folded by the assumed period p, are fitted by a cubic spline function $f(\phi)$ (Appendix B). The phase of an observation at time t_i for an assumed period $\tilde{p}$ is given by

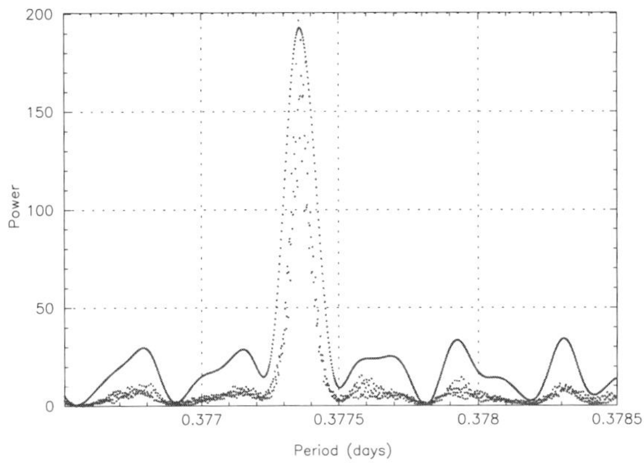

Figure 7.10. The results of the AOV and DFA analysis for the RRc star HIP 8163. The DFA results (the smooth distribution) have been scaled to the AOV data (the more scattered distribution). The peak shows the period of this 11th magnitude star

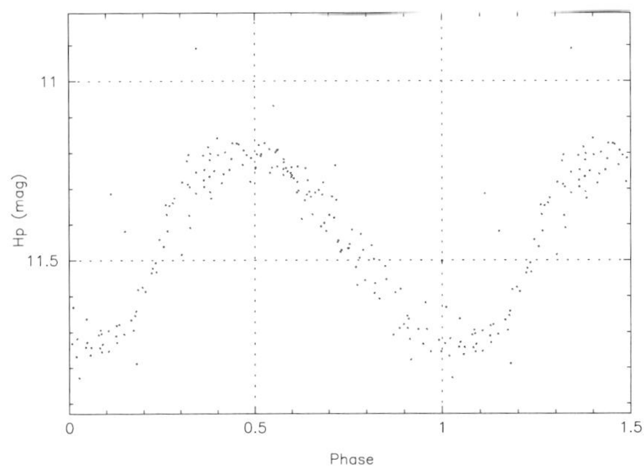

Figure 7.11. The data for HIP 8163 folded with the period derived from the AOV analysis

$\phi_i = \mathrm{mod}((t_i - t_0)/p, 1)$, where t_0 is an arbitrary reference time, preferably taken close to the mean time of all observations used in the determination. In order to obtain near continuity between phase $\phi = 0.0$ and phase $\phi = 1.0$, the data are repeated such that they cover three complete cycles for the spline fit, of which for the further analysis only the central coverage is used.

An adjustment of the period by Δp will shift the phase of observation i by:

$$\Delta\phi_i = -\frac{t_i - t_0}{\tilde{p}}\frac{\Delta p}{\tilde{p}}. \tag{7.10}$$

The change in response expected for a change in phase will in first order depend on the derivative of the fitted light curve, which leads to the following relation:

$$O_i - f(\phi_i) = -\frac{t_i - t_0}{\tilde{p}}\frac{\Delta p}{\tilde{p}}\left(\frac{\partial f(\phi)}{\partial\phi}\right)_{\phi=\phi_i}. \tag{7.11}$$

This can be solved by linear least squares, to give a period correction and formal accuracy of the period, as well as a standard deviation for the curve fitting, which may in some cases be an indication of the presence of multiple periods (see for example Koen 2001). The solution may need to be iterated a few times to take into account the way the estimated period can affect the shape of the light curve. Figure 7.12 shows the formal errors determined this way on the periods of over 200 Cepheids as based on the *Hipparcos* photometric data. They follow the relation $\sigma p = 10^{-5}p^2$, where the dependence on p^2 follows directly from Eq. 7.11, and the factor 10^{-5} is specific for the Cepheid light curves.

7.3.1.5 *Eclipsing binaries*

The methods described above tend to have rather limited success for eclipsing binaries, in particular for systems with highly elliptical orbits, which have asymmetric light curves. For these cases the poor epoch coverage of the data

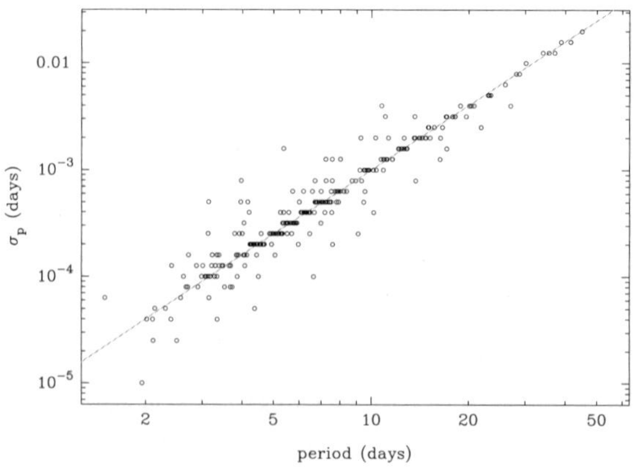

Figure 7.12. Formal errors on periods as determined from the *Hipparcos* data for over 200 Cepheids

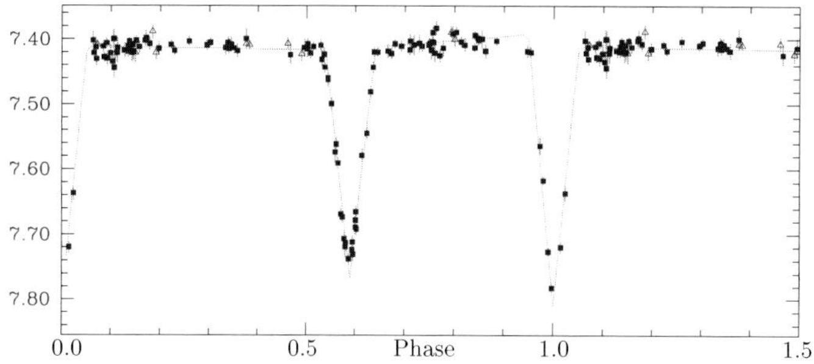

Figure 7.13. The light curve of the eclipsing binary HIP 270, which, due to the ellipticity of its orbit, defeated all automatic period-searching routines employed in NDAC and at Geneva Observatory

appears to leave no other solution than to examine the data manually. This was the case for the newly discovered binary HIP 270, shown in Fig. 7.13. The DFA method (and any other Fourier-analysis based method) fails because of the strong asymmetry in the light curve and the sharp minima. The AOV method fails mainly because of the sharp features at the eclipse minima. Solving these type of variables in an automated manner will be a major challenge for the variability analysis in the Gaia mission, where around 10^4 such cases might be expected (see Chapter 14).

7.3.2 Amplitudes for Unsolved Variables

Even when the period cannot be detected, which is often the case for small amplitude stars, the actual amplitude can still be estimated from the distribution of the magnitudes and their formal errors. For simplicity, a sinusoidal variation is assumed for those stars:

$$m_i = m_0 + m_1 \cos \phi_i, \tag{7.12}$$

where m_1 is the amplitude that has to be estimated. The variance of the observations m_i is approximated by:

$$\langle (m_i - m_0)^2 \rangle = m_1^2 \frac{\int_0^{2\pi} \cos^2 \phi \mathrm{d}\phi}{\int_0^{2\pi} \mathrm{d}\phi} = 0.5 m_1^2. \tag{7.13}$$

The observed variance in the observations also contains the noise from the observational errors. Thus, taking into account the standard errors on the individual measurements σ_i through weights $w_i = \sigma_i^{-2}$, and defining $\Delta m_i = m_i - \widetilde{m_0}$, the amplitudes of variations can be estimated as follows:

$$\widetilde{m_1} = \sqrt{2 \frac{(\sum_i \Delta m_i^2 w_i)/(n-1) - 1}{\sum_i w_i/(n-1)}}, \tag{7.14}$$

where n is the number of observations. The amplitudes of the variations equal $2\widetilde{m}_1$ and have been checked against the amplitudes obtained from the fitting of sinusoidal functions for a large number of small-amplitude periodic stars. No systematic discrepancies were detected. These intrinsic dispersions are included in the unsolved variables table that forms part of the *Hipparcos* data release (ESA 1997).

7.4 Newly discovered variables

Table 7.5 provides a summary of the variable star investigations that were completed for the published catalogue (ESA 1997). Since then a few other studies have tried to further investigate some of the unsolved low-amplitude variables in the *Hipparcos* data (Koen and Eyer 2002; Kallinger and Weiss 2002; Percy et al. 2002).

The discovery power of the *Hipparcos* photometry for periodic variables is demonstrated through comparing the histograms of periods as known for catalogue stars before the mission with the histogram based on the analysis of

Table 7.5. Summary of the variable stars investigations

Stars variable or possibly variable	11597	(8237 new)
Periodic Variables	2712	(970 new)
Cepheids	273	(2 new)
δ Sct & SX Phe	186	(9 new)
Eclipsing binaries	917	(343 new)
Other types	1238	(576 new)
Non-periodic and unresolved	5542	(4145 new)
Not investigated (small amplitudes	3343	(3122 new)

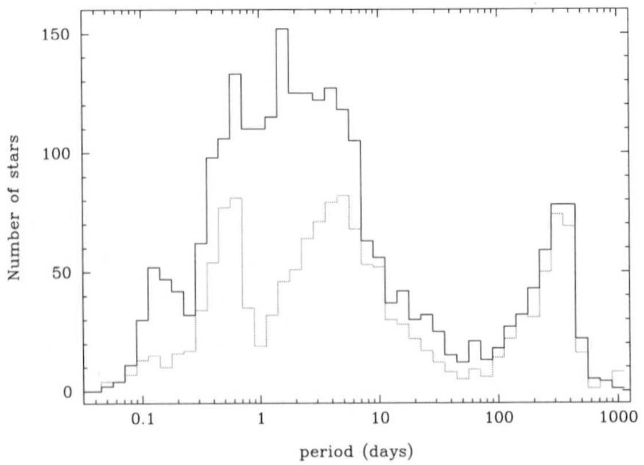

Figure 7.14. The distribution of periods for stars in the *Hipparcos* catalogue from information available before the mission (dotted line) and as a result of the analysis of the mission data (solid line)

the *Hipparcos* photometric data (Fig 7.14). Unlike ground-based observations, there is for the *Hipparcos* data no problem detecting periods close to one day. Many of those have been detected among the Ap, late A- and early F-type stars (see also Section 5.3.2, in particular concerning γ Dor stars), as well as SPBs.

PART IV

HIPPARCOS ATTITUDE MODELLING

The accuracy of the reconstructed attitude reflects in the formal errors on the astrometric parameters, and the level of abscissa-error correlations between neighbouring stars. Through understanding of the dynamical behaviour of the satellite, the accuracy of the reconstructed attitude can be improved, and is therefore of great relevance for obtaining the highest accuracies and best overall statistical properties for the main scientific products of the mission: the astrometric parameters of the 118 000 target stars. The pointing-reconstruction accuracy and requirements of the *Hipparcos* mission far-exceeded anything seen before. A further study of the attitude data can therefore provide insight into issues ranging from the Earth magnetic field and satellite magnetic moments, through statistics on external hits, to the overall rigidity of the satellite. Building on these experiences, a fully-dynamic attitude reconstruction procedure is developed and applied to the *Hipparcos* data, resulting in much reduced levels of in particular systematic noise in the reconstructed attitude.

Part IV covers three Chapters. In Chapter 8 the theoretical background for describing the rotations of a rigid body in space, affected by internal and external torques, is described. Chapter 9 focuses on the calibration through observations of the solar radiation and magnetic torque components. Chapter 10 presents the application of these torque studies in a novel approach for the on-ground reconstruction of the satellite attitude: the fully dynamic attitude model (FDA).

Chapter 8

A FREE-FLOATING RIGID BODY IN SPACE

8.1 Dynamics of a rigid body in space

Any displacements of a satellite, being a free-moving rigid body in space, are naturally split between translational and rotational movements: the translations are described by the orbital dynamics (Section 1.2.3), the rotations are covered by the attitude description. Understanding the rotations from a theoretical point of view is the main topic of this Chapter, which presents the theoretical background for describing the rotations of a rigid body in space, affected by internal and external torques. This includes the gyro-induced, gravity gradient, solar radiation and magnetic torques.

8.1.1 The Basic Equations

While the mean force over the satellite will affect its orbit, the imbalance with respect to its Centre of Gravity (COG) will cause torques $\boldsymbol{N}$, affecting its angular momentum $\boldsymbol{L}$. In the inertial reference system (subscript i) we have:

$$\left(\frac{\mathrm{d}\boldsymbol{L}}{\mathrm{d}t}\right)_i = \boldsymbol{N}. \tag{8.1}$$

The description of the evolution of the angular momentum of a rigid body is simpler in body coordinates (subscript b), which are obtained with the transformation:

$$\left(\frac{\mathrm{d}\boldsymbol{L}}{\mathrm{d}t}\right)_i = \left(\frac{\mathrm{d}\boldsymbol{L}}{\mathrm{d}t}\right)_b + \boldsymbol{\omega} \times \boldsymbol{L}, \tag{8.2}$$

where $\boldsymbol{\omega}$ is the angular-velocity vector of the body. Thus, in the body reference frame Eq. 8.1 becomes:

$$\left(\frac{\mathrm{d}\boldsymbol{L}}{\mathrm{d}t}\right)_b = \boldsymbol{N} - \boldsymbol{\omega} \times \boldsymbol{L}. \tag{8.3}$$

The angular momentum of a rigid body is related to the angular-velocity vector and the inertia tensor I through:

$$\boldsymbol{L} = \mathsf{I} \cdot \boldsymbol{\omega}. \tag{8.4}$$

Implementing this in Eq. 8.3, and assuming the inertia tensor to be constant, we obtain the Euler equation, which describes the rotation of a free-moving rigid body:

$$\mathsf{I}\frac{\mathrm{d}\boldsymbol{\omega}}{\mathrm{d}t} = \boldsymbol{N} - \boldsymbol{\omega} \times \mathsf{I}\boldsymbol{\omega}. \tag{8.5}$$

For a detailed discussion of these relations the reader is referred to any book on classical mechanics, for example Goldstein (1980).

8.1.2 A Spinning Satellite

In the case of *Hipparcos* and other survey satellites, the ω_z component of the angular-velocity vector represents the spin rate, and is much larger than the ω_x and ω_y components. In this situation, an approximate, simplified solution for the Euler equations can be obtained in case of spin-synchronous torques, such as due to solar radiation (see also van Leeuwen et al. 2002). Ignoring, for the moment, the torques altogether, a number of characteristic parameters for the dynamical system can be defined. The first set of parameters concern the ratios of the diagonal elements of the inertia tensor:

$$\begin{aligned} f_x &\equiv \frac{I_{zz} - I_{yy}}{I_{xx}} \\ f_y &\equiv \frac{I_{xx} - I_{zz}}{I_{yy}} \\ f_z &\equiv \frac{I_{yy} - I_{xx}}{I_{zz}}. \end{aligned} \tag{8.6}$$

These three relations are not independent, as $f_z = -(f_x + f_y)/(f_x f_y + 1)$. Similarly, the diagonal elements I_{xx} and I_{yy} of the inertia tensor can be related to an assumed value of I_{zz}, and measured values of f_x and f_y:

$$\begin{aligned} \frac{I_{xx}}{I_{zz}} &= \frac{1 + f_y}{f_x f_y + 1} \\ \frac{I_{yy}}{I_{zz}} &= \frac{1 - f_x}{f_x f_y + 1}. \end{aligned} \tag{8.7}$$

These ratios rather than the individual values of the diagonal elements in the inertia tensor play the more important role in determining the satellite dynamics.

The characteristics of the inertia tensor for a rotating satellite are often expressed in $\lambda \equiv 1 + \sqrt{-f_x f_y}$. Instead, we use $\gamma \equiv \lambda - 1 = \sqrt{-f_x f_y}$. Then, in its simplest approximation, Eq. 8.5 can be given as:

$$\begin{aligned}\frac{\mathrm{d}\omega_x}{\mathrm{d}t} &\approx N_x/I_{xx} - f_x\omega_y\omega_z,\\ \frac{\mathrm{d}\omega_y}{\mathrm{d}t} &\approx N_y/I_{yy} - f_y\omega_x\omega_z,\\ \frac{\mathrm{d}\omega_z}{\mathrm{d}t} &\approx N_z/I_{zz},\end{aligned} \tag{8.8}$$

which, in the absence of external torques, has an exact solution:

$$\begin{aligned}\omega_x &= \omega_1\cos(\bar{\omega}_z\gamma t) - \omega_2\sin(\bar{\omega}_z\gamma t),\\ \omega_y &= \frac{\gamma}{f_x}\Big[\omega_2\cos(\bar{\omega}_z\gamma t) + \omega_1\sin(\bar{\omega}_z\gamma t)\Big],\\ \omega_z &= \omega_3,\end{aligned} \tag{8.9}$$

where the vector $\omega_1, \omega_2, \omega_3$ defines initial rates at a reference time $t = 0$ for the x, y, z axes respectively. The satellite will nutate with a frequency of $\bar{\omega}_z\gamma$, where $\bar{\omega}_z$ is the nominal rotation rate around the spin axis. The difference between the actual rotation rate ω_z and the nominal rate $\bar{\omega}_z\gamma$ introduces a negligible second order correction. The positional nutation amplitudes are given by:

$$\begin{aligned}T_x &= f_x\frac{\omega_y}{\gamma\bar{\omega}_z}\\ T_y &= -f_y\frac{\omega_x}{\gamma\bar{\omega}_z},\end{aligned} \tag{8.10}$$

where T_x and T_y are the offsets in satellite coordinates between the actual and the mean spin axis position, which itself is fixed in space. Thus, the spin axis of the satellite describes, as a result of the nutation, an ellipse in body coordinates, and a rotating ellipse in space coordinates. Given the values presented in Table 8.1, we find a nutation period for *Hipparcos* of just over 12 hours. In that configuration, an amplitude of 1 arcsec s^{-1} in, say, ω_x would, following Eq. 8.10, correspond to an amplitude of about 20 arcmin in T_y. The mission requirements set the allowed range in position at 10 arcmin and in rates 5 arcsec s^{-1}. In reality, the external torques very much disturb this idealized picture

Table 8.1. Measured, assumed and derived properties of the inertia tensor for the middle of the *Hipparcos* mission

Parameter	Value	Property	Parameter	Value	Property	Units
f_x	-0.24357	Measured	I_{zz}	459.420	Assumed	kg m^2 s^{-2}
f_y	0.12597	Measured	I_{xx}	533.667	Derived	kg m^2 s^{-2}
f_z	0.12132	Derived	I_{yy}	589.405	Derived	kg m^2 s^{-2}
λ	1.17516	Derived	$\bar{\omega}_z$	168.75	Fixed	arcsec s^{-1}

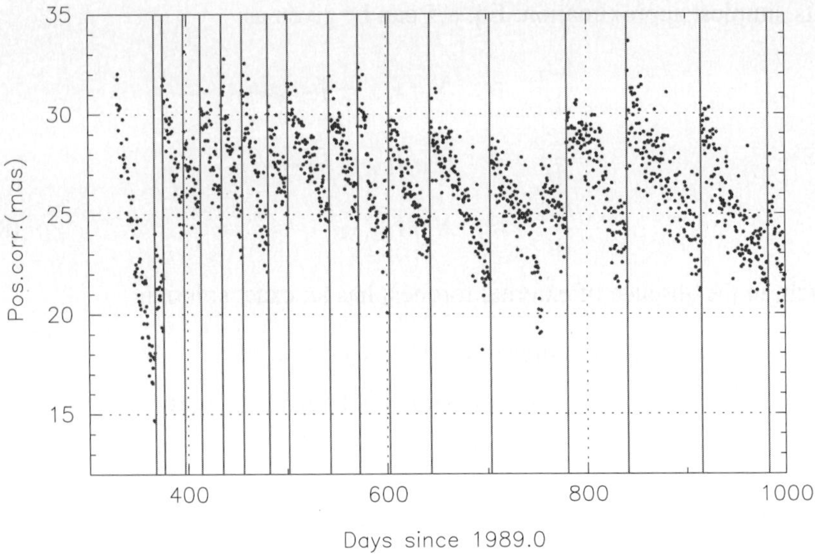

Figure 8.1. Calibration results for the correction of the position of the chevron slits relative to the centre of the main grid. The vertical lines represent instances of refocusing, compensating for the gradual drift in focal length of the telescope. The discontinuity at day 755 is the result of a problem with the thermal control electronics, which led to loss of some control over the payload temperature

for a relatively slow rotating satellite like *Hipparcos*, and no exact, analytical solutions are available for the attitude modelling (see also Bois 1986, 1987).

8.1.3 The Attitude Reference System

The reference system for the description of the *Hipparcos* attitude is fixed to properties of the payload and therefore not aligned with the principal moments of inertia of the satellite, or centred on the centre of gravity of the satellite, COG. The difference between the centre of gravity and the centre of mass is negligibly small, so although centre of gravity is used throughout this text, substituting centre of mass makes no significant difference to the results. The relation between the (x, y) plane of the satellite reference system and the "sky" is defined by the way the sky is projected through the two fields of view onto the focal plane of the telescope. One can turn this around and see it as the focal plane being projected onto two locations of the sky. These projections are continuously moving along a direction that is (almost) perpendicular to the lines on the main grid. These will therefore not provide information on the position of the grid perpendicular to its motion. The chevron slits of the star mapper detector have been specially designed to do so instead. Thus, the two projections on the sky of the tips of these chevron slits provide two well-defined positions on the sphere, and can be used to define the track of a great circle (see Fig. 2.4). The reference positions for the star mapper slits are defined as the mean positions of the four slits in each group. This great circle thus defined will be referred to as the scan circle. The directions of the centres of the preceding and following fields of view are defined at the crossings of the positions halfway between grid-lines 1344 and 1345 with the scan circle, about 53 arcmin from the reference position on the chevron slits. The variation in this distance is at the level of 0.01 arcsec, and is calibrated to an accuracy of about 1 to 2 mas (see Fig. 8.1). It is affected by the adjustments of the telescope to the space environment, causing slow changes in focal length, and refocusing exercises, correcting in discrete steps the slow drifts.

With the scan circle in place, and two reference points identified on it, the coordinate system can now be defined: the x axis is defined as the bisector of the optical axes for the two field of view positions, the z axis is defined perpendicular to the scan circle, and the y axis completes the right-handed reference system (see Fig. 1.8).

8.2 The internal torques and inertia tensor

Hipparcos was subjected to two types of internally generated torques: thruster firings (see Section 13.3) and gyro-induced torques. The satellite was equipped with rate-integrating gyros, which were part of the on-board attitude control system. The input and spin axes of these gyros remain largely fixed within the satellite reference frame: actual displacements of the input axes provide measurements of the satellite's rotation rates, and by re-adjusting the input axes every $16/15$ seconds to their starting positions, they appear to be fixed. Thus, the coupling between the rotation $\boldsymbol{\omega}$ of the satellite and the angular momentum $\boldsymbol{h}$ of the gyros will cause an internal torque:

$$\boldsymbol{N}_G = \sum_k \boldsymbol{h}_k \times \boldsymbol{\omega}, \tag{8.11}$$

where the sum is over the active gyros, which varied during the mission from 2 to 5. In the default configuration the resulting torque $\boldsymbol{N}_G$ was close to zero (gyros 1, 2 and 4 or 5 active). Details on gyro characteristics, input axes and angular momentum vectors are provided in Section 13.2, Table 13.1.

Internal torques can in principle be observed from specific correlations between the observed accelerations and inertial rates of the satellite. A difficulties in determining the angular momentum vectors of the gyros is caused by an ambiguity with the determination of some of the off-diagonal elements in the inertia tensor: they both produce the same relations between rates and accelerations. The gyro-torque components $h_y\omega_z$ on x and $-h_x\omega_z$ on y will produce near constant internal torques on these axes, similar to the off-diagonal elements of the inertia tensor in the cross product in Eq. 8.5: $-\omega_z^2 I_{yz}$ on x and

$\omega_z^2 I_{xz}$ on y. The components $-h_z\omega_y$ on x and $h_z\omega_x$ on y are in practice difficult to distinguish from the cross terms in the Euler equation: $(I_{yy} - I_{zz})\bar{\omega}_z\omega_y$ and $(I_{zz} - I_{xx})\bar{\omega}_z\omega_x$ respectively. Information can be extracted from $h_x\omega_y$ and $-h_y\omega_x$ on z, given the high accuracy level of the reconstructed along-scan attitude. But here too, there is an ambiguity with the inertia-tensor elements: $\omega_x\omega_z I_{yz}$ and $-\omega_y\omega_z I_{xz}$. Thus, uncertainties remain in the off-diagonal elements of the inertia tensor and the angular-momentum vectors of the gyros.

Changes in the configuration of the active gyros helped to determine h_x and h_y, at least in a relative sense for an activated gyro. Inactive gyros were subjected to despin exercises; when, for example, this involved gyro-3, the alignment of the angular-momentum of this gyro with the z axis could clearly be observed from the changes in the spin rate during spin-up and spin-down, as can be seen in Fig. 8.2. As no effects are observed on the spin rates for the x and y axes in these situations, it can be assumed that the angular momentum of gyro 3 is directed almost entirely along the z axis. Similar observations showed that gyros 4 and 5 do not have significant angular-momentum components along the z axis.

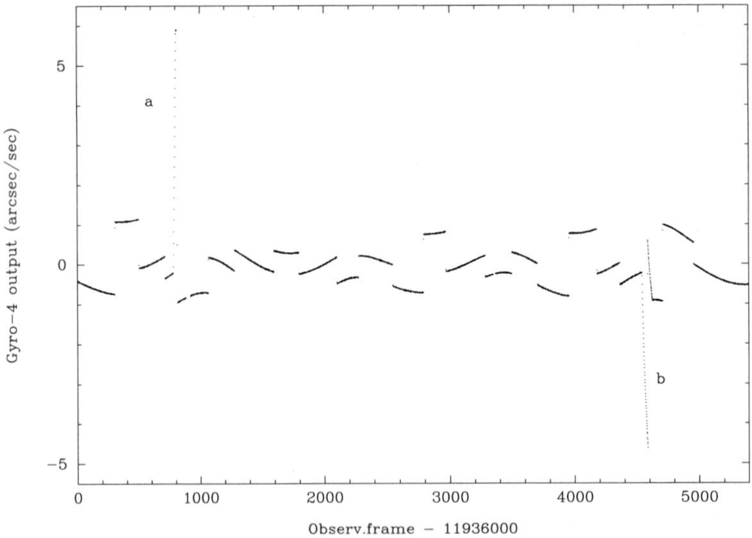

Figure 8.2. An example of the influence on the rate of change of the inertial rates caused by a gyro de-storage exercise. The readings of gyro 4 are shown, calibrated to rates relative to the nominal scan velocity. The redundant gyro 3 is spun up at position a. Its spin axis is aligned with the input axis of gyro 4 (parallel to the satellite's z axis), which is noted in the accelerations. At position b the redundant gyro 3 is spun down, causing a deceleration. The discontinuities in the rates are the result of thruster firings, which also compensated the changes in the satellite's angular momentum that resulted from the de-storage exercise

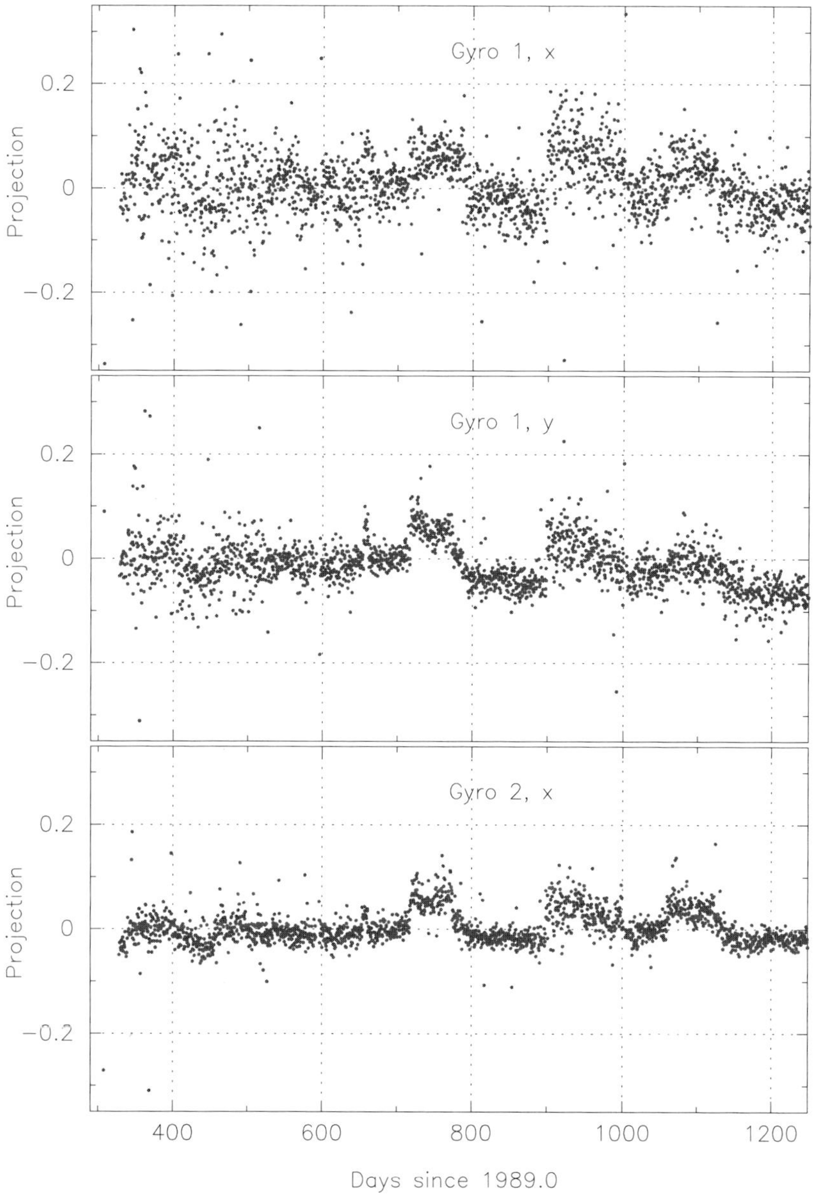

Figure 8.3. Relative variations in the responses of gyros 1 and 2: (Fig. 8.3) over part of the mission. Synchronized discontinuities in the responses are observed at times of switching on or off gyro 4 (in redundant mode) at days 655, 661, 718, and 789, and the change of tank used to supply the cold gas for the thruster firings at day 900. Temperature changes are also observed to affect the responses (around day 450 and day 1100)

The orientations of the input axes of the gyros are easier to determine, and, as the input axes are perpendicular to the angular momentum vectors, this provides some constrains on these vectors. Detailed examination of the gyro-orientation calibration results indicate some unexplained variations at a level of 1 part in 10^3 to 10^4. Some of these variations are systematic and observed for example on the three main components of gyros 1 and 2: (Fig. 8.3). These look like being related to other activities taking place in the spacecraft, such as the running of an additional gyro, the switching between the gas tanks and periods of relatively low average temperature. It seems possible that small changes in the power supply caused these variations.

The most serious consequence of the uncertainties in the determination of the off-diagonal elements of the inertia tensor is the way these affect predicting the gravity gradient torques, in particular when the satellite was observing at relatively low altitude. These uncertainties, combined with problems determining the magnetic torques and the Earth-albedo induced torques, mean that towards perigee no accurate prediction of the environmental torques can be made.

8.3 External torques acting on the satellite

The motion of a free-floating rigid body in space is sensitive to the minutest of torques acting on it, the most important of which are:

- the solar radiation pressure;
- the gradient of the gravitational field of the Earth over the satellite;
- the coupling between the magnetic moment of the satellite and the Earth magnetic field;
- radiation pressure due to reflected and scattered light from the Earth;
- hits by external particles.

The solar radiation pressure is proportional to the inverse squared distance between the satellite and the Sun. All Earth-related torques are proportional (or roughly proportional in the case of scattered light) to the inverse of the distance from the centre of the Earth to power 3, and therefore decrease rapidly with increasing altitude above the Earth. External hits by interplanetary dust particles and space debris act in the same manner as thruster firings, except that their presence has to be deduced from the science data. The detailed motion of the satellite can further be affected by its thermal adjustments, resulting from temperature gradients caused by rotation, eclipses of the Sun by the Earth, and friction heating during perigee passages, effects that can cause violations of the principle assumption of a free-floating rigid body.

8.3.1 Solar-radiation Torques

8.3.1.1 General description of radiation torques

The solar energy flux F_e is the dominant source of torques for the satellite at altitudes above 10 000 km. The flux, integrated over all wavelengths, is to within 0.3 per cent given by (Spence 1978):

$$F_e \approx 1358 R^{-2} \mathrm{W\ m}^{-2}, \tag{8.12}$$

where R is the distance from the Sun measured in AU at the time of observation. The pressure exerted by the radiation flux will create a torque on the satellite, depending on the position, orientation, and optical properties of each element of the satellite exposed to the radiation. The radiation can be absorbed, mirror-like (specular) reflected, or scattered (diffuse reflection), depending on the characteristics of the surface element. The mean momentum flux P along the normal to the Sun's direction is given by:

$$P \equiv \frac{F_e}{c} = 4.5298 \cdot 10^{-6}\ \mathrm{N\ m}^{-2}, \tag{8.13}$$

where $c = 299\ 792\ 458\ \mathrm{m\ s}^{-1}$ is the speed of light.

We consider a surface element $\mathrm{d}A_i$, with its normal unit vector $\boldsymbol{n}_i$. The direction of the Sun is given by the unit vector $\boldsymbol{s}$. The coefficients $C_{a,i}$, $C_{s,i}$ and $C_{d,i}$ give the fraction of the solar flux being absorbed, reflected and scattered, respectively, by this element. In an equilibrium situation (temperature of the surface element remains constant), the three coefficients add up to one. The inner product $\boldsymbol{n} \cdot \boldsymbol{s}$ is always positive for surface elements exposed to solar radiation, and represents the angle of incidence θ:

$$\cos\theta_i = \boldsymbol{n}_i \cdot \boldsymbol{s}. \tag{8.14}$$

Thus, $0 \leq \theta \leq 90°$ applies.

The force on a surface element $\mathrm{d}A_i$ due to absorbed radiation is derived as (Spence 1978; Fantino 2000):

$$\mathrm{d}\boldsymbol{f}_{a,i} = -PC_{a,i}\cos\theta_i \boldsymbol{s}\mathrm{d}A_i. \tag{8.15}$$

The reflected radiation has a direction $-\boldsymbol{s} + 2\boldsymbol{n}_i\cos\theta_i$. The specular radiation force can thus be given by:

$$\begin{aligned} \mathrm{d}\boldsymbol{f}_{s,i} &= PC_{s,i}\big[(\boldsymbol{s} - 2\cos\theta_i\boldsymbol{n}_i)\cos\theta_i - \cos\theta_i\boldsymbol{s}\big]\mathrm{d}A_i \\ &= -2PC_{s,i}\cos^2\theta_i\boldsymbol{n}_i\mathrm{d}A_i. \end{aligned} \tag{8.16}$$

In scattering, the radiation is distributed proportional to $\cos\phi$, where ϕ is measured relative to $\boldsymbol{n}_i$. Integrating the reflected radiation over all angles gives:

$$\mathrm{d}\boldsymbol{f}_{d,i} = -PC_{d,i}\left[\frac{2}{3}\cos\theta_i\boldsymbol{n}_i + \cos\theta_i\boldsymbol{s}\right]\mathrm{d}A_i. \tag{8.17}$$

The total force $\mathrm{d}\boldsymbol{f}_{t,i}$ under equilibrium conditions, i.e. $C_{a,i} + C_{s,i} + C_{d,i} = 1$, is now given by:

$$\mathrm{d}\boldsymbol{f}_{t,i} = -P\big[(1 - C_{s,i})\boldsymbol{s} + 2(C_{s,i}\cos\theta_i + 1/3C_{d,i})\boldsymbol{n}_i\big]\cos\theta_i \mathrm{d}A_i. \quad (8.18)$$

The torque resulting from the radiation on the surface element is given by:

$$\mathrm{d}\boldsymbol{N}_i = \boldsymbol{r}_i \times \mathrm{d}\boldsymbol{f}_{t,i}, \quad (8.19)$$

where $\boldsymbol{r}$ is the position of the element in the satellite reference frame, relative to the COG. The total torque is obtained by summation or integration over all exposed surface elements.

8.3.1.2 The torque caused by the solar panels

In the specifications for *Hipparcos* the absorption coefficients C_a are given. The reflection was considered either fully specular or diffuse: in each case the absorption and the relevant reflection coefficient add up to one. The main elements are:

- the solar panels: $C_a = 0.75$, specular reflection;
- the lower platform: $C_a = 0.96$, diffuse reflection;
- the ABM nozzle, outer surface: $C_a = 0.92$, diffuse reflection;
- the ABM nozzle, inner surface: $C_a = 1.00$, no reflection;
- the lateral surfaces: $C_a = 0.96$, diffuse reflection;
- excess-heat radiators: $C_a = 0.13$, specular reflection;
- top cover: $C_a = 0.12$, specular reflection.

With these values and the outer dimensions of the satellite, a reasonably accurate picture can be obtained of the expected solar-radiation torques.

We take as an example the solar panels of the satellite. Each panel measures 1.29 by 1.69 m, and is attached, on the shorter side, to the satellite at a distance of 0.965 m from the axis through the COG, which is situated 0.76 m above the plane containing the three panels. For a surface element $\mathrm{d}A_i$ on the Sun-lit side of a panel, the vectors $\boldsymbol{s}$ and $\boldsymbol{n}$ are given by (see Fig. 1.8, page 22, for the definitions of Ω and ξ):

$$\boldsymbol{s} = -\begin{bmatrix} \cos\Omega\sin\xi \\ \sin\Omega\sin\xi \\ \cos\xi \end{bmatrix}, \quad \boldsymbol{n} = \begin{bmatrix} 0 \\ 0 \\ -1 \end{bmatrix}. \quad (8.20)$$

With the optical properties specified above, the force on this surface element is given by:

$$\mathrm{d}\boldsymbol{f_i} = P\cos\xi \begin{bmatrix} 0.75\cos\Omega\sin\xi \\ 0.75\sin\Omega\sin\xi \\ 1.25\cos\xi \end{bmatrix} \mathrm{d}A_i. \tag{8.21}$$

The position vector of the surface element as seen from the COG is given by:

$$\boldsymbol{r}_i = \begin{bmatrix} -x_i \\ -y_i \\ -z_{\mathrm{COG}} \end{bmatrix}, \tag{8.22}$$

producing a torque given by:

$$\mathrm{d}\boldsymbol{N}_i = P\cos\xi \begin{bmatrix} -1.25y_i\cos\xi + 0.75z_{\mathrm{COG}}\sin\Omega\sin\xi \\ -0.75z_{\mathrm{COG}}\cos\Omega\sin\xi + 1.25x_i\cos\xi \\ -0.75x_i\sin\Omega\sin\xi + 0.75y_i\cos\Omega\sin\xi \end{bmatrix} \mathrm{d}x_i\mathrm{d}y_i. \tag{8.23}$$

The integration is done for the panel aligned with the x axis, for which the integration boundaries are ± 0.645 m in y and 0.9546 m to 2.6446 m in x:

$$\boldsymbol{N}_a = P\cos\xi \begin{bmatrix} 1.24\sin\Omega\sin\xi \\ -1.24\cos\Omega\sin\xi + 4.90\cos\xi \\ -2.94\sin\Omega\sin\xi \end{bmatrix}. \tag{8.24}$$

The same result is obtained when assigning to the vector $\boldsymbol{r}_i$ the position of the COG of the solar panel $(1.80, 0, -0.76)$, and to $\mathrm{d}A_i$ its total surface area (2.18 m^2). That simplification is used to calculate the torques on the other two solar panels (b, c), for which the positions are $(-0.9, \pm 1.56, -0.76)$ m. The torques on those panels are given by:

$$\boldsymbol{N}_{b,c} = P\cos\xi \begin{bmatrix} \pm 4.25\cos\xi + 1.24\sin\Omega\sin\xi \\ -1.24\cos\Omega\sin\xi - 2.45\cos\xi \\ \mp 2.55\cos\Omega\sin\xi + 1.47\sin\Omega\sin\xi \end{bmatrix}. \tag{8.25}$$

For the sum over the torques from the three solar panels, we find:

$$\boldsymbol{N}_{a+b+c} = P\cos\xi \begin{bmatrix} 2.48\sin\Omega\sin\xi \\ -2.48\cos\Omega\sin\xi \\ 0 \end{bmatrix}, \tag{8.26}$$

which is measured in the satellite reference frame. In the inertial reference frame, rotating the satellite reference frame by $-\Omega$, we get:

$$\boldsymbol{N} = P\cos\xi \begin{bmatrix} 0 \\ -2.48\sin\xi \\ 0 \end{bmatrix}, \tag{8.27}$$

which is equivalent to a constant torque on $\bar{y}$ of -5.6μNm. This compares with an observed value of -6.1μNm for the entire satellite. Thus, ideally the only contribution the solar panels make directly to the satellite torque is the constant torque on $\bar{y}$. All further modulation is the result of the radiation pressure on the lateral surfaces and the way these are affected by the shadows from the solar panels. The presence of the ABM nozzle further complicates the calculations.

8.3.2 The Gravity Gradient Torques

The gravity gradient torque $\boldsymbol{N}_{\rm GG}$ results from the gradient of the gravitational potential V across the body of the satellite, and is defined as:

$$\boldsymbol{N}_{\rm GG} = \int \boldsymbol{r} \times \nabla V \, {\rm d}m, \tag{8.28}$$

where $\boldsymbol{r}$ gives the position in the satellite reference frame for the mass element ${\rm d}m$. Following Spence (1978), and integrating over all mass elements for the potential associated with a purely spherical Earth, gives the first approximation of the gravity gradient induced torque $\boldsymbol{N}_{GG0}$ for the satellite:

$$\boldsymbol{N}_{\rm GG0} = \frac{3GM_\oplus}{R_S^3}\left[\hat{\boldsymbol{R}}_S \times \left(\mathsf{I} \cdot \hat{\boldsymbol{R}}_S\right)\right], \tag{8.29}$$

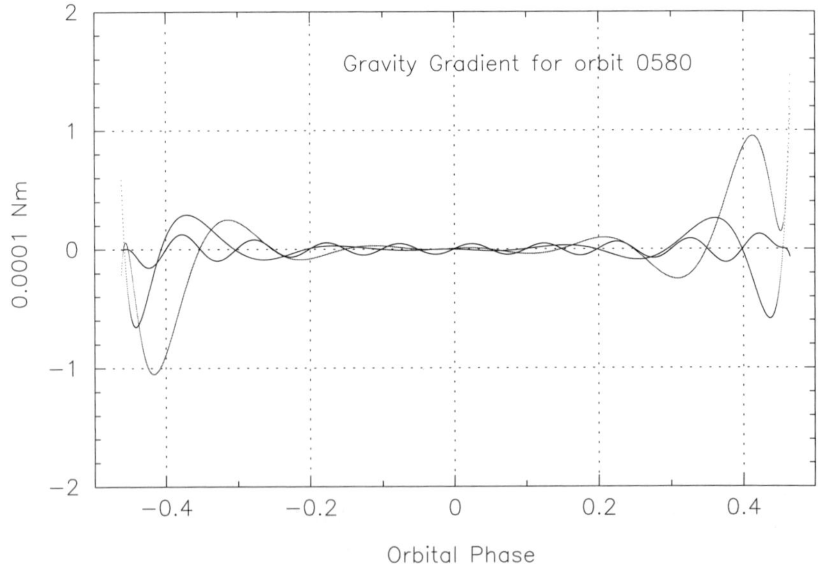

Figure 8.4. The gravity gradient torques over orbit 580 (20 July 1990). Clearly visible is the increase towards perigee and the modulation with the rotation of the satellite. The modulation phase shifts due to the changing direction of the Earth as seen from the satellite over its orbit

where $\hat{\boldsymbol{R}}_S$ is the unit length vector indicating the geocentric direction to the origin of the satellite reference frame, R_S is the geocentric distance of the satellite, G is the gravitational constant and $M_\oplus$ is the mass of the Earth. Thus, the gravity gradient torque produced by a spherical mass distribution decreases with the inverse of the third power of R_S. An example of the gravity gradient over an orbit is shown in Fig. 8.4.

The possible effects of the oblateness of the Earth on the gravity gradient was investigated by Fantino (2000); Fantino and van Leeuwen (2003). These contributions, for which no exact but only an approximate description was obtained, decrease with the fifth power of R_S, and are generally very much smaller than the spherical term. These higher order contributions will therefore be ignored. The results from those papers are summarized in Table 8.2.

8.3.3 Magnetic Moment

The interaction between the magnetic moment $\boldsymbol{m}$ of a satellite and the Earth's magnetic field $\boldsymbol{B}$, as seen in the satellite reference frame, causes a torque:

$$\boldsymbol{N}_\mathrm{M} = \boldsymbol{m} \times \boldsymbol{B}. \tag{8.30}$$

We define a set of geotropic coordinates in the same way as we defined earlier the heliotropic coordinates: thus, the reference meridian for this system goes through the instantaneous spin axis of the satellite and the direction of the centre of the Earth as seen from the satellite. The magnetic field of the Earth is described as a function of distance and equatorial coordinates (see below) of the observer, and the geotropic coordinates provide a direct link between this description and the satellite coordinates. The angle Ω_G provides the link

Table 8.2. Typical amplitude of the gravity gradient torque on *Hipparcos* at perigee, at 10 000 km altitude ($h^\star$) and at apogee, according to the spherical Earth model and to the oblate potential model. The units are Nm

position	axis	spherical potential model	oblateness contribution
	x	$2 \cdot 10^{-4}$	$2 \cdot 10^{-5}$
perigee	y	$1 \cdot 10^{-4}$	$1 \cdot 10^{-5}$
	z	$1 \cdot 10^{-4}$	$2 \cdot 10^{-5}$
	x	$2 \cdot 10^{-5}$	$3 \cdot 10^{-7}$
$h^\star$	y	$1 \cdot 10^{-5}$	$2 \cdot 10^{-7}$
	z	$8 \cdot 10^{-6}$	$3 \cdot 10^{-7}$
	x	$1 \cdot 10^{-6}$	$3 \cdot 10^{-9}$
apogee	y	$6 \cdot 10^{-7}$	$2 \cdot 10^{-9}$
	z	$4 \cdot 10^{-7}$	$3 \cdot 10^{-9}$

between the satellite reference frame and the magnetic field $\bar{B}$ in the (quasi) inertial reference frame:

$$B = \begin{bmatrix} \cos\Omega_G & \sin\Omega_G & 0 \\ -\sin\Omega_G & \cos\Omega_G & 0 \\ 0 & 0 & 1 \end{bmatrix} \bar{B}. \tag{8.31}$$

The magnetic moment is the primary source of magnetic disturbance torques. Effects due to eddy currents and hysteresis are considerably less significant. The main magnetic torque decreases with the intensity of the magnetic field, that is, with the inverse of the third power of the geocentric distance. Eddy currents, on the other hand, are proportional to the square of the magnetic-field intensity, therefore decreasing much more rapidly.

As far as the geomagnetic field is concerned, it is customary to distinguish between the "main" field and the "disturbance" field (Plett 1978):

1 the main field (or core field) is the field produced and maintained, like that of an electric dynamo, by some energy source within the Earth. This contribution dominates the field from the Earth's surface up to about four Earth radii.

2 following Tsyganenko (1990), the disturbance field can be separated into the "crustal" (or anomaly) field, which is produced by the crustal magnetisation and is responsible for features called magnetic anomalies, and the "external source" field, which is due to current systems in the Earth's magnetosphere and is strongly affected by the solar wind interactions with the magnetosphere which dominate the main field beyond four Earth radii.

According to the modular principle (Tsyganenko 1996), the geocentric magnetic field $\bar{B}$ can be represented as the sum of the main field (or internal field) $\bar{B}_I$ and the disturbance field (or external field) $\bar{B}_E$ produced by the extraterrestrial electric current systems:

$$\bar{B} = \bar{B}_I + \bar{B}_E, \tag{8.32}$$

thus enabling the two terms to be modelled separately.

The geocentric internal geomagnetic field $\bar{B}_I$ can be conveniently expressed as the gradient of a scalar potential U which, because it satisfies Laplace's equation $\nabla^2 U = 0$, can be described by means of spherical harmonics:

$$U(r,\theta,\phi) = a_\oplus \sum_{n=1}^{k} \left(\frac{a_\oplus}{r}\right)^{n+1} \sum_{m=0}^{n} \left(g_n^m \cos m\phi + h_n^m \sin m\phi\right) P_{nm}(\theta), \tag{8.33}$$

where $a_\oplus$ is the equatorial radius of the Earth; g_n^m and h_n^m are called the *Gaussian coefficients*; r, θ, ϕ are the geocentric distance, co-elevation, and east longitude from Greenwich; and $P_{nm}(\theta)$ are the *associated Legendre functions* of the second kind, of degree n and order m. The Gaussian coefficients are calibrated every 10 years and presented as the International Geomagnetic Reference Field (IGRF). In the current study we use the representation as given by Barraclough (1985).

The components of $\bar{\boldsymbol{B}}_I$ in spherical coordinates (in the instantaneous equatorial system) are given by Plett (1978):

$$\begin{aligned}
\bar{B}_r &= \frac{-\partial U}{\partial r} = \sum_{n=1}^{k} \left(\frac{a_\oplus}{r}\right)^{n+2} (n+1) \sum_{m=0}^{n} (g_n^m \cos m\phi + \\
&\qquad h_n^m \sin m\phi)\, P_{nm}(\theta), \\
\bar{B}_\theta &= \frac{-1}{r}\frac{\partial U}{\partial \theta} = -\sum_{n=1}^{k} \left(\frac{a_\oplus}{r}\right)^{n+2} \sum_{m=0}^{n} (g_n^m \cos m\phi + \\
&\qquad h_n^m \sin m\phi)\, \frac{\partial P_{nm}(\theta)}{\partial \theta}, \\
\bar{B}_\phi &= \frac{-1}{r\sin\theta}\frac{\partial U}{\partial \phi} = \frac{-1}{\sin\theta} \left(\frac{a_\oplus}{r}\right)^{n+2} \sum_{m=0}^{n} m\,(-g_n^m \cos m\phi + \\
&\qquad h_n^m \sin m\phi)\, P_{nm}(\theta).
\end{aligned} \tag{8.34}$$

The position of the satellite with respect to the Earth is obtained from the orbit files as described by Dalla Torre and van Leeuwen (2003). This relates the three components of the magnetic field to the origin of the satellite coordinate reference frame. These are transformed to the reference epoch in which the attitude of the satellite is described, to obtain the components of the magnetic field in the satellite reference system.

The secondary effects of the interplanetary field and the magnetic field carried by the solar wind have been derived by Tsyganenko (1996), and have been described for application to the *Hipparcos* data by Fantino and van Leeuwen (2003), using the Tsyganenko T96_01 model. Figure 8.5 shows the magnitude of the two fields in satellite coordinates as a function of the orbital phase for orbit number 95.

The orbit described by *Hipparcos*, together with the inclination of the magnetic field with respect to the polar axis of the Earth, offers a way of directly recognizing magnetic-field related torque components: since the magnetic field is tilted with respect to the polar axis of the Earth, its properties depend not only on elevation above the Earth, but also the rotation phase of the Earth. Since *Hipparcos* completed roughly two orbits in one day (see Fig. 1.6, page 19), the features of the magnetic field at the position of the spacecraft on successive odd or even orbits were approximately the same, while often clear differences exist between the odd and even orbits (Fig. 8.6). The nominal scanning law introduces in addition a regular 57-day variation in the direction of the

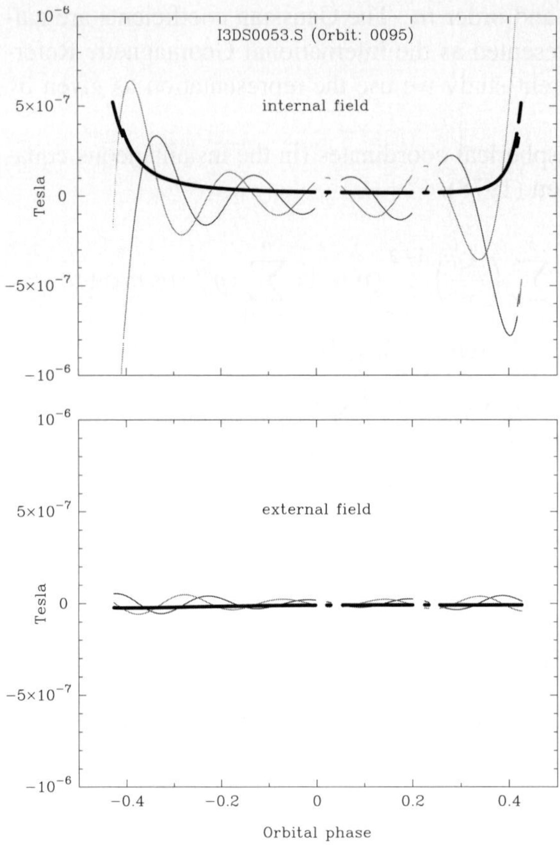

Figure 8.5. Magnetic field at the position of *Hipparcos* in satellite coordinates as a function of orbital phase ($\pm$ 0.5 at perigee, 0.0 at apogee) split into internal contribution (top) and external contribution (bottom): the thin curves represent the field components on x and y; the thick curve is the z component. The units are Tesla

magnetic field as seen in satellite coordinates: any such periodicity in the torque components is related to the precession of the spin axis, and is a clear sign of magnetic torques.

8.3.4 External Hits and Thruster Firings

Sudden changes in the along-scan velocity can be observed down to a detection limit of around 2 mas s^{-1}. This is equivalent to an angular momentum change of $4.4 \cdot 10^{-6}$ kg m^{2}s^{-1} and in energy of $2.1 \cdot 10^{-14}$ kg m^{2}s^{-2}. Suppose a particle hits the satellite at a velocity of 30 km s^{-1} and at a distance of 1.5 m from its COG. If the particle gets absorbed by the satellite, then a lower limit for the mass of the particle can be estimated:

$$M \approx \frac{\mathrm{d}\boldsymbol{L}}{\boldsymbol{a} \times \boldsymbol{v}}, \tag{8.35}$$

where $\boldsymbol{a}$ gives the position of the hit on the satellite surface, and $\boldsymbol{v}$ the velocity vector of the particle in the satellite reference frame. For the assumed velocity,

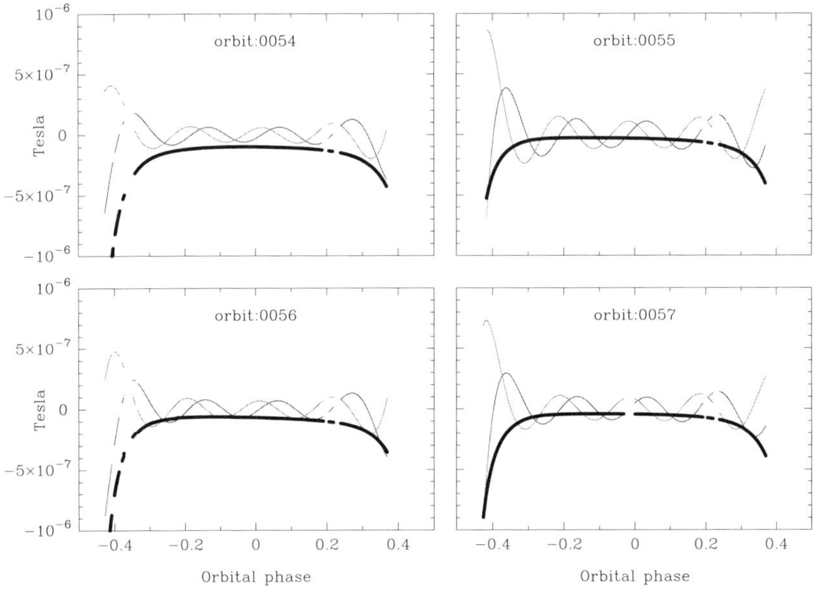

Figure 8.6. Magnetic field at the position of *Hipparcos* in satellite coordinates as a function of orbital phase ($\pm$ 0.5 at perigee, 0.0 at apogee) over 4 successive orbits: the thin curves represent the field components on x and y; the thick curve is the z component. The units are Tesla

impact position and limit on the observable angular momentum changes, this gives a lower limit on the mass of the particle of approximately 0.001 mg. That is all that is needed to disturb noticeably the rotation velocity of the satellite. It is therefore not surprising that of the order of 80 hits have been recorded for this mission, where the statistics indicate that still many more smaller hits have remained undetected. The largest hits, creating velocity discontinuities at a level of a few arcsec s^{-1}, were detected from examinations of the gyro data. Discontinuities that were not due to thruster firings were marked in the preparations of the data for the reductions. The medium-size hits, down to a level of about 0.1 $arcsec^{-1}$, were mostly detected in the star-mapper based attitude reconstruction. The smallest hits showed up in examinations of the abscissa residuals. All hits were treated in the data analysis in the same manner as thruster firings, except that for the smallest hits the rate discontinuity was only introduced for the spin axis and in the final along-scan attitude solution using the abscissa data.

Most sudden rate changes, however, were intended and caused by thruster firings. These thruster firings always took place at the start of a frame (2.133 s data collection period), and lasted up to 0.5 s. Their aim was to keep rotation rates and error angles within specified limits from the nominal attitude values. The thrusters were fed from one of two cold-gas tanks. Halfway the mission the first tank was empty and swapped for the second tank. The use of the cold-gas reflects in changes in the inertia tensor.

8.4 Non-rigidity

The clearest sign of non-rigidity of the spacecraft are the scan-phase jumps of up to 120 mas that can take place after abrupt thermal changes to the spacecraft, such as following the start or end of an eclipse, or following a low perigee passage. The phase jumps are recorded by the payload, which implies that the payload has rotated with respect to the mean body of the satellite, as there should be no change in angular momentum. The only way this is possible is through a counter movement of another part of the satellite. It has been suggested in Vol. 2 of ESA (1997) that these jumps are the result of thermal adjustments of the solar-panel hinges, and current observations at least partly justify that conclusion.

In the preparation of the published data only a small number of phase jumps had been identified, primarily following the end of eclipses. The new reduction has investigated the data systematically for phase jumps, and identified around 1500 with amplitudes between 7 and 120 mas. Their distribution as a function of rotation phase of the satellite is strongly directional, in particular for the positive jumps which are associated with heating up of the satellite (Fig. 8.7).

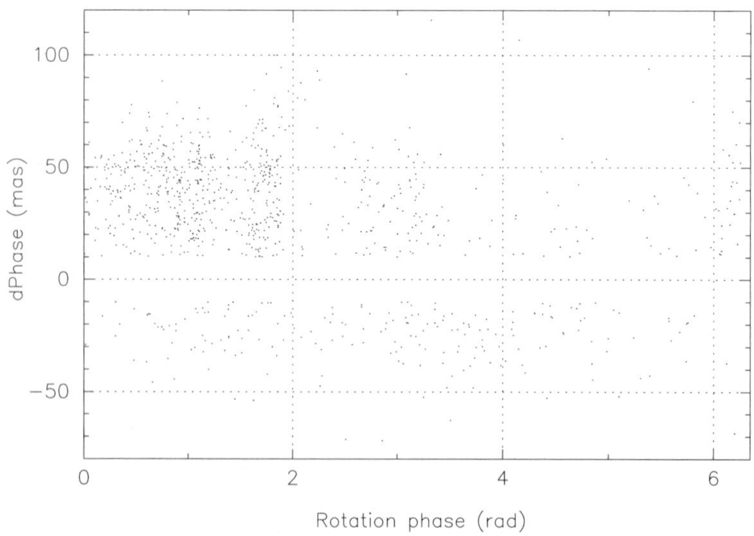

Figure 8.7. The distribution of 954 scan-phase jumps as a function of the rotation phase of the satellite. The positive jumps are associated with heating up, the negative jumps with cooling down of the satellite. A strong asymmetry in the distribution of the positive jumps suggests that exposure of one of the side panels to sunlight could be an indirect reason for these jumps

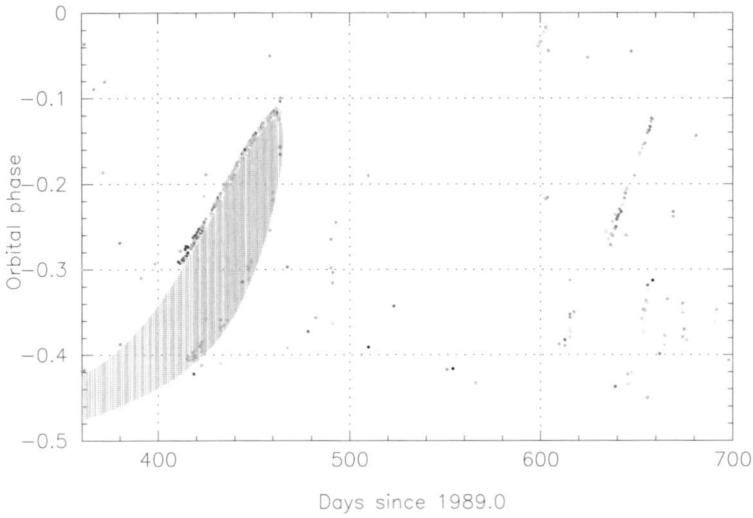

Figure 8.8. Orbit and orbital phase distribution of phase jumps for part of the mission. The grey area indicates eclipses, causing the satellite to cool down. Positive jumps concentrate after eclipses, but can also occur under different, as yet not fully understood circumstances. Negative jumps tend to occur shortly after the start of an eclipse

These jumps occur very soon after the end of an eclipse, and must therefore occur in an area directly exposed to the Sun. The thermal record of the satellite, as derived from the on-board clock, shows that the inner parts of the spacecraft experience temperature changes such as due to eclipses only about 2 hours later, and in a much less abrupt manner.

There are also, however, phase jumps which are much more strictly bound to the rotation phase of the satellite, and which can occur many hours after an eclipse, or even when no eclipse is present at all. Some of these events may be associated with conditions during the perigee passage. Both types of jumps can be observed in Fig. 8.8. Identification of these jumps allows the attitude-reconstruction software to insert appropriate degrees of freedom in the solution. Doing so removes a major source of systematic, correlated errors from the final abscissa residuals.

Chapter 9

THE TORQUES ON HIPPARCOS AS OBSERVED OVER THE MISSION

9.1 Relation between attitude and torque reconstruction

In the new reduction of the *Hipparcos* data the attitude reconstruction has been developed as a reconstruction of the underlying torques rather than a modelling of the error angles, as was done for the published data (Chapter 10). However, this process could only be started from a preliminary reconstruction of torques as based on the old, published solutions. From the old attitude solutions the underlying torques, as well as the inertia tensor characteristics, are derived from a comparison between rates and rate changes as a function of time, implementing all elements of the Euler equation for a rotating rigid body (Eq. 8.5). The torques as derived in the new solution provide corrections for the inertia tensor, and more direct input for the torque analysis.

In the following analysis, the underlying external torques are assumed to be a continuous function up to its second derivative in time. There are two exceptions to this assumption that we know off: possible abrupt changes in the satellite's magnetic moment due to switching on or off of electrical equipment (heaters primarily), and eclipses. The generally short pen-umbra phases of eclipses have been rejected from the data for this reason. They showed more complications in modelling than could be supported by the available data.

9.1.1 The Inertia Tensor Calibration

The ratios of the diagonal elements in the inertia tensor are easy to recognize and calibrate:

$$\begin{aligned}\dot{\omega}_x &\approx N_x/I_{xx} - f_x\omega_y\omega_z,\\ \dot{\omega}_y &\approx N_y/I_{yy} - f_y\omega_x\omega_z,\end{aligned} \qquad (9.1)$$

where f_x and f_y are defined in Eq. 8.6, and are the parameters we wish to calibrate. The specific correlations with the rates ω_y and ω_x in these relations decouple them from the fitting of the torques.

A similar calibration is made for two of the off-diagonal elements in the inertia tensor, I_{xz} and I_{yz}, which can be partially resolved from the observed rates and accelerations on the z axis:

$$\dot{\omega}_z \approx \frac{N_z}{I_{zz}} - f_z\omega_x\omega_y - \frac{I_{xz}}{I_{zz}}\omega_x\omega_z + \frac{I_{yz}}{I_{zz}}\omega_y\omega_z. \tag{9.2}$$

These elements can only be resolved partially because, as was described earlier, also the gyro angular momentum introduces such dependencies on ω_x and ω_y.

A similar situation exists for the element I_{xy}, which in principle can be calibrated from the rates and accelerations on the x and y axes, adding the following terms to Eq. 9.1:

$$\begin{aligned} \dot{\omega}_x &\approx \ ... + \frac{I_{xy}}{I_{xx}}\omega_x\omega_z, \\ \dot{\omega}_y &\approx \ ... - \frac{I_{xy}}{I_{yy}}\omega_y\omega_z, \end{aligned} \tag{9.3}$$

but, with typically $I_{xy}/I_{xx} \approx 0.01$, the accuracy of this calibration is rather poor.

9.1.2 Torque Fitting Procedures

The first step in the torque fitting is the subtraction of the contribution from the gravity gradient. The uncertainties in the off-diagonal elements of the inertia tensor affect the accuracy with which the gravity gradient can be predicted, in particular close to perigee. As here also the magnetic torques are relatively strong and not accurately predictable, and there is a likely torque contribution from the Earth's albedo, it is not possible to reverse the issue and use the gravity gradient torques to estimate the off-diagonal elements in the inertia tensor. Attempts to do so (by bringing in relevant degrees of freedom in the torque fitting) have all failed, probably because of all the complications mentioned above.

The remaining signals are fitted with magnetic-field specific dependencies, to recover the magnetic moment of the satellite, and harmonics of the rotation period of the satellite. The latter cover the contributions from the solar radiation torques and any other unresolved issues, such as the Earth's albedo torques and inaccuracies in the representation of the magnetic and gravity gradient torques.

9.2 Solar radiation torques

9.2.1 Low-frequency Components

The solar component in the torques acting on the satellite can easily be recognized, as it is dominant for most of its orbit. After subtracting the gravity gradient, the only other (small) torque components left around apogee are the magnetic torques, which will be discussed below. As an example, Fig. 9.1 shows the observed torques over two full rotations around apogee (folded with the rotation phase), after subtracting the gravity gradient contributions. The torques are shown both in the satellite coordinates and the inertial reference frame.

The torques as observed in the inertial reference frame clearly expose the three-fold symmetry of the satellite. For example, the constant offset on the inertial y axis is mainly caused by the positions of the three solar panels with respect to the COG, while the torque modulations are related to the shadows of those panels on the main body of the satellite (see Section 8.3.1). Given a clear three-fold symmetry of the satellite, this leads to an initial approximation of the solar radiation torques as follows:

$$\bar{\boldsymbol{N}}_{\mathrm{R}} \approx \begin{bmatrix} 0 \\ b_0 \\ 0 \end{bmatrix} + \sum_{n=1}^{m} \begin{bmatrix} a_n \sin 3n\Omega \\ b_n \cos 3n\Omega \\ c_n \sin 3n\Omega \end{bmatrix}, \tag{9.4}$$

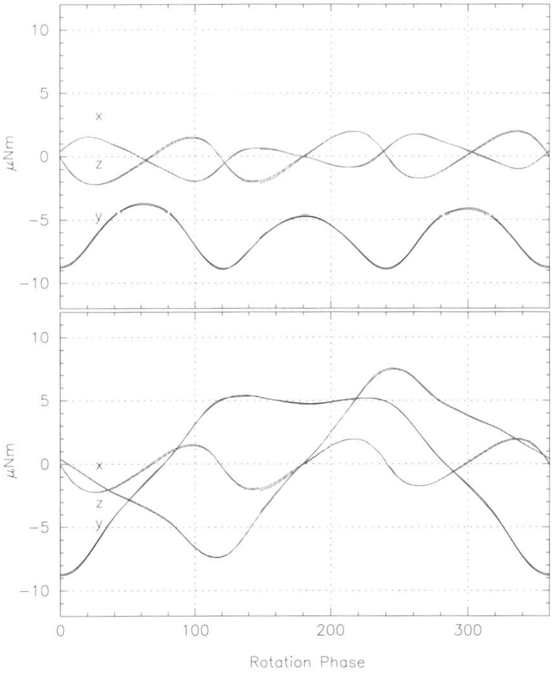

Figure 9.1. The solar radiation torques for two full rotations of the satellite around apogee. The top graph shows the torques in the inertial reference frame, clearly exposing the three-fold symmetry of the satellite. The bottom graph shows the torques in the satellite reference frame, as observed and applied in the Euler equation

where a_n, b_n and c_n are the characteristic coefficients to be determined. $\bar{\boldsymbol{N}}_{\mathrm{R}}$ refers to an inertial reference frame ($\bar{x}$,$\bar{y}$,z) in which z is parallel to the nominal spin axis, $\bar{x}$ is on the great-circle through z and the direction to the Sun as seen from the satellite and $\bar{y}$ completes the orthogonal triad. The satellite coordinate system (x,y,z) can be obtained from the inertial coordinate system ($\bar{x}$,$\bar{y}$,z) through a rotation around z by the angle Ω. Hence the solar radiation torques $N_{\mathrm{R}x}$ and $N_{\mathrm{R}y}$ around the satellite x and y axes are given by the following approximate expressions:

$$\begin{aligned} N_{\mathrm{R}x} \approx b_0 \sin\Omega \quad & +\tfrac{1}{2}\textstyle\sum_{n=1}^{m}(a_n - b_n)\sin(3n-1)\Omega \\ & +\tfrac{1}{2}\textstyle\sum_{n=1}^{m}(a_n + b_n)\sin(3n+1)\Omega, \\ N_{\mathrm{R}y} \approx b_0 \cos\Omega \quad & -\tfrac{1}{2}\textstyle\sum_{n=1}^{m}(a_n - b_n)\cos(3n-1)\Omega \\ & +\tfrac{1}{2}\textstyle\sum_{n=1}^{m}(a_n + b_n)\cos(3n+1)\Omega. \end{aligned} \tag{9.5}$$

Fitting only these harmonics is insufficient, and other harmonics need to be included too, not all of which necessarily reflect (only) the solar radiation. The complete set of fitted coefficients is given in Table 9.1. All coefficients showed, when the relative noise levels are considered, variations on a time scale of 5

Table 9.1. Harmonic coefficients in the torque fitting. ".": not used; "s": mainly solar; "m": probably magnetic

axis	x		y		z	
	cos	sin	cos	sin	cos	sin
0	m	.	m	.	s	.
1	s,m	m	s,m	m	m	m
2	s	s	s	s	s	s
3	s	s	s	.	s	s
4	.	s	s	s	s	.
5	s	s	s	s	.	.
6	.	s	s	.	.	s
7	.	s	s	.	.	.
8	.	s	s	.	.	.
9	.	s	s	.	.	s
10	.	s	s	.	.	.
11	.	s	s	.	.	.
12	.	s	s	.	.	s
13	.	s	s	.	.	.
14	.	s	s	.	.	.
15	.	.	s	.	.	s

to 10 days and in amplitudes up to 10 per cent. The variations on the main solar-radiation components for the z axis (3Ω, 6Ω, 9Ω) are clearly correlated. This would indicate either a variation in the force (but variations in the solar radiation are very much smaller) or in the structure of the satellite. The latter appears the more likely explanation: anti-correlated variations are observed on the main solar-radiation torque components for the x and z axes (see Fig. 9.2). Examination of the x and y coefficients together, transforming them back to the inertial reference system, clearly identifies these and other variations in the large constant offset in the inertial $\bar{y}$ torque. As was shown in Section 8.3.1, this component originates mostly from the solar panels, which are also the most likely elements of the satellite to show variations in positioning.

The simplest variation one can apply to the position of a solar panel is a tilt around its hinges. This would affect the vector $\boldsymbol{n}$ in Eq. 8.20. The observed variations are up to a level of about 10 per cent. This can be obtained by tilting one of the three solar panels by about 2.5°. Although it is not clear at this moment whether this was realistically possible, there seem to be few if any other possibilities available to explain these observations. A varying tilt of one

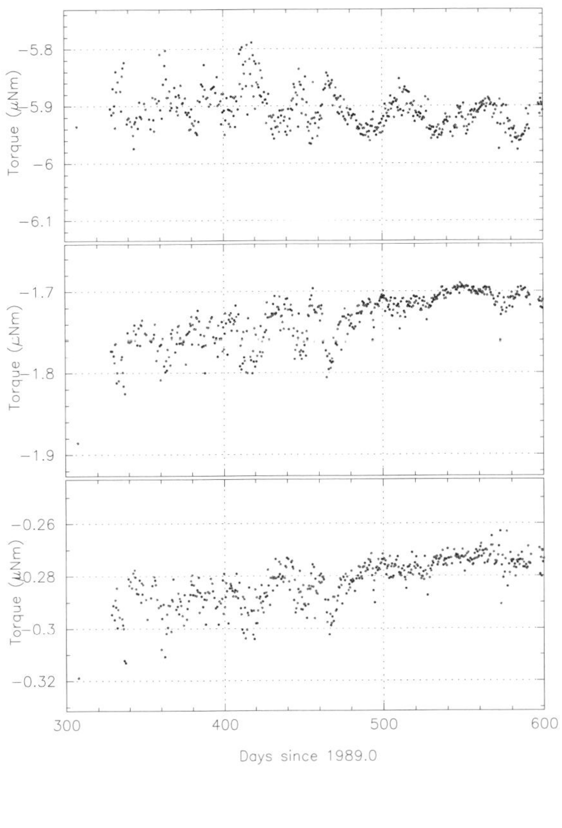

Figure 9.2. The evolution of three components of the solar radiation torques. From top to bottom: $\sin\Omega$ in x, $\sin 3\Omega$ and $\sin 6\Omega$ in z. Note the similarity of the variation features in the lower two graphs, and the opposite behaviour in the top graph

of the solar panels may also explain the reflection in the torques on the z axis, as this would affect the shadows on the lateral surfaces of the satellite.

There is one other type of variations that can be identified: those that are strongly correlated between either even or odd numbered orbits. This is a reflection of the asymmetry of the Earth's magnetic field, and is therefore a clear indication of the limitations of the magnetic-moment fitting. These effects will be dealt with in Section 9.3.

One other interesting coefficient is the constant on the y axis, as shown in Fig. 9.3. A constant torque on the x or the y axis can have two possible origins: an internal torque, such as caused by the gyros or a radiator, or through $\sin\Omega$ and $\cos\Omega$ dependencies in the solar torques (Eq. 9.4). A constant torque in y in the satellite reference system in particular requires a significant $\sin\Omega$ term on $\bar{x}$ and $\cos\Omega$ term on $\bar{y}$, if it originates from solar radiation. The systematic decrease of this coefficient, in particular during the early part of the mission, is not reflected in any other solar radiation component. The expected modulation with the varying distance between the Earth and the Sun is also not visible. Similarly, gyro-induced torques are not observed to vary this way. A remaining possibility may be the radiating of excess power by the satellite, which became gradually less as the power supply from the solar panels dwindled. The main radiator is positioned on the x axis, and would thus create a torque on the

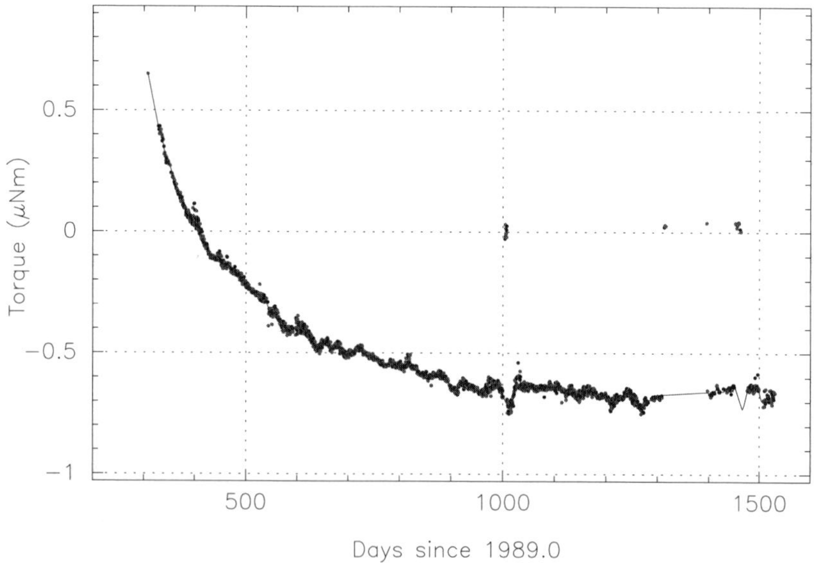

Figure 9.3. Evolution of the constant torque on the y axis. Changes in a constant torque indicate an internal rather than external origin, such as radiation of excess power. Different values observed during Sun-pointing periods (like around day 1000) could be due to the increased power from the solar panels due to a higher temperature

y axis. The changes observed during Sun-pointing mode could in that case reflect the higher efficiency of the solar panels over those periods.

9.2.2 High-frequency Components

The residuals left after the fitting described above are accumulated as functions of rotation and orbital phase of the satellite. Here we are interested in the residuals as a function of rotation phase, which reveal the higher harmonics in the solar radiation pressure. These higher harmonics reflect, for example, the effects of shadows from the solar panels on the lateral surfaces of the satellite. These surfaces contain various structures, such as the apertures of the two fields of view, excess-heat radiators, and the thermal cover for the beam combiner. In detail, these effects depend on the actual solar aspect angle, ξ, which varies within a range of 10 arcmin around its nominal value of 43°. This variation is not, as one may expect, random, but in fact highly systematic (Fig. 9.4). This makes it harder to recover a possible $\Delta\xi$ dependence of the high-frequency torques, with the range of ξ values relatively small and unevenly distributed.

Data from two satellite rotations (just over four hours) are used in the accumulation of the high-frequency residuals, using any orbit with sufficient coverage over that time span, but excluding Sun-pointing data. For each orbit a solution is made for the coefficients of Table 9.1, using all available data

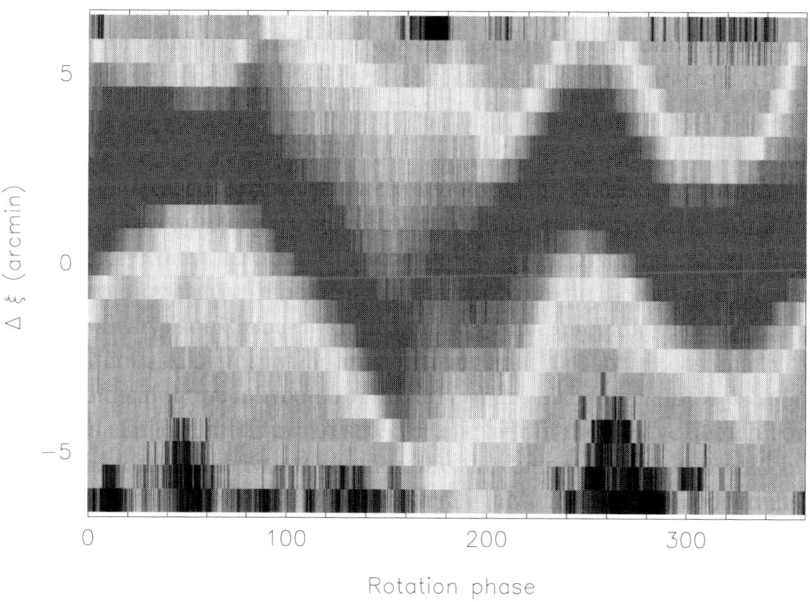

Figure 9.4. The distribution of the solar aspect angle as a function of the rotation phase of the satellite, as accumulated over the mission. The distinctive trends result from torques acting on the satellite

between orbital phases -0.3 and $+0.3$. The residuals relative to this model, when accumulated as a function of the rotation phase of the satellite, reflect the high-frequency components in the solar radiation torques. To ensure a minimum disturbance from magnetic torques, only the residuals for the 4 hours of observations around apogee are accumulated. The accumulation is done in bins, and is further resolved according to the instantaneous values of ξ. For each rotation-phase bin the accumulated values are fitted as a second-order function of ξ. The mean values, slopes and curvatures are fitted with harmonics in Ω, up to order 40 in x and y, and order 80 in z. The difference reflects the much higher resolution of the along-scan attitude solution, based on the main-grid observations, than the across-scan solutions, which is based on the star mapper chevron-slits observations. In order to eliminate correlations between these fits and those of the low-frequency coefficients, the two solutions are iterated: in subsequent iteration steps the best estimate of the high-frequency component is subtracted before the low-frequency components are fitted. This iteration converges rapidly.

The solutions for the x and y axis do not reveal much detail due to their relatively low accuracy, but the solution for the z axis is full of details, some of which can be identified with features on the lateral surfaces of the satellite (Fig. 9.5). The noise level on these solutions is also of interest: in general the standard deviation of the points per bin (after fitting) is relatively high. The

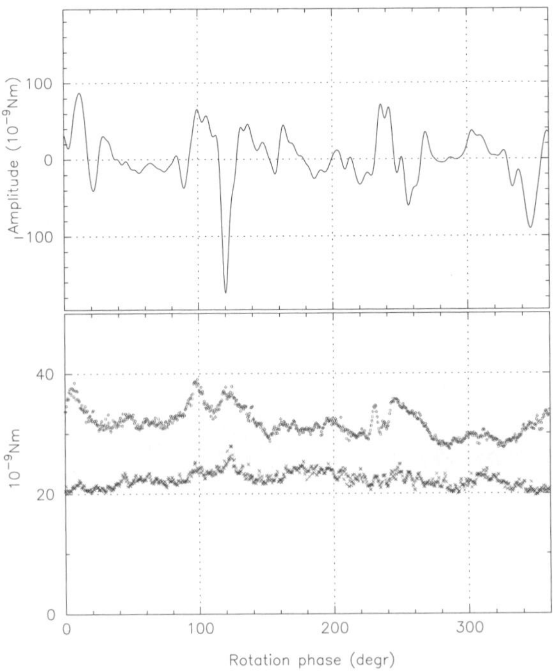

Figure 9.5. Top graph: the high-frequency component of the solar-radiation torque around the z axis, as derived from the accumulated residuals over the first 33 months of the mission. Bottom graph: the standard deviations for the binned data used for deriving the top graph. The bottom set of points is for an interval where the low-frequency components in z are very stable, the top set of points for an interval with large variations in the low-frequency components, as shown in Fig. 9.2

height of the noise level is, however, strongly correlated to the variation level of the low-frequency torque components. This is as expected if these variations are due to tilting variations for one or more of the solar panels.

9.2.3 The Power Spectra of the Solar Radiation Torques

Adding up the contributions from the low and high frequency components in the solar radiation torques, we can derive the power spectra of these torques on the three satellite axes: (Fig. 9.6). The multiples of 3Ω in z and $(3n-1)\Omega$ and $(3n+1)\Omega$ in x and y dominate as expected (Eqs. 9.4 and 9.5). The main importance of the power spectra is, however, to establish the requirements for the attitude modelling: the relation between variations in amplitudes of torques and positions. In first approximation the amplitude of the positional variations caused by harmonic n with amplitude A_n for the z axis is given by:

$$a_n = \frac{A_n}{I_{zz}(n\bar{\omega}_z^2)}. \tag{9.6}$$

With A_n expressed in μNm and a_n in mas, the following approximate relation is obtained:

$$a_n = \frac{6.5 \cdot 10^5}{n^2} A_n. \tag{9.7}$$

Thus, the highest frequency in z ($n = 80$, $A_n \approx 8 \cdot 10^{-4}$) contributes variations in scan phase (which constitute the positional variations caused by torques on the spin axis) with an amplitude of $a_n \approx 0.08$ mas. Assuming the coverage of a full cycle with two spline-node intervals (see Chapter 10), gives a node interval of around 48 s, equivalent to 4.5 telemetry formats. The average node interval used in the attitude solutions is 6 formats. Inserting the high-frequency solar-radiation torque components in the initial torque model takes out some of the stress of the torque fitting, in particular around very sharp features as observed at $\Omega = 120°$ for the z axis (Fig. 9.5).

9.2.4 Eclipses and Sun Pointing

During eclipses and Sun-pointing observations there are effectively no solar-radiation torques, exposing more clearly the effects of the remaining torques: gravity gradient and magnetic (Fig. 9.7). The absence of significant solar-radiation torques shows also in the noise level of the torque fitting, which is systematically lower during Sun-pointing periods. This again could be an indication of the "variability" of the shape of the satellite, which would be of little or no influence in Sun-pointing mode.

The situation is much more complicated during eclipses, and in particular during the pen-umbra phase. A few things are all happening at the same time: rapid temperature changes, changes in the solar-radiation torques, and,

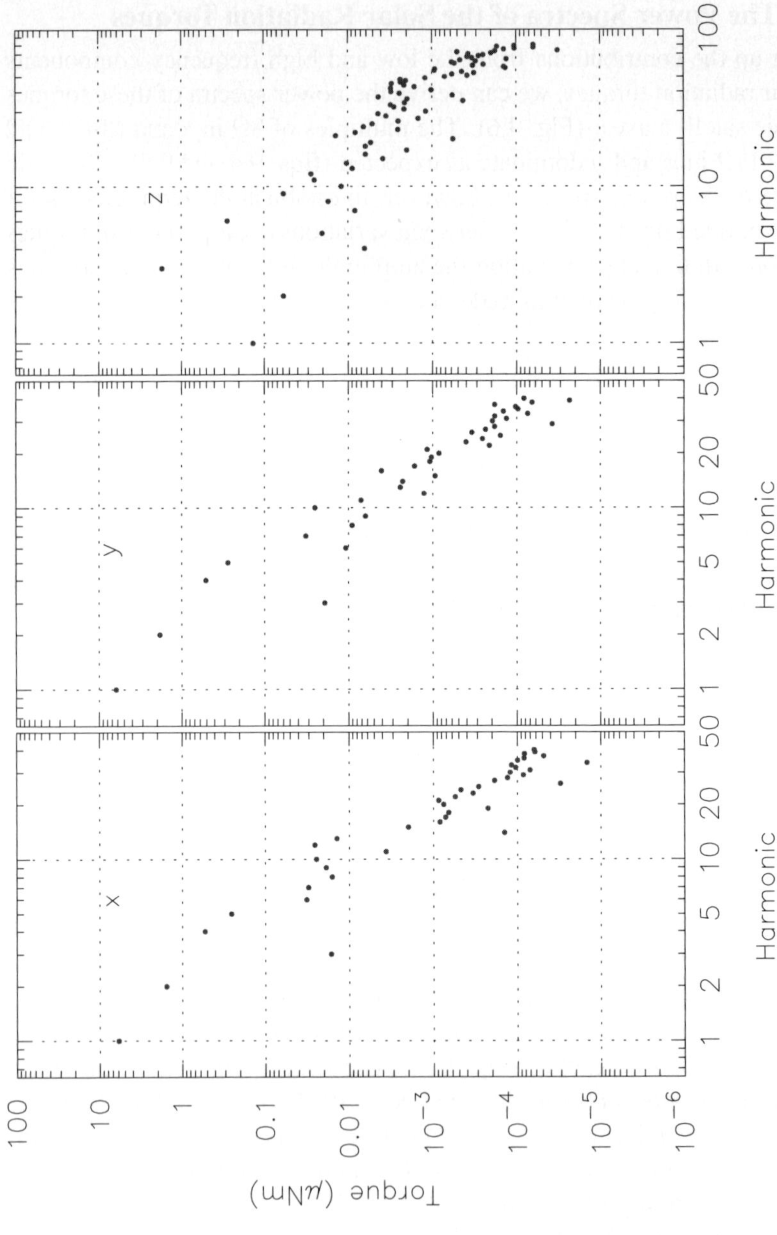

Figure 9.6. Power spectra of the solar radiation torques as registered on the satellite axes. The higher harmonics for the x and y axes are suppressed due to the limited accuracy and resolution provided by the star mapper measurements. On the z axis the resolution is generally approximately 10 to 20 times higher

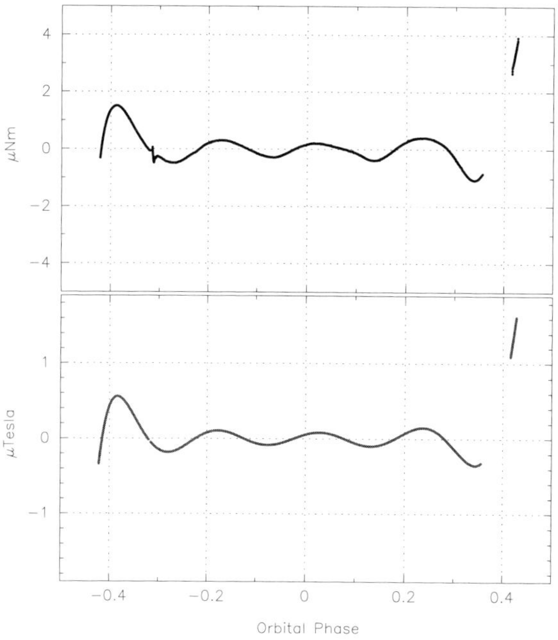

Figure 9.7. Top graph: The observed torques on the z axis over a Sun-pointing data set (orbit 1571). No solar radiation torques are present, and the gravity-gradient torque has been subtracted. The remaining variations reflect the magnetic torque, in particular due to the magnetic moment $N_z = m_y \cdot B_x$, which emerges from a comparison with the magnetic field B_x (bottom graph)

it seems, changes in the magnetic moment of the satellite, in particular along the z axis. The exact timing of an eclipse and the changes in solar radiation received during the pen-umbra phase depend on characteristics of the Earth's atmosphere conditions on the rim. The data available to model what may be going on during the pen-umbra phase is very limited, and the safest way out is to eliminate data obtained during or close to (short) pen-umbra phases. Over longer pen-umbra phases, in particular when no umbra phase is reached, the data could still be fitted.

9.3 Magnetic torques and the remaining torque variations

The first indication of the presence of a significant magnetic moment contributing to the satellites torques was obtained from an analysis of the behaviour of the $\sin\Omega$ and $\cos\Omega$ terms for both the x and z axes. These coefficients showed a regular variation with the period of the precession of the satellite's spin axis, 57 days. In other words, observed were coefficients for which the amplitude depend on the orientation of the spin axis of the satellite. This excludes solar-radiation pressure, while gravity gradient torques are well defined. There was one other telltale sign: on a small scale these coefficients showed systematic differences between odd and even numbered orbits. Due to the orbital period of just over ten hours, odd and even orbits have their apogee on nearly opposite sides of the Earth. It is only the magnetic field of the Earth that can introduce such effects, and we are thus observing the torque caused by the interaction between the magnetic moment of the satellite and the Earth's magnetic field. Introducing first the simple, and later the more advanced model of the magnetic field (see Section 8.3.3 and Fantino 2000; Fantino and van Leeuwen 2003), the magnetic moment was calibrated per orbit. Unfortunately, there is not always a sufficient amount of data available for a reliable calibration, so instead a calibration over intervals of 30 orbits (13.3 days) was introduced (Fig. 9.8). It is clear from residual behaviour that this solution is not ideal: correlated behaviour between odd and even orbits remains often present. However, the time scale of the magnetic moment variations appears to be too short for the calibration data available to resolve it completely. They are, however, also too large to attribute them to variations in the magnetic field. Although the data shown in Fig. 9.8 do not provide the small details on the magnetic moment variations, they do show that magnetic moments are far from constant, while formal errors and other noise characteristics show that these variations are mostly real. In subsequent fittings for solar-radiation torque components the magnetic moment is assumed fixed at the relevant value shown in Fig. 9.8.

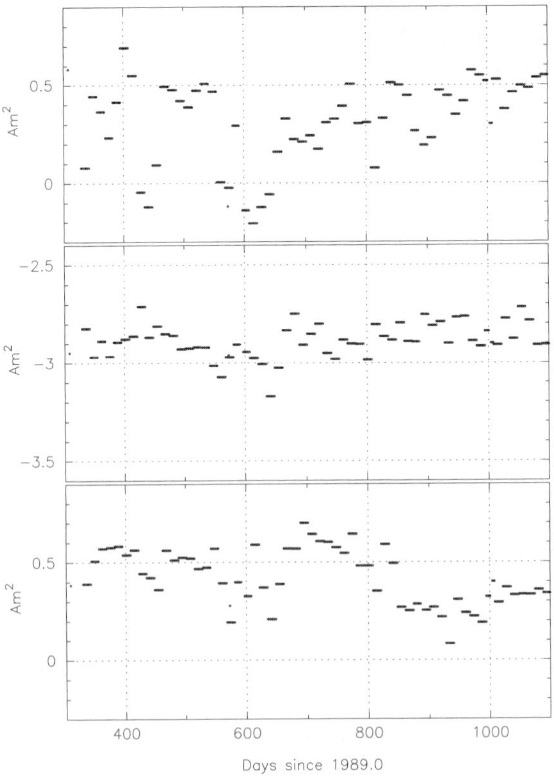

Figure 9.8. The magnetic moment of the satellite as observed over intervals of 30 orbits (just over 13 days). From top to bottom: the magnetic moment for the x, y and z axes. The magnetic moment of the satellite was clearly varying with time, in particular on the x and z axes, with some variations showing a long-term systematic trend

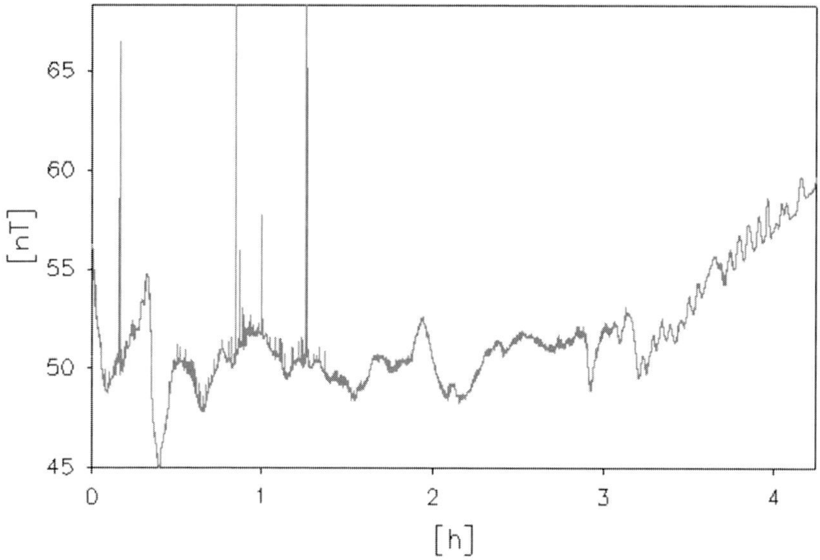

Figure 9.9. Variations of the magnetic field over a time interval of 4 hours, roughly between $5.3 \cdot 10^4$ and $6.4 \cdot 10^4$ km from the Earth surface, as observed by the magnetospheric German space mission *EQUATOR-S* (Baumjohann et al. 1999)

Next to the magnetic moment, also the magnetic field is varying. Variations of the field intensity up to 20 nT over a few minutes have been measured on the equatorial plane between 53 and 64 thousand kilometres from the Earth surface by the German space mission *EQUATOR-S* (Baumjohann et al. 1999) (Fig. 9.9). It is often difficult to model the magnetic field in sufficient detail, in particular considering the effects of the external field. The *Hipparcos* mission took place during a solar maximum, and solar wind variations, which affect the external field (Section 8.3.3), were considerable.

Some of these variations may be behind systematics observed in the residual torques. The correlations between residual torques for even or odd numbered orbits again suggest a magnetic field origin. This is shown in Fig. 9.10 for 6 successive orbits from the first week of June 1991. Several discrete features can be distinguished, some of which are repeated between either odd or even numbered orbits. It is, unfortunately, not possible to turn the *Hipparcos* observations around into corrections for the magnetic field: it is impossible to derive both the size and the direction of such corrections from this kind of data.

9.4 Predictability of the environmental torques

The predictability of the environmental torques depends to a large extent on two major factors: the overall rigidity of the spacecraft and the predictability of its magnetic moment. For *Hipparcos* both these elements do not fulfil the

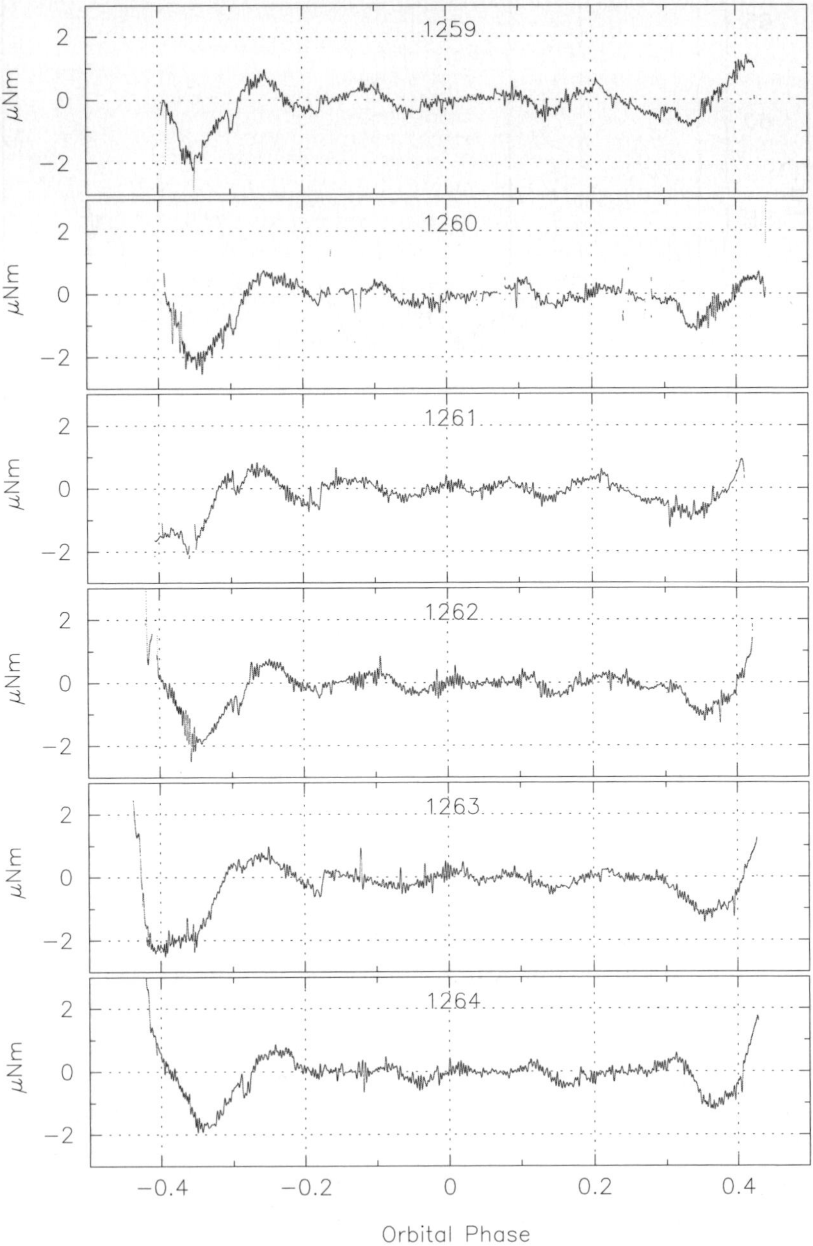

Figure 9.10. The residuals in the z torques after torque fitting for 6 successive orbits, showing the systematics and the correlations between odd or even numbered orbits. These correlations indicate a magnetic-field related origin for these residuals

requirements for torque prediction at an accuracy level required for the reconstruction of the along-scan attitude. The lack of rigidity of the spacecraft was exposed in scan-phase jumps and, what very much looks like, movements of the solar panels. Neither effect would, however, be noticeable for almost any satellite but *Hipparcos*. The scan-phase jumps are in a way the worst and the easiest: they had to be identified manually, and about 1500 have been identified, but once they are identified, their presence can be accounted for in the attitude modelling. The possible tilting of the solar panels makes the solar-radiation coefficients in the attitude modelling unpredictable to the required accuracy, but are otherwise probably fairly harmless.

The magnetic torques was a more serious problem. The character of the observed variations in the torque-fitting residuals suggests quite sudden changes. The same may happen with the magnetic moment of the satellite. These changes require modelling in the along-scan attitude with models that are unaware of these events. This can lead to modelling errors and correlated residuals in the abscissae. However, as has been shown in Section 3.2 and in Fig. 2.11, the abscissa-error correlations are observed to be at a very low level for the new reduction.

Chapter 10

FULLY-DYNAMIC ATTITUDE FITTING

10.1 Outline of the method

If we can consider the satellite to be a rigid body, then its motions are governed by the Euler equation and the underlying external and internal torques, which mostly can be considered as continuous functions of time. If, under these circumstances, we try to model the attitude in positional space with cubic splines, then we may well violate in our description the natural kinematics of the satellite. If on the other hand we try to model the underlying torques as an approximate continuous function through, for example, a cubic spline, and derive inertial rates and error angles through integrating the dynamical model of the satellite, then we may obtain a more realistic and accurate attitude reconstruction. This is the basic philosophy behind the fully dynamic model for the attitude fitting.

The first experiments on recovering dynamical information from the *Hipparcos* attitude reconstruction started early in the mission, followed by the first attempts to implement these (van Leeuwen et al. 1992). These early experiments were limited to the star mapper based attitude, and even at that level not implemented in full. It was only after the publication of the *Hipparcos* data (ESA 1997) that these studies were further developed (Fantino 2000), until a full end-to-end reduction pipeline based on an implementation of the fully dynamic attitude modelling was obtained (Fantino and van Leeuwen 2003; van Leeuwen and Fantino 2003). In what follows we shall refer to this method as the Fully Dynamic Attitude (FDA).

The principle assumption of the FDA is the existence of continuous torques affecting the satellite as a rigid body in space. In the preceding two chapters we have seen that there are some limitations to this assumption, but on the whole it holds quite well. The FDA then works as follows:

- Divide the data stream into uninterrupted intervals, i.e. intervals with (nearly) continuous data coverage. Short interruptions (less than 1 minute) can in most cases be bridged by the model;
- Create a reference torque description for the interval based on calibrations of the external and internal torque parameters;
- Identify within each such interval the thruster firings and other rate- or positional-discontinuities (hits, phase jumps), dividing the interval into sub-intervals;
- Define a reference time near the centre of each sub-interval; these will be the starting times for integrations;
- At the reference time, derive from information like the gyro data and the Real Time Attitude Determination (RTAD), starting values for the rates and error angles;
- Using the Euler equation and assumed starting rates, integrate the torques to provide rate estimates over each sub-interval;
- Integrate the rates from the starting values of the error angles to provide error angles over each sub-interval;
- Compare the predicted values of rates or error angles with observed values and fit the differences with fourth or fifth order splines;
- Derive from the fitted splines corrections to the torque model and to the starting values of the rates and error angles;
- Iterate when necessary.

Before describing in more detail how the different data streams are used in the model, and how each contributes, we first have a look at the mechanisms for fitting the residuals and deriving the corrections to the model parameters, and how interruptions are handled.

10.1.1 Fourth and Fifth Order Splines

The assumptions presented above lead to the following solution:

- Represent the torque corrections by a third order exact or B-spline;
- Represent rate corrections (gyro data) by a fourth order spline, the first derivative of which is a third order spline; insert an additional degree of freedom for the starting-rate corrections in each coordinate and for each sub-interval;

- Represent positional variations by a fifth order spline, of which the second derivative is the third order spline of the torque corrections; add for each sub-interval two degrees of freedom to account for corrections to the starting positions and rates.

The idea of a fifth-order spline may seem alien or frightening, but one should realize that for an exact solution over a given interval a fifth-order spline uses only two more degrees of freedom than a third-order spline with the same number of nodes. Details on these spline functions can be found in Appendix B.

10.1.2 Thruster Firings, Hits and Scan-phase Discontinuities

The implementation of a thruster firing, external hit, or a phase discontinuity in the FDA involves little more than defining interval boundaries, as each interval has its own starting values for rates and error angles. In cases where a phase jump or hit is registered close to a thruster firing, a remaining stretch of data can be too short to allow a reliable solution, and the affected data is rejected. The majority of phase jumps were small with respect to the accuracy of the star-mapper based attitude reconstruction. Most phase jumps are for that reason only incorporated in the final, IDT-based along-scan attitude.

10.1.3 Eclipses

Short pen-umbra phases of eclipses are excluded from the data, starting ten seconds before the beginning of the eclipse, and lasting to ten seconds into the umbra phase. There was often also much data disruption after the start of an eclipse due to the non-rigid behaviour of the satellite (scan-phase discontinuities). The amount of data affected is relatively small, and would be of little value due to the level of complexity in the attitude fitting compared to the number of data points available. The latter part of umbra phase posed in general no problems.

10.2 The integration engines

The FDA model is based on an integration of torques over the Euler equation (Eq. 8.5) and over the inertial rates to represent the error angles with respect to the nominal attitude. It is thus essential that these integrations do not introduce systematic errors themselves, as this would make the method fail. The accuracy of the integrations was tested by integrating backwards and forwards to arrive at the same starting point to within a specified numerical accuracy, much higher than the potential accuracy of the attitude reconstruction. The following two sections provide some details of the specific aspects for the two integrations. Each integration is performed over a time interval of 2.1333 s, or one frame at a time, providing reference values for rates and angles at the centre of the frame.

10.2.1 Integrating over the Euler Equation

The first step in the integration over the Euler equation is to create a continuous representation of the external torques for the integration interval. The torques, rates and error angles for any stage of the attitude reconstruction are only preserved for a single reference point per observing frame (2.1333 s). Integrations require knowledge at a higher resolution, for which a third-order polynomial fit to the torques is made. This fit, in each of the three coordinates, uses the estimates from the preceding frame, the starting frame of the integration, the end frame of the integration, and the frame after the end frame, four points in total. With these polynomial fits, the external torques are available at any time needed for the integration.

The integration interval is sub-divided in four equal-length sub-intervals, which was observed to be sufficient for the accuracies required. For each sub-interval the following steps are carried out:

- Calculate the external torques halfway the sub-interval: mid-point;
- Determine the cross-term contribution as based on the starting values of the rates, provisionally extrapolated to mid-point;
- Derive the gyro torques based on the total angular momentum of the active gyros;
- Use these data to re-evaluate the rates at the mid-point;
- Re-calculate the cross-term;
- Integrate across the sub-interval to obtain the rates at the end of the sub-interval.

The main problem in the integration over the Euler angles is the way it is affected by the changes in the rates, through the cross-product and the gyro-induced torques. The sensitivity to the accuracy of the rate estimates was further tested by introducing one extra loop, backwards and forwards again through each sub-interval. The effect of this was well within the required accuracy levels. To obtain a numerical accuracy for the integrations of 0.1 mas after 400 s, the acceleration estimates need to be free from systematic errors down to a level of 10^{-6} mas s^{-2}, or to approximately one part in $3 \cdot 10^6$, which is achieved by the method described above.

10.2.2 Integrating the Rates to Error Angles

The integration over the rates follows a similar pattern, but here the integration interval is divided in 17 sub-intervals to ensure sufficient numerical accuracy. The integration of the rates to corrections for the error angles relative to the nominal attitude is described in detail in Appendix A.

10.2.3 The First Torque Model

The first approximation of the attitude model comes from three different data sources. The torque model is constructed from reconstructed torque coefficients as obtained over the mission, adjusted to the positions of the Earth relative to the Sun, and the satellite relative to the Earth, and the pointing of the satellite axes. The starting rates are obtained from the gyro data, fitting with a second-order polynomial the 80 measurements around the centre of each thruster-firing interval. If the interval is very short, and fewer than 40 observations are available, the data are fitted to a linear function of time. The fitted values for the three gyros at the centre of the interval are transformed to rotation rates in the satellite coordinates. These transformation coefficients have been determined from comparisons between reconstructed rates (based on star mapper and IDT data) and observed gyro rates (see Section 13.2).

The starting values for the error angles are in a similar manner obtained from the Real Time Attitude Determination (RTAD), which is the attitude according to the on-board computer for the time of observation. Here the interval chosen is 70 frames, fitted with a third-order polynomial. Fitting the error angles could be difficult in the presence of large updates instigated from the ground to assist the on-board attitude convergence. These cases have been "corrected" by applying the error-angle updates to all preceding data, thus bringing the data from before the update in line again (Fig. 10.1).

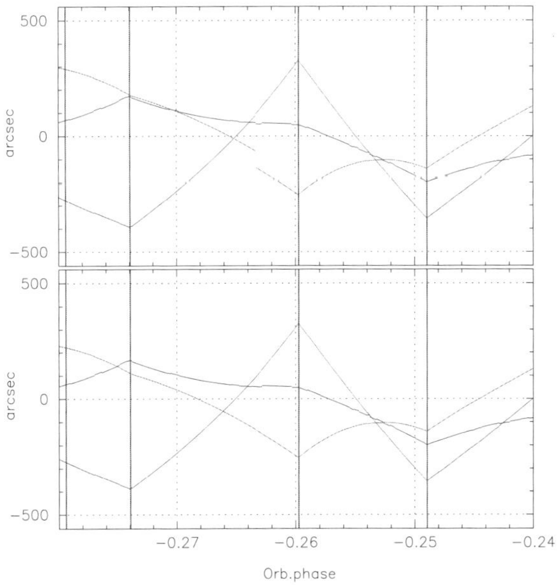

Figure 10.1. An example of an error-angle update from the ground station at orbital phase ≈ -0.2633 (top graph). The graph shows the evolution of the three error angles over a time interval of just over 25 min for orbit 145 (8 January 1990). The update is "removed" by applying it in retrospect to all preceding data (bottom graph)

10.3 Implementing the gyro data

10.3.1 Correcting Gyro Data for Drift

Before the gyro data can be used, they have to be corrected for drifts. Three regular types of drift are observed in the gyro data: the drift zero-points, slowly changing over the mission, trends at a time scale of hours (Fig. 10.2), reflecting the thermal conditions outside the spacecraft, and spin-synchronous drift modulations (Fig. 10.3). These drifts and the way they are derived are described in more detail in Section 13.2, page 335.

The mean gyro-drift-modulation curves have been determined on the basis of all suitable material collected over the mission, and in subsequent reductions these curves rather than the individual calibrations are used to correct the data. The mean drift and trend over an orbit was still calibrated and applied individually per orbit.

10.3.2 Modelling the Rate Corrections

Every step in the attitude reconstruction consists of modelling the differences between a set of observations and their predicted values based on the assumed attitude model. The modelling parameters are used to improve the attitude model: the torque model and the starting rates and angles at the centre of

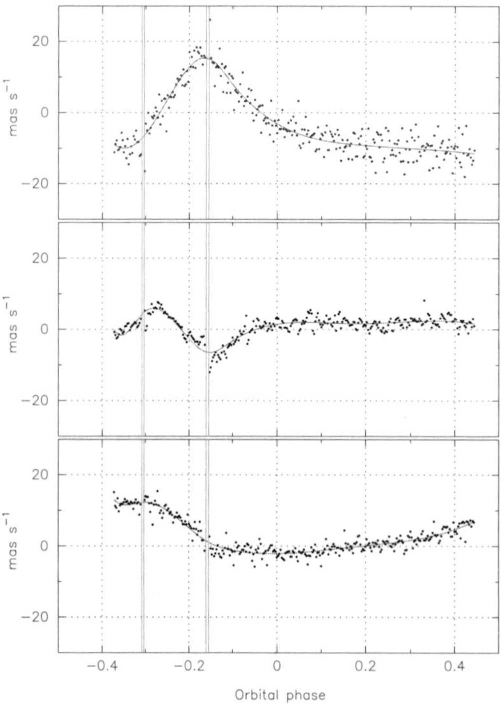

Figure 10.2. Trends in the gyro drift over an orbit, reflecting temperature changes due to an eclipse (indicated by the interval between the vertical lines). Each dot represents the average over 25 consecutive observations. The graphs show from top to bottom: gyro 1, gyro 2 and gyro 4

each thruster-firing interval. In the case of using the gyro data as observations, only torques and starting rates are corrected.

The rates-to-gyros transformation matrix G is used to derive predicted values for each gyro measurement. The differences between the predicted and the observed measurements are collected over five telemetry frames, 10.67 s. There is no information contained in those data at a higher resolution. In addition, at full resolution the interpretation of the data is often hampered by the digitisation of the measurements for the telemetry stream.

Per gyro, the data are fitted with a fourth-order spline function, with separate zero points added for all but the first thruster-firing interval. This can be seen in Fig. 10.4 for a relatively short stretch of data. The nodes in the spline are spaced according to the noise on the gyro data, and according to the orbital phase of the data. Data with a lower noise level can be used to extract more detailed information. Closer to perigee more nodes are needed to accommodate the more rapidly changing, and more difficult to accurately predict, torques. For the example shown in Fig. 10.4 the solution contained 2831 observations for 150 unknowns, of which 56 represent the starting-rate corrections for the thruster-firing intervals, and 94 the spline function (fourth order with 89 nodes). The gyro noise in this example is among the lowest observed

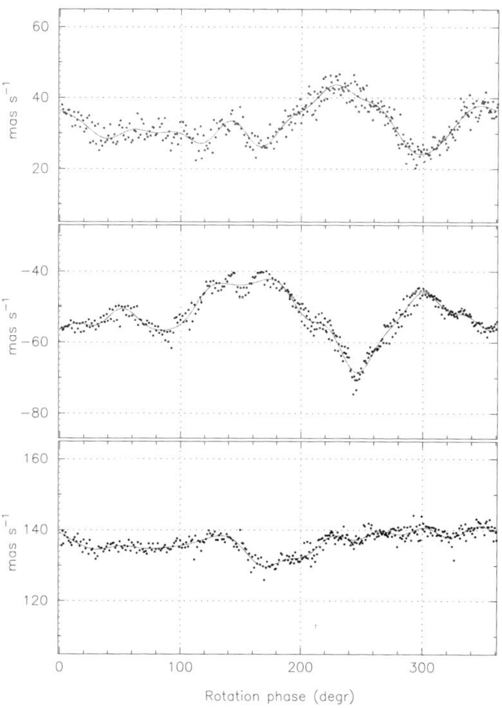

Figure 10.3. The modulated component of the gyro drifts as observed for orbit 1264. These drifts persisted almost unchanged over the mission. Their origin is still unknown. Each dot represents the average over 25 consecutive observations. From top to bottom: gyro 1, gyro 2 and gyro 5

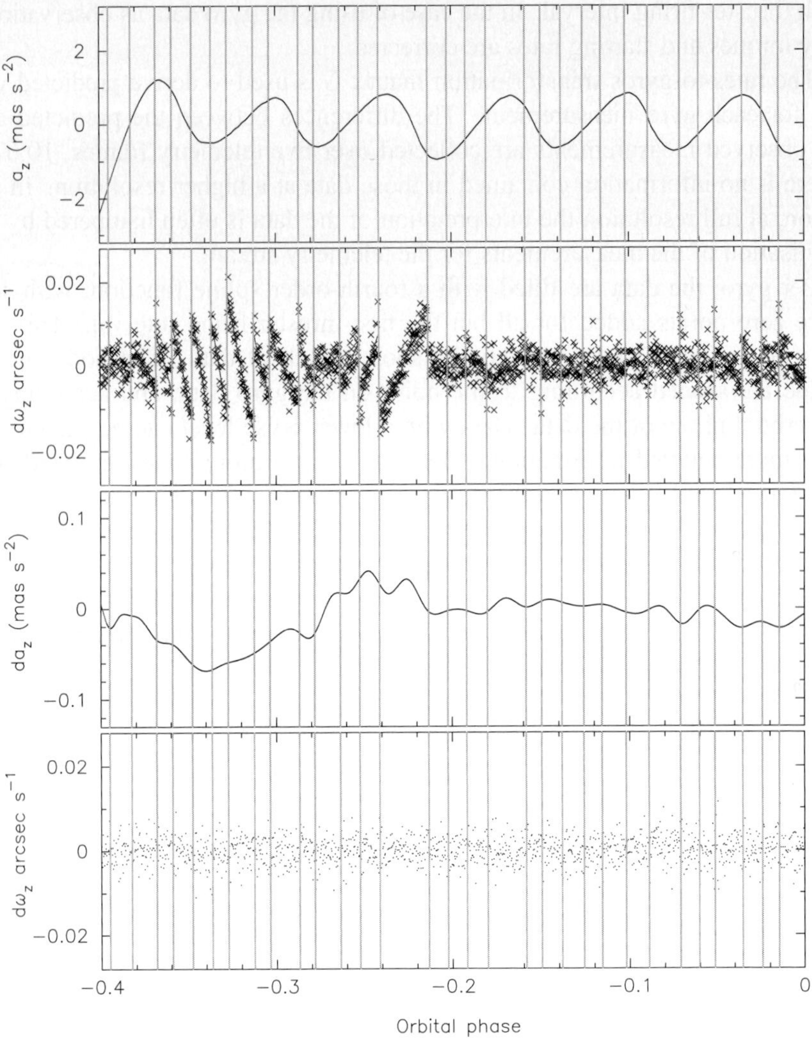

Figure 10.4. The first stages in the attitude modelling, from top to bottom: the *a priori* torque model; the residuals of the gyro data with respect to this model, and the fitted curve; the torque correction as derived from fitting the gyro residuals; the gyro residuals after correcting the torque model. Instances of thruster firings are shown by grey lines. The data are for the z axis and gyro 5 in orbit 1264. The torques are shown expressed in the resulting accelerations (1 mas s^{-2} $\approx 2.4\mu$Nm)

during the mission, so in general the number of nodes was lower, the intervals longer, and there have been more observations per estimated parameter.

The first derivative of the fitted function, a third order spline function without any discontinuities, is used to correct the torque model. The rate corrections are evaluated at the centre of each thruster-firing interval, and applied accordingly. In both cases the inverse transformation G^T has to be applied to the results before applying them to the model. No iteration is needed. The new attitude model is used as input for the star mapper data reduction and is the reference model for the star-mapper based attitude reconstruction.

10.4 Implementing the star mapper data

10.4.1 Characteristics of the Star Mapper Data

The star mapper of the *Hipparcos* satellite consisted of two sets of unevenly spaced slits, one set perpendicular to the scan direction, and one set at ± 45 degrees (Fig. 2.4, page 47). The unevenly spaced slits (Fig. 10.5) made it possible to locate any transit to a specific reference line on the grid. The light behind the entire slit system was split in a blue and a visual pass band, and sampled at 600 Hz. At the start of the mission, a count rate of 10 at 600 Hz corresponds roughly with a 9th magnitude star. Stellar images from either field of view and through either of the slit groups were all recorded in the same data stream. This data stream, in its entirety, formed the input for the *Tycho* mission (Hoeg et al. 1997, 1998, 2000b,a; Fabricius et al. 2002), and provided B − V photometry for the main mission.

The star mapper data stream for the main mission consists of strings of 250 samples cut from the *Tycho* data stream in ten fixed units of 25 samples. The

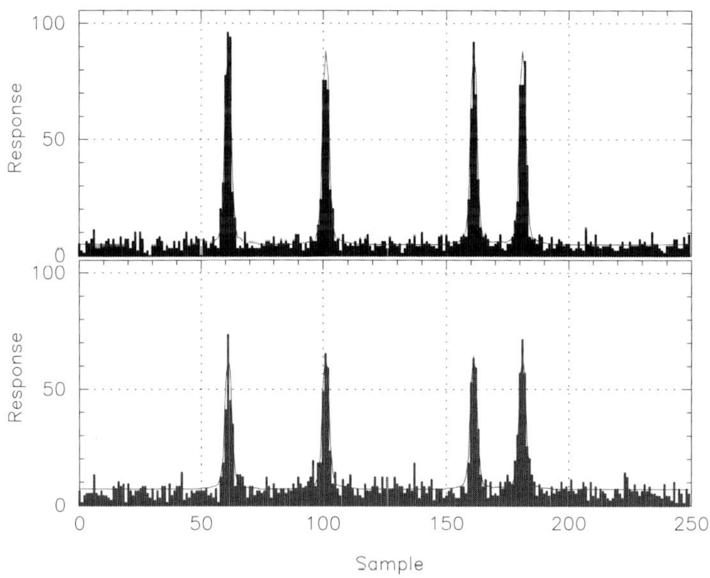

Figure 10.5. A transit through the vertical star mapper slits for HIP 41168 (V = 7.12, B = 7.00) at the start of the mission. Top: B counts, bottom: V counts

cuts were made roughly centred on the time of a transit as derived from the star catalogue and the on-board attitude, the RTAD. When the RTAD had not properly converged, these cuts could contain only part or none of the information of the star. The same string could also contain part or all of the transit from one or more different stars, situated in either the same or the alternative field of view, and coming from the same or the alternative slit group.

The star mapper detectors were very sensitive to the radiation environment, creating at times a background signal that was too high to even detect the brightest of stars (Daly et al. 1994) (Fig. 10.6), and leading to loss of on-board attitude control. The disentanglement and analysis of the star mapper data stream was therefore far from simple, and ultimately only a relatively low fraction of the transits (depending mainly on the intensity of the background) could be preserved for use in the attitude reconstruction.

The analysis of the star mapper data involves detection of the signal, which is not always trivial in the presence of a relatively high background, followed by fitting the response and transit time using a set of single-slit response functions. Different sets of response functions are required for the two pass bands, the two slit groups, the two fields of view and the upper and lower branches of the slits, 16 combinations in total. Each single-slit profile was carefully calibrated from the science data. The spacing of the slits is based on the (in-flight) calibrated grid geometry and the instantaneous scan velocity, derived from the

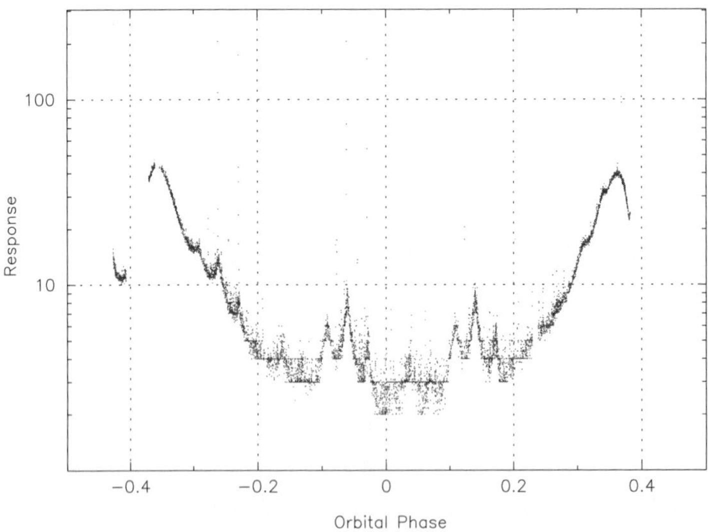

Figure 10.6. The background in the V-channel for orbit 75 (December 1989). The peaks towards perigee are due to the radiation belts. The double peaks around apogee are the transits of the two fields of view through the galactic plane. Transits through the ecliptic plane (zodiacal light) can also be distinguished. Digitisation of the response occurs at low photon-count levels

attitude model. For the vertical slits the scan velocity is simply the rate around the z axis, but for the inclined or chevron slits the scan velocity depends on the field of view and whether the transit goes through the upper or lower branch. The vertical velocities for the two fields of view are given by:

$$\begin{aligned} \omega_p &= \omega_y \cos(\gamma/2 + \mathrm{d}\gamma) - \omega_x \sin(\gamma/2 + \mathrm{d}\gamma), \\ \omega_f &= \omega_y \cos(\gamma/2 - \mathrm{d}\gamma) + \omega_x \sin(\gamma/2 - \mathrm{d}\gamma), \end{aligned} \tag{10.1}$$

where $\gamma/2$ is half the basic angle, and $\mathrm{d}\gamma = 0.0125$ rad represents the offset of the inclined slits from the centre of the main grid. Secondary offset corrections, relating to the actual ordinate of a transit, are small and ignored at this stage. The effective scan velocity for a transit through the inclined slits is now:

$$\begin{aligned} \omega_{i,p} &= \omega_z \mp \omega_p \\ \omega_{i,f} &= \omega_z \mp \omega_f, \end{aligned} \tag{10.2}$$

where the plus sign applies to the lower branch and the minus sign to the upper branch of the slit group.

The local background for a star mapper transit is derived from the mean of the 18 samplings around the median of the intensity distribution for the 250-sample extract. Where possible, the actual background applied is based on Kalman filtering of accumulated results to reduce the noise level.

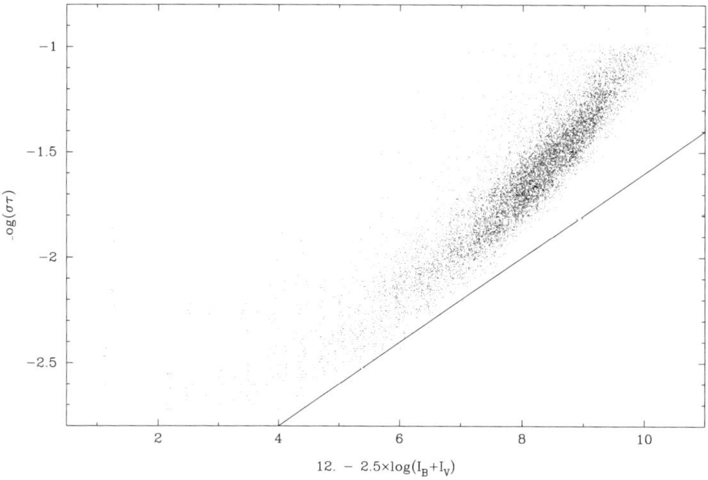

Figure 10.7. The formal standard errors (σ_T, in arcsec) of the transit time determinations as a function of the total signal intensity, transformed to an approximate magnitude scale. The graph shows results for 9400 reduced star mapper transits obtained over a period of 9 hours in January 1990. The line represents the development of accuracies when only affected by the photon noise of the signal

Intensities and transit times are estimated in an iterative loop rather than simultaneously: the intensities by fitting the responses for the four slits to the single-slit response profiles:

$$\begin{aligned} R_{\mathrm{B},i} - b_{\mathrm{B}} &= I_{\mathrm{B}} \cdot f_{\mathrm{B},i,\tau} + \epsilon_{\mathrm{B},i}, \\ R_{\mathrm{V},i} - b_{\mathrm{V}} &= I_{\mathrm{V}} \cdot f_{\mathrm{V},i,\tau} + \epsilon_{\mathrm{V},i}, \end{aligned} \tag{10.3}$$

where $R_{\mathrm{B},i}$ and $R_{\mathrm{V},i}$ are the observed responses in B and V, b_{B} and b_{V} the backgrounds, $f_{\mathrm{B},i,\tau}$ and $f_{\mathrm{V},i,\tau}$ the response curves at sample position i and for the reference position τ, and I_{B} and I_{V} the estimated intensities. Only data points for which the expected response is at least 10 per cent of the maximum response are used in the solution. More than one profile may be fitted at any one time, accounting for double stars and accidental superpositions of transits.

The transit times are obtained by fitting the differences between the observations and the fitted response profiles to the time derivative of the profiles:

$$\begin{aligned} R_{\mathrm{B},i} - b_{\mathrm{B}} - I_{\mathrm{B}} \cdot f_{\mathrm{B},i,\tau} &= \mathrm{d}\tau\, I_{\mathrm{B}} \frac{\partial f_{\mathrm{B},i,\tau}}{\partial i} + \epsilon_{\mathrm{B},i}, \\ R_{\mathrm{V},i} - b_{\mathrm{V}} - I_{\mathrm{V}} \cdot f_{\mathrm{V},i,\tau} &= \mathrm{d}\tau\, I_{\mathrm{V}} \frac{\partial f_{\mathrm{V},i,\tau}}{\partial i} + \epsilon_{\mathrm{V},i}. \end{aligned} \tag{10.4}$$

These equations are used to solve $\mathrm{d}\tau$ simultaneously for the B and V counts. Applying the reference position correction to the response curves completes the iteration cycle.

The final identification of transits is based on the agreement between the responses I_{B}, I_{V} and the expected responses based on the photometric calibration for the specific orbit and the magnitude and colour index given in the catalogue. Large-amplitude variables are excluded from the star mapper data stream. The formal errors on the transit times, when transformed to angular displacements, are generally between 0.10 and 0.01 arcsec (Fig. 10.7).

10.4.2 Attitude-model Corrections from the Star Mapper Data

The differences between the observed and predicted star mapper transit times form the input for the next stage in building the attitude model. The corrections to the model are expressed in fifth-order splines with specific zero and first order terms for each thruster-firing interval. The star mapper observations for the vertical slits describe directly the corrections for the z axis, but those for the inclined slits are a combination of the three axes, depending on field of view and upper or lower branch. If the functions $g_x(t)$, $g_y(t)$ and $g_z(t)$ represent the corrections on the three axes, then the residual star mapper transit positions are described by:

$$\mathrm{d}p = f_1 g_x(t) + f_2 g_y(t) + g_z(t), \tag{10.5}$$

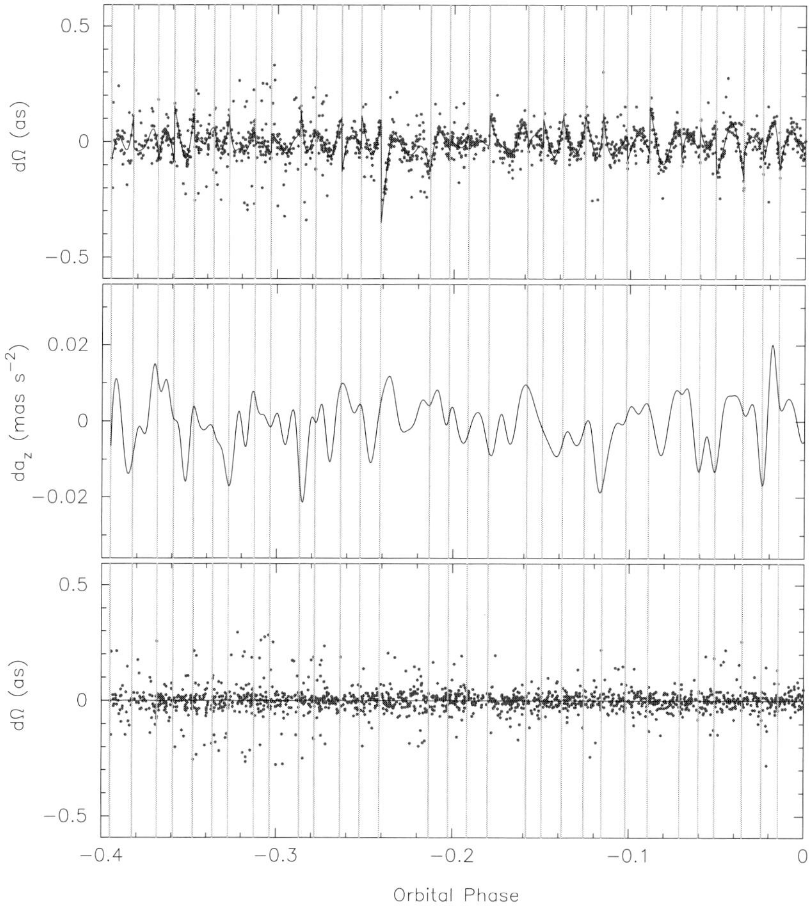

Figure 10.8. The second stage in building the attitude model, incorporating the gyro data. Data for the vertical slits are shown. Top: The transit residuals with respect to the first stage model, and the fitted curve. Middle: The acceleration corrections for the z axis as derived from the fitted curve. Bottom: The same residuals, after implementing the corrections to accelerations and zero points in rates and angles. The grey lines show the instances of thruster firings

where f_1 and f_2 are zero for the vertical slits, and functions of γ and $\mathrm{d}\gamma$ in the same way as described above in Eq. 10.1 and Eq. 10.2 for the inclined slits. Part of this fit can be seen in Fig 10.8.

This solution requires three to four iterations before the corrections for the x and y axes settle down. The extraction of the attitude-model corrections operates along the same lines as for the gyro data: the torques are corrected by the second derivatives of the three fitted functions $g_x(t)$, $g_y(t)$ and $g_z(t)$. At the centre of each interval corrections to the starting rates and angles are obtained and applied.

10.4.3 Instrument Parameters

The star-mapper geometry, measured with respect to the scan direction and the position of the main grid, is not fixed over the mission, and required calibrations. A complete set of parameters was fitted during preliminary test runs of the reduction software. Algorithms and methods used for these calibrations (simultaneously with the attitude fitting) are presented in Appendix D. For later runs most parameters were fixed on values obtained by fitting the collected calibration values as a function of time. Up to three parameters were re-determined: the basic angle as applicable to the star mapper slits, the grid rotation, and the separation between the vertical and inclined slits. A fourth parameter, the separation of the star-mapper grid from the main grid, which is affected by the focus of the telescope, is obtained as part of the next step in the attitude modelling. This parameter, shown in Fig. 10.9, is determined from the mean difference in scan phase between the star mapper determination and that of the main grid.

10.4.4 Accuracy of the Star Mapper Based Attitude Model

Tests carried out during the mission confirmed that the star-mapper based attitude fulfilled the basic requirements of the mission, with standard errors below 0.1 mas (Donati and Sechi 1992). This limit was set by the methods used in the great-circle reduction, where transits were projected on a reference great circle, based on the star mapper model of the attitude. In the current study,

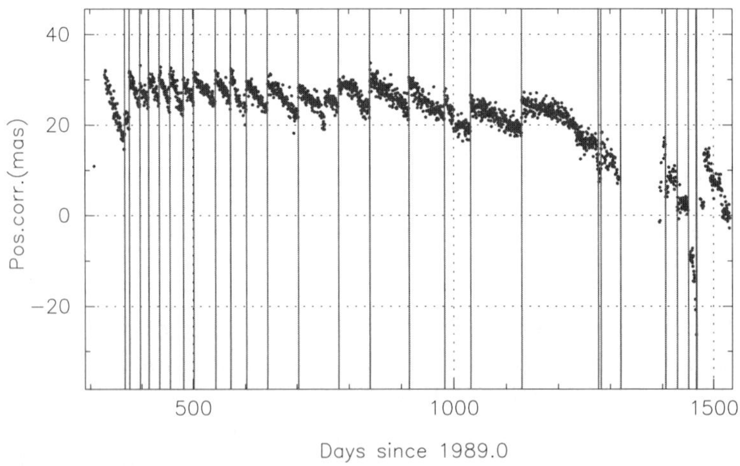

Figure 10.9. The correction to the nominal position of the chevron slits over the mission, as determined from a comparison between the along-scan attitude based on the star mapper data and as based on the main grid data. The vertical lines represent refocusing events. Compare with Fig. 2.12

these projections are no longer required, and the accuracy is no longer a critical parameter. However, observations of the performance of the star-mapper based attitude showed that although for most of the time the required limits were easily reached, at times of high background uncertainties were often higher and would have caused additional noise on the abscissa residuals in the published data.

10.5 Implementing the IDT transit data

The preceding steps in building the attitude model provide a reference frame against which the IDT-based along-scan attitude is fitted. This is a critical step for the *Hipparcos* reductions, as it provides the reference frame for the abscissa measurements, which form the input for the astrometric parameters. The preceding steps ensured that this final adjustment can be made relative to a reference frame with proper dynamical characteristics: based on a description of the torques by continuous functions, and integrations of those torques over the Euler equations. This stage of the solution is as critical for the final accuracies in the catalogue as is the stability of the basic angle. The along-scan attitude solution, when performed correctly, is the tool that provides the link between the observations made in the two fields of view, and thereby makes it possible to determine absolute parallaxes for the whole sky. In the *Hipparcos* mission this issue was partly resolved by the use of an Input Catalogue (ESA 1992), which ensured that the density of selected stars on the sky was reasonably even. It was only later realized how critical evenness in weight (or integrated intensity over an area of sky) is for the success of the mission.

10.5.1 Characteristics of the IDT Data

The input for the final along-scan attitude corrections is the same as the input for the astrometric parameters, namely the modulation phase β_3 and its formal error as defined in Eqs. 2.33 and 2.47, Section 2.2. These accuracies reflect the accumulated photon count of a transit rather than the magnitude of the star, to account for the quite substantial variations in observing time.

The IDT was considerably less sensitive to background radiation than the star-mapper detectors, and in general the background signal was low (Evans et al. 1992). The IDT measurements are sensitive, however, to the presence of other stellar images, either due to duplicity or due to accidental superposition from the other field of view. The combined signal of two or more stars can still be fitted with the five-parameter model, but the relation between the phase of the first-order modulation and the position of the star becomes a complex function of the relative offset along the scan direction and the relative brightness of the other star (Chapter 4). Such data can generally not be used for the along-scan attitude determination. Data affected by the presence of other images can be recognized from a comparison between the observed and expected values of the parameters β_4 and β_5, which represent the relative amplitude and phase of the second harmonic in the modulated signal (Eq. 2.33 on page 50), and show up through a χ^2 test with the model of the optical transfer function, as described in Section 12.1, Eq. 12.2.

10.5.2 Attitude-model Corrections from the IDT Data

The differences between the observed and predicted transit times for the IDT observations form the input data for the along-scan corrections to the attitude model. The solution mechanism is very similar to that used for the star

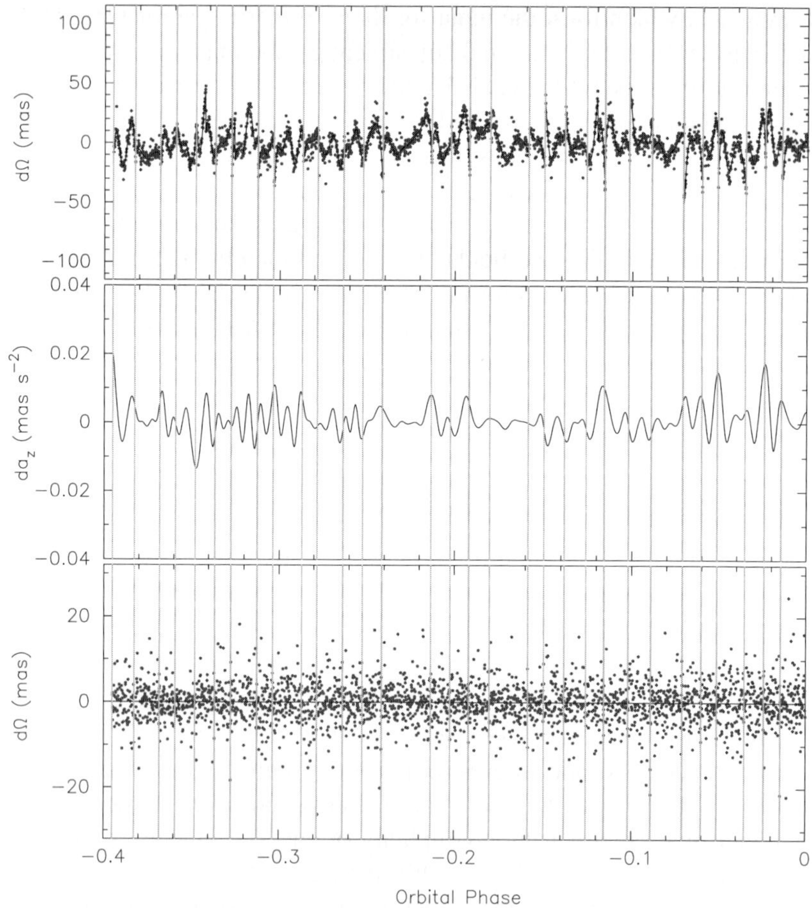

Figure 10.10. The third stage in the attitude modelling: the along-scan attitude corrections using the IDT data. Top graph: the transit residuals with respect to the star-mapper based attitude model and the fitted curve; middle graph: the torque corrections derived from the fitted curve; bottom graph: the residuals after applying the corrections to the attitude model

mapper data, and although only the torques and zero points for the z axis are updated, the attitude is at each iteration step re-determined on all three axes. An additional feature at this stage of the reductions is the introduction of the scan-phase discontinuities and a number of smaller hits to the data, playing the same role as thruster firings. Figure 10.10 shows an example of the fitting process. The along-scan attitude fitting is iterated about 4 to 5 times, the last two of these iteration steps include the fitting of corrections to the instrument parameters (see below).

The data points, transit time differences for individual frame transits, are combined over a telemetry format (five telemetry frames, 10.6 s), for each field of view. The formal error assigned to a transit-time difference is based on the Poisson error on the transit time and the accuracy of the predicted position (Fig. 10.11). When combining different frame transits of the same star, the second error will be correlated and does not average out. As is clear from Fig. 10.11, the formal errors have a considerable range.

The spacing of the nodes is quite critical: too few, and there is not enough freedom in the fit to represent the actual changes taking place, and too many will lead to modelling of noise, Poisson noise as well as noise on the astrometric parameters for the reference positions. The latter may cause the attitude to adapt itself to astrometric parameters which are not the optimal solution for the data, and thus cause perpetuating errors. The average node interval chosen is 6 formats, or just over 1 minute. Over that time the satellite rotated 3 degrees.

Perpetuating errors can occur for fields with a high concentration of bright stars, creating a lot of (local) weight in the along-scan attitude solution. To solve this problem, which when uncorrected violates a basic principle of the mission, we need artificial adjustments of weights.

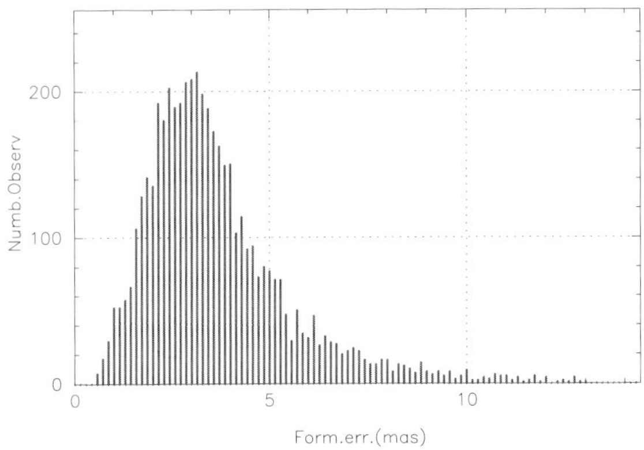

Figure 10.11. A histogram of the formal errors on the weighted mean abscissa residuals per format and per field of view. The data is for orbit 409

10.5.3 Weight Distribution Between the two Fields of View and Iterations

The formal errors of the weighted mean abscissa residuals that form the input for the along-scan attitude correction show a considerable spread, reflecting fluctuations in number and brightness of stars in the *Hipparcos* catalogue as a function of position on the sky. It is essential for the creation of absolute parallaxes that the weights of the contributions of the two fields of view at any given time do not differ too much (see also Section 3.2). Only that way, will residuals, associated with corrections of the astrometric parameters, have a chance of being recovered. This may have been a problem for the core of the Pleiades cluster observations in the published catalogue: the stars in the centre of the cluster appeared to have a parallax that was about 0.3 mas larger than the halo of the cluster (see also Section 5.2.2).

The way in which weights have been re-adjusted is as follows. For every interval between two nodes of the spline function, the total weight from each field of view is determined. If the ratio in total weight between the two fields of view for the node interval is above 2.72, the highest weight is reduced such that a weight ratio of 2.72 is obtained. The value used here has been determined empirically from the results of the iterations, and is effectively a compromise between the speed of convergence (for which a small value works better) and the residual noise left in the data (which is reduced with a higher value). The reduction factor is applied to the individual abscissa normal points when used as input to the solution. One of the effects of the weight reduction is that

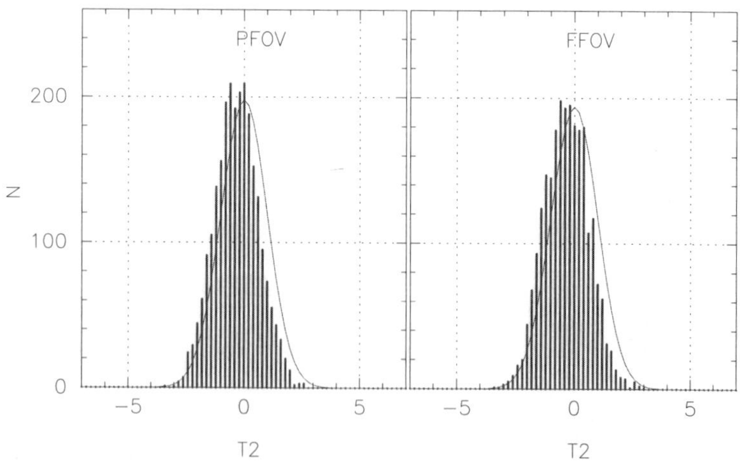

Figure 10.12. Histograms of the F2 statistics of the mean abscissa residuals in the two fields of view. The distributions are offset from the normal distribution due to application of weight adjustments, needed to ensure evening out of residuals between the two fields of view

the distribution of residuals obtained after the model fitting is distinctly non-Gaussian (Fig. 10.12).

Only a weight ratio of one could optimally re-distribute the residuals, but this would introduce considerable attitude noise, as it will reduce all data to the lowest accuracy. With a maximum weight ratio of 2.72 (sigma ratio of 1.65), iterations are needed to ensure the re-distribution of residuals. This has been observed in the iteration steps through comparing the old and new solutions. Where the old solution was dominated by attitude noise, the memory of the old solution was gradually lost in the iterations. It is also observed in the convergence of the normalized formal errors on the field-transit residuals, the distribution of which converges gradually to Gaussian through the iteration steps (Fig. 10.13).

10.5.4 Scan-phase Discontinuities

A scan-phase discontinuity is like a jump in phase (Fig. 1.12, page 34), and as such is directly related to the rigidity of the satellite: as the total angular momentum of the satellite is not affected, a jump has to be the result

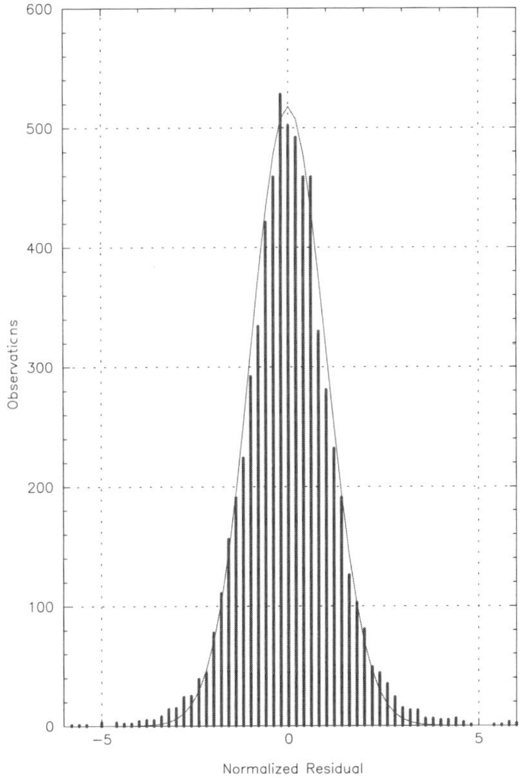

Figure 10.13. The distribution of normalized errors on the field transit residuals after the second iteration. The curve shows the Gaussian distribution for $\sigma = 1$ and the same total number of observations. The data are for orbit 409

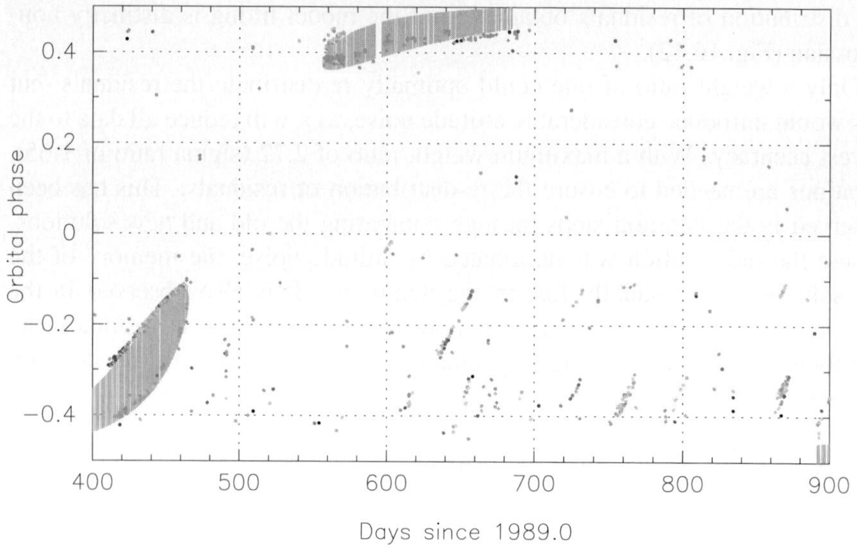

Figure 10.14. The distribution of scan-phase discontinuities over 500 days of the mission. Every dot represents a recorded jump. The grey areas are times of eclipses. Jumps are clearly concentrated towards the start and end of eclipses, but also occur systematically away from those events. As the orbital period of the satellite was close to five complete rotations, the rotation phase of the satellite for a given orbital phase changed little from one orbit to the next. Jumps tend to concentrate around a fixed rotation phase of the satellite, causing the high levels of correlation in the diagram

of one part of the satellite moving with respect to some other part. A clear indication of which part may be moving comes from the temporal distribution at which jumps take place: a large fraction is directly linked to eclipses, when the satellite moves into, or out of, the Earth shadow. These jumps appear to be immediate reactions to the strong temperature changes taking place at these instances (Fig. 10.14). Jumps are otherwise found to be concentrated around (but not restricted to) a relatively narrow interval in the rotation phase of the satellite (Fig. 10.15). The rotation phase, by definition, represents the direction of the solar radiation seen from the satellite. Both aspects point to an external element of the satellite, and the most obvious candidate is any one of the three solar panels, as was also identified in the earlier preliminary investigations on this subject. We can estimate the amplitudes of the possible displacements from the sizes, masses and inertia moments involved. Each solar panel measures 1.69 by 1.39 m and has a mass of 6.363 kg. They are connected (on the shorter side) to the spacecraft at 0.9546 m from the spin axis. Assuming an even mass distribution, the inertia moment of a single solar panel around the spin axis is 23.15 kg m^2. The inertia moment around the spin axis for the entire satellite is 459 kg m^2, giving a ratio of 1 : 19 for a single solar panel versus the

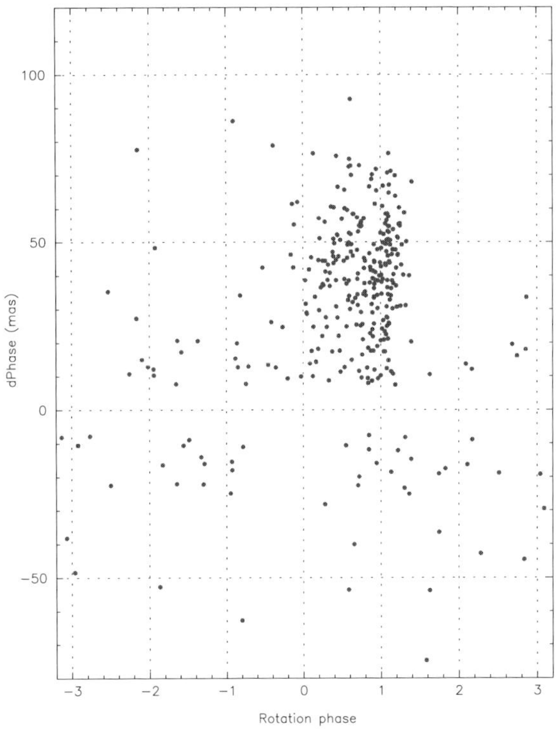

Figure 10.15. The distribution of satellite-rotation phase against the size of the discontinuity, for scan-phase discontinuities that are not directly associated with the start or end of an eclipse. The strong concentration in phase for all positive jumps (associated with heating-up) is an indication that the specific direction of the solar radiation played a role in causing these jumps. For negative jumps (cooling) there is no such preference observed

rest of the spacecraft. The rotational displacement of a solar panel required to give a 40 mas rotation of the payload is therefore just over 0.75 arcsec. At the point of attachment of the solar panels, a 0.75 arcsec rotation amounts to a shift of 3.4 μm. Thus, very small discrete displacements in the solar panel positions with respect to the spacecraft are sufficient to cause the observed scan-phase discontinuities.

Jumps tend to be negative at the start of an eclipse, and positive after, which may be expected when heating up and cooling down cause opposite movements. Jumps do occur also in a systematic manner away from eclipses, and are noticeably more frequent in the few hours after the perigee passage (Fig. 10.14). During the perigee passage the outer layers of the Earth atmosphere can affect the temperature of the solar panels, and possibly in some cases even the position.

The overall effect of phase jumps on the published data is difficult to assess. The amount of data potentially directly affected can be estimated from the number of jumps detected: around 1500. The time interval around a jump for which the attitude reconstruction is affected, if the presence of the jump is not incorporated in the attitude model, is about 150 s. Over the mission this would amount to approximately 3 to 4 per cent of the data. However, it is dif-

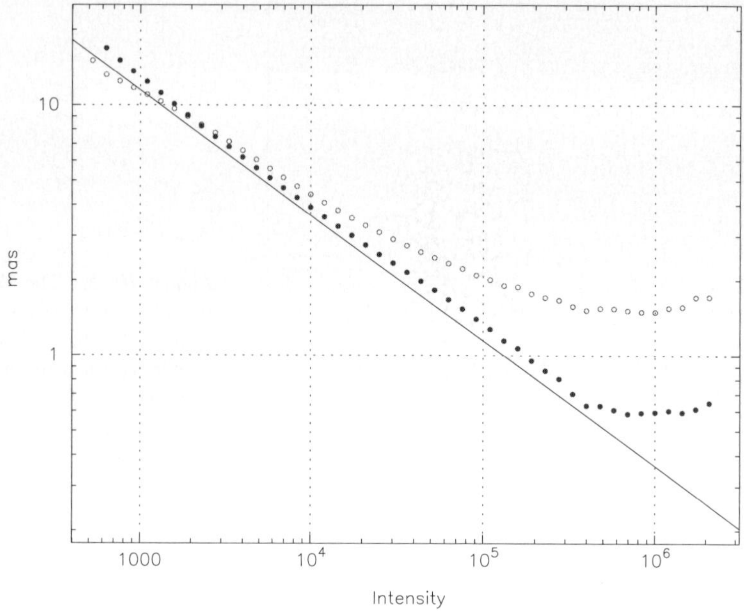

Figure 10.16. The reduction in the abscissa dispersions due to, amongst others, the detection of scan-phase jumps shows in a comparison between results before removal (open symbols) and at the end of the fourth iteration (filled symbols), when nearly all these defects had been detected and incorporated in the modelling. A further difference is visible for the low intensities, where in the earlier stages of the reduction the cut-off criterion for poor-quality measurements had been set too low, creating artificially low dispersions. The slope of the diagonal line represents the contribution of the Poisson-noise

ficult to estimate how these jumps may have affected the along-scan attitude reconstruction in the great-circle reduction due to the way the data are projected on a reference great circle. As a result, local problems in the scanning can replicate between different rotations of the satellite and between different parts of the scan. The affected time may therefore be much larger, with the effects of a phase jump repeating at intervals of the basic angle, 58°. Such error correlations are observed in the statistics for abscissa residuals (van Leeuwen and Evans 1998) for intervals up to eight times the basic angle. This "spreading out" of scan-phase problems is for example displayed by the systematic effects observed in the abscissa residuals in orbits 721 to 796 (covering days 638 to 660 in Fig. 10.14). These orbits are all affected by phase jumps that are not directly associated with an eclipse. The abscissa residuals for these orbits in both the FAST and NDAC solutions show a relatively high level of systematics, spread over all rotation phases, and in some cases resembling phase jumps. Contrary to the assumptions for the published data, statistics from the new reduction clearly indicate that these phase jumps did impose a significant,

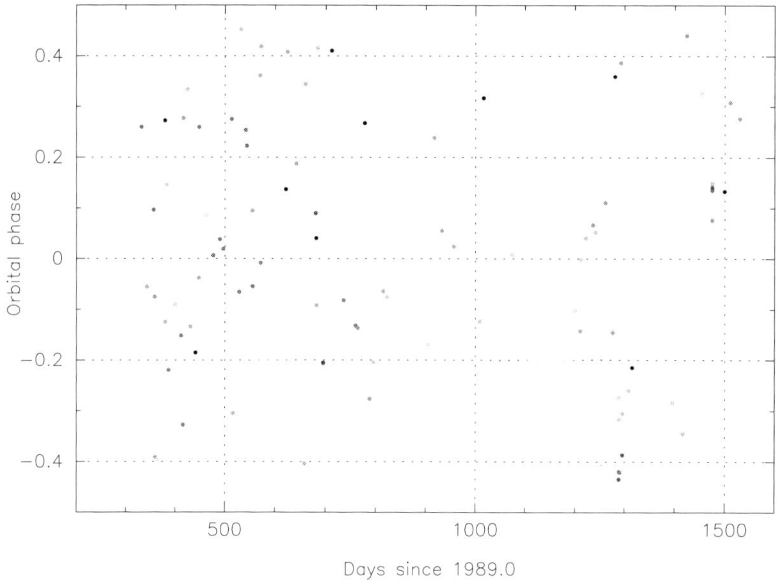

Figure 10.17. The distribution of hits as observed for the satellite from the attitude data, as a function of orbital phase and epoch. The hits plotted range from 0.2 mas s^{-1} upwards

non-Gaussian, noise component on the abscissa residuals. This is shown most clearly in the noise levels of the abscissa residuals before and after detecting and taking care of these events (Fig. 10.16).

10.5.5 External Hits

The first hits of the spacecraft were recognized by the NDAC team during initial inspections of the gyro data. A hit of the spacecraft causes a discontinuity in its inertial rates. Such discontinuities were also caused by thruster firings, but the instances of the firings were known from the satellite telemetry, and firings always took place in a designated time interval. Inertial rate discontinuities not associated with thruster firings and not in the designated time interval could safely be attributed to external hits. The largest of these caused rate changes at a level of a few arcsec s^{-1} and four of these have been recorded over the mission. In the NDAC solution these events were incorporated in the attitude model in the same way as thruster firings; in the FAST reductions the gyro data were not analysed and the affected data have most likely been rejected at some stage in the reductions.

Many more smaller hits took place, but these could not be recognized from the gyro data because of noise levels. Smaller hits do show up, however, in the behaviour of the abscissa residuals with respect to the star-mapper based attitude reconstruction, as shown by the example in Fig. 1.13 (page 35). Some

80 of these hits have been recorded, none of which had been incorporated in the reduction of the *Hipparcos* data. As with the phase jumps, the fraction of time directly affected is relatively small, about 0.4 per cent, but due to propagation in the great-circle reduction, more data may have been indirectly affected. There is no obvious dependence on orbital phase in the occurrence of hits (Fig. 10.17), which suggest that they are predominantly caused by particles not trapped in the gravitational field of the Earth. Among the strongest (and most likely real) hits, 13 are above 4 mas s^{-1}, 33 between 2 and 4, 33 between 1 and 2 and 11 between 0.2 and 1.0 mas s^{-1}.

The actual energy transfer associated with a hit producing a $\Delta\omega = 5$ mas s^{-1} rate jump is surprisingly small. With an inertia moment around the spin axis of $I_{zz} = 459$ kg m^2 and a nominal scan velocity of 168.75 arcsec s^{-1}, the change in rotational energy is about 9 nJ. The change in angular momentum provides a lower limit of the typical particle mass involved:

$$m \leq \frac{\Delta\omega I_{zz}}{v \cdot r_{\max}}, \tag{10.6}$$

where v is the velocity of the particle, and $r_{\max}$ the maximum arm-length of the impact position (about 0.95 m). At typical velocities of around 30 km s^{-1}, the masses of the particles involved are small (10^{-3} to 10^{-4} mg). However, the shape of the satellite is such that the torque for the spin direction caused by a particle hit will have been relatively inefficient in most cases, and the actual particle masses involved in these hits are likely to be larger than indicated by the angular-momentum transfer for the spin axis. The same uncertainty of where a particle may have hit, and how much it changed the rates on the other two axes (where due to the 100 times higher noise levels the measurements are too unreliable), means that we cannot obtain reliable statistical information from these events.

10.5.6 Torque Disturbances

Torque disturbances are observed around apogee during or shortly after times of high solar activity (Fig. 10.18). These may be related to the position of the bow-shock of the Earth's magnetic field, which can go down to the altitude of geostationary orbits in the presence of strong solar winds. The main effect on the data is a requirement for a significant increase in the number of nodes in the spline function, which can weaken the solution. In the NDAC and FAST data analyses increases in the number of nodes were determined automatically on the basis of the dispersion of the data. In the published data, the data sets concerned do not appear to be badly affected.

Considering that the torque modulation on the spin axis is primarily the result of a variation of Sun-lit and shadow areas on the body of the spacecraft, and solar variations are at least two orders of magnitude smaller than the variations

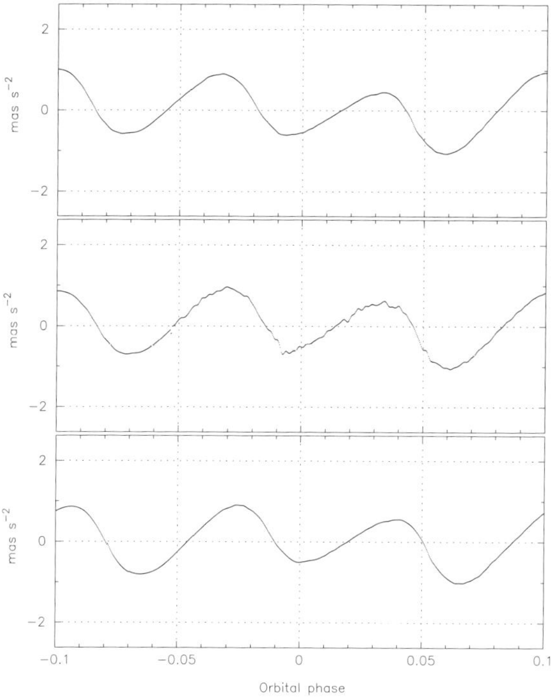

Figure 10.18. Reconstructed torques around the spin axis for one rotation of the satellite around apogee, expressed in equivalent accelerations. From top to bottom: orbits 1317, 1318 and 1320. The torques in orbit 1318 are significantly disturbed, possibly due to solar wind following high solar activity

observed here, an explanation for these phenomena may have to be found in positional variations of the solar panels, even though this appears unlikely.

Problematic attitude behaviour was also encountered around the start and end times of eclipses, when solar radiation torques change rapidly. It is difficult to retrace in the *Hipparcos* data how the NDAC and FAST solutions coped with these situations, as the same orbits are also affected by the scan-phase discontinuities. In both reduction chains the abscissa residuals for these orbits show well above average levels of abscissa-error correlations.

10.5.7 Instrument Parameters

The instrument parameters describe the relation between the physical coordinates on the grid, in particular the grid lines, and relative positions on the sky. The representation used here is that of a two-dimensional third order polynomial in the grid coordinates, with supplementary coefficients in the colour $(V - I)$. The instrument parameters are solved for differentially, i.e. as corrections to an earlier solution.

The instrument model depends quite strongly on the (partial) inclusion of the partially observed stars and is for that reason also uniquely linked to the medium-scale distortion maps (Section 2.4.2). The partially observed stars are either entering or leaving the field of view, and therefore cover the very edges of the grid, where the distortions tend to be largest. Thus, full inclusion, as was done by NDAC, or complete exclusion, as done by FAST, produces noticeable differences in the model, most of which are compensated in the medium-scale distortion map (Fig. 2.13, page 61). In the new reduction, only observations of partially-observed stars observed over at least 10 of the 16 interlacing periods (Section 2.2.3) have been included in the reductions. Thus, the instrument parameters found are somewhere between the FAST and NDAC solutions. An overview of the instrument parameters and their evolution over the mission is presented in Section 12.2.

An unavoidable feature of the instrument-parameter solution is the correlation between the first- and third-order parameter determinations. This situation is controlled in the new reduction by applying a spline fit to the third-order parameters obtained in the early test runs of the reductions. These spline fits included, where necessary, discontinuities for re-focusing instances. For the published data, FAST also applied constraints on the third-order parameter, but did this instead by applying fixed values over intervals in time. The effect of controlling the third-order parameters is to significantly reduce the variation on the first-order parameters: the scale and rotation of the grid (Fig. 10.19).

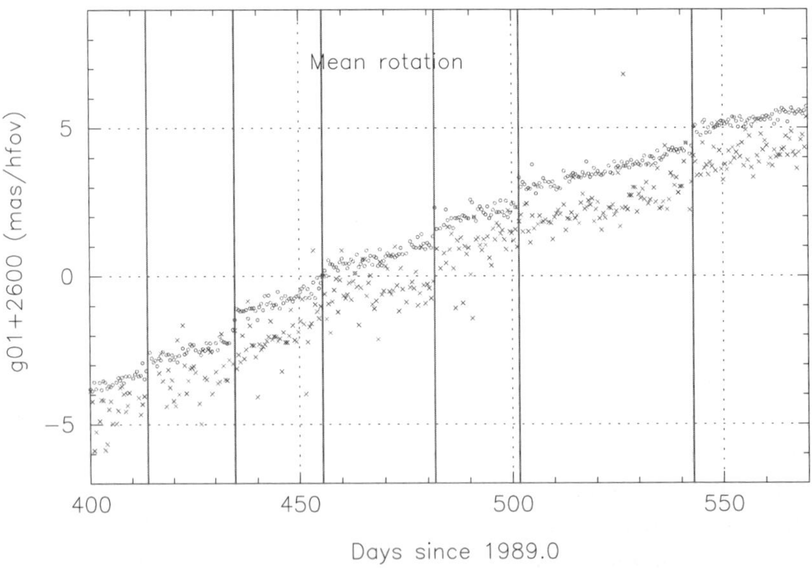

Figure 10.19. The mean rotation of the main grid as measured in the new (circles) and old NDAC (crosses) instrument parameters. The much reduced noise on the new determination is partly due to restricting the third-order terms in the solution, partly to the generally reduced noise of the attitude solution. This diagram covers about half a year of data from early to mid 1990

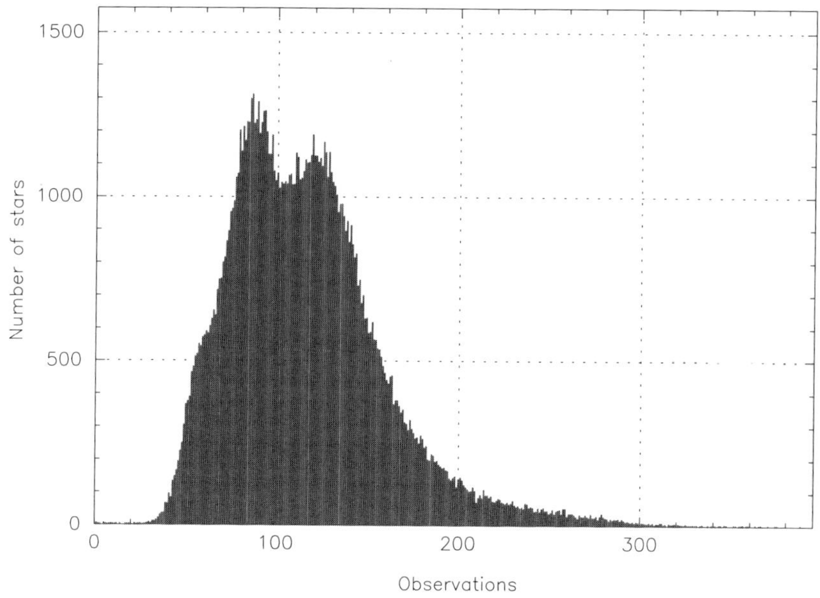

Figure 10.20. A histogram of the number of field transits per star. The bi-modal distribution results from the scanning strategy, but is much less pronounced than for orbit transits

10.5.8 Creating Field Transits

An important difference between the presentation of the published data and the current study is the level of time-resolution for the abscissa measurements. The published data used the concept of orbit abscissae: combining the frame transit data from all field transits for a star as observed over an orbit. To create orbit abscissae required projecting data on a reference great circle, introducing the possibility of projection errors whenever the reconstruction of the position of the spin axis as based on the star mapper data was not reliable. This could in particular happen at times of a high background signal. This procedure was, however, essential for the first reduction of the raw data: the initially large errors on the astrometry of the stars necessitated a combined solution of the along-scan attitude and the abscissa residuals, a process referred to as the great-circle solution (van der Marel 1988; van der Marel and Petersen 1992).

The new reduction uses the published catalogue as a starting point. The errors in the astrometric parameters could therefore be considered as small and contributing only white-noise. This allowed the de-coupling of the along-scan attitude fitting from determining the final abscissa residuals. The field transit now becomes the natural unit for those residuals. The field transit accumulates the abscissa residuals, after fitting the final along-scan attitude, for all frame transits associated with one pass of a star through a field of view. Resolving the astrometric data to this level gives a much better chance of detecting and eliminating data affected by an accidental parasitic transit from the other field of view. It also makes it possible to associate with a transit the total integrated intensity as a direct measure for its precision, and to preserve other critical information such as the mean offset of the instantaneous field of view of the IDT, the field of view for the observation, the ordinate at the time of the transit, and its distribution in time. In total, 13 678 866 field transits are collected for 117 956 stars, just under 116 observations per star. Figure 10.20 shows the distribution of the number of field transits per star, a distribution which reflects the scanning strategy through its double peak and extended tail, but these features are much less pronounced than for orbit transits (see Fig. 58 in van Leeuwen 1997). The number of observations compares with about 3.5 million orbit transits in the published data.

10.5.9 Accuracy of the Along-scan Attitude Model

The accuracy of the final along-scan attitude is difficult to assess. The precision can be derived from the residuals in the astrometric solutions for the brightest stars. In Fig. 2.16 (page 65), showing the dispersion of the abscissa residuals, the attitude noise is exposed through the dispersion of the abscissae residuals for the brightest transits. The distribution for these residuals is, however, not exactly Gaussian as the attitude dispersion will fluctuate. Further experiments indicate an average along-scan attitude noise level of about 0.7 mas at field-transit level, and a possible range from 0.5 to 1.0 mas. In comparison, the noise level on the orbit abscissae in the published data is 2 mas. With 3.5 field transits per orbit transit, this is equivalent to 3.7 mas per field transit, giving a factor five improvement for the brightest stars. However, this is precision, not necessarily accuracy.

One other way of assessing the accuracy of the attitude reconstruction is by means of the self-consistency of the data. One test of self-consistency is the recovery of the Poisson statistics for the abscissa residuals in the astrometric solutions of all stars fainter than about $\mathrm{Hp} = 4.5$. Another indicator is the distribution of normalized residuals for the field transits over one orbit: once all astrometric solutions have fully converged, these residuals will show a pure Gaussian distribution with $\sigma = 1$. As iterations progressed, this became increasingly the case (Fig. 10.13).

10.6 Conclusions

The FDA has, in its application to the *Hipparcos* data, provided an attitude modelling which is, through imposing physical constraints, considerably more stable than pure numerical approaches. The physical constraints exposed at the same time those instances where the rigidity assumption was violated, the so-called phase jumps. The implementation of the model provided an easy and physically meaningful way of accounting for these instances. Together with the implementation of a much more stringent control on the weight distribution between the observations from the two fields of view, these are the most important improvements in the new reduction that allowed obtaining the much improved astrometric data for stars brighter than magnitude 8 in the catalogue. However, the modelling of the satellite attitude could not be brought to a level where torques could be sufficiently accurately predicted purely based on a study of the external and internal torques. It appears that next to the discrete instances of non-rigidity shown by the satellite, there probably were also less easily noticeable small distortions. Some of these could be long-term distortions, that appear to show through systematic variations in the observed values of the solar radiation torques. The solar torques were further complicated by small variations in the solar aspect angle. This would cause variations in the exact location of the shadows of the solar panels on the body of the satellite, and would thus affect the solar-radiation induced torques. The magnetic moment, its probable variability and the localized variations of the Earth's magnetic field also complicated the torque model well beyond the possibility of accurate prediction. The FDA model was therefore strictly limited to two assumptions only: rigidity of the satellite and continuity of the torques affecting the satellite. Even though in some detail both these assumptions were violated from time to time, overall they provided a workable, stable and generally accurate modelling of the satellite attitude.

The extensive calibration exercise exposed some still unexplained phenomena, in particular concerning the spin-synchronous component of the gyro drifts. This component, with a specific signature for each gyro, had characteristics that seem to contradict each other. The long-term gyro drifts show a slow and partially delayed reaction on solar radiation variations due to eclipses. The spin-synchronous drifts on the other hand show sharp and accurately repeated features, which in amplitude are also linked to solar radiation variations (see also Section 13.2). A possible solution for this apparent contradiction as due to a mechanical distortion appears unlikely. A mechanical distortion could change the projection of the rotational velocity on the gyro input axes, which would mimic a drift, as the rotational velocity is nearly constant. It would require "bending" by about ± 18 arcsec of the spacecraft platform, a local displacement of the order of 20 to 50 μm to create the observed variations of ± 18 mas s^{-1} in the gyro drifts. It is unclear whether this might at all have been possible. A further complication of this possible explanation is that the gyros were suspended on the underside of the payload platform, which excludes direct exposure to Sun light.

PART V

SUMMARY OF SELECTED SPACECRAFT AND PAYLOAD CALIBRATION RESULTS

Calibration parameters monitor the health of the spacecraft and payload, and show some of the environmental strains the satellite was subjected to during the mission. The *Hipparcos* environment was dominated by passages through the Van Allen Belts approximately every 5 hours, bursts of high solar activity, and temperature fluctuations resulting from low perigee passages and long eclipses.

Chapter 11

THE MISSION TIMELINE

11.1 Trend analysis and anomalies

Payload and spacecraft monitoring are essential tools in the data analysis of an astronomical survey instrument. They may alert, for example, the data-processing about changed conditions which could require an adaptation in the instrument model. An example of how this situation occurs in the *Hipparcos* data processing concerns the basic angle. The initial assumption is that the basic angle is constant over the time interval of one orbit. The value of the basic angle as determined per orbit is monitored over the mission. At some instances the derived value is significantly larger, or in one case smaller, than expected from the trend. Each orbit showing this kind of discrepancy is investigated in more detail: the date and time of the measurements are checked in the European Space Operations Centre (ESOC) operation reports for possible anomalies, and the data is reduced in interactive mode. In nearly all cases when a drift of the basic angle is observed, this can be related to a hardware event in the spacecraft and/or the payload. In subsequent processing, data sets affected this way are no longer fitted with a fixed, but with an evolving basic angle, significantly improving the quality of the reduced data for those orbits.

This section provides a summary of events noted in the *Hipparcos* data from the trend analysis of the payload and spacecraft calibration parameters, and from information provided in bi-weekly operations reports supplied by ESOC over the mission. It identifies events and conditions that may have, and are often known to have, affected the analysis of the science data. As such, it provides a reference for identification of data points that may possibly be affected by unfavourable circumstances. It also exposes some of the problems the data analysis for this mission had to cope with.

Throughout this section, the timing of events is by orbit number and orbital phase, where the orbital phase runs from -0.5 (perigee) to 0.0 (apogee) to $+0.5$ (perigee). The relation between time and orbit number is approximated by a fourth-order polynomial $orb = \sum_{k=0,4} a_k t^k$, where t is measured in years of 365.25 days, counted from 1991.25. Table 11.1 lists the values of the coefficients. The value for a_1 is equivalent to an average orbital period of 10.651 hr.

11.1.1 Focusing

It became clear during the commissioning phase that the optimal focus for the two fields of view differ significantly, and that a compromise situation had to be adopted. The optimal focus is determined by the relative amplitude of the first harmonic of the modulated signal on the main grid (Fig. 11.1). Changes in the focus happened as a result of the payload adjusting to the space environment.

Rapid changes in the telescope focus can affect the science data. This happened during anomalies in the thermal control electronics (see below), or when a refocus command took place during data gathering. As can be seen from Table 11.2, the latter situation occurred at least 4 times during the mission.

Table 11.1. The coefficients of the fourth-order polynomial used to fit the orbit number as a function of time. The time parameter is expressed in years since 1991.25

Coeff	Value	Form.acc.	units
a_0	1159.0794	±0.0005	orbit
a_1	823.0119	±0.0009	orbit yr^{-1}
a_2	0.2081	±0.0011	orbit yr^{-2}
a_3	0.0051	±0.0007	orbit yr^{-3}
a_4	0.0015	±0.0005	orbit yr^{-4}

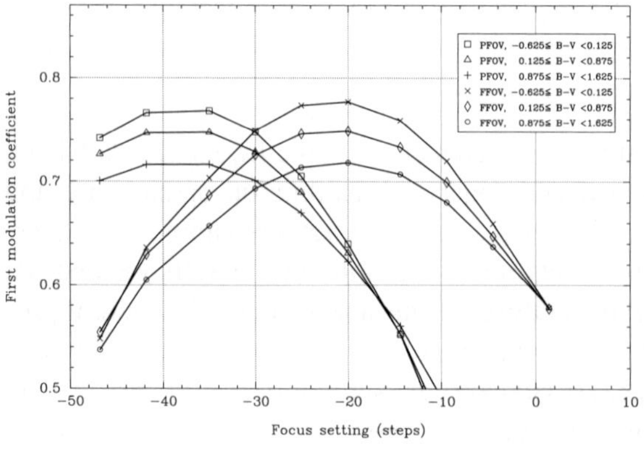

Figure 11.1. The determination by ESOC at the start of the mission of the optimal focus for the two fields of view by means of the relative amplitude of the first harmonic in the modulated signal. The maximum modulation showed no colour dependence, but a strong dependence on the field of view

Table 11.2. Instances of refocusing over the mission, identifying the time, orbit number and orbital phase. Notes: 1. refocus took place during gathering of science data; 2. time estimated, not logged; 3. no science data gathered around that time

Year	Day	Hr	min	Days-1989.0	Orbit	Orb.Phase	Note
1990	3	15	19	367.138	134	+0.38	
1990	12	11	54	375.996	154	+0.32	
1990	33	9	23	396.891	201	+0.38	
1990	50	6	10	413.757	239	+0.37	
1990	71	5	19	434.722	287	-0.42	
1990	91	23	59	455.499	333	+0.38	
1990	118	4	38	481.693	392	+0.38	
1990	138	4	24	501.683	437	+0.40	
1990	179	9	18	542.888	530	+0.21	1
1990	208	10	9	571.923	596	-0.38	
1990	239	1	16	602.553	665	-0.39	
1990	279	16	35	643.191	756	+0.16	1
1990	339	17	35	703.233	891	+0.42	2
1991	51	14	29	780.103	1065	-0.38	
1991	111	22	27	840.435	1201	-0.44	
1991	186	20	30	915.354	1369	+0.39	
1991	254	9	31	982.897	1522	-0.40	
1991	304	0	34	1032.524	1633	+0.45	
1992	36	10	39	1129.944	1853	+0.03	1
1992	183	11	37	1276.984	2185	-0.49	
1992	188	14	6	1282.088	2196	+0.01	1
1992	312	21	34	1406.399	2476	+0.30	
1992	335	11	2	1428.960	2527	+0.17	
1993	6	18	0	1466.250	2611	+0.26	3
1993	7	14	12	1467.092	2613	+0.16	3
1993	7	21	31	1467.397	2614	-0.15	3
1993	7	23	2	1467.460	2614	-0.01	3
1993	34	10	22	1493.932	2674	-0.32	
1993	57	9	1	1516.876	2725	+0.43	

11.1.2 Basic-angle Drifts

Temporary loss of thermal control for the payload would generally result in changes in focus and basic-angle drifts. The most frequent of these events remained, due to a coincidence of circumstances, unnoticed in the processing of the published data: when unexpected signals started appearing in the telemetry from the satellite, the on-board computer was shut down and reloaded by the Mission Operation Centre (MOC). This process generally took a few hours, during which time there was no thermal control of the payload. Consequently, the payload cooled down, which reflected in the basic angle value obtained after observations were resumed. Because these data sets generally covered only part of an orbit, they were not considered for the first-look processing in Utrecht, which left most, if not all, of these events undetected. Table 11.3 lists details on orbits affected by a drift of the basic angle. In a few cases also the following orbit was still affected, and in one case the restarting of the on-board computer took place during the second half of the orbit, allowing the recovery

Table 11.3. Orbits during which the basic angle drifted. Notes: CBS: on-board computer reloaded; RTU: Thermal control unit anomaly; V: Power-supply anomaly; u: unidentified; or: operations report as supplied by ESOC; ar: Anomaly report

Year	Day	Days-1989.0	Orbit	or	ar	Note
1990	4	368	137	3	20	CBS
1990	89	453	329	8	26	RTU
1990	117	481	393	10	29	CBS
1990	243	607	675	18	38	CBS
1990	321	685	852	23		CBS
1991	49	778	1060	29		CBS
1991	88	817	1150	31	47	CBS
1991	126	855	1235	33	49	CBS
1991	202	931	1407	38		V
1991	308	1037	1645			u
1991	314	1045	1663			u
1991	356	1086	1754	48		CBS
1992	39	1134	1860	51		CBS
1992	46	1142	1879	51		CBS
1992	83	1177	1960	54		CBS
1992	126	1220	2058	56		CBS

of the thermal control to take place during the perigee passage (orbit 1456), and no basic-angle drift is observed for that particular event.

The drift in the basic angle is fairly simple to incorporate in the calibrations. This is not the case, however, for the related changes in the scale and the modulation parameters (the OTF, see Section 12.1), the effects of which can be observed for the one or two orbits with the largest of the basic-angle drifts. The worst case is the first data set after the resumption of the mission towards the end of 1992. Observations were re-started well before the payload had returned to its nominal operations temperature, which can be observed in a very substantial basic-angle drift and also problems in the calibration of other instrument parameters. Data from this orbit (number 2450) have not been used in the new reduction.

11.1.3 Gyro Configuration

Four main periods of gyro configuration are distinguished. Only for the first two of these have the gyro data been used in the data analysis, but in all four cases the configuration of active gyros has been implemented in the satellite dynamical model. The four intervals are:

- **(1)** Orbits 1 to 533, active gyros: 1, 2 and 4;
- **(2)** Orbits 535 to 2274, active gyros: 1, 2 and 5;
- **(3)** Orbits 2450 to 2576, active gyros: 2, 3 and 5;
- **(4)** Orbits 2581 to 2579, active gyros: 2, 3.

Table 11.4. Orbits during which gyro de-storage exercises took place. Start and end times are given in orbital phase

Orbit	Start	End	Gyro	Active gyros	Orbit	Start	End	Gyro	Active gyros
152	-0.230	-0.036	3	1,2,4	763	-0.315	-0.080	4	1,2,5
170	-0.123	0.050	5	1,2,4	781	-0.346	-0.152	3	1,2,5
260	-0.157	0.025	3	1,2,4	844	-0.253	-0.064	3	1,2,5
278	-0.133	0.090	5	1,2,4	846	0.031	0.247	4	1,2,5
391	-0.272	-0.080	3	1,2,4	920	-0.076	0.123	3	1,2,5
406	0.110	0.322	5	1,2,4	1006	-0.376	-0.182	3	1,2,5 (+4)
465	-0.113	0.098	3	1,2,4	1078	0.033	0.222	3	1,2,5 (+4)
467	0.067	0.287	5	1,2,4	1186	-0.199	0.045	3	1,2,5
533	0.131	0.410	5	1,2,4	1303	-0.205	-0.005	3	1,2,5
544	0.013	0.206	3	1,2,4	1666	0.178	0.500	3	1,2,5
636	-0.029	0.163	3	1,2,5	1667	-0.500	-0.245	3	1,2,5
702	-0.275	-0.084	3	1,2,5					

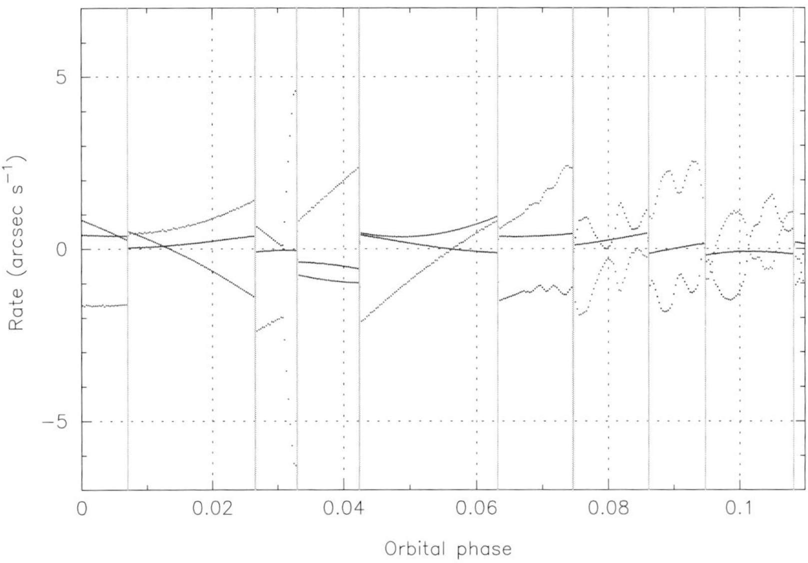

Figure 11.2. An attempt to de-store gyro 4 after its breakdown earlier in the mission. The de-storage takes place between phase 0.031 and 0.033. Initially all gyros are well behaved, but after phase 0.066 the spin rate of gyro 4 starts to vary. As the spin axis of gyro 4 lies in the (x, y) plane, these changes are recorded by gyros 1 (crosses) and 2 (circles). The irregular behaviour of the spacecraft that resulted from the problems with gyro 4 led to a temporary loss in on-board attitude control. The vertical lines represent thruster firings. The data comes from orbit 846

The active gyros are the only ones for which the readouts are communicated to the ground for use in the data processing.

Changes to the gyro configuration affect the torques acting on the satellite, and therefore the dynamical modelling. Gyro configuration changes took place for two reasons:

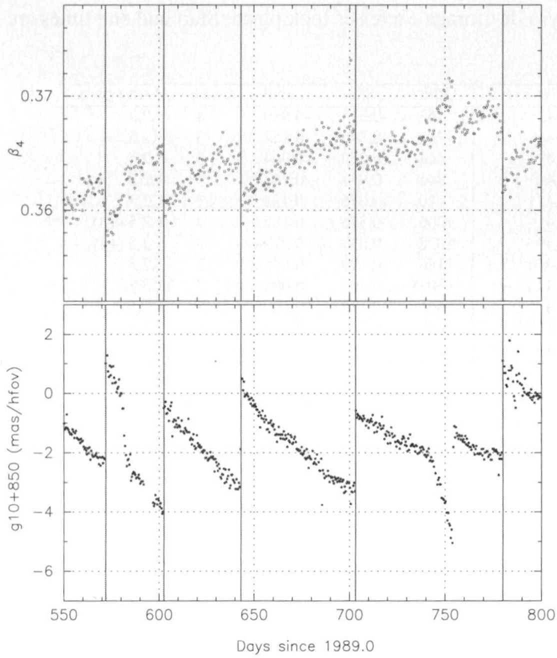

Figure 11.3. Problems with the Thermal Control Electronics (TCE) of the payload reflect in focal length variations of the telescope, from there in the scale (lower graph) and signal modulation (upper graph). The changes towards day 755 are the result of TCE1 breakdown and its replacement by TCE2. The changes around day 580 (orbits 617 and 618) are not associated with any known events

- Gyro de-storage, during which a redundant gyro was spun-up, tested, and spun down (see Fig. 8.2, page 224). These periods usually lasted for two to three hours (Table 11.4);
- Gyro failures and replacements. Due to radiation damage to the electronics, various breakdowns of gyros occurred during the mission.

In addition to the de-storage events and configuration changes, testing of gyro 4 took place over two periods of one week (orbits 783 to 796) and two months (orbits 925 to 1085) around the end of 1990 and the start of 1991. During the second of those periods twice a de-storage on gyro 3 took place, implying all five gyros were active at that time.

The breakdown of a gyro could also badly affect the science data, as this was often accompanied by changes in the angular momentum of the gyro (Fig. 11.2). These changes caused difficulties to model torque variations on the spacecraft and errors in the derived inertial rates. This in turn resulted in poor attitude convergence or complete attitude loss (see also Section 13.2.3, page 340).

11.1.4 Thermal Control

A tight thermal control of the payload was a crucial requirement for the mission: without it changes in the optical system could quite easily affect the science data in a very destructive manner. Still, on a few occasions temperature

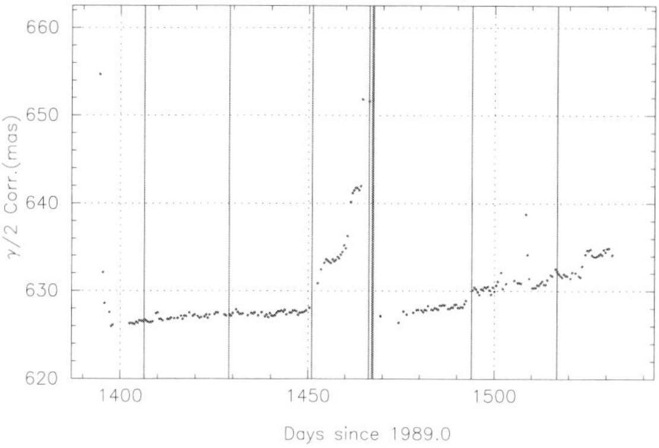

Figure 11.4. The variation in the basic angle for the final three months of the mission. Due to failed heaters, thermal control was poor and substantial drifts of the basic angle developed on a few occasions

control for the payload was partly lost due to failure of the thermal control unit. This was not always immediately noticed, and as a result associated drifts are observed in many of the calibration parameters. The failure of the payload thermal control unit (TCE1) that started on day 741 was first noticed by Schrijver in the First-Look processing of selected data in Utrecht. The failure reflects clearly in the changes in scale and modulation amplitude as recorded by the various calibrations (Fig. 11.3). It was rectified on day 755 by replacing TCE1 with the redundant TCE2, after which normal conditions returned.

An earlier fairly sudden change in focus, around day 580, can not be identified with any hardware or other problems on the spacecraft. It does not behave in a way expected for the thermal control electronics: after about one day the normal evolution of the focus is restored, but the focus position is not returned to its expected value. This focal change may possibly have a non-thermal cause. The most severe changes in focus resulted from the deteriorating thermal control towards the end of the mission (Fig. 11.4). The evolution of basic angle and scale parameters is fairly chaotic for these last three months of the mission. Still, even for that time in the mission the changes in focus were sufficiently slow for the focus to be assumed constant for most orbits.

The thermal control of the payload (the thermal insulation and the heaters) was able to compensate for the changes in heat received as a result of the rotation of the satellite: spin-synchronized modulations as observed for some spacecraft parameters are not detected for any of the payload parameters, even towards the end of the mission.

11.1.5 Sun Pointing

The Sun-pointing mode formed a safety backup for the operations: implemented automatically when it appeared that the attitude control of the satellite was badly lost. This happened on day 1001, as a result of an error made in the error-angle corrections uploaded from the ground station. Four days later observations were resumed in Sun-pointing mode, until on day 1008 the nominal scanning mode was resumed. As a result, the data for orbits 1570 to 1578 are in Sun-pointing mode. Data obtained over those orbits do not contain information on parallaxes (see Chapter 2).

The remaining two time intervals over which the satellite observed in Sun-pointing mode were containment measures to deal with very difficult situations for the on-board attitude control as a result of gyro failure. The first of these periods covered orbits 2264 to 2274, the second covered orbits 2581 to 2608.

Sun-pointing operations naturally did not show any of the spin-synchronous modulations as observed in normal mode for some of the spacecraft calibration parameters.

11.1.6 Star Mapper Background

The star mapper background level was seriously affected by radiation levels outside the satellite. The *Hipparcos* mission took place during a solar maximum period, and the unplanned orbit took it through the Van Allen belts twice every 10.6 hours. Background levels for the star mapper varied as a result considerably, and could be on occasion so high that no stars could be seen at all (Fig. 11.5). This affected the on-board attitude control and the mission's science return, but also provided input for studies on the evolution of the radiation activity of the Van Allen belts (Daly et al. 1994; Wicenec and van Leeuwen 1995).

The radiation activity was linked, but not in a simple way, to solar activity as observed from the ground. At times of very high activity the entire orbit of the satellite was contained in a high radiation environment. For this reason, many data sets between orbits 1300 and 1340 contained little or no useful data.

11.1.7 Various Other Events

Various other "events" left their mark on the mission. The first to do so was the rotation by 5 arcminutes of the main grid, well beyond the expected range. This was only properly recognized after the first data had been successfully analysed in the First-Look processing. It required a correction to the pointing of the IFOV of the sensitive area of the IDT, which was applied on 23 January 1990. Observations obtained before that date, in particular for stars in the corners of the field of view, suffer from poor centring. The most conspicuous effect of the grid-rotation correction was observed in some of the photometric calibration parameters, for example the dependence on the transit ordinate of the amplitude of the second harmonic in the modulated signals (Fig. 11.6), and the response gradients for the IDT photometric calibrations (Figs. 12.11 and 12.12).

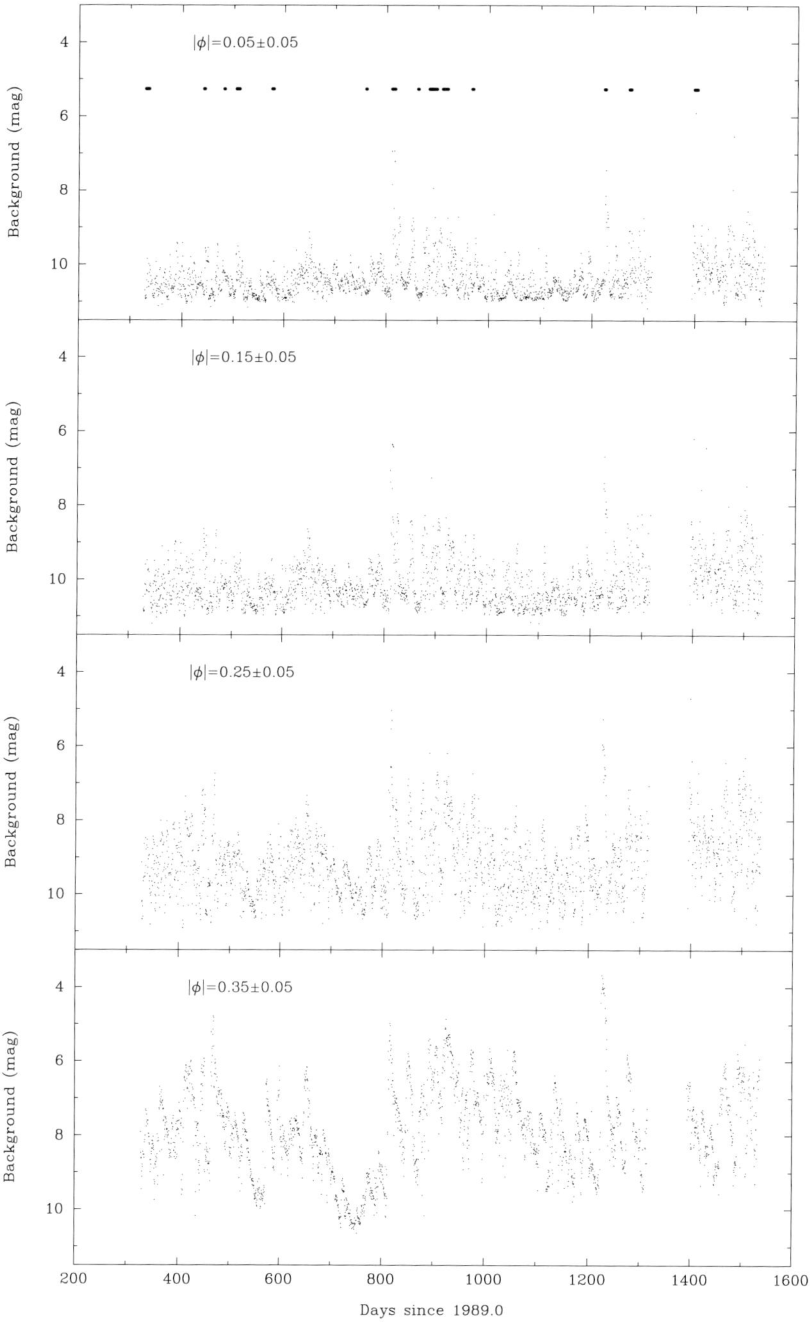

Figure 11.5. Star mapper background levels in the V channel at different orbital phases over the mission. The orbital phases ϕ range from -0.5 at perigee to 0.0 at apogee to 0.5 at perigee. The top graph also indicates events of high solar activity (just below the mag 5 level)

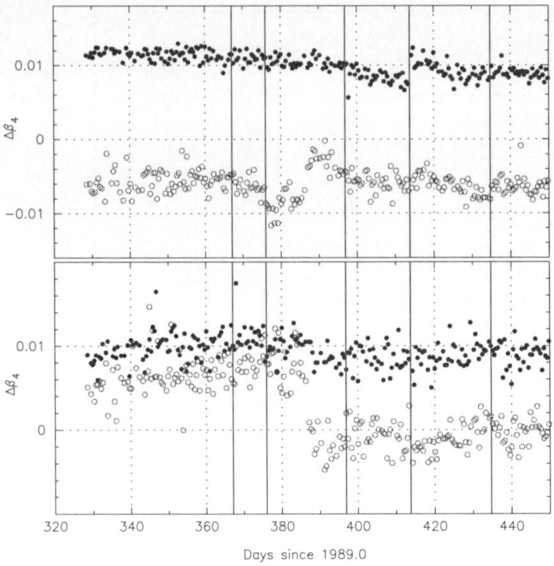

Figure 11.6. The effect of the correction for the grid rotation in the dependence of the second-order harmonic amplitude on the ordinate z of a transit. Top graph: z dependence, bottom graph: z^2 dependence. The correction for the grid rotation in the pointing took place around day 387. The open and closed symbols refer to data for the two fields of view

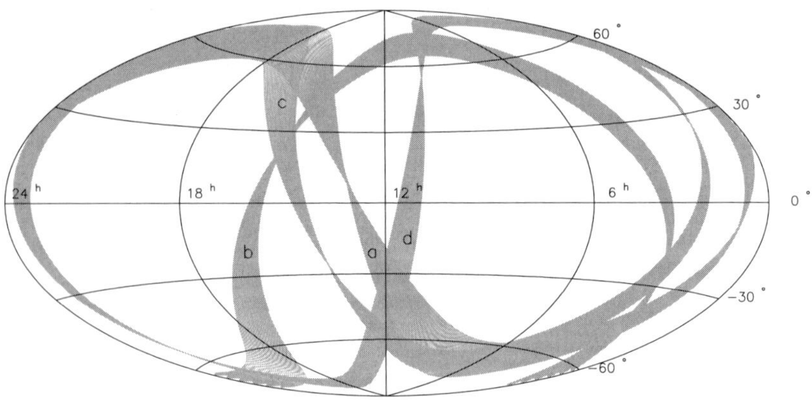

Figure 11.7. The sky coverage of 4 periods of data lost due to errors or system faults. See text for identification

The second event that affected the mission was a fault in the implementation of the analogue recording for very bright stars. While the signals for most stars were recorded in photon counting mode, those brighter than 2nd magnitude should have been measured in analogue mode. As such cases were rather rare, it took till mid January before it was realized that the synchronization of the switching between photon counting and analogue mode was wrong, and took place a few samples too early. This corrupted the data for a fainter star as well as for the bright star. It was possible to selectively delete damaged samples and still use some of the affected data, but after mid January 1990 the analogue

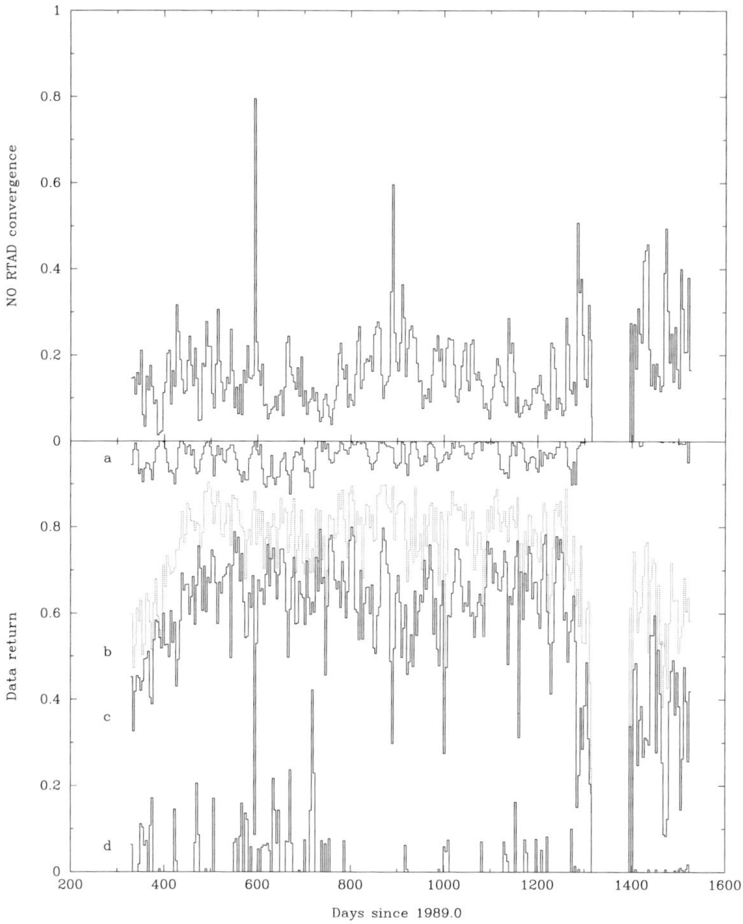

Figure 11.8. The data return summary over intervals of 4 days (close to 9 orbits). The top graph shows the fraction of time lost due to "no attitude convergence". The bottom graph shows between the upper boundary and curve "a" the fraction of time lost due to occultations during data coverage. Curve "b" (dotted) shows the maximum possible data return (ground-station coverage minus occultations). Curve "c" shows the actual data return. The difference between "b" and "c" is the same as the curve in the upper graph. Curve 'd' shows the fraction of data contained in data sets of less than 1200 frames, usually too short to be included in the final results. The difference between curves "a" and "b" shows the fraction of time lost due to 'no ground station coverage'

mode was switched off, and all stars were measured in photon counting mode. This means that for the brightest stars the signal becomes partly saturated and distorted. These signals are generally only reliable for astrometry, noisy for photometry, and close to useless for detection of visual double stars.

Four to five days of observations in August 1990 were lost (orbits 641 to 651) due to a faulty attitude initialization, indicated by "a" in Fig. 11.7. Two days were lost in September 1991 due to an uplink command error (orbits 1564 to 1569) (indicated by "b" in the Figure). This forced the satellite into safe-mode, i.e. Sun-pointing, where observations were obtained for the following 5 days. A tape fault March 1992 meant that two days of observations were lost (indicated by "c"), and for the introduction of software patches to enable observations with damaged or no-longer functional gyros 3 days were lost in July 1992 ("d"). The satellite was put into safe-mode (Sun-pointing and spun up) for August, September and October of 1992, before a final 6 months of observations.

11.2 Data coverage and detector response

The science return of the mission was primarily determined by ground station coverage and on-board attitude control convergence. A relatively minor role was played by occultations of one of the fields of view by the Earth or the Moon. By summarizing the data returns over four-day periods, covering almost exactly 9 orbits, a good impression can be obtained of science return as a function of time over the mission. This is shown in Fig. 11.8. On average the science return over the mission was around 60 per cent, which has to be seen in the context of operations being forced to take place in a highly unfavourable orbit. The, for such circumstances, high return was obtained through major efforts by all teams involved: ESOC, the data reduction consortia and the project team and project scientist at the European Space Technoly Centre (ESTEC).

Mission response was further affected by the reduction in the response of the main detector to nearly 50 per cent over the mission. This reflects in the astrometric data for the faintest stars, and in particular those that are variable.

Chapter 12

PAYLOAD CALIBRATIONS

The payload calibrations reveal information on the instrument's health during the mission. These calibrations were monitored during the mission on a regular basis by the so-called First-Look processing (see for example Schrijver and van der Marel 1992), and also as part of the NDAC and FAST full-scale data reductions. Also after completing all reductions useful information can be extracted from the calibration data. When compared with the spacecraft calibrations (Chapter 13), one can for example draw conclusions on the thermal sensitivity in the payload.

Here we summarize results for the three most prominent calibrations:

- The optical transfer function: describing the phase and amplitude of the second harmonic in the modulated signal obtained by a single star passing over the main grid, relative to the phase and amplitude of the first harmonic in that signal;
- The large-scale grid distortion: describing the calibration of the (along-scan) positional corrections on the grid;
- The photometric calibration: describing the relation between the observed fluxes and calibrated magnitudes, as well as the reconstruction of the pass band for the main grid.

In all cases, these calibrations were carried out as a function of position on the grid and colour of the star. They are closely related, as they are all the result of the imaging characteristics of the telescope, and its evolution over the mission. One may therefore expect to encounter the same event (like a thermal control disturbance) standing out in all or most of these calibrations, as was shown in the case of basic-angle drifts by van Leeuwen and Fantino (2005). Variations

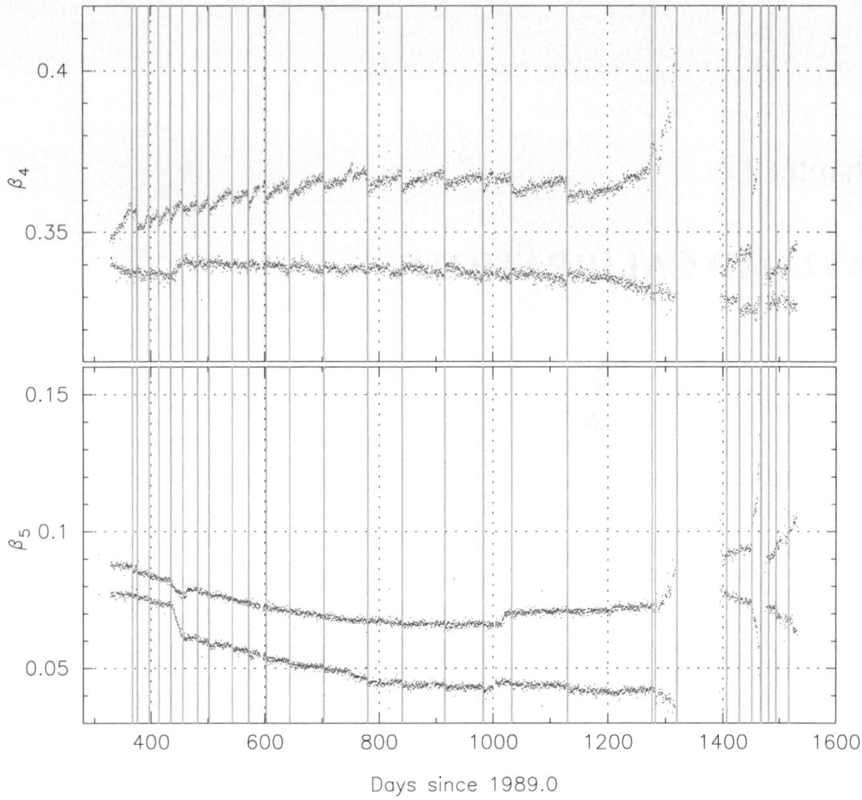

Figure 12.1. Evolution of the parameters β_4 and β_5, describing the relative amplitude and phase of the second harmonic in the modulated signal, over the mission. The vertical lines represent changes in focal plane position. The discontinuity visible at day 755 is due to a thermal-control failure and change of thermal control electronics. The top curve in each graph refers to the following, the lower curve to the preceding FOV

in focal length of the telescope for example are reflected in most of the calibration results presented here. The optimal focus of the *Hipparcos* telescope was slightly different for the two fields of view, and a focal position was selected for operations that provided the best return for both fields: maximizing the amplitude of the first harmonic of the modulated signal (see Chapter 2 and Section 11.1.1, Fig. 11.1, page 288). Thus, both fields are usually slightly out of focus. As focal changes took place due to mechanical adjustments of the payload, the focus for the preceding field of view improved, while for the following it deteriorated. A refocusing, as described in Section 11.1.1 corrected this. The new focal position would favour the following field of view, so that on average both fields were equally well-focused (see also Fig. 2.8, page 55).

12.1 The optical transfer function

The modulated signal produced by the passage of a stellar image over the main grid (Fig. 2.4, page 47) is well represented by a five-parameter model as described in Section 2.2.1, page 46. The relative width of the transparent lines of the main grid was chosen such that a second harmonic would be present in the signal, sufficiently strong to be used in the disentangling of transits for double and multiple stars (Chapter 4). For the analysis of those complex images it is therefore essential that the relation between the first and second harmonic in the modulated signal can be described and calibrated in a simple model. This model is referred to as the optical transfer function or OTF.

12.1.1 The Model

The OTF model describes the relative phase and amplitude of the second harmonic in the modulated signal, β_4 and β_5 (Eq. 2.33, page 50), as a 3rd order function of the positional coordinates on the grid (w, z) (10 coefficients), a first- and second-order dependence on the star colour index $c \equiv \big[(\mathrm{V} - \mathrm{I}) - 0.7\big]$, and the two colour-position cross terms $w \cdot c$ and $z \cdot c$:

$$\begin{aligned} \tilde{\beta_4} &= f_4(w, z, c) \\ \tilde{\beta_5} &= f_5(w, z, c) \end{aligned} \tag{12.1}$$

Initially all 14 parameters are calibrated over the whole mission. However, as the noise on the higher-order parameters in a single determination is clearly much larger than the actual variations of these parameters, those higher-order parameters have been fitted as a function of time over the mission and implemented as such. This also removes a substantial natural correlation between the first and third order parameters, and therefore significantly reduces the noise on those determinations. The two colour-position cross terms have been handled in the same manner. Separate solutions apply for the preceding and following FOVs.

There are two applications of this model, both concerning double stars. The first application uses the calibration as a means for detecting disturbed (double or multiple) images. The majority of those are due to actual double stars, but some are the result of an accidental superposition by an image from the other field of view. For each transit (as described in Section 2.2.2: Eq. 2.33, page 50) the observed values of β_4 and β_5 and their covariance matrix are compared with the predicted values, to provide a χ^2 value for the likelihood of the transit image being the result of a single point source. The covariance matrix only concerns the 4 elements of the matrix 2.44 (page 54) that refer to β_4 and β_5:

$$\chi^2 \equiv \begin{bmatrix} \delta\beta_4 & \delta\beta_5 \end{bmatrix} \begin{bmatrix} R_{4,4} & R_{4,5} \\ R_{5,4} & R_{5,5} \end{bmatrix} \begin{bmatrix} \delta\beta_4 \\ \delta\beta_5 \end{bmatrix}, \tag{12.2}$$

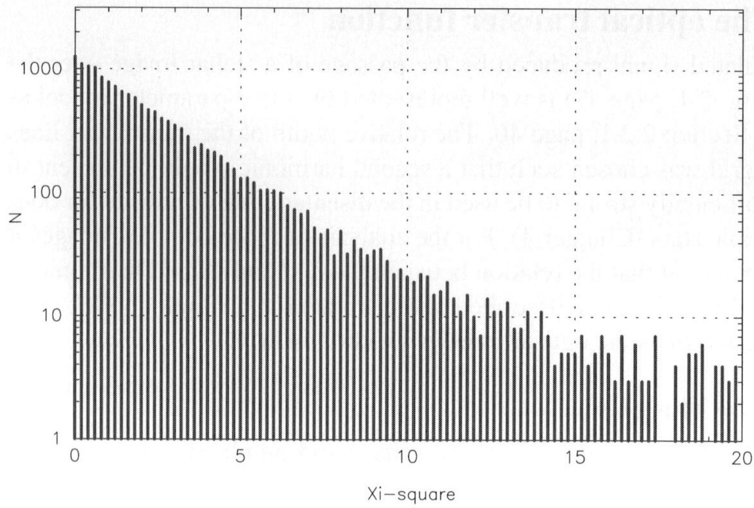

Figure 12.2. An example of the χ^2 distribution for the observed second-harmonic parameters β_4 and β_5, with respect to the predicted values as derived from the OTF model. The data are for orbit 53, November 1989

where $R_{i,j}$ are elements of the matrix R_β as defined by Eq. 2.44, and $\delta\beta_{4,5}$ represent the differences between the observed values and those predicted by the OTF calibration model. A typical example of the observed distribution of χ^2 values for single measurements is shown in Fig. 12.2.

The second application is part of the preparation for the double-star analysis. To ease that analysis, the OTF calibration model is used to eliminate the nominal phase difference between the first and second-order harmonics. The model predictions $\tilde{\beta}_4$ and $\tilde{\beta}_5$ represent the predicted relative phase offset for the second harmonic given by:

$$\tilde{\phi}_2 = \arctan(\tilde{\beta}_5/\tilde{\beta}_4), \tag{12.3}$$

and the predicted nominal amplitude:

$$\tilde{R} = \sqrt{\tilde{\beta}_4^{\,2} + \tilde{\beta}_5^{\,2}}. \tag{12.4}$$

The normalized second-harmonic parameters are now defined as:

$$\begin{aligned} \hat{\beta}_4 &= [\ \ \beta_4 \cos(2\tilde{\phi}_2) + \beta_5 \sin(2\tilde{\phi}_2)]/\tilde{R}, \\ \hat{\beta}_5 &= [-\beta_4 \sin(2\tilde{\phi}_2) + \beta_5 \cos(2\tilde{\phi}_2)]/\tilde{R}. \end{aligned} \tag{12.5}$$

This transformation prepares β_4 and β_5 for the double star analysis as described in Chapter 4, where also their observed distributions are shown.

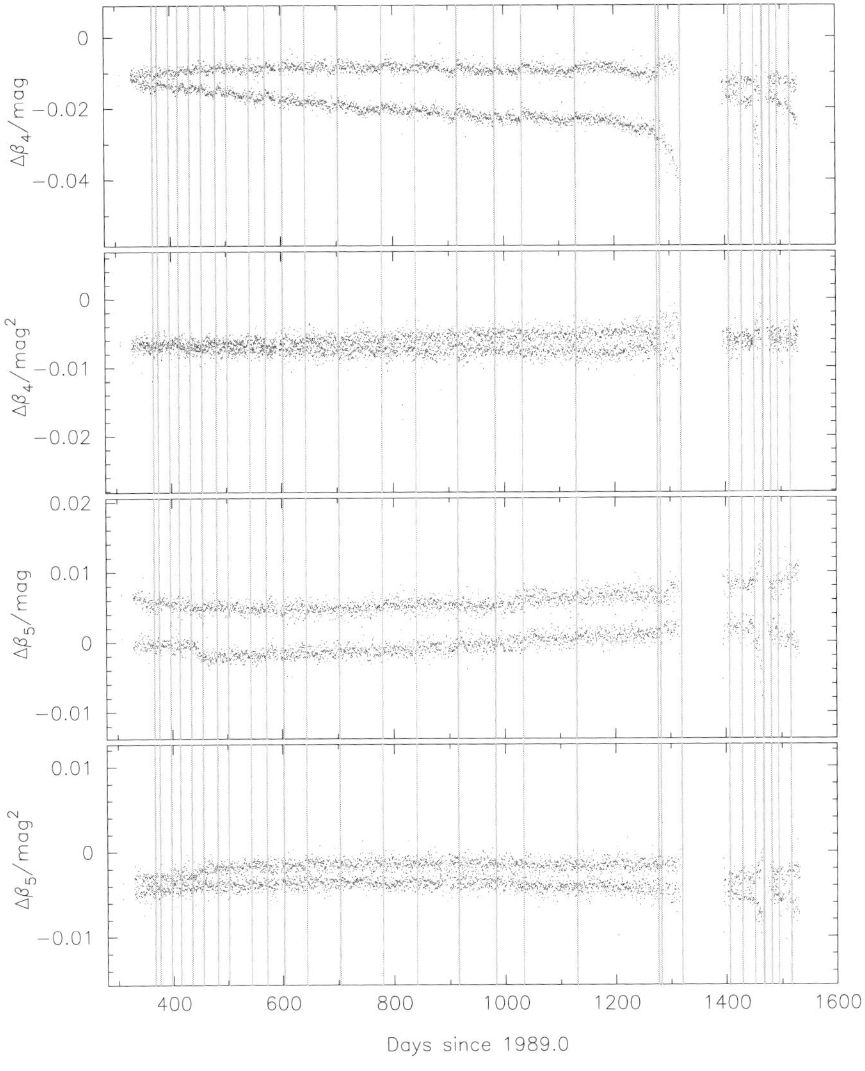

Figure 12.3. Evolution over the mission of the first and second-order colour dependence in the parameters β_4 and β_5. The top two graphs refer to the linear and quadratic colour dependence in β_4, and the bottom two graphs similarly to β_5. The two curves in each graph refer to the following and preceding FOVs. The grey lines indicate instances of refocusing the telescope. Only the linear colour coefficient for β_4 shows a minor dependence on focus

12.1.2 Parameter Evolution

The evolution of the OTF model parameters reflects the adjustments in the telescope over the mission due to the settling of the satellite in the space environment. It also shows reflections of periods with long eclipses, around days

450 and 1020 (Fig. 12.1). The β_4 values for the following FOV are particularly sensitive to focusing disturbances, showing as discontinuities at refocusing instances.

Most important for the processing of double stars (Chapter 4) is the colour dependence of the modulation coefficients. When processing a close double star, usually only a mean colour is known, which then is assumed to be the colour of both components. A strong dependence on colour would make this approach fail in many cases. The colour dependencies are, however, quite small, as is shown in Fig. 12.3.

12.1.3 Conclusions on the OTF Calibration

The OTF calibration has thus far only covered the large-scale distortions, and judging by the results, this is probably sufficient. The presence of a small-scale distortion can, however, not be ruled out, as similar scale distortions have also been observed in the geometric calibrations. These distortions may, for the brightest binary systems, have even some minor effect on the accuracy and the reconstructed double-star parameters. From the residual statistics it appears, however, that the model applied has left no more than negligible residual effects for the majority of stars.

12.2 Large-scale geometric-distortion calibration

The geometric distortions of the grid are defined as the systematic differences, as a function of observation-position on the grid, or as a function of colour of the observed star, between the predicted and observed abscissae for a transit of a single point source. The predicted abscissae assume an ideal grid and substrate on which the grid was printed, i.e. a perfect spherical surface fitting the focal surface of the telescope, and a grid printed on it as projected from a tangential plane at the optical axis of the telescope. The distortions are split in three groups, each of which has its own characteristics: (1) the small-scale distortions are related to the printing of the individual scan fields of the modulating grid (see Fig. 2.4, page 47), (2) the medium-scale distortions, related to the distortion of the substrate on which the grid was printed, and (3) the large-scale distortions, related to the overall telescope optics and their mountings in the payload. The first two of these are highly stable over the mission, and their characteristics are presented in Sections 2.4.2 and 2.4.3 respectively, page 60. Here we present further details on the large-scale geometric calibration only, for which some characteristics are directly related to the OTF and to some extent to the photometric calibrations.

The large-scale distortions affect the main grid as well as the star mapper grid, but due to the very different characteristics of these two grids, their distortions have to be treated separately.

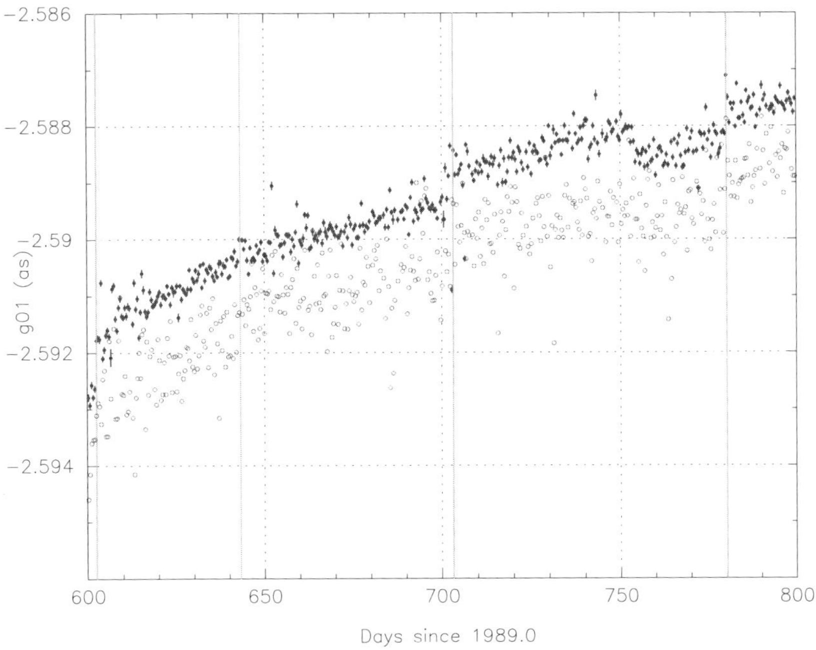

Figure 12.4. Detail of the evolution of the mean grid rotation as recorded in the large-scale distortion parameters for the main grid. The small vertical bars refer to the new reduction, with constrained third-order parameters. The small circles refer to the original NDAC solutions, without parameter constraints. The vertical lines indicate instances of refocusing. The discontinuity around day 755 is due to a switch of the thermal control electronics, and can now be easily recognized in the calibration data. Other (small) discontinuities are due to refocusing, indicated by the grey vertical lines

12.2.1 The Main Grid

The large-scale distortions of the main grid have been represented by a two-dimensional third-order polynomial in position and linear colour dependencies in along- and across-scan directions. Following the NDAC conventions, the large-scale distortions are described by the mean values over the two fields of view (coefficients indicated by "g"), and half of the differences (coefficients indicated by "h"), such that the sum of the two represents the preceding field of view, and the difference ("g"−"h") the following. As an example, the coefficient "g12" stands for the mean distortion proportional to the first order along-scan and second-order across scan position. Coordinates used in the calibration have been normalized by the half-width of the field of view, so coefficients show the maximum possible effect they may have at the edge or the corners of the field of view.

As was done for the OTF calibration described above, the third-order positional and colour terms were, after initial calibrations, fixed through fitted

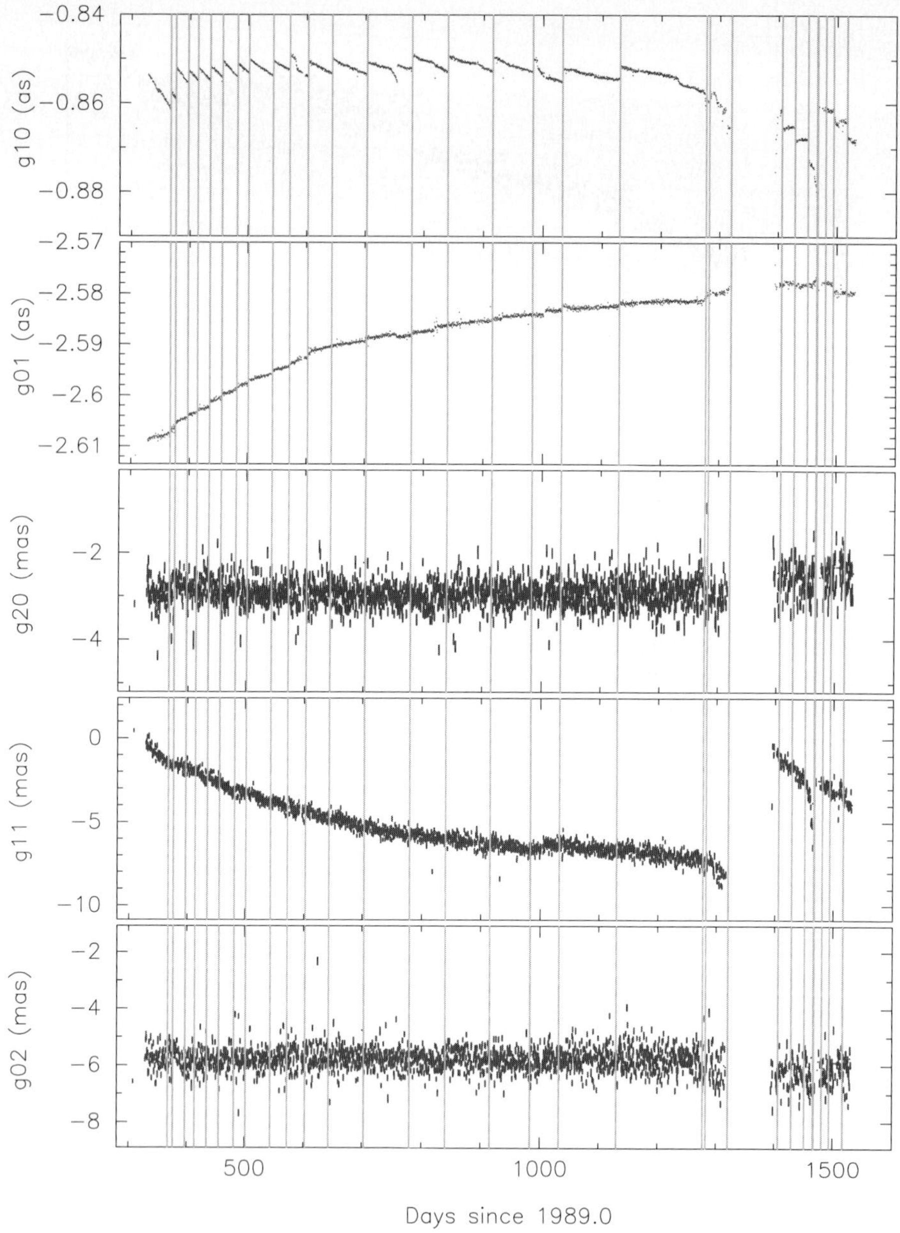

Figure 12.5. Evolution of the mean values of the first- and second-order terms in the large-scale distortion model for the main grid

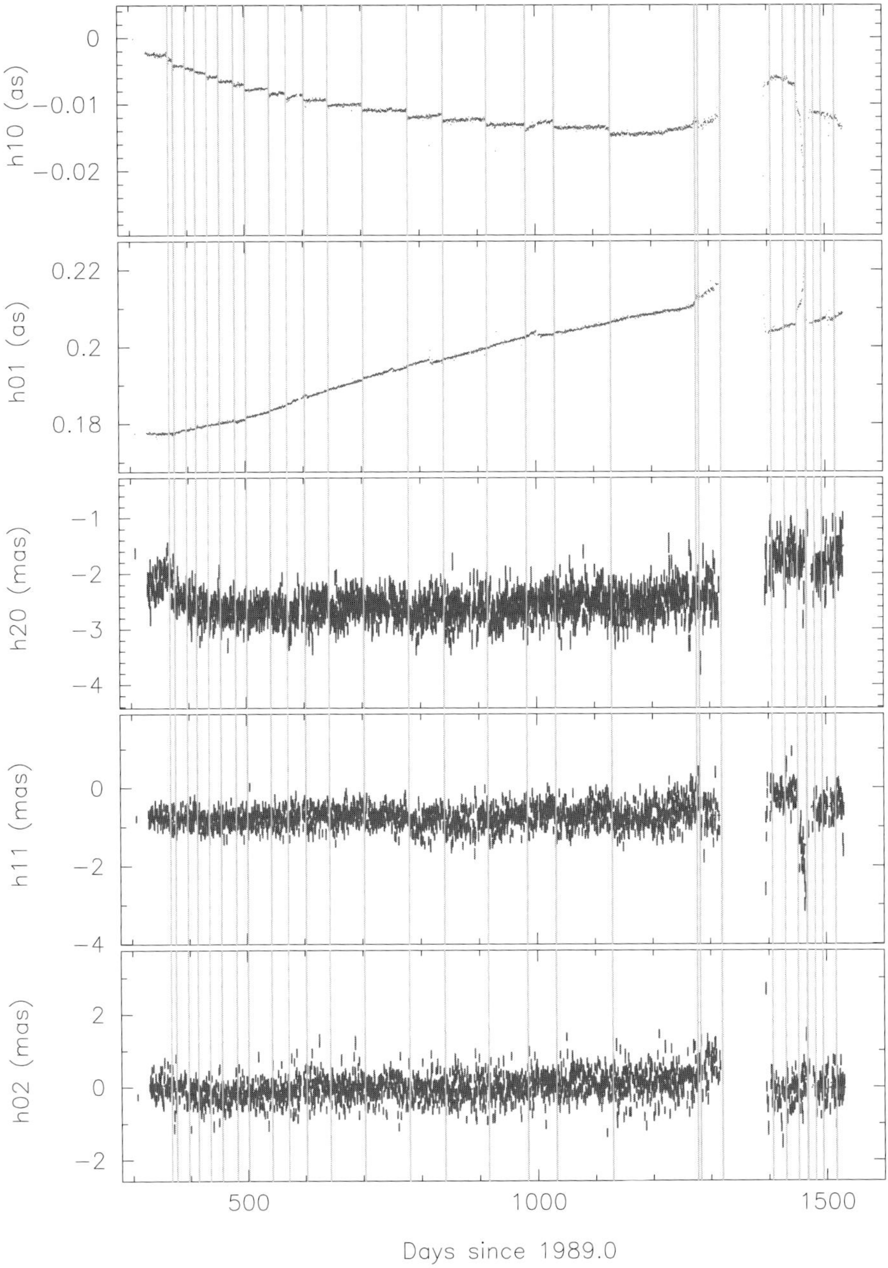

Figure 12.6. Evolution of the differences for the first- and second-order terms in the large-scale distortion model for the main grid. The grey lines indicate the instances of refocusing of the telescope, reflecting mainly in the parameter h10, the difference in scale between the two fields of view

spline functions. These functions take into account discontinuities that occurred in some coefficients at times of refocusing, or as a result of thermal control anomalies. The uncertainty of an individual determination of these coefficients appears to be considerably larger than their actual variation over the mission. Fixing these coefficients also removes the strong correlations with the first-order coefficients and leaves those determinations considerably less noisy. The lower noise level improves the sensitivity to the detection of mission anomalies, and therefore the overall understanding of the mission. An example is shown in Fig. 12.4 for the mean grid rotation. The evolution of all first and second order positional parameters is shown in Fig. 12.5 for the mean values and Fig. 12.6 for the differences.

The large-scale geometric distortion parameters have been related to a simplified telescope model by Lindegren et al. (1992) and in Chapter 5 of Volume 2 of (ESA 1997). Rotations and translations are applied to this model to see how these reflect in the large-scale distortion parameters. A comparison with the observed instrument parameters provides a record over the mission of a set of more "physical" parameters, like mirror rotations and translations. These movements are found to be all at a level of 0.1 μm day^{-1} or less, and getting smaller as the mission progressed.

12.2.2 Chromaticity

The calibration of the chromaticity in the geometric distortions in the new reductions was part of the instrument-parameter determinations. With the positional reference system already well-determined, there was no longer a potential instability caused by this inclusion. In the published data the chromaticity corrections could only be determined as part of the sphere reconstruction (see Chapter 16 of Vol. 3 of (ESA 1997)). That determination was, as a result, very limited in resolution, with the chromaticity determined as a linear function of time over the mission. The new reduction provides a single determination for most data sets. These have subsequently been fitted with a cubic spline as a function of time. The results are shown in Fig 12.7. The main problem with the chromaticity correction is the application to the very red stars, which generally are too rare to seriously affect the calibration. A simple linear extrapolation over colour index is all that can be done.

12.2.3 Conclusions on Geometric Distortions

The overall success of the corrections for the three levels of geometric grid distortions is provided by the level of detail visible in the small-scale distortions, and by the very low noise levels on the abscissae residuals for the brightest stars. It is unlikely that any significant improvements can still be made to the available corrections as applied to in new reduction. The improved stability

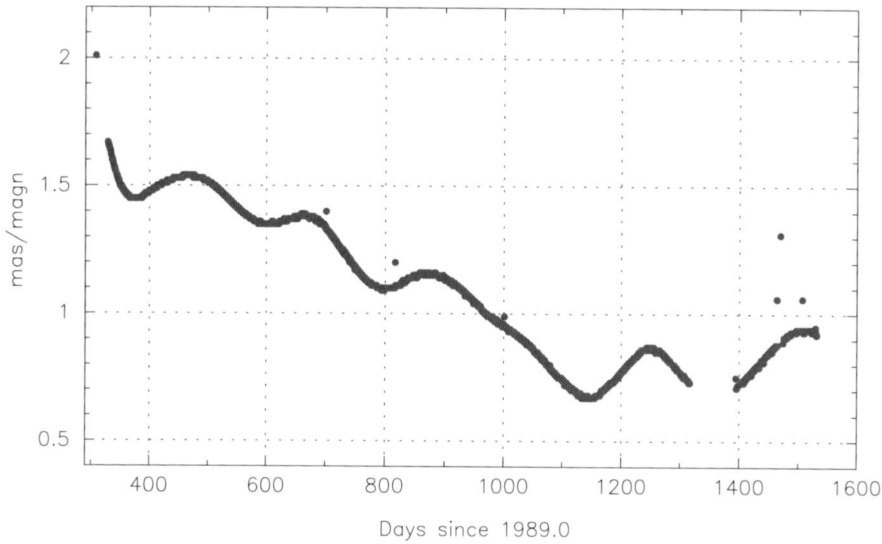

Figure 12.7. Evolution over the mission of the mean chromaticity over the two fields of view. These results agree broadly with the corrections as applied by NDAC up to day 1200, but for the last three months of the mission they are systematically higher

of the model through the constraining of third-order parameters also reveals the generally high level of stability of the geometric-distortion parameters.

12.2.4 Star Mapper Geometry

The geometric calibration of the star mapper grids has been included in the attitude reconstruction as derived from the star mapper transits. During the first reduction iterations, seven parameters were included in the calibration model. Four of these were eliminated through either a fixed value or slowly evolving values. Thus, the rotation difference between the two fields of view, the difference in relative positions of the vertical slits between the fields of view, and the curvature of the vertical slits were all fixed. This left only the grid rotation, the mean position of the vertical slits with respect to the chevron slits, and the effective basic angle for the star mapper slits. Complex optical distortions meant that the effective basic angle for the star mapper detector was about 10 mas smaller than for the main grid, but showing variations over the mission that were 1.5 to 2 times larger than for the main grid. A fourth instrument parameter was determined as part of the final stage of the attitude reconstruction, where, using main grid observations, corrections to the attitude are determined relative the star mapper based attitude reconstruction. This provides the calibration of the relative position of the star mapper grid with

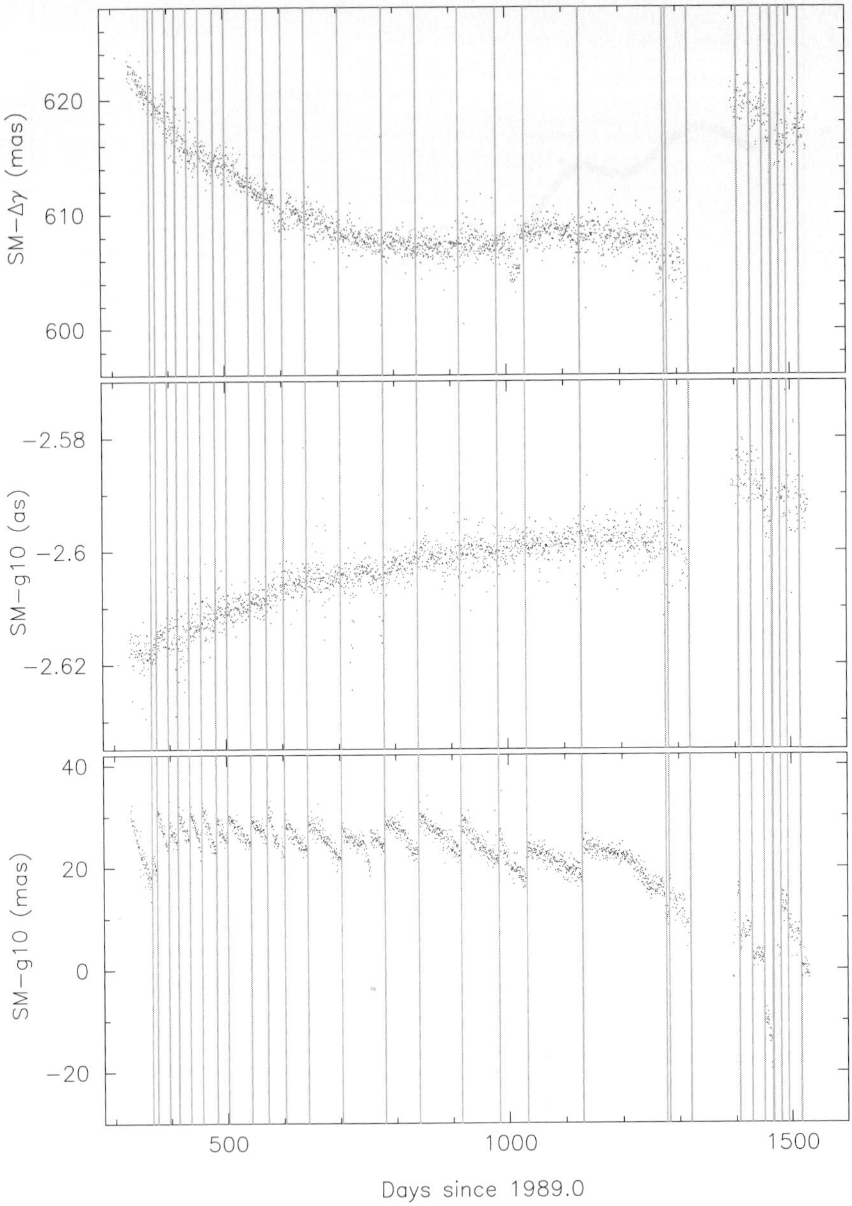

Figure 12.8. Three parameters in the star mapper geometry calibration (top to bottom): the basic-angle correction, the mean grid rotation, and the correction to the assumed distance of the chevron slits crossing point and the centre of the main grid. The discontinuities in the lower graph result mostly from refocusing of the telescope

respect to the main grid. This is mainly an indication of the scale variations in the focal plane, and therefore sensitive to refocusing. The evolution of some of these parameters is shown in Fig. 12.8.

The small-scale distortions, due mainly to grid printing, have been corrected using the ground-based calibration file, after verification that corrections were indeed representative for the observed residuals.

12.3 Photometric calibrations

The photometric calibrations did not require a new reduction, and what is presented in this section is a summary of the calibration methods and results as obtained by NDAC (Evans et al. 1992).

12.3.1 The Main Grid or IDT Photometry

The main-grid, or IDT, photometry was based on the calibration of the mean intensity, β_1, and amplitude of the first harmonic, β_2, as a function of position in the field of view and star colour, for each of the fields of view. These two parameters are also referred to as the DC and AC *Hipparcos* photometry. The relations between these parameters and a pseudo-intensity I for a single star are given by Eqs. 4.1 and 4.2, page 114. Thus, β_1 provides a measurement of the intensity of the star or stars visible in the instantaneous field of view, and the background. Section 4.1.1 shows how β_2 can be affected by duplicity. This parameter is also affected by a bias, in particular when the modulation amplitude is low. Apart from these specific complications, the calibrations of the two parameters follow the same model. This model specifies the responses as a function of field of view, position on the grid and colour index of the star.

12.3.1.1 The model

The photometric calibration model consists of 5 elements:

- Parameters X_1 to X_3 representing the mean response and its linear gradients over the field of view;
- Parameters X_4 to X_6 representing the pass band corrections up to third order in $(V-I)-0.5$;
- Parameters X_7 and X_8 representing the variation of the pass band over the grid, and given by the linear cross terms of position and colour;
- Parameters X_9 to X_{14}, representing a linear spline as a function of distance from a fixed reference position on the grid, the so-called radial model parameters;
- Parameter X_b, representing the background contribution for the DC photometry.

All coordinates were expressed in units of the half-width of the field of view, i.e. in units of 0.45 degrees. Coefficients for these parameters are fitted through linear least squares, and are in most cases part of a running solution.

12.3.1.2 Running solutions

To improve the stability of the photometric calibration for the more difficult to determine parameters, each parameter estimate was effectively done through its own Kalman filter. Using the Householder formalism for processing the least squares solutions, a new solution included the upper-triangular information array of the previous solution as additional observations (see Appendix C), with the information for each parameter down weighted according to its typical variation time scale and the noise level on its determination based on an individual orbit. The weights applied for the preceding solution could be 0, 0.7 or 0.98. A weight of zero was applied to all parameters after a long data interruption, always for well determined parameters like mean response and response gradients, and after refocusing for those parameters that are sensitive to focal setting. A weight of 0.98 was used for the most difficult to determine parameters, such as the higher order colour, the correction of the pass band, and the radial corrections. The remaining parameters received a weight of 0.7.

12.3.1.3 Background determinations

The background modelling consisted of two parts: an empirically predicted background and a constant offset. In the empirical model the following contributions have been included:

1 Detector related background signals, at a few counts s^{-1};

2 Galactic glow, which covers the accumulated light from faint background stars, which in some parts of the galactic plane reached a level of 20 counts s^{-1};

3 Zodiacal light, i.e. Sun light reflected by dust in the ecliptic plane, with a peak level of 30 counts s^{-1};

4 General radiation induced photon counts, in particular during the crossings of the Van Allen belts, peak levels above 100 counts s^{-1}.

Unfortunately, there is a strong correlation between complexity (or difficulty to predict) of the background contribution and its potential strength. The modelling of components 2 to 4 made use of whole, or sections of, data sets in which the on-board attitude had not converged, and the orientation of the satellite axes was not known with sufficient accuracy. As the satellite was not aware of this, the observations in main detector continued, but in general there

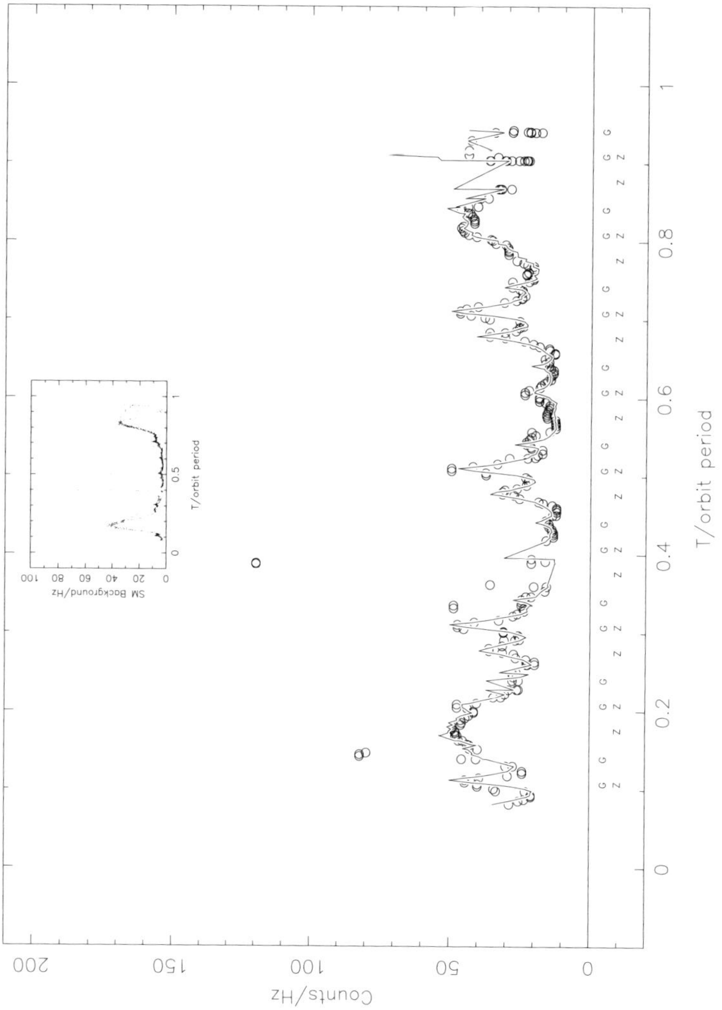

Figure 12.9. The observed (dots) and modelled (line) background signal in the IDT photometry. The transits of the ecliptic plane (zodiacal light) are indicated by "Z", transits of the galactic plane (Galactic glow) by "G". The inset shows the background for the star mapper detectors over the same orbit, with the signal dominated by radiation

were no stars captured in the instantaneous field of view, which therefore measured mostly the background. Many such sections of "data" were investigated, to ultimately provide empirical models of the galactic glow and the zodiacal light. Both of these produce contributions from the two fields of view. Radiation induced background, on the other hand, has no relation to fields of view, as it does not enter through the telescope optics, but randomly as it penetrates the payload. However, there was available the radiation profile as derived from the star mapper detectors, and all that was needed was a scaling with the much smaller radiation effects as observed in the main detector. Figure 12.9 shows an example comparison between the model and observations for an orbit without on-board attitude convergence.

Overall, background contributions varied mainly between 10 and 50 to 100 counts s^{-1}. Halfway the mission 100 counts s^{-1} was approximately equivalent to the mean response of a mag. 12 star, and 2 per cent of the response of a mag. 9 star.

Although the background contribution is an addition in intensity space, it is not advisable to carry out the photometric calibrations in that space, but instead in magnitude space. Following Evans et al. (1992), we express the observed quantity β_1 as:

$$\log(\beta_1) = \log(I_s) + \log(1 + \frac{I_b}{I_s}) \approx \log(I_s) + 0.434\frac{I_b}{I_s} + 0\left(\frac{I_b^2}{I_s^2}\right). \qquad (12.6)$$

We can now express the differences between the observed β_1 values and the calibration magnitudes as a function of a set of instrument parameters X_i and the background contribution:

$$-2.5\log(\beta_1) - H\text{p} \;=\; \sum a_i X_i - 1.086\left(\frac{I_b}{I_s}\right). \qquad (12.7)$$

The coefficient a_1 represents the zero point in the response, or, in intensity scale, the ratio between the observed and reference intensities. As all other contributions are much smaller, we can approximate:

$$I_s \approx 10^{-0.4(Hp+a_1)} = 10^{-0.4(8+a_1)} \cdot 10^{-0.4(Hp-8)}, \qquad (12.8)$$

with which we can now eliminate the signal contribution in the background coefficient and introduce:

$$a_b = -1.086 \cdot I_b 10^{0.4(8+a_1)}, \qquad (12.9)$$

representing the background contributions, to give:

$$-2.5\log(\beta_1) - H\text{p} \;=\; \sum a_i X_i + a_b \cdot 10^{0.4(Hp-8)}. \qquad (12.10)$$

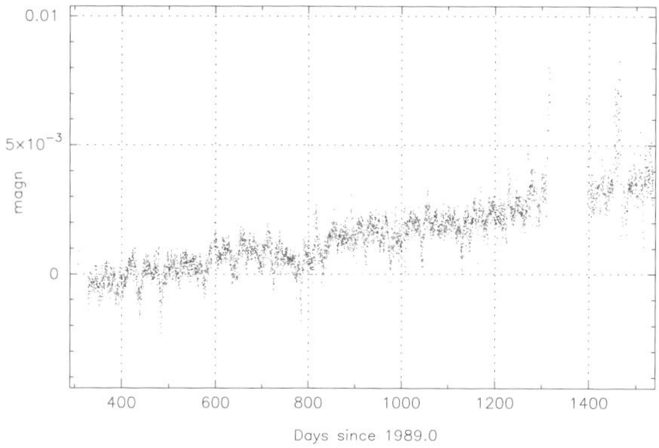

Figure 12.10. Evolution of the coefficient a_b, representing the background in the IDT photometry, over the mission

The actual background can then be derived from the calibrated values of a_1 and a_b.

To ensure that indeed a background signal is measured, only the calibration contributions of stars fainter than magnitude 8 or 9 included the background coefficient. In addition, the calibrated signal is kept small by first subtracting the *a priori* empirical background model from the counts. The background correction then becomes a constant offset correction to the model, generally small enough to be treated in the differential format described above. Figure 12.10 shows the evolution of this background correction (expressed in magnitudes) over the mission. There is an indication from correlations between features in this graph and the graph of the star mapper background (Fig. 11.5) that there were still residual radiation effects in the background signal. It also appears that the *a priori* model gradually slightly under estimated the background towards the end of the mission.

12.3.1.4 Mean response and response gradients

The coefficient a_1, representing the mean detector response, deteriorated quite dramatically over the mission, as was described in Chapter 7 and shown in Fig. 7.2, page 201. Its direct effect on the use of the photometric data is described in Chapter 7. Of interest here are the parameters describing the variation of the flux over the field of view. This variation was dominated by two features: linear gradients along (Fig. 12.11) and across scan (Fig. 12.12), and a circular symmetric feature. The evolution of the linear coefficients shows a number of features also encountered in various other instrument parameters. Highly noticeable is the adjustment on day 387 of the pointing of the instantaneous field of view to the grid rotation (see also Fig. 11.6, page 296). The focus corrections and thermal anomalies are clearly visible for the following field of view for β_2, very similar to what is observed for β_4 (Fig. 12.1) and for the first-order modulation amplitude $M1$ (Fig. 2.8). What is not observed,

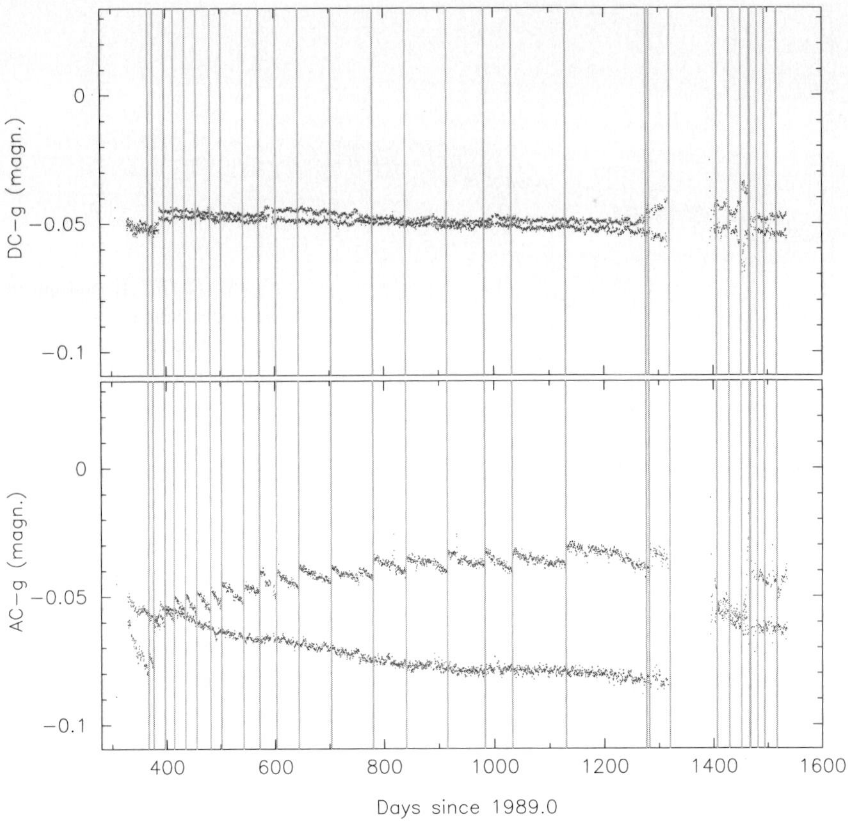

Figure 12.11. The response dependence on the along-scan position of a transit, for β_1 (DC-photometry, top graph) and β_2 (AC-photometry, bottom graph). The two sets of points represent the preceding and following fields of view, with in both graphs the following field of view showing the higher values at the start of the mission. The grey lines show the instances of refocusing of the telescope

however, is the feature around day 450 which is clearly present in Fig. 12.1 as well as in the evolution of the basic angle.

12.3.1.5 The circular-symmetric response variation

After subtracting the zero point and linear gradients of the photometric response, the remaining residuals show a clear circular feature (Fig. 12.13). This feature was fitted with a linear spline as a function of distance from the centre, as determined from the accumulated residuals of a large number of data sets. Over the mission this centre was not observed to shift significantly. To avoid a "singular" point at the centre of the circular feature, the spline function was used only to represent data outside a given radius. The mean response

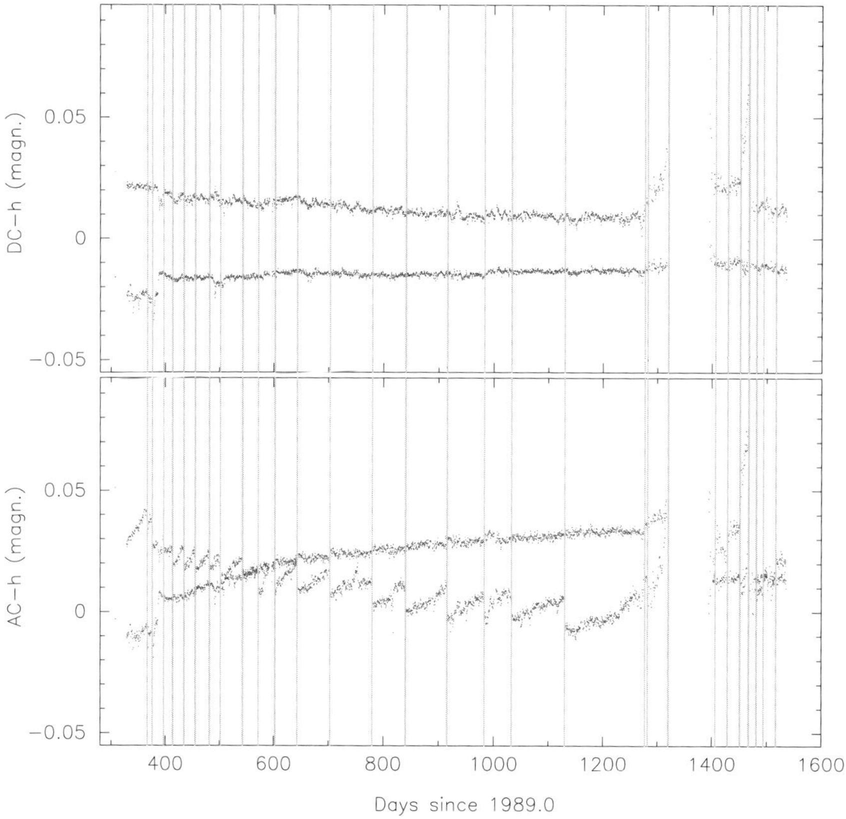

Figure 12.12. The response dependence on the across-scan position of a transit, for β_1 (DC-photometry, top graph) and β_2 (AC-photometry, bottom graph). The two sets of points represent the preceding and following fields of view, with in both graphs the following field of view showing the lower values at the start of the mission. The grey lines show the instances of refocusing of the telescope

X_1 served as the reference response for the centre of the field. The spline nodes were placed at 0.2 units intervals (where one unit is equivalent to half the field width), starting at 0.1 unit from the centre. The response correction for a star was determined according to its position on the grid with respect to the nodes, and was linearly interpolated between the two nearest nodes. Figure 12.14 shows how this contour evolved between the start of the mission and day 1300. There was insufficient data available to investigate a possible colour dependence of this feature. Observations in the far corners of the field of view, beyond the radius of 1.3, were corrected using a linear extrapolation of the corrections at 1.1 and 1.3 units.

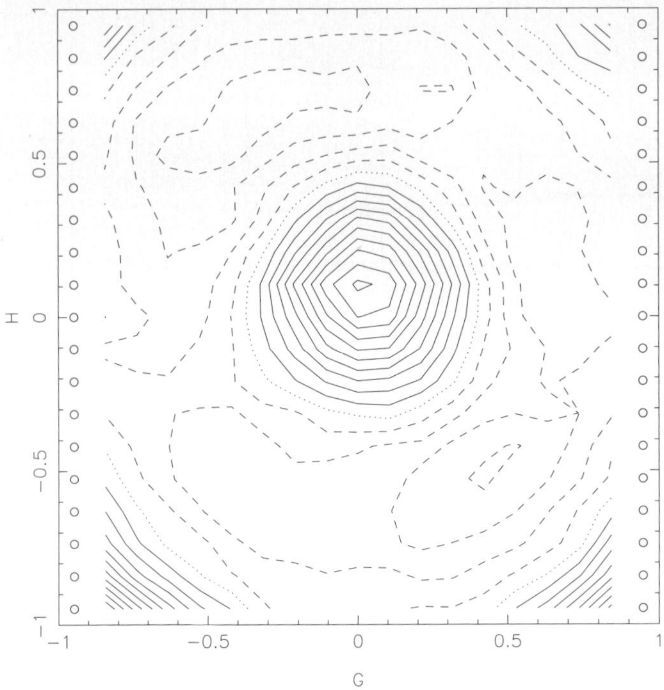

Figure 12.13. The mean residuals in the photometric calibration after subtracting the zero point and linear features, showing clearly the circular characteristic of the remaining response variations

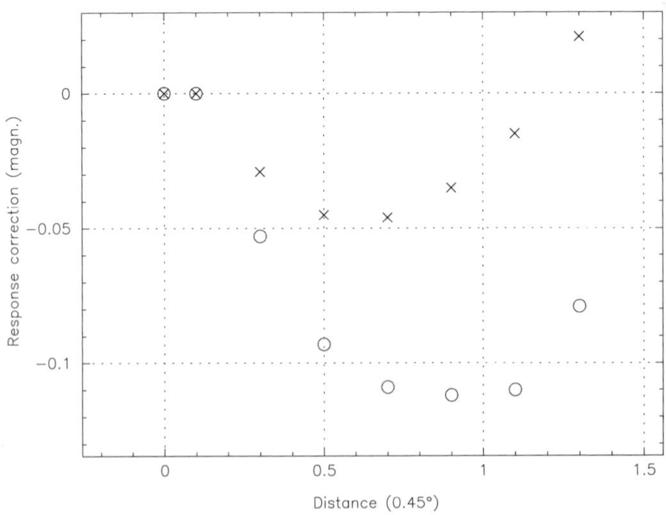

Figure 12.14. The fitted contour of the circular response feature at the start of the mission (crosses) and at day 1300, circles

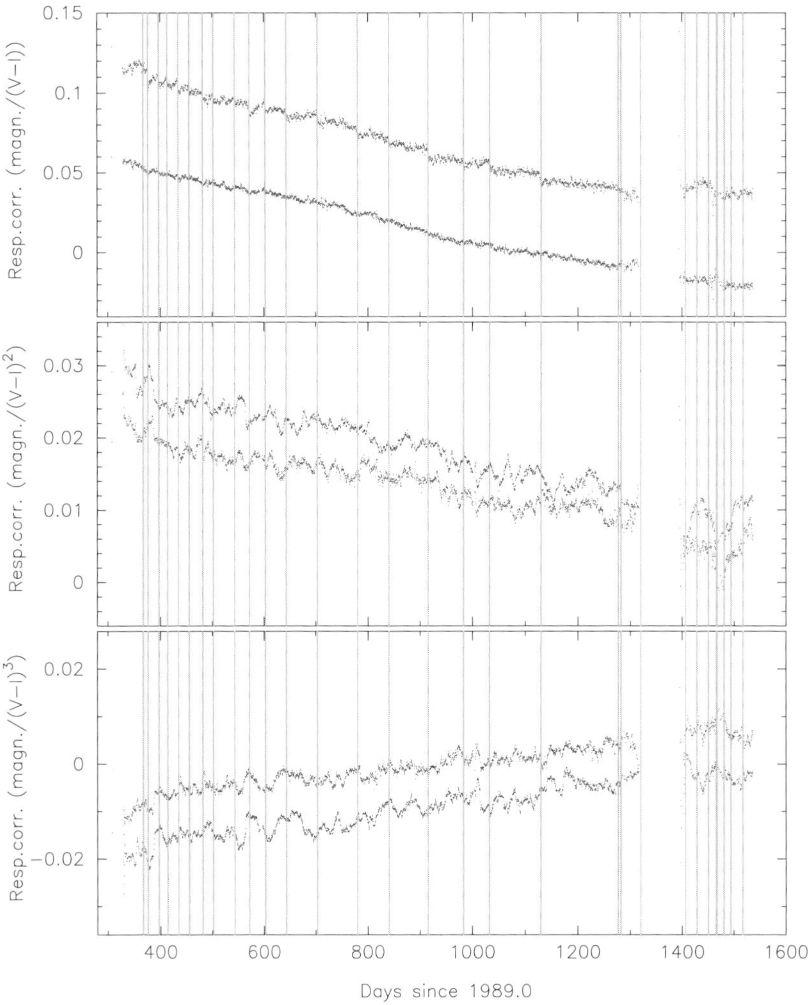

Figure 12.15. The colour coefficients in the photometric calibration model for the DC and AC photometry as obtained in the preceding field of view. From top to bottom: the linear, quadratic and third order terms. In the top two graphs the upper curve and in the bottom graph the lower curve refers to the AC photometry. The grey lines show the instances of refocusing of the instrument, events that, as could be expected, did not affect the colour coefficients

12.3.1.6 The pass band variations

The pass band used for the photometric reductions is a reference pass band, which was approximately equal to the actual pass band for the DC photometry in the preceding field of view, around day 1100. The characteristics of the colour coefficients and of the colour gradients across the field of view all show that the actual pass band at any time of the mission was a function of field of

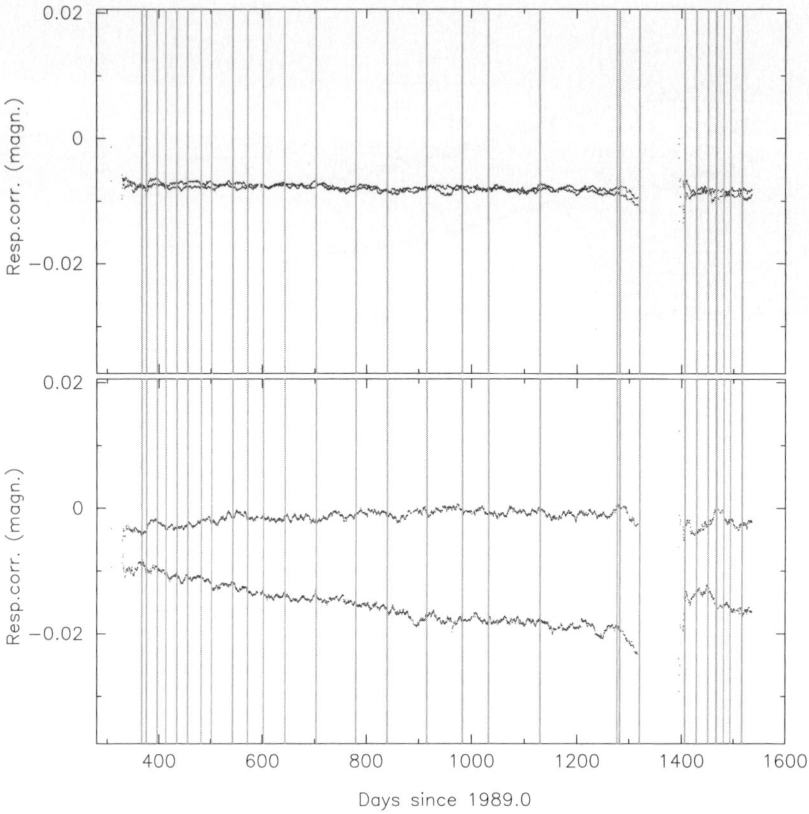

Figure 12.16. The cross terms of colour index and along-scan coordinate, showing a systematic shift of the pass band along the scan direction. The upper graph shows the data for the DC, the lower graph for the AC photometry, for both fields of view. The grey lines show the instances of refocusing the telescope, which had no significant effect on these dependencies

view, position in the field of view, and detector: the AC or the DC photometry. The colour-dependent parameters in the calibration model ensured that all data could be calibrated to the reference pass band. The evolution of those parameters shows the extent of the variation in the pass band, mainly caused by radiation darkening of the optical elements in the Image Dissector Tube assembly. The colour coefficients for the preceding field of view are shown in Fig. 12.15, demonstrating a quite significant difference between the AC and DC pass bands, as well as a significant quadratic term for both AC and DC photometry. The data for the following field of view look very similar, but showing small systematic offsets with respect to the data for the preceding field of view. Thus, also between the two fields of view differences in effective pass band existed.

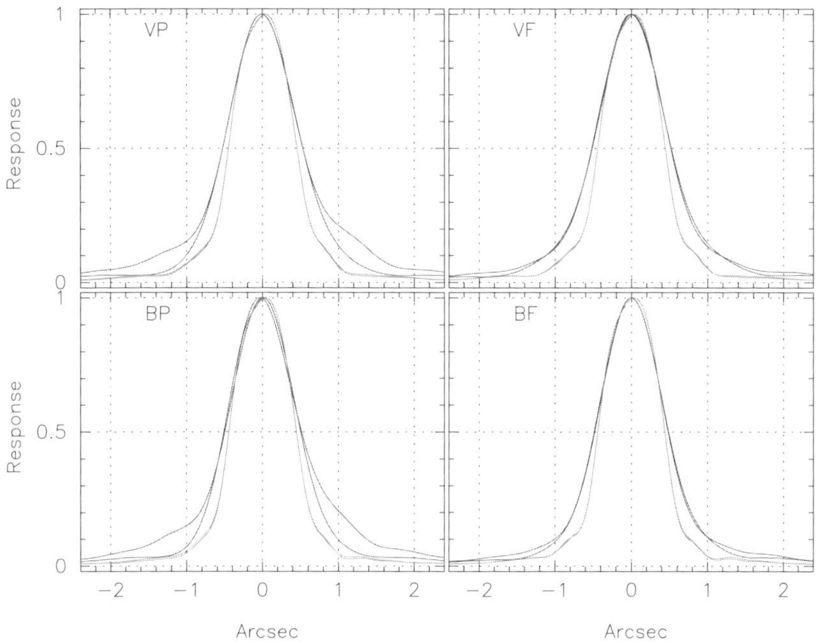

Figure 12.17. The single-slit response functions for star mapper transits. The narrower central curves represent the vertical slits, the wider outer curves the inclined slits. Curves are drawn for upper and lower branches. "B" and "V": B and V band respectively. "P" and "F": Preceding and following fields of view

The cross terms between colour index and along and across scan coordinates show that the transmission gradients noted earlier are in fact colour dependent. The relevant coefficients for the along-scan-colour correction are shown in Fig. 12.16. A comparison with Fig 12.11 shows the strong similarity between the evolutions of these coefficients. In both cases their value for the DC photometry remains effectively constant over the mission, apparently unaffected by radiation darkening that affected the overall transmission. There is also little or no difference between the two fields of view, indicating a common optical element as the source of the effect. The same applies, though somewhat less stringent, to the gradients across scan.

12.3.2 The Star Mapper Photometry

12.3.2.1 The single-slit response functions

The star mapper data was very different in characteristics from the IDT data. The star mapper recorded all transits from both fields of view, irrespective of any pre-selection criteria. The light was received for the entire grid (two slit groups: the vertical and inclined slits, see Fig. 2.4, page 47), and subsequently

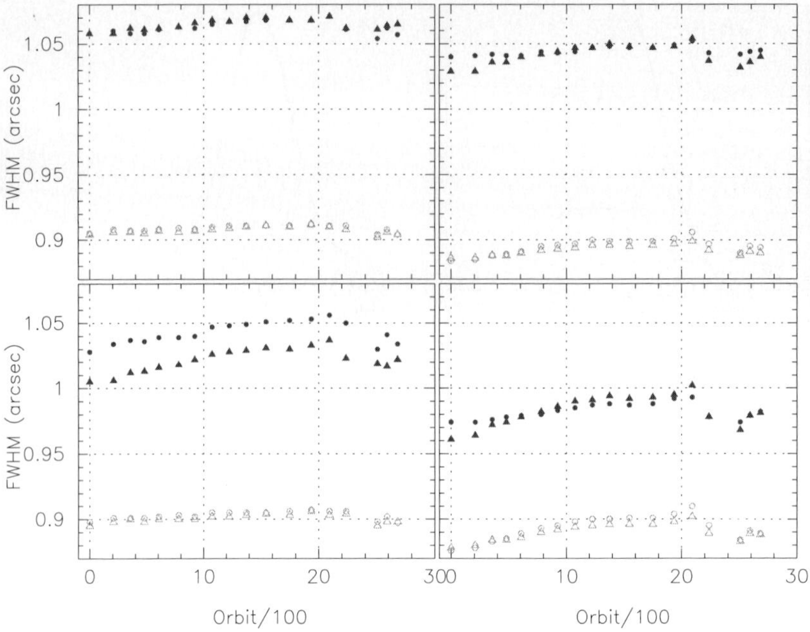

Figure 12.18. The evolution of the FWHM values of the SSRFs over the mission. Left: Preceding FOV, right: Following FOV; top: V, bottom: B. Lower curves: vertical slits, upper curves: inclined slits. Circles: lower branch, triangles: upper branch

split, using a dichroic prism, into a V and B channel, where photo-multipliers integrated the received flux over 1/600 s intervals. The transits over the four star mapper slits in a slit group were fitted with single-slit response functions (Single-Slit Response Function (SSRF)) to establish the response. Separate SSRFs were used for transits according to field of view, slit group, upper or lower branch, and channel (B or V), 16 different possibilities (Fig. 12.17).

As all instrument parameters, the SSRFs evolved over the mission. Collecting calibration data over intervals of about 160 orbits, the FWHM value for each profile was evaluated (Fig. 12.18). Determinations towards the end of the mission are generally noisy due to the poor coverage. Still, also here some typical characteristics seen in other instrument parameters can be recognized, in particular concerning the changes that took place over the three-months interruption in 1992.

12.3.2.2 The star mapper photometric calibration

Table 12.1 summarises the responses in the different channels for a 10th magnitude star (in either B or V) with $V - I = 0.5$. These values have been obtained as part of the star mapper photometric calibration. This model describes the observed intensities of transits through the star mapper slits as (for

Table 12.1. Responses in counts at 600 Hz at the start of the mission in the star mapper detectors for a 10th mag. star with $V - I = 0.5$

Channel	resp.	Channel	resp
VP-vert	3.665	VF-vert	3.911
VP-incl	2.613	VF-incl	2.766
BP-vert	5.122	BF-vert	4.937
BP-incl	3.946	BF-incl	3.710

the B channel):

$$I_{obs}/10^{-0.4(B-10)} = a_0 + a_1 C + a_2 C^2 + a_3 z + a_4 z^2 + a_5 zC, \quad (12.11)$$

where $C = (V - I) - 0.5$ and z is the ordinate of the transit, normalized such that it ranges between -1 and $+1$. Thus, Table 12.1 gives the coefficient "a0" as measured at the start of the mission.

The star mapper photometric calibrations were performed on groups of orbits. This provided a more stable calibration than could be obtained from a single orbit, while the shorter-term variations in the calibration parameters could not be significantly recovered: the star mapper data is generally quite noisy. Altogether, 329 calibrations were obtained covering nearly 2300 orbits. As for the IDT, the transmission of the optics changed over the mission, though for the star mapper the loss was only around 10 per cent over the mission, and much less colour dependent (Fig. 12.19). The way the response "recovers" after the three months the satellite spent in hibernation, in a way similar to the recovery of some of the instrument parameters, indicates that the loss in response is more related to the changes in the SSRFs. As shown above, these all showed a systematic increase of FWHM over the same period for which the response decreased.

As for the IDT, the star mapper pass bands are reference pass bands for the calibrations, and a slightly different pass band applied for each channel and at different epochs of the mission. These differences show up in the colour terms a_1 and a_2 being different from zero. There appears to be little or no gradient over the mission in the colour coefficients. The pass band deviations that cause the colour coefficients are largest in the B channel, where corrections are opposite for the two fields of view (Fig. 12.20). They are larger for the vertical than for the inclined slits, which may be due to a dependence on the along-scan coordinate, which would affect the measurements for the inclined slits, but not those of the vertical slits. A similar effect is noted for the across-scan response gradients and their colour dependence (see below).

With the entire optical chain shared between the two fields of view, it is not clear where the differences between the fields of view may originate, but its

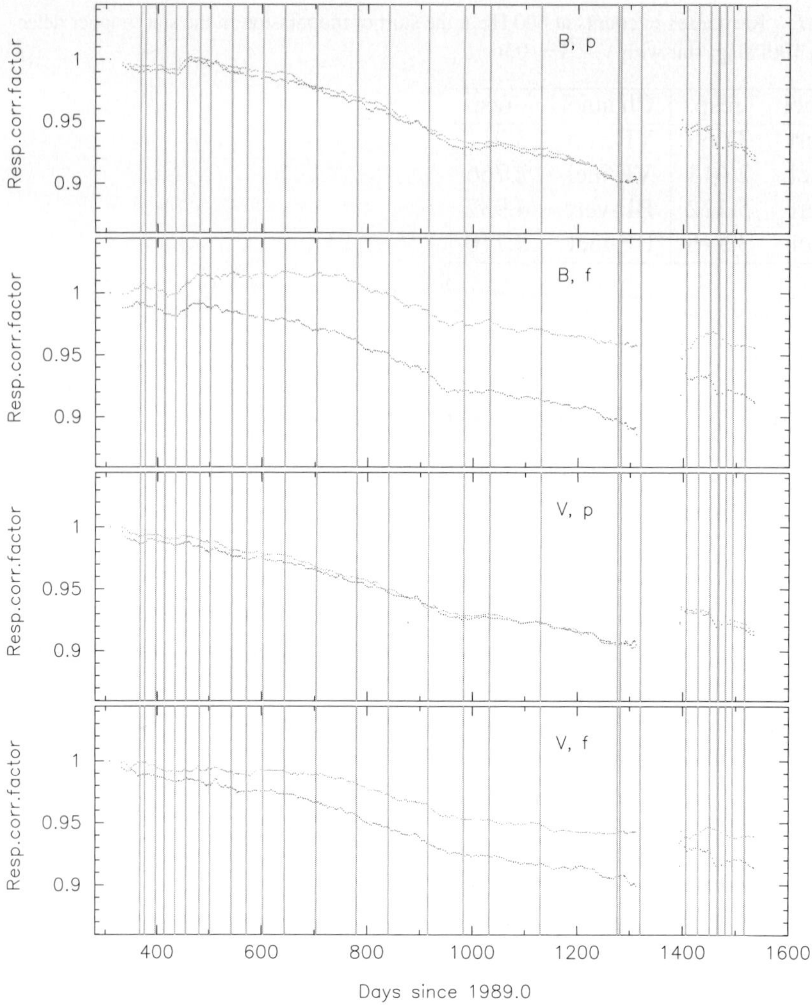

Figure 12.19. Changes in response for the different channels, as measured over the mission. The responses have been normalized to the value measured for the first data sets. From top to bottom: B in the preceding field of view; B in the following field of view; V in the preceding field of view; V in the following field of view. The two curves in each graph represent transits through the vertical and inclined slit groups. The grey lines show the instances of refocusing the telescope

nature, affecting primarily the B band, suggests that it may have to do with the near-UV cut-off in the optics, which in that case would need to be slightly different for the two halves of the beam combiner. All other mirrors are fully shared by the two beams. It is also noted from the SSRF profiles in the preceding field of view that the images are asymmetric in the across-scan direction

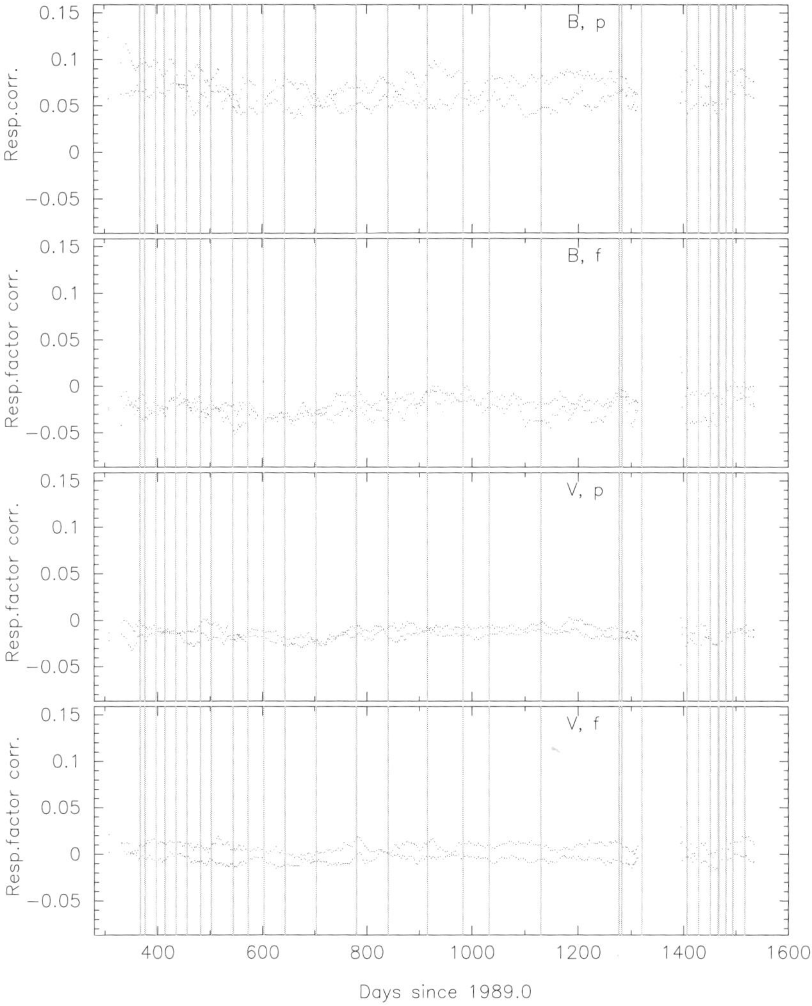

Figure 12.20. Differences in response as a function of the colour index V − I − 0.5, for the different channels, as measured over the mission. This is coefficient a_1 in Eq. 12.11. From top to bottom: B in the preceding field of view; B in the following field of view; V in the preceding field of view; V in the following field of view. The two curves in each graph represent transits through the vertical and inclined slit groups. The grey lines show the instances of refocusing the telescope

(Fig. 12.21). There appears to be a "ghost" image about 1 arcsec below the main image (preceding in the upper branch transits and following for the lower branch), and at about 10 per cent of the intensity of the main image. The SSRFs have not been resolved according to star colour, which otherwise might have

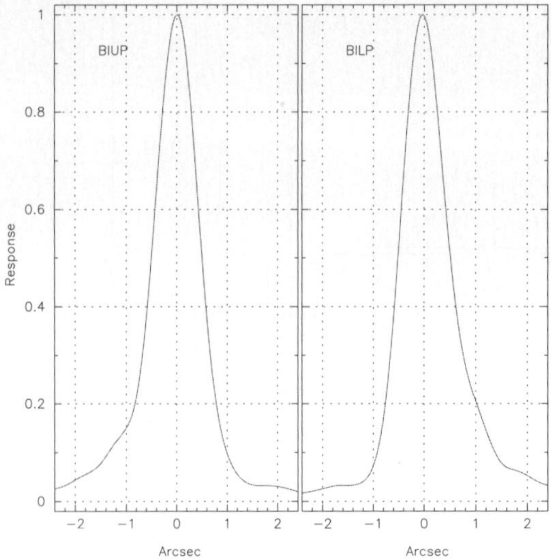

Figure 12.21. Asymmetry in the SSRFs for inclined slits in the preceding FOV, B channel. The additional hump preceding for the upper (left), and following for the lower branch (right), is also clearly visible for the inclined V channel, but not for the vertical slit transits

thrown some light on these and subsequent colour-related aspects of these calibration results.

As for the IDT, a transmission gradient is observed as a function of the ordinate, with a similar colour dependence associated with it (Fig. 12.22). This effect is much more strongly present in the B than in the V channel, and is more pronounced for the vertical than the inclined slits.

12.3.3 Conclusions on Photometry

The *Hipparcos* photometric reductions have been far from simple, with changes in pass bands taking place over the entire mission. For extreme-colour stars, mainly those of very red colour indices, the calibrations have been hampered by lack of reference stars, as most stars in that regime tend to be variable. A major ground-based support effort was organized by Michel Grenon during the mission, and made it possible to cross-check calibrations for extensions to large colour index values. In principle it was possible to have included a new photometric analysis using the epoch colour information for the red stars as provided by (Pourbaix and Jorissen 2000; Pourbaix and Boffin 2003), but it was felt that this would not provide enough improvement to justify the work involved.

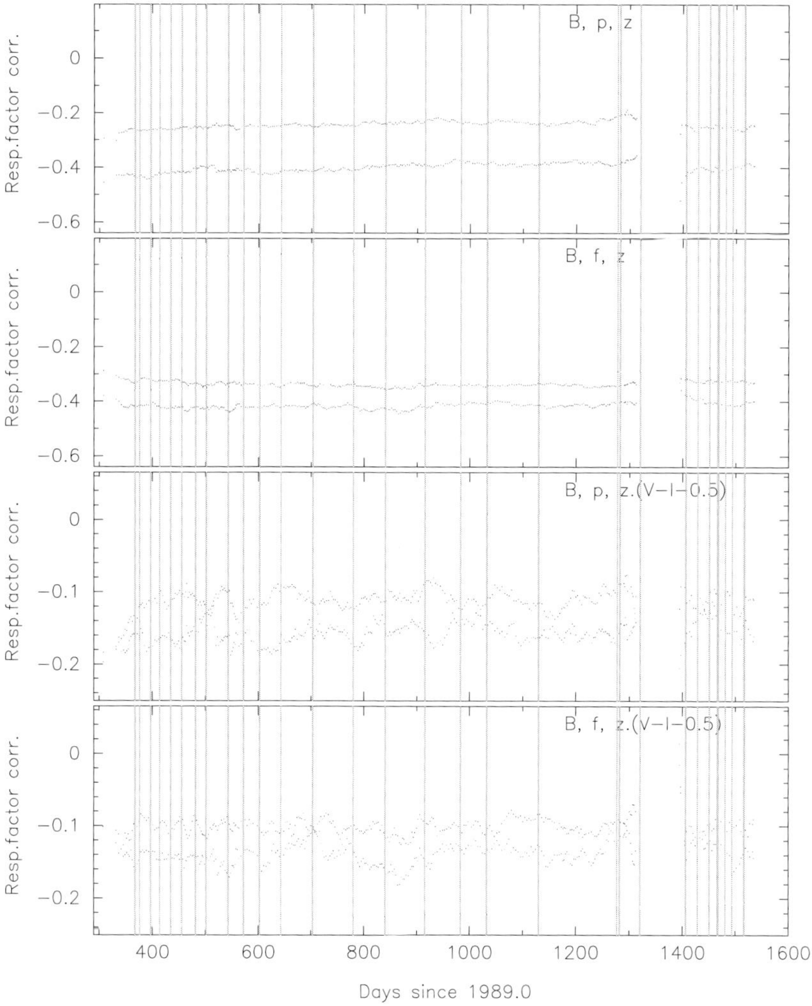

Figure 12.22. The dependence on ordinate and ordinate times colour of the responses in the B channel. From top to bottom: B in the preceding field of view, z (ordinate) dependence; B in the following field of view, z dependence; B in the preceding field of view, z and colour dependence; B in the following field of view, z and colour dependence. The upper and lower curves in each graph represent transits through the inclined and vertical slit groups respectively. The grey lines show the instances of refocusing the telescope

Chapter 13

SPACECRAFT-PARAMETER CALIBRATIONS

13.1 The on-board clock

All on-board measurements were regulated by signals derived from the on-board clock. The stability of these timings can be checked by comparing the on-board times with the ground-station times. *Hipparcos* had very limited memory for on-board data storage, and its operations depended on continuous contact with a ground station for immediate downloading of observations. The telemetry was sent to the ground station in telemetry formats, covering 10.667 s of operations (spacecraft and payload). Each format received by the ground station, was time-tagged using a signal from the ground-station clock. With the orbital position of the satellite generally known to within a few hundred meter, the time delay between satellite and ground station could be calculated to the sub-micro second level, thus providing a direct comparison between on-board and ground-station clocks. For obvious reasons, the ground-station clock was considerably more stable than the on-board clock, and drifts between the two could generally be assigned to the on-board clock. What was not always stable, however, was the ground-station time delay: the time passed between the signal reaching the antenna and it being time-tagged. This applied in particular to the Goldstone ground station.

In the following sections the input data, calibration methods and calibration results of the on-board clock are reviewed.

13.1.1 Calibration Data

The calibration-data stream consists of one measurement every 10.667 s for every telemetry format received from the satellite, independent of the observing status of the payload. The analysis shows an intrinsic noise for the on-board clock of $\sigma \approx 0.3\mu$s, and very much larger systematic drifts . Moreover, dis-

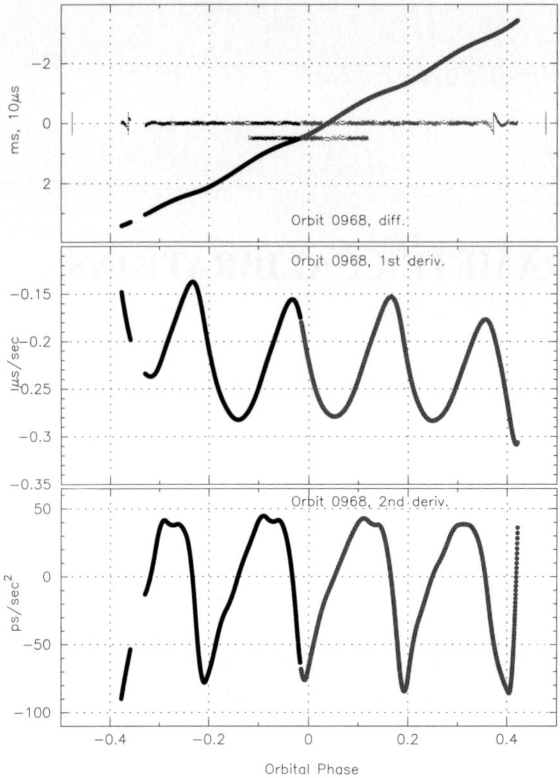

Figure 13.1. On-board clock-drift analysis for orbit 968, in which no eclipses occurred and the perigee passage was relatively high. Top graph: the evolution of the drift over the orbit, and the residuals for the spline fit (around zero) and the harmonic fit (around apogee, offset). Middle graph: The first derivative of the spline fit to the clock drift, showing the temperature variations, here dominated by the rotation of the satellite. Bottom graph: The second derivative of the spline fit, showing the heat intake (negative values) or loss (positive values), depending on the orientation of the satellite

crete jumps in the clock differences are often observed when there is a change of ground station, showing that the ground-station delay times as supplied were not accurate and in some cases not even constant. Before any fitting of the data could be attempted these jumps needed to be removed. This was done by fitting a second order polynomial through a small amount of data around the jump (which could bridge a short gap), using an additional offset for the data after or before the jump. All data before or after the jump were subsequently corrected for the additional offset. Jumps in the timing are most frequently observed in those data received and time-tagged by the Goldstone ground station, where jumps also occurred without ground-station changeovers. This may have been due to switching on/off of electrical equipment in the same power circuit as the time-tagging equipment.

The observed time differences represent the integrated effect of the on-board clock drift, the actual drift is given by the time derivative of the time-differences curve. If we assume that clock drifts are related to operational temperatures, for which there is ample evidence as will be shown later, then the second time derivative of the time difference curve, which shows the changes in the drift, will also show the changes in local temperature or the heat intake or

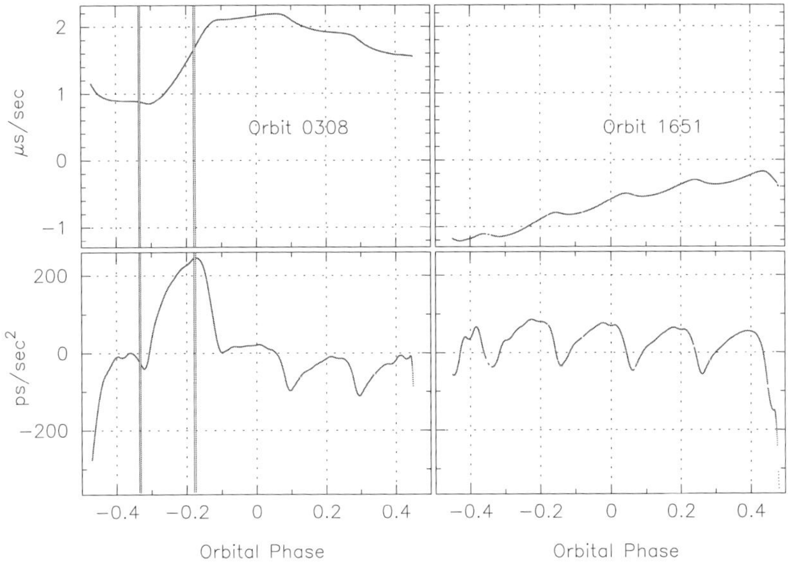

Figure 13.2. The first and second derivatives of the on-board clock drift for orbits 308 (long eclipse as indicated by the vertical lines) and 1651 (friction heating during low perigee passage). The temperature (1st derivative) and heat-intake (2nd derivative) are clearly affected by these events. The eclipse in orbit 308 also shows that there is some delay in the reaction of the spacecraft on the changes in the exposure, which reflects the insulation blankets applied to the spacecraft

loss. The main sources of heat intake for the satellite were the solar radiation (sometimes interrupted by eclipses and generally affected by the rotation of the satellite and the varying distance between Earth and Sun) and friction experienced during perigee passages, in particular when perigee height was relatively low (450 to 470 km) (Dalla Torre and van Leeuwen 2003). In the following we will refer to the on-board clock as OBC, and the ground-station clock as GSC.

To accommodate the various features affecting the OBC, the clock differences (OBC-GSC) were fitted by means of a fourth order spline (see App. B), so that the second time derivative still provides a continuous function up to its first time derivative. In addition, the differences for the two hours around apogee were fitted with a third-order polynomial and harmonics of the satellite's rotation phase up to order 5. An example of a "quiet" situation is shown in Fig. 13.1. For this orbit there is only a very short eclipse during the perigee passage, the satellite is most of the time Sun-lit. The perigee height for this orbit is too high for friction to have a significant impact on the spacecraft temperature. The clock-drift variations are therefore dominated by the rotation of the satellite, and reflect the positioning of the clock on the space-craft platform: close to one of the side panels. Whenever this panel is exposed to the Sun, there is an effective intake of heat for the area around the clock. Of further interest are the discontinuities in the residuals relative to the spline fit. These mark the start and end time of so-called orbit records. Each orbit record provided the parameters for reconstructing the position and velocity vectors of the spacecraft over a specified interval of time. This representation was based on Keplerian orbit parameters and Chebichev polynomials for final adjustments (see for further details van Leeuwen and Penston 2003). The discontinuities are typically of the order of several μs, representing positional discontinuities in the orbit representation of a few km. These discontinuities occurred only close to perigee, and did not significantly affect the data. It would in principle be possible to correct for these local inaccuracies using the OBC-GSC comparisons.

Eclipses and low perigee passages significantly affected the spacecraft temperature, which was directly reflected in the OBC behaviour. This can be observed in Fig. 13.2, where we show the OBC data for orbit 308 (long eclipse) and orbit 1651 (low perigee, no eclipse). In the presence of long eclipses, the spacecraft will be taking up heat for most of the time when exposed by the Sun, while following low perigee passages it loses heat, despite being Sun-lit. The heat-loss during and after the eclipse shows that there is a delayed as well as an almost immediate reaction to the eclipse: heat loss starts to decrease almost immediately after the satellite comes out of eclipse, but still continues for about 30 min. It takes a further 2 hours before the spacecraft is warming up again.

13.1.2 Long-term Drifts

To investigate the long term drifts of the OBC, reference data from the polynomial fit around apogee, for each orbit with sufficient coverage, was collected. Because this fit included harmonic coefficients, the polynomial fit should show mainly the averaged values one would get after removing the spin-synchronous modulations. Figure 13.3 shows the behaviour of the on-board clock for a reference time at apogee in each orbit for which data is available. The drift of the on-board clock accumulated to about 40 s over the mission. The derivative of the drift, shown in the middle graph, provides the clearest indication that its main dependence is temperature. As a comparison is shown a curve representing the exposure factor of the satellite. This exposure factor is defined as:

$$E = (1/R)^2 \cdot (T_o - T_e)/T_o, \qquad (13.1)$$

where R is the distance Sun–Earth (in AU), T_o the orbital period, and T_e the length of time spent in eclipse during the orbit, all at the time of observation (see also van Leeuwen 1997). The factor E thus defines the normalized amount of solar radiation received by the satellite over an orbit. The curve thus produced is scaled and shifted to fit the observed clock-drift values, leaving only secondary variations to explain. Clearly recognizable are the periods with no

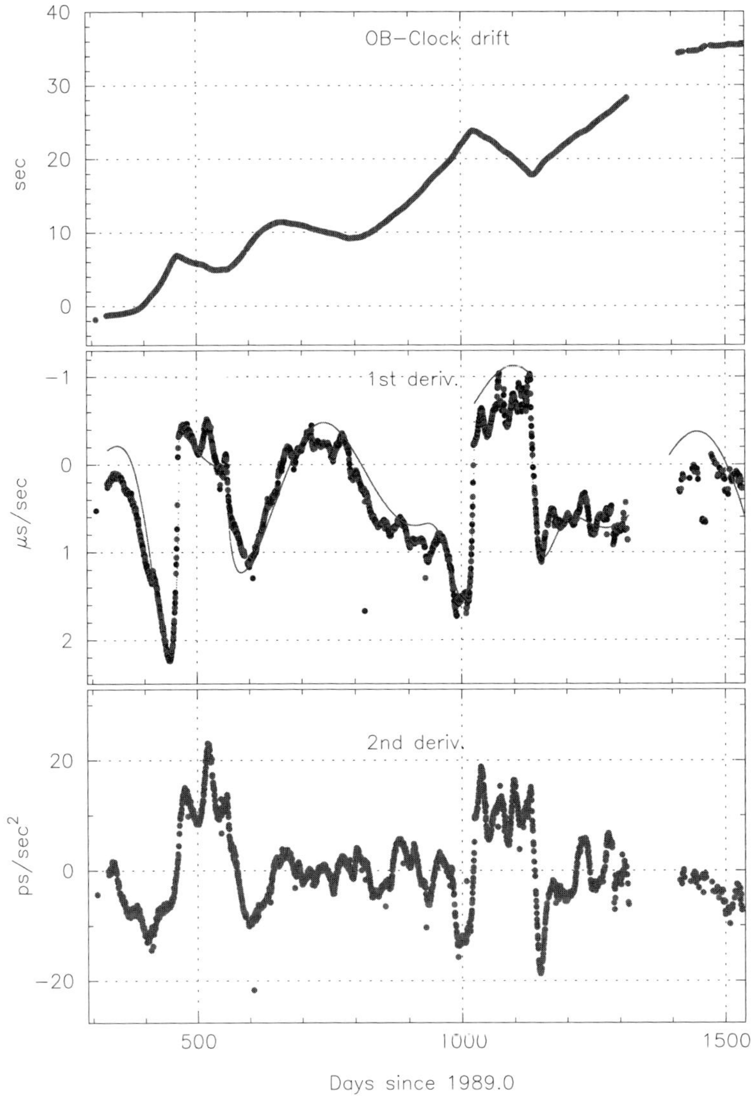

Figure 13.3. The on-board clock drift as measured at apogee over the mission. Top graph: accumulated time difference; Middle graph: the mean clock drift at apogee, compared with the satellite exposure factor (see text); Bottom graph: the mean change in clock drift, showing the heat intake(negative values) or heat loss, depending on conditions in the orbit

eclipse, around days 500 and 1100. Also clearly represented are the results of the variation in radiation intensity due to the ellipticity of the Earth orbit. The remaining variations appear to have a typical timescale about equal to the precession period (57 d) of the satellite's spin axis.

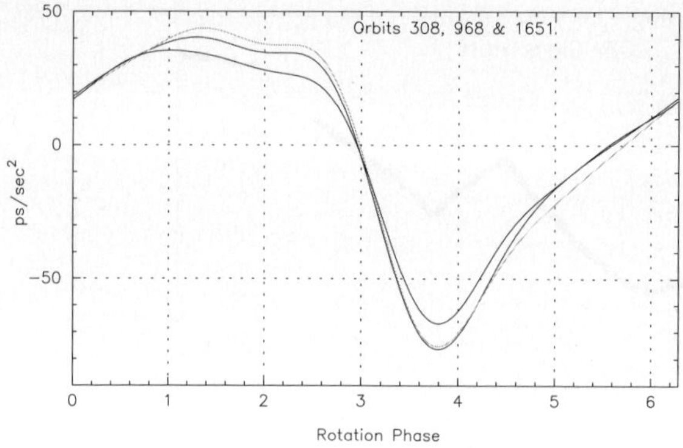

Figure 13.4. The spin-synchronous component in the second derivative of the clock drift for the three orbits shown above: 308, 968 and 1651. Despite quite significant differences in operational conditions, the curves show only small variations, most noticeably a smaller amplitude for orbit 308 (cold conditions)

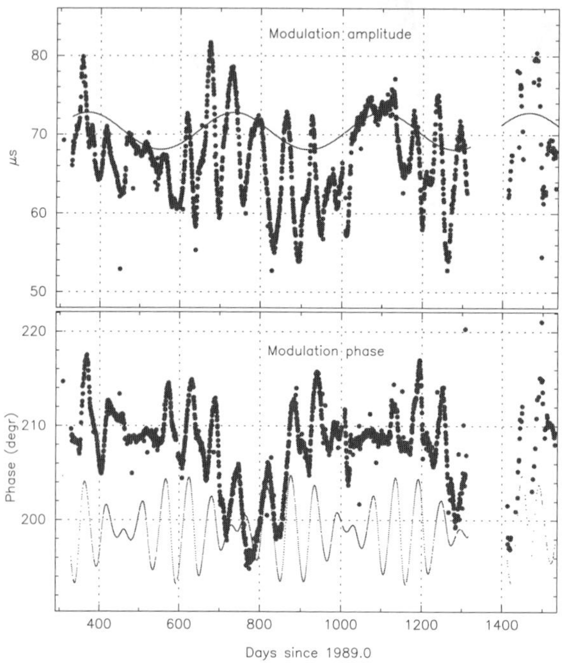

Figure 13.5. The evolution of the amplitude (top) and phase (bottom) of the first harmonic in the spin-synchronous component of the OBC drift. The amplitude variations reflect the varying distance between Earth and Sun (smooth reference curve), but show on top of that strong modulations. The same modulations are observed in the phase. The smooth curve in the lower graph shows the beat pattern produced by periods of 57 and 584 days (see text)

Although all photon count processing took place using the on-board clock timings, the relevant proper ground-station times were assigned to all measurements and used for all ephemerides calculations (satellite orbit and all solar-system object observations).

13.1.3 Spin-synchronous Drift Modulation

As was shown above, the on-board clock drift has a long term as well as a spin-synchronous component. The latter is fairly stable, but shows some dependence on the mean temperature (or drift level) of the OBC (Fig. 13.4).

The spin-synchronous component is not observed when the satellite spin axis is pointed directly towards the Sun, which happened a few times during the mission as part of a recovery operation. Operations of this kind are referred to as Sun-pointing mode observations.

There are variations in total amplitude and phase of the modulation, the origins of which are still unclear, but the character of which is intriguing. Both amplitude and phase show a modulation with time which appears to be affected by two periodic phenomena: the precession of the satellite's spin axis (57 days) and the alignment period of the Sun, Earth and satellite at apogee (584 days) (Fig. 13.5). The two phenomena together determine to some extent the conditions during the perigee passage of the satellite: perigee height and pointing direction of the spin axis during the perigee passage. However, at times when there are no eclipses, this modulation is no longer observed. Also, the amplitudes of the modulations of these two parameters when eclipses do take place seems to be unaffected by the lengths of the eclipses. The double-beat modulation curve is shown as comparison in the bottom graph of Fig. 13.5.

13.1.4 Conclusions on the On-board Clock

The on-board clock has provided us with a very precise record of the temperature variations in the spacecraft. It should be noted here that the spacecraft was much less thermally insulated or controlled than the payload, so these variations represent an extreme rather than a typical behaviour, a behaviour that is not reflected in any way in payload calibration parameters.

13.2 Gyro characteristics

Table 13.1. Gyro orientation and angular momentum values

	Input axes			Angular momentum (kg m^2 s^{-1})		
Gyro	x	y	z	x	y	z
1	−0.541	+0.891	0	0.01157	0.00701	0
2	1.056	0	0	0	−0.01410	0
3	+0.526	+0.869	0	0	0	−0.00134
4	0	0	+1.056	−0.01157	+0.00701	0
5	0	0	−1.056	−0.01157	+0.00701	0

Hipparcos was equipped with five rate-integrating gyros, of which in normal operations three would be active and two redundant. Three of the gyros had their input axes in the plane scanned by the payload, and two gyros had their input axes perpendicular to that plane, along the spin axis of the satellite. The main task of the gyros was to provide readings of the inertial rates around the satellite axes for the on-board closed-loop attitude control system. For the on-ground processing the gyro readings provided rate estimates for the star mapper data processing and in the new reduction they provide the first step in building the fully dynamic attitude model (Chapter 10). The orientation and angular momentum details are shown in Table 13.1. The orientation parameters also contain a factor to correct the default scaling from gyro readouts to inertial rates. For rate integrating gyros the input and angular momentum axes remain effectively fixed within the satellite reference frame. A rotation around the input axes is measured every second, and applied in reverse to correct the orientation of the angular momentum vector to its reference position. These movements, when coupled with the rotation of the satellite, will cause a torque (see Chapter 9).

The behaviour of the gyros during the mission has had much influence on the data products. Most noticeably the breaking down of nearly all gyros forced the operations over the final four months of the mission to operate under non-optimal conditions, leading to the collection of only short stretches of data during that period. In the following sections the noise and drift properties of the gyros are summarized, and some examples are shown for what happens when a gyro breakdown takes place.

13.2.1 Noise and Drift Properties

All gyros behaved differently during the mission, for both noise characteristics and drift. Gyro drift represents a zero point in the measurements of the inertial rates by the gyro. With a nearly constant rotation rate around the spin axis, it was difficult to unambiguously distinguish drift on gyros in the (x, y) plane with a slight misalignment towards the spin axis. However, most of the observed variations in the drift would be hard to explain from misalignments and should be real drift variations. Figure 13.6 shows noise and drift behaviour for the active gyros over the mission. Noise levels varied between gyros by more than a factor ten. At the lowest noise levels recorded (0.01 arcsec s^{-1}) the digitisation of the readings became the limiting factor, and the data were in weight nearly compatible with the star mapper data; at the highest noise levels (above 0.1 arcsec s^{-1}) the data were effectively worthless.

The gyro drifts were determined after the attitude reconstruction from a comparison between the reconstructed inertial rates and the observed gyro readings. The gyro orientations were reconstructed in the same process. Gyro drifts were observed to be affected by eclipses, in the same manner as the on-board clock. The drifts also show a very clear and persistent modulation component reflecting the rotation phase of the satellite. This component is further described below. Orientation components were observed to be constant within the measurement accuracy, and the amplitudes of "misalignments" are generally less than one part in a thousand.

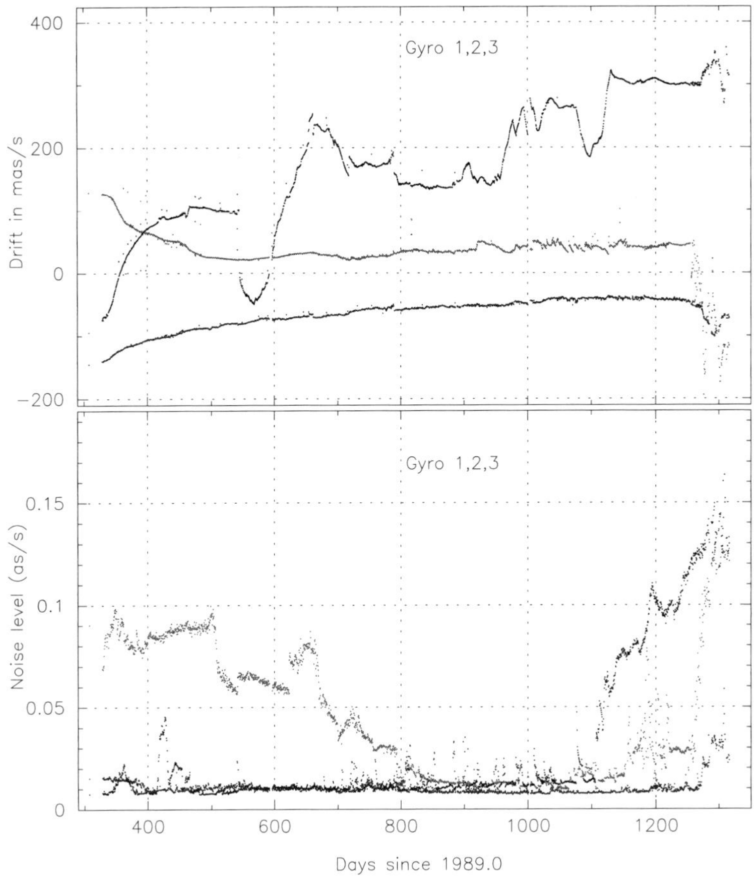

Figure 13.6. Drift (top) and noise (bottom) on the operational gyros (1,2,4 or 5) over the mission (a change from gyro 4 to 5 took place on day 545)

Although there is no sign in the gyro drift of the kind of changes observed in the on-board clock over the mission, there are similarities in the variations observed over a single orbit. The specifics vary between the gyros, and an example is shown in Fig. 13.7 for gyros 4 and 5, showing the same orbits as used to illustrate the on-board clock drifts in the preceding section.

13.2.2 Rotational-Drift Modulation or Reference Frame Distortions?

A most puzzling feature in the gyro drifts is the presence of a rotation-synchronous modulation, and in particular the presence of quite high harmonics. These modulations are observed to repeat quite accurately all through the mission and are only not present when the satellite observes in Sun-pointing

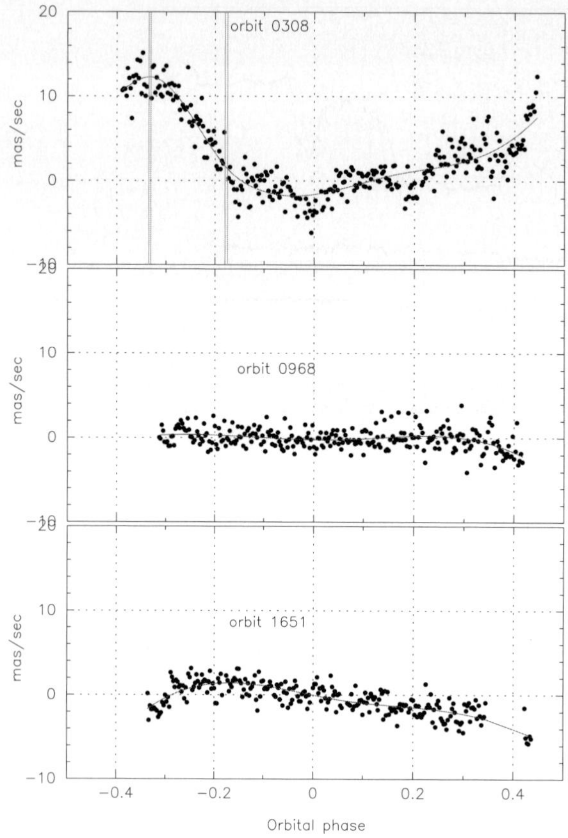

Figure 13.7. Changes in the drift for gyro 4 (orbit 308) or 5 (orbits 968 and 1651) over the orbit. Features similar to what was observed for the on-board clock can be seen. From top to bottom: Presence of a long eclipse (between vertical lines); A relatively undisturbed orbit; An orbit with very low perigee passage

mode. The rotation phase, together with the solar aspect angle, both as defined in the heliotropic angles, provide a direct measure of the direction of the Sun as seen from the satellite. For Sun-pointing mode observations the solar aspect angle of the spin axis is zero degrees, while under normal observing conditions it is 43.0 ± 0.15 degrees.

The modulation features shown in Fig. 13.8 repeat accurately over the entire mission. There are two aspects of this modulation that together are difficult to explain. The first is the presence of high harmonics. As we have seen for the on-board clock as well as in the reaction to eclipses of the gyro drifts, temperature changes as perceived within the spacecraft are much smoothed out, leaving no room for sharp features as observed for Gyro 2 at phase 248 degrees for example. Secondly, an examination of the amplitudes of the most prominent feature (the same minimum at 248 degrees for Gyro 2) shows a reflection of the changes in solar radiation received as a result of the ellipticity of the Earth orbit, which would indicate that solar-radiation is the reason behind these drifts (Fig. 13.9).

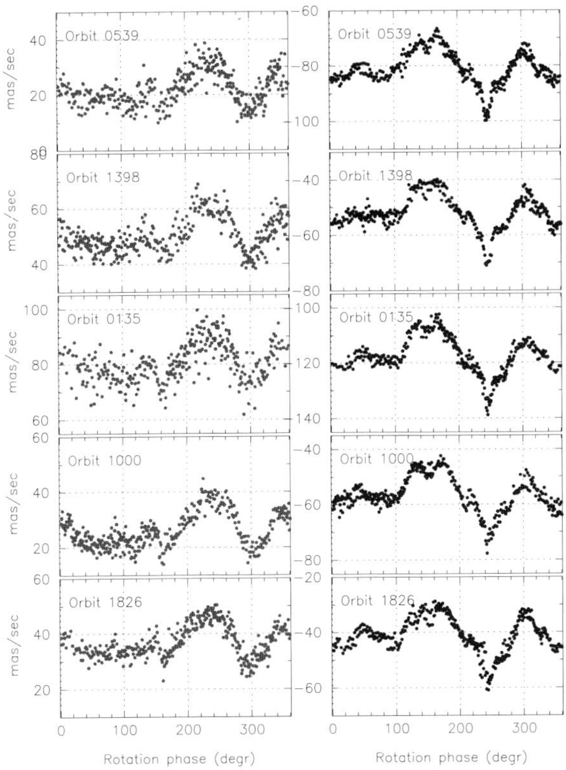

Figure 13.8. Spin synchronous drifts for gyros 1 (left) and 2 (right) over 2.5 years of the mission. The top two orbits are at maximum distance Earth-Sun, the remaining three at minimum distance

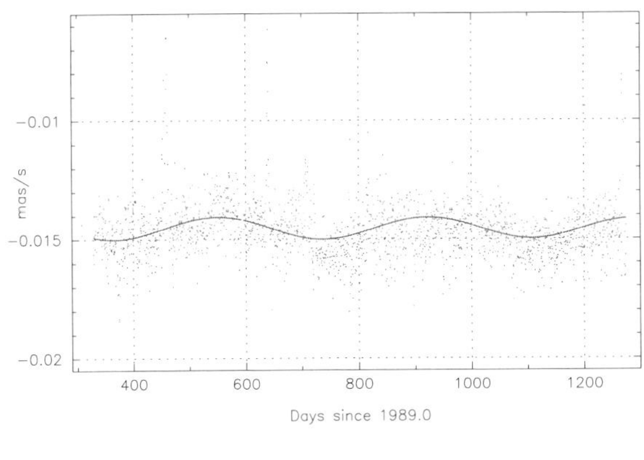

Figure 13.9. The variation in amplitude of the most prominent feature in the spin-synchronous drift modulation for Gyro 2, compared with the expected variation if due to solar radiation

The high-frequency character of some of the observed features hints at a source on, or very close to, the outside of the satellite, where exposure changes are much more pronounced. What may in that case be affected is the actual mounting of the gyro box (suspended under the payload platform, quite close to the outside edge of the satellite).

13.2.3 Gyro Break Down

The gyro breakdowns are generally attributed to a failure of the circuitries controlling the gyro motors. Changes in rotation rates result in torques on the satellite, which led to rapid inertial rate changes which could no longer be followed through the star mapper transits. As a result, on-board attitude control was lost, and in the worst cases the satellite resorted to its safety mode: moving the spin-axis in the direction of the Sun. Data around these events is nearly always lost. An extreme example of such event is shown in Fig. 11.2, page 291.

13.3 Thruster firings and the Centre of Gravity

13.3.1 Thrusters: Positions and Actions

Hipparcos was equipped with 6 sets of cold-gas thrusters for attitude control. Each set consisted of an active and a redundant cluster. The z-thrusters were situated in what was supposed to be the plane close to the COG, but this had shifted due to the still full tank of the ABM. The x and y thrusters were situated on the lower platform. The positions and nominal firing directions of all thrusters are given in Table 13.2, in a coordinate system centred on the nominal position of the COG. The torques caused by each thrusters is given by:

$$\boldsymbol{T}_i = \boldsymbol{p}_i \times \boldsymbol{d}_i F_i, \tag{13.2}$$

where i indicates one of the six thrusters, $\boldsymbol{p}_i$ the position of the thruster, $\boldsymbol{d}_i$ the direction of the thrust, and F_i the force caused when activated.

Activation of thrusters was controlled by the on-board closed loop attitude control system. The lengths of thruster firings were quantized in units of 1/75 s, with a minimum of initially 4, later 0 or 2 units, and a maximum of around 40 units. Through thruster activations the actual attitude of the satellite (positioning and rotation rates) was kept within pre-defined limits from the nominal

Table 13.2. The positions and directions of the 6 active thrusters

Thruster	Position in mm			Direction		
	x	y	z	d_x	d_y	d_z
+x	0	-945	-860	0.000	0.000	-1.000
-x	0	+945	-860	0.000	0.000	-1.000
+y	889	0	-860	0.000	0.000	-1.000
-y	-1003	0	-860	0.000	0.000	-1.000
+z	625	-906	-25	0.309	0.951	-0.008
-z	428	-1042	-25	-0.978	0.208	-0.008

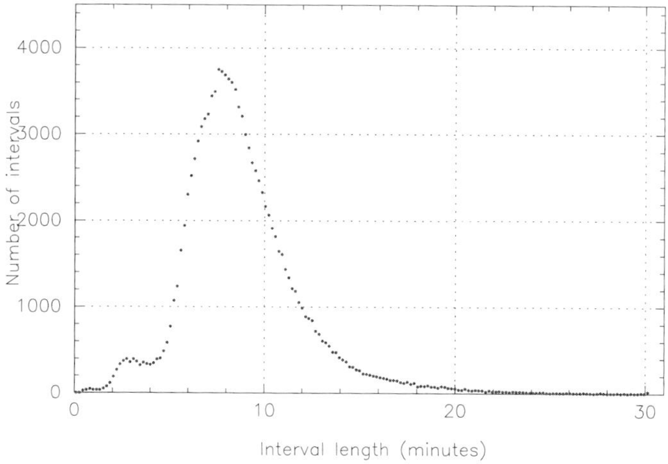

Figure 13.10. A histogram of all thruster-firing intervals over the mission. The distribution of short intervals originates from near-perigee observations

attitude. Activation was triggered by the position and/or the rate for at least one of the axes approaching one of these limits. The on-board computer calculated, based on its current estimates of the satellite's position and rates, the actuation times required to move the satellite back to its nominal position. This was initially done for all three axes simultaneously, but after day 800 small actuations on the z axis were not applied.

The thruster-firing strategy incorporated a simplified model of the solar radiation torques to extrapolate the current attitude and estimate the required thruster firings at any one time. Ultimately, this model determined the average lengths of the intervals between two firings. This interval length is an important parameter in the attitude modelling, as has become apparent in Chapter 10. Firing criteria had to be adjusted close to perigee due to the increased amplitudes of the gravity gradient and magnetic torques, a situation that was not foreseen in the original mission planning. Figure 13.10 shows the distribution of the actual thruster-firing lengths over the mission.

13.3.2 Thruster-firing Calibration

In the calibration of the thruster-firing torques, the length of the firings (measured in units of 1/75 s) are compared with the resulting change in velocity:

$$\mathrm{d}\omega_{i,j} = a_{0,j} + a_{1,j} \cdot t_j, \tag{13.3}$$

where t_j is the actuation length and j identifies the thrusters: individual calibrations are needed for positive and negative firings, as they refer to different thrusters. An example of a calibration is shown in Fig. 13.11. The slopes a_1

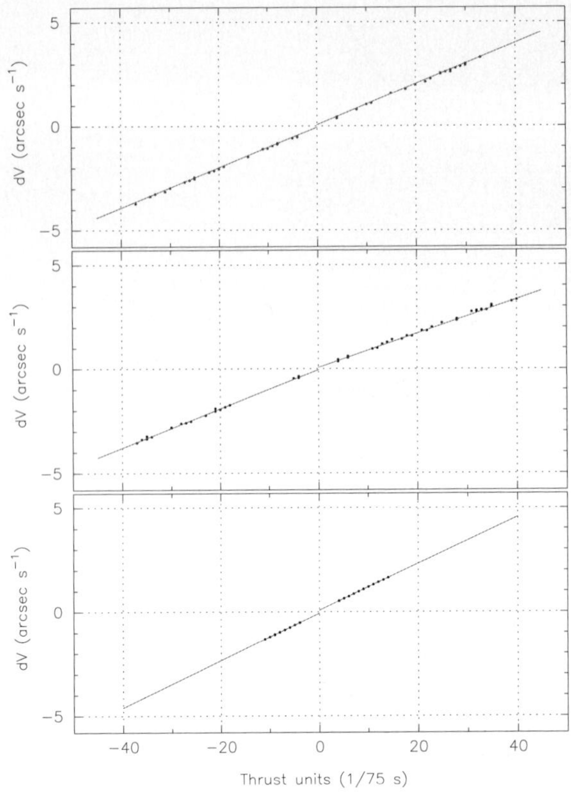

Figure 13.11. The calibration of the relation between thruster-firing lengths and the resulting changes in rotation velocities. Thruster-firing lengths come in units of 1/75 s. The data are for orbit 68

of these relations are used to derive the force exerted by each thruster when activated. For example, for the positive firings on the x axis we have:

$$T_{+x} \approx \frac{75 a_{1,+x} I_{xx}}{206264.8 \cdot p_{+x,y} d_{+x,z}}, \tag{13.4}$$

where $p_{+x,y}$ is the y component of the position vector for the $+x$ thruster, and similarly, $d_{+x,z}$ is the z component for the thrust-direction vector for the x thruster.

Figure 13.12 shows the evolution of the thruster force over the mission for all 6 thrusters. The force increases as over the mission the gas tanks empty. This is the result of the pressure compensator in the gas tank, which did its job slightly “better” than required. The first gas tank was nearly empty when, on day 900, it was replaced by the second tank. The pressure supplied by the second tank was considerably noisier than for the first tank. This may have had to do with the positioning in the spacecraft of the two tanks: the first tank was situated in an area that was less likely to be affected by large temperature fluctuations (behind one of the wide side panels, in the shadow of a solar panel) than the second tank (behind one of the small side panels,

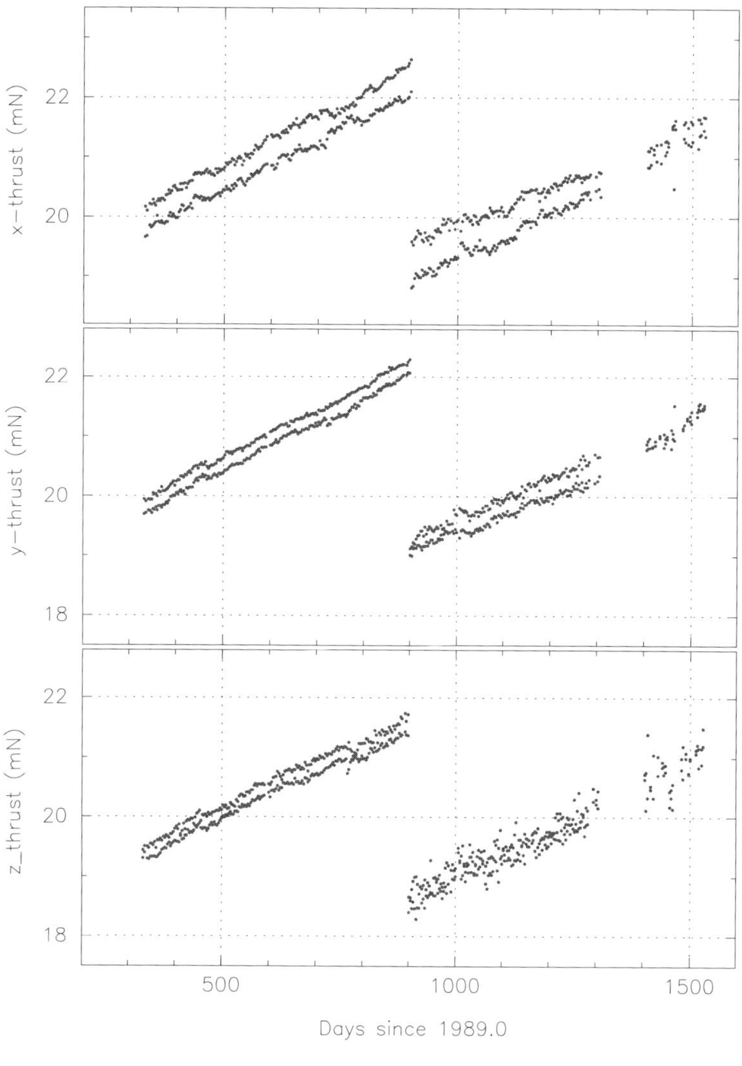

Figure 13.12. Calibrated force values for the thrusters over the mission. The two sets of points in each graph represent firings on the positive and negative thrusters. The slopes are due to the depletion of the cold-gas (see text). The discontinuity on day 900 is caused by changing the cold-gas supply to the second tank. Slope offsets between day 718 and day 788 for the x thrusters are due to the redundant running of gyro 4 (see text). Data for up to 7 orbits is used for each data point to decrease the noise

which could become fully exposed to the Sun). The data for the x thrusters between days 718 and 788 show the effect of the slight non-linearity of the relation between the thruster-firing lengths and the resulting velocity change, as over that interval gyro 4 was running redundantly (see above). The change

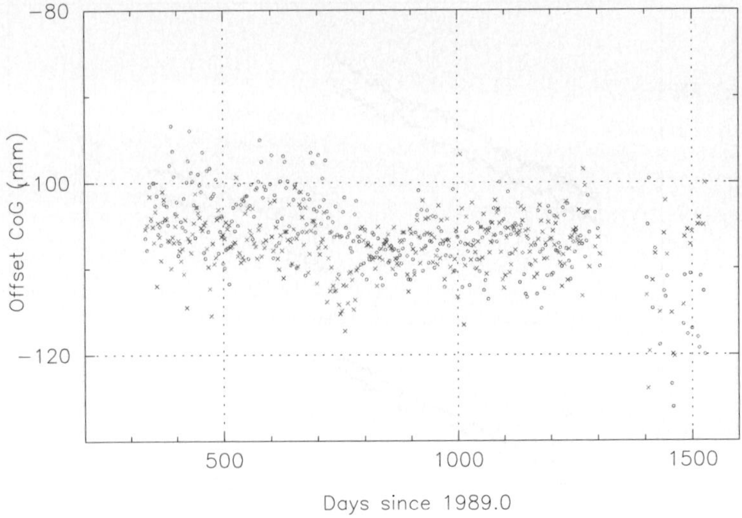

Figure 13.13. The offset of the COG from the plane containing the z thrusters, as calibrated from their torques on the x (circles) and y (crosses) axes. Data for up to 7 orbits are combined for each data point

in firing strategy for the z thrusters took effect around day 800, decreasing the average firing length and allowing firings on the z axis to be skipped, which again reflects in an apparent decrease of the thruster force.

The zero point (bias) a_0 is about 0.008 arcsec s^{-1} for all thrusters, which is equivalent to a thruster firing effectively lasting 1 ms (or 0.073 units) longer than intended. There is a slight non-linearity in the slopes, such that the zero-point tends to be larger for shorter than for longer thruster firings. This shows up on at least two occasions:

- when the shortest firing length on the z thrusters was reduced from 4 to 1 units, the slopes decreased and the zero points increased;
- when gyro 4 was running in addition to the three nominal gyros over the last few months of 1990, the additional torques on the x and y axes resulted in more and larger, negative firings and fewer and smaller positive firings on the x axis, causing a decreased zero point and increased slope for the negative firings, and an opposite effect for the positive firings.

13.3.3 The Position of the Centre of Gravity

In addition to the coefficients of Eq. 13.3 two more coefficients are required for the rate changes on each of the x and y axes to allow for the shifted position of the COG. This shift increases the torque of the z thrusters on the x and y

axes, and by calibrating these dependencies the actual position of the COG is reconstructed. As these contributions only depend on the sign of the z thrust, these coefficients are solved for in a combined solution for the positive and negative firings on the x or y axes. To calculate the offset of the COG, we assume the force exerted by the z thrusters to be known from the calibration. The equivalent of Eq. 13.4 now becomes:

$$\begin{aligned} p_{z,z} &\approx \frac{75a_{x,-z}I_{xx}}{206264.8F_z d_{x,y}}, \\ p_{z,z} &\approx \frac{75a_{y,+z}I_{yy}}{206264.8F_z d_{y,x}}, \end{aligned} \tag{13.5}$$

where the first equation gives the most significant contribution on the x rates (from negative firings on z), and the second on the y rates (from positive firings on z). The skipping of firings on the z axis after day 800 provided a much better defined zero point for these calibrations, as is clear from the reduced noise for the period between days 800 and 900. After day 900, the noise increased again due to the use of the second cold-gas tank. Figure 13.13 summaries the results of these calibrations over the mission, placing the COG 108 ± 3 mm below the plane containing the z thrusters, and therefore approximately 133 mm below its nominal position.

13.3.4 Predictability of the Velocity Jumps

An important aspect of the thruster firings is their reproducibility. This can, to some extent, be read from the noise on the calibrations. Without taking into account non-linearity of the calibration relations and a possible temperature dependence (the thruster nozzle and/or the gas tank), a noise level of about 4 mas s^{-1} is observed for the z axis. This compares with reconstructed-rate accuracies of much better than 1 mas s^{-1}.

13.3.5 Thruster-firing Anomalies

The on-board control of the thruster firings did not operate properly any longer when very large updates to the attitude angles were up-linked from the ground station. In extreme cases, these updates put the satellite at more than the allowed distance from the nominal attitude. This triggered thruster firings, but as it would take some time for the satellite to return within the permitted boundaries, the condition for thruster firings was not eliminated, and firings would take place every frame, backwards and forwards by small amounts, until the satellite did reach the required limits. Obviously, such intervals would not produce any useful data. The statistics of such events have not been incorporated in the histogram of the thruster-firing interval lengths (Fig. 13.10).

PART VI

THE NEXT GENERATION

The *Hipparcos* mission provided a big step forward in the astrometric determination of stellar distances, but on a galactic scale the area covered remained very small. The next generation in astrometric missions aim at improving the dynamic range of parallax measurements from the 1 mas level to the 20 micro arcsec level, or even better. This will extend the horizon from a few hundred parsec to tens of kiloparsec, covering most of the visible parts of our galaxy, covering some globular clusters, and providing the first statistical parallaxes for the LMC and SMC.

Chapter 14

GAIA

14.1 Introduction

Building on the unique experienced gained with the *Hipparcos* mission, and using the same principle of measurement to obtain absolute parallaxes, the *Gaia* satellite project intends to extensively map the Galaxy in positions and velocities, covering more than one billion objects down visual magnitude 20. As *Hipparcos* 30 years earlier, the *Gaia* project is ambitious and will be ground breaking in many areas (Luri et al. 2004; Bailer-Jones 2005). It will affect nearly all aspects of astronomy, from solar-system studies, through stellar structure and evolution, galactic dynamics and reconstruction of the history of our Galaxy, to fundamental physics and cosmology.

The current *Hipparcos* study has already been contributing to the data processing approach for the *Gaia* mission, testing and applying aspects of the iterative astrometric solution planned for *Gaia*. In this section various instrumental and data processing aspects of the *Hipparcos* and *Gaia* missions are compared, with an emphasis on where parallels can assist in solving the rather large data processing task for *Gaia*. We start with a brief summary of the instrument and mission plan, and then look at the various data processing challenges ahead.

14.2 The spacecraft and payload

The instrument consists of a 35 m focal length telescope with two apertures, folded within the 3.5 m diameter body of the satellite. Projected on the sky, the two fields of view, measuring 0.67 by 0.69 degrees for the astrometric field detectors, will be 106.5 degrees apart. In comparison, the *Hipparcos* astrometric fields of view were 0.9 degrees square, and 58 degrees apart. A separation closer to 90 degrees means an easier segregation of the parallax contributions than for *Hipparcos*: as the parallax factor along-scan is proportional to the rotation phase through $\cos\Omega$ (see Section 3.3, page 85), a phase difference of 90 degrees provides fully uncorrelated parallax factors for the to fields of view. However, a basic angle of 90 degrees also fits exactly four times in a single rotation of the satellite, and would therefore create major solution-instability problems (see Section 1.1.4.3, page 14).

A second parameter in the scanning that is difficult to optimize is the solar aspect angle of the spin axis, set at 43 degrees for *Hipparcos*, and at 45 degrees for *Gaia*. The solar-aspect angle ξ effectively defines three distinct "scan-quality" regions on the sky: the ecliptic plane, the turn-over, and ecliptic pole regions. The most homogeneous sky coverage is obtained in the polar regions. The turn-over regions (at ecliptic latitudes of $\pm(\pi/2 - \xi)$) receive the most observations, but many of these observations are at 90 degrees from the direction of the Sun, and contain therefore little or no parallax information. The ecliptic plane region is receiving the poorest scan coverage, with restricted directions and long interruptions between successive scans. An increase of the solar aspect angle would therefore be good for the scan, as it increases the well-covered polar regions, and decreases the poorly covered ecliptic-plane region. However, a larger solar aspect angle also decreases the minimal aspect angle between the apertures and the direction of the Sun, which causes problems with stray-light protection. In fact, *Gaia* is equipped with a 10 m diameter shield to protect the spacecraft from direct solar radiation. A larger solar-aspect angle requires a larger shield, which soon becomes impractical in size. The solar-aspect angle is therefore a compromise between what may be desirable and what is technically feasible. The shield is further used to accommodate the solar panels for the energy provision of the satellite.

The solar shield protects the spacecraft from large temperature variations, the kind of variations that have been observed in the *Hipparcos* spacecraft. The *Gaia* telescope with its multiple mirrors and long focal length, is very sensitive to the stability of its optical bench, which has to be kept under strict thermal control. Keeping direct Sunlight well away from the payload, through the solar shield, is a necessity to achieve this. The solar shield also provides a very simple reaction of the spacecraft on solar radiation torques: in a nearly-inertial reference frame, these depend only on the direction and distance of the Sun as seen from the spacecraft. For *Gaia* there will be no modulated torques resulting from solar radiation as was observed for *Hipparcos*.

Both apertures of the telescope focus on the same focal plane, consisting of 106 specially manufactured Charge-Coupled Device (CCD) detectors, 102 of which are used for direct gathering of the science data (Fig. 14.1). Each CCD has 4500 pixels along scan (covering 0.074 degrees) and 1966 pixels across scan (covering 0.097 degrees), and measures 45 by 59 mm. The pixels are three times longer than they are wide (10 by 30 μm), which reflects the much higher

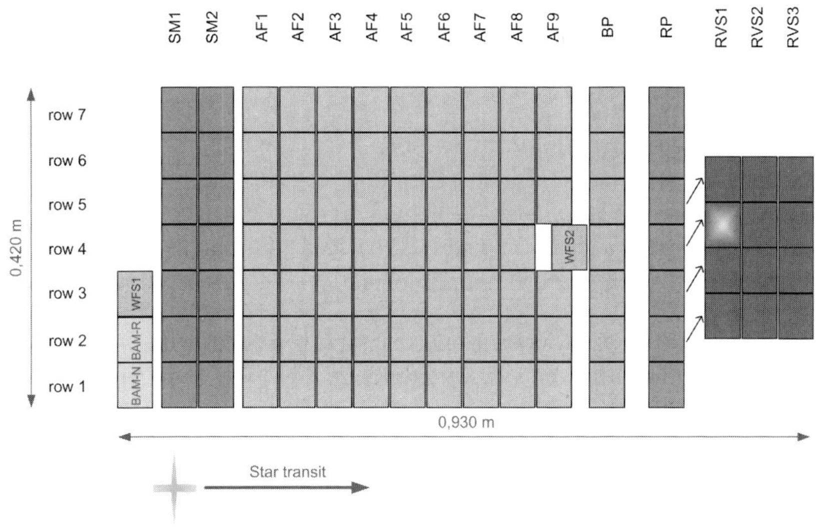

Figure 14.1. The configuration and functionality of the *Gaia* focal plane assembly (courtesy of EADS-Astrium)

spatial resolution of the telescope along scan. Still across scan the pixels are 0.177 arcsec, and during a single CCD transit (4.42 s) of a star the across-scan movement of the satellite needs to be small enough to keep the stellar image within a few pixel heights. Thus, transverse motions of the satellite have to be restricted to no more than 0.5/4.42 = 1.2 arcsec s^{-1}. The *Hipparcos* transverse motions were restricted to ± 5 arcsec s^{-1}, and the increase in requirements by two orders of magnitude puts very high demands on the on-board attitude control mechanism. For most CCDs the transit data will be binned across scan, so that only along-scan information is retained. For bright stars, however, the full information is retained, and an across-scan position can also be derived. This helps a little in improving the astrometric data for objects near the ecliptic plane, where the across-scan measurements provide details in directions that cannot be reached with the along-scan measurements. This was never possible for the *Hipparcos* mission, where the modulating grid did not allow a sensible measure of the across-scan position.

The *Gaia* focal plane contains 4 "types" of detectors:

1 The Star Mapper (SM) CCDs, one column of 7 CCDs for each field of view. These CCDs are used for object detection and qualification, defining a transit time, transit ordinate and intensity estimate;

2 The Astrometric Field (AF) CCDs, 9 columns of 7 CCDs each. These CCDs provide the astrometric data and broad-band photometry. The first column, AF1, is used to confirm detections from the SMs, and to establish the local

scan velocity. This may also create the possibility to handle the so-called Near-Earth Objects, a few of which may move significantly during a transit of the focal plane;

3 The dispersion-spectra photometry CCDs. Two columns of 7 CCDs each, with a dispersion prism and filter in the light beam just before reaching the detectors. The Blue Photometry (BP) column covers a wavelength range of 330 to 680 nm, spread over 37 pixels. The Red Photometry (RP) column covers 640 to 1000 nm, spread over 35 pixels.

4 The spectroscopy CCDs. A grating spectrograph is placed in the light beam, providing a spectrum for the wavelength range 847 to 874 nm, spread over 1035 pixels.

All CCDs operate using the Time-Delayed Integration (TDI) mode. The charges on the CCDs follow the images as they cross the CCD, and are read out when the final row of pixels is reached. Special operation modes are considered for bright objects, integrating only over a fraction of the CCD. The integration time interval affects the reference time for a transit as derived from the CCD readouts, and how it relates to the reconstruction of the along-scan attitude. A full transit over a CCD takes 4.4 s, over which time interval attitude variations are smoothed out. A reduced integration time interval will produce a different average over the attitude variations, which needs to be taken into account when using these measurements for the reconstruction of the along-scan attitude.

14.3 The mission plan

The *Gaia* mission will perform a 5 to 6 year survey, in coverage very similar to that of *Hipparcos* (described in Section 1.2.6, page 21): the satellite's spin axis describing a precessional motion around the direction of the Sun, and the satellite rotating around this spin axis once every 6 hours. The longer rotation period (which was 2.13 hours for *Hipparcos*) results in a significantly lower number of transits through the field of view per object per time interval. In fact, *Gaia* will over its mission measure fewer of these field-of-view transits per object than *Hipparcos* did over a period of just over three years of observations. The slower rotational velocity allows for a higher spatial resolution in the focal plane, which is required for the scientific goals of the mission.

Gaia will operate at the Lagrangian point $L2$ of the Sun-Earth system. This position is situated opposite the Sun as seen from the Earth, at 1.5 million km distance. Around it a satellite can describe a nearly stable Lissajous orbit. The advantage of such an orbit is that eclipses of the Sun by the Earth (and the associated strong temperature changes) can be avoided. Also avoided are occultations of the field of view by the image of the Earth. There are still the estimated 50 occasions of partial eclipses by the Moon, which may block up to 15 per cent of the Sun light. The many interruptions that affected the *Hipparcos* data stream are thus avoided. The downside of the L2 orbit is the long time it takes to get there (it currently takes one month), and the power required to relay the data to the ground station. The transfer period allows for only limited testing, as the observing conditions are still by no means fulfilled. It is, for example, only for short periods per day possible to keep the satellite properly orientated towards the Sun, and to maintain contact with it from the Earth.

Gaia will thus produce a sparser but much less interrupted data stream than was received from *Hipparcos*. For the latter the data stream was further interrupted by the (unplanned) perigee passages. The sparser data stream will affect the variability analysis of the photometric data, being less sensitive to the detection of short periods (a few hours to a few days) than *Hipparcos*, and being equally insensitive to the detection of intermediate-length periods (a few days to around 80 days). The longer periods will be covered a little better due to the longer mission length, but for stars with periods longer than about 100 days (Miras for example) the light curves are often not very stable, and the mission will provide insufficient coverage to track the more detailed behaviour of these stars. Eyer and Mignard (2005) have carried out a pilot study on the possibilities to recover variability periods from the *Gaia* photometric data.

A major difference with respect to the *Hipparcos* mission is the selection of stars: while *Hipparcos* operated using an elaborately prepared input catalogue (see for example Turon et al. 1992), *Gaia* will in principle observe any object brighter than $V = 20$ it detects, provided it has time and space for the observation. This will be a problem in particularly dense regions of the sky: some of the densest areas near the galactic plane and the cores of globular clusters. In both these cases the star density can get higher than the maximum number of images the on-board software can handle for detection and tracking. It has been estimated (Robin et al. 2003, 2004; Robin 2005) that there are over 10^9 objects in the sky that would thus qualify for being observed by *Gaia*. Some of these will have only a short visibility "life": it is estimated that *Gaia* will observe around 20 000 novae during its mission. The nova detection is part of the science alert activities, which will be used to activate ground-based observations for particularly interesting temporary phenomena observed by *Gaia*. Other objects that may qualify for science alerts are stars that appear to be going through some of the very rapid stages of stellar evolution. These stages are naturally poorly populated, but can provide insight in particularly difficult to study stages of stellar evolution.

Gaia will operate with one or two ground stations. A satellite at the $L2$ position is generally visible for only 8 to 14 hours per day, depending on geographic latitude and time of the year. While *Hipparcos* required continuous ground-station contact to enable its observations, *Gaia* will operate autonomously for most of the time, and relay the accumulated observations once of twice per day to the ground station. In case there is only one ground station, there will not always be sufficient time available to download all accumulated data, and some data losses will occur.

14.4 The astrometric data reduction

The astrometric data for *Gaia* are of the same kind as for *Hipparcos*: they are in effect transit times over a detector. In the case of *Hipparcos* the detector was a modulating grid, for *Gaia* it is the data accumulated for a passage over one CCD. These transit times have to be transformed into rotation phases on the sky, and the link that supplies this transformation is the reconstruction of the along-scan attitude. The along-scan attitude ultimately represents the reconstruction of the actual rotational motion and phase of the satellite within an inertial reference frame. As the reference frame itself is built up from the data of the mission, and those same data are being used as reference points for the attitude reconstruction, it requires some iterations between the reconstruction of the astrometric catalogue and the reconstruction of the along-scan attitude before the whole system will converge. This procedure is referred to as the global iterative solution, and is in many ways equivalent to the new astrometric reduction of the *Hipparcos* data as presented in this volume.

Considerations that applied for the new reduction of the *Hipparcos* data will also apply to the *Gaia* data:

- For the converged system to produce absolute parallaxes it is essential that the weight distribution over the sky of stars contributing to the iteration is very homogeneous; the presence of localized high-weight areas on the sky can cause ill-defined local parallax offsets;

- For the attitude reconstruction to operate reliably, the dynamical motion of the satellite needs to be well-understood; poorly understood peculiarities in the rotational motion of the satellite can lead to localized correlated errors in the abscissa residuals, which can accidentally accumulate to local errors in the astrometry, as was observed for the published catalogue;

- All large-scale and small-scale distortions by the detectors need to be calibrated to the highest possible accuracy as part of the iterative solution.

The homogeneity requirement means that the number of stars that can be used in the global iterative solution is effectively limited by the average stellar densities towards the galactic poles. The typical area-size for weights is set by the number of unknowns per time interval as used in the along-scan attitude: the time interval defines an arc-length covered, and the number of unknowns in the solution. When this solution is a spline function, each unknown will roughly correspond to an interval between two nodes. This interval then gives the typical area diameter for the weight smoothing.

The interval size is also affected by any discrete action (thrusters) to offset the effects of external torques, in this case mainly solar radiation. In order to maintain a near-zero across-scan velocity (to keep stellar images following a single pixel row on a CCD), activations of thrusters or similar devices may be required every 1 to 2 seconds. Such short intervals would imply a typical area size of 1 to 2 arcmin only.

14.5 The photometric data reduction

The photometric data reduction consists of two very distinct elements: the broad-band photometry derived from the SM and AF CCDs, and dispersion spectra photometry derived from the BP and RP CCDs. The broadband photometry will be used as a homogeneous reference magnitude over the sky and over a large dynamic range (14 magnitudes). It will be the most sensitive source for variability detection on various time scales. The dispersion photometry has two aims: to provide the astrometric reductions with the colour information needed to correct residual chromaticity effects in the image positions, and to provide the data analysis with information on astrophysical parameters of the objects observed.

14.5.1 The Broad-band or *G* Photometry

The first step in the analysis of the broad-band photometry concerns the extraction of one or more intensity-sensitive parameter(s) from the image as recorded during the CCD transit. This is obtained through the fitting of a Point Spread Function (PSF), which in the case of the *Gaia* data analysis becomes a Line Spread Function (LSF): on-board processing of the data for all but the brightest stars involves adding-up of counts within a 6-pixel wide and 6 or 18 pixel long window to give a one-dimensional window in the along-scan direction. The longer windows, in rows 2, 5 and 8 are intended to be used for image reconstruction through superimposing images obtained in different scan directions, in some ways similar to the reconstruction of double and multiple stars as implemented for *Hipparcos*, though without the phase-ambiguity problem.

The few fully resolved windows for the brighter stars allow for a PSF reconstruction, those for the fainter stars will have to do with a reconstructed LSF. On a large scale, the PSF and LSF are expected to be a function of position in the focal plane, in a different way depending on the field of view. They are also expected to be a function of the spectrum of the observed object, a dependence which may be represented as a function of one or more colour indices, derived from the dispersion spectra measurements. On a small scale the PSF and LSF are affected by local CCD characteristics, and may have to be resolved down to the level of an image height or across-scan coverage (12 pixels). These local CCD characteristics can range from variations in quantum efficiency, through variation in spectral response to Charge Transfer Inefficiency (CTI). Most of these effects can also be expected to have a time dependence. In case of the CTI, for which the main cause is radiation damage of the CCDs, these variations are likely to be in the form of discrete steps rather than a continuous evolution. The accurate reconstruction of the PSF and LSF is important for reaching the desired astrometric and photometric goals of the mission.

Calibration of the LSF and PSF (as functions of time) is part of the outer loop in the iterative calibration, which together with the inner loop establishes a stable internally defined photometric system. The inner loop describes the transformations required to relate the intensity parameters derived from the transit images to a self-consistent system of intensities. For this purpose, stars that serve as calibration standards are selected from the *Gaia* data stream. Initially the selection of these standards will be rather poor, in the sense that the selection is likely to contain a number of variable stars, and the calibration values are not yet properly determined. Through systematic iterations both problems are gradually removed, and an internally calibrated system is established. This system then provides the reference data for the calibration of the PSF and LSF, which are then used to produce a new, preferably improved, set of image parameters. The new image parameters are then subjected again to the inner-loop calibration procedure.

Given the very substantial numbers of transits involved in this process, it should be expected that at least the large-scale aspects of these calibrations will converge quickly. For the small scale aspects this is not necessarily the case. The across-scan window size is just over 2.12 arcsec, or 1/1172 times the height of the field of view. At a nominal scan velocity of 60 arcsec s^{-1}, an area with the height of a window will see on average about 21 000 stars per day per field of view. Of those, there will be around 1800 stars brighter than 16th magnitude, of which possibly only 10 per cent is useful as calibration objects. It will therefore require most likely the accumulation of many weeks of data (in the form of residuals relative to the assumed calibration model) to obtain the required resolution for the small-scale calibration. It has to be established during the mission what will be the optimum time intervals to be used for these calibrations.

After establishing the internally calibrated photometric system, it needs to be connected to an absolute system using ground-based spectrophotometric, spectroscopic and photometric observations. This calibration will effectively connect the *Gaia* photometric system to the overall standard Vega. The number of standards required for this calibration is at this moment not yet established, but estimated to be of the order of 100. The photometry ultimately released for

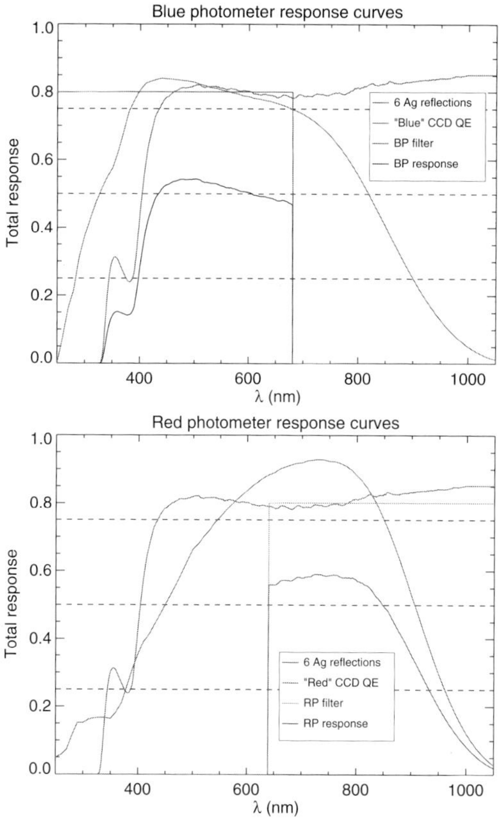

Figure 14.2. Summary of the response curves that determine the overall response for the BP (top) ad RP (bottom) dispersion prism data by *Gaia* (courtesy of A. G. A Brown)

the *Gaia* mission will therefore be on a well established absolute scale, ready for further astrophysical research.

The internally calibrated photometry will be used for detection and analysis of variability. In first instance the detection of variability will be tuned primarily to establishing a selection of apparently non-variable stars from which standard stars can be selected for the photometric calibrations. The analysis of variability for the *Gaia* data will be more difficult than for the *Hipparcos* data: the longer rotation period of the satellite means that the same stars are re-appearing about three times fewer in one of the fields of view. This affects the detection of short periods (of length up to a few hours). Only the extreme-short periods, which may be relevant for the brighter white dwarfs, are in some cases still detectable from the variations over the individual CCD transits.

14.5.2 The Dispersion Spectra

As is shown in Fig. 14.1, two rows of CCDs are equipped with prisms and filters to provide low-resolution spectra in predominantly blue ($330 < \lambda <$

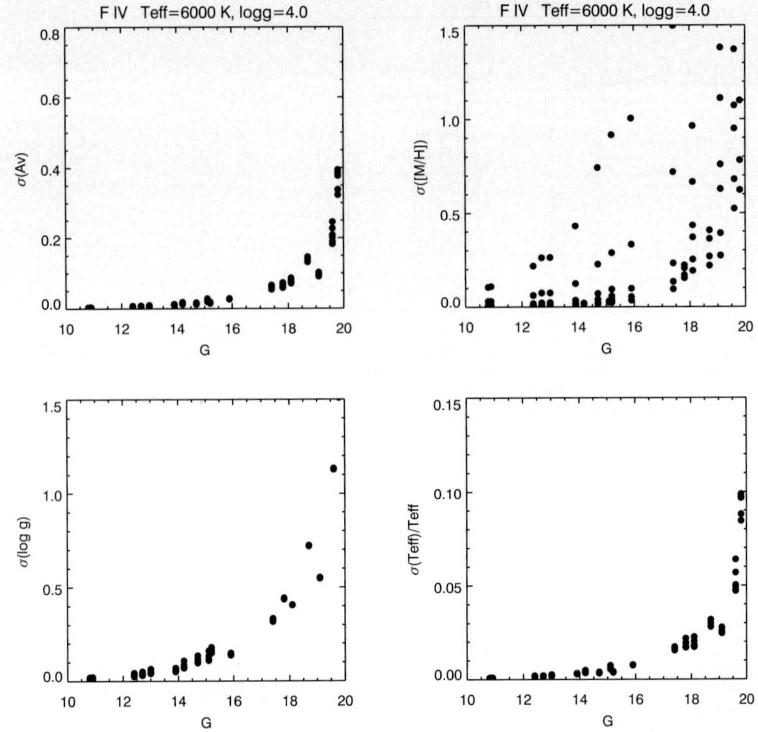

Figure 14.3. Accuracy estimates, as obtained from experiments with simulated spectra, for the reconstruction of astrophysical parameters and reddening, illustrated here for an F star of luminosity class IV. Top: reddening and metallicity, bottom: surface gravity and effective temperature. (courtesy of C. Jordi)

680 nm) and red ($640 < \lambda < 1050$ nm). Figure 14.2 shows the expected responses for rather idealized filter profiles. The dispersion of the spectra varies non-linearly with wavelength, from 3.3 nm per pixel at 330 nm to 28.5 nm per pixel at 690 nm in BP, and from 6.3 nm per pixel at 630 nm to 15.3 nm per pixel at 1050 nm in RP. There are in addition variations expected in the dispersion as a function of across-scan position. The variations will need to be calibrated in flight, as they tend to be temperature dependent, and will complicate the processing of the BP and RP photometry considerably.

Extensive experiments have been carried out to assess the effectiveness of the spectral information in deriving astrophysical parameters, following a similar exercise for photometric pass bands by Jordi et al. (2006). These showed that the current design is close to optimal for its type, and compatible with the multiple pass band design described by Jordi et al. (2006). An example of the

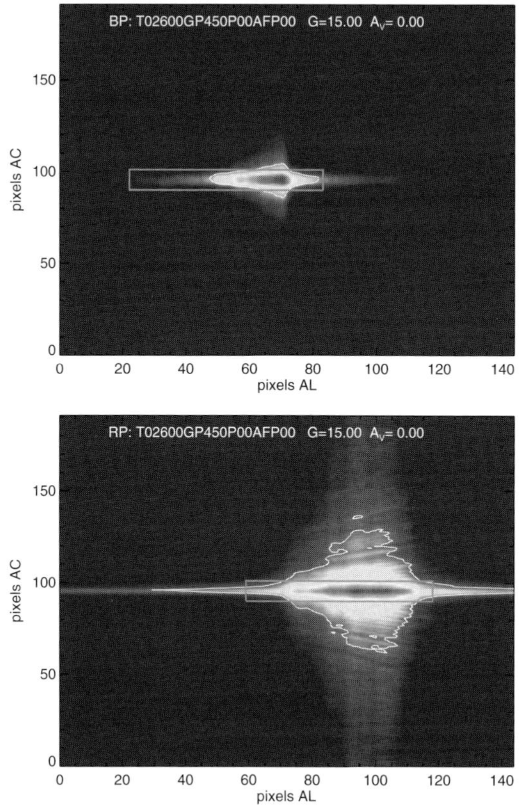

Figure 14.4. Simulated dispersion spectra for an un-reddened $15^{\rm th}$ magnitude M6V star. Top: in BP, bottom: in RP. The windows used for reading out the spectra are shown by the narrow rectangular boxes (courtesy of A. G. A Brown)

expected accuracies for the reconstruction of astrophysical parameters using the dispersion spectra is shown in Fig. 14.3.

The spectra, of which an example is shown in Fig. 14.4, are sampled in a window. The position of this window is derived on-board, based on the initial detection of the object in one of the SM CCDs, and the subsequent confirmation in the AF1 CCD, combined with the knowledge of the current value of the scan velocity and any large-scale sky-to-field distortion. On-board all responses in the across-scan direction are added up to provide a one-dimensional spectrum, which is down-loaded for the data processing.

The data processing for these spectra will involve several aspects:

- Calibration of the wavelength scale, which may be done using objects with clear spectral lines (Emission-line stars, QSOs) or stars with otherwise very clear spectral features (Hβ line for early A-type stars). The wavelength scale will be a function of transit ordinate and will not be linear;
- Relating spectra measured with different dispersions to single reference spectra;

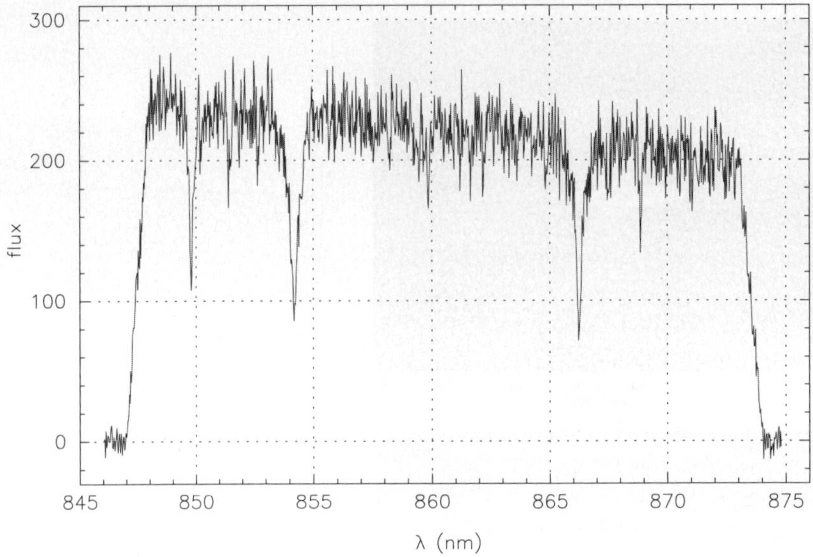

Figure 14.5. The simulation of the spectrum of a V = 10 G0V star as it will be observed by the *Gaia* spectrograph. The simulations were obtained by Carine Babusiaux using the GIBIS software

- Calibrating the variations of the response functions across the field of view;
- Calibrating the overall response as a function of the across-scan coordinate and as a function of time;

Only objects with sufficient signal to noise ratio may be used in these calibrations; the limiting magnitude probably around 16 to 17 in G, leaving a relatively small number of calibration stars.

The dispersion spectra are, more than the AF broad-band photometry, affected by crowding, simply because their "windows" take up 5 to 6 times more space.

14.6 The spectroscopic data reduction

The primary goal of the spectroscopic measurements is to provide radial-velocity data for the tracer sample to be used in the study of the galactic structure: the Red-Clump stars (see for example Babusiaux and Gilmore 2005). These stars are at an absolute magnitude close to 1.0. The study needs to cover an area of some 20 kpc from the Sun, the target magnitude for these stars still to be observed with the spectrograph is 17.5. This may not be possible to reach with the proposed hardware configuration, however, even a limit of 17.0 would still provide a radius of nearly 16 kpc.

The spectroscopic data are obtained in the near-infra red, over a spectral range of 847 to 874 nm (Fig. 14.5). This covers a mixture of strong and weak lines from a range of chemical elements, and in particular the CaII triplet, which will be the main provider of information on radial and rotational velocities. The spectrograph provides a resolution of $\lambda/\Delta\lambda = 11500$. The dispersive element is a grating.

The focal plane for the spectrograph covers an area of 0.22 by 0.39 degrees, and is an extension of the focal plane of the astrometric instrument. Due to the smaller-size field of view, there will on average be only about 40 field transits per object for the spectroscopic data. In the focal plane is situated an array of 3 (along-scan) by 4 (across-scan) CCDs, operated, like the astrometric instrument, in TDI mode. The spectral images will be read out in windows of 1104 pixels along scan and 10 pixels across scan. These windows will be preserved at full resolution only for stars brighter than magnitude 7, ignoring stars brighter than magnitude 4.75. A small fraction of the fainter stars, selected for calibration purposes, will also be preserved at full resolution. For stars between magnitudes 7 and 10 the data will be co-added across scan, giving a single vector of 1104 data points. For images fainter than magnitude 10 the along-scan samples will in addition be co-added in groups of 3, giving a lower resolution but higher signal-to-noise ratio. In fact, at full resolution, as applied to the brighter stars, the spectra will be, at a pixel width of 0.026 nm, slightly oversampled, while at for the faint stars the under-sampling is about a factor two. Stars may be observed down to magnitude 16.5 to 17.0.

The processing of these data is organized around three principle tasks:

- Calibration of the Radial Velocity Spectrograph (RVS) characteristics;
- Extraction and cleaning of spectra;
- Derivation of the physical parameters.

For each of these tasks a brief summary is provided below.

14.6.1 Calibration of the RVS Characteristics

The calibration of the spectroscopic data will follow a similar global iterative procedure as described for the astrometric and photometric reductions above. The spectrographic measurements have to be self-calibrating, while those stars recognized as stable will be selected as standard stars for the calibration. Some of stars will be selected already before the start of the mission, using ground-based observations. To be able to use stars as calibration sources, they need to be characterized and identified as stable, for which a calibration of the characteristics of the spectrograph is required. These characteristics are derived from the selected stable source, and the whole procedure needs to iterate to converge. Convergence can only be reliably obtained when the number of calibration sources is large with respect to the number of parameters to be calibrated, and when those parameters are no more than slowly varying as a function of time, so that long time-spans can be used in single calibration solutions. This is in particular the case for the spectroscopic data, as only relatively bright stars can effectively be used in the calibration, and of these there are only small numbers available.

The instrument parameters that require calibration for the spectrograph are:

- The photometric response, its linearity and saturation level;
- The across-scan characteristics of the Point-Spread Function;
- The along-scan characteristics of the Point-Spread Function;
- The geometric calibration relating a wavelength zero point to a position on the CCD as predicted based on the astrometry and the satellite attitude;
- The wavelength dispersion law.

The last two of these will be the most difficult, depending for example on payload temperature, and affected by radiation damage of the detectors. These calibrations will mainly use single G and K stars which are abundant, not too faint intrinsically, and have relatively simple spectral characteristics and a clear presence of the CaII triplet.

14.6.2 Extraction and Cleaning of Spectra

The main complication for the extraction of the spectra is caused by the overlapping of the windows used to extract the spectra, and the differences in resolution associated with different windows. The cleaning procedure involves amongst others extensive calculations for the background determination, determined using both the observations and predicted background transits, for which the information is derived from the astrometric and photometric catalogues produced by *Gaia* (which reach about 3 magnitudes fainter than the spectrograph measurements). To recover the spectra of the faintest stars, first all of the brighter stars need to be extracted, analysed and their influence subtracted from the images of the fainter stars. It is in particular for this reason that an accurate reconstruction of the across-scan response profile is crucial, as well as the geometric calibration, which together with the astrometric and attitude data will provide accurate predictions of where on the CCD a disturbing spectrum was exactly situated at the time of observation, and thus what its contribution is expected to be to other observations.

Only after all cleaning and calibration is completed can the spectra be added to provide the input for the determination of physical parameters. For the brightest stars the co-adding may not be required, and spectra can be examined individually for, for example, radial-velocity variations. Adding the spectra will be a complicated job, as sampling of the spectra will be different for each transit due to the exact position on the CCDs with respect to the TDI time intervals, as well as due to variations in dispersion and wavelength scale across scan. It is therefore important for the further examination of those combined spectra that the "combination history" is preserved, so that it can be fully incorporated where relevant in the analysis. These effects naturally are most relevant for the brighter objects.

A major complication of the procedure described above may come from cosmic rays, which are particularly difficult to identify in the spectra folded across-scan on-board the satellite.

14.6.3 Derivation of the Radial Velocities

The determination of radial velocities will be primarily based on cross-correlation with synthetic spectra. The choice of the most relevant spectrum may not always be obvious from the data available, in particular for the faintest objects. Here the astrometric and photometric data together can assist in narrowing down the choice of possible spectra. In some of these cases fits with several synthetic spectra may be applied, and data examined for all those fits together. In general, however, the radial velocities will be determined from the correlation peak derived from the comparison between the observed and synthetic spectrum.

In the correlation comparison for the brighter stars it is important to incorporate the sampling history of the observations, and to incorporate the same sampling in the synthetic spectrum. The multiple observations of the spectra provide in principle an oversampled, and therefore potentially higher resolution combined spectrum. However, as individual samples provide each their own peculiar integration over a small part of the spectrum, the results of the combined spectrum can be different for different combinations of the data, and these differences need to be accounted for in creating the synthetic spectrum.

In the case of double stars and other objects with radial velocity variations, the radial velocity will need to be examined as a function of time based on the combined spectra per epoch. This will probably only be possible for a relatively small selection of bright stars.

14.7 Data-analysis challenges

The data-analysis challenges posed by *Gaia* are considerable, ranging from the size and complexity of the data stream, through dealing with crowded areas, to the detail required for most of the calibrations.

14.7.1 The Size and Complexity of the Data Stream

The *Gaia* mission will register an average of 85 transits through the field of view for approximately 1 billion objects. Each transit involves 12 CCDs: 1 SM, 9 AF, BP and RP, which leads to a total of just over 10^{12} individual CCD transits produced by the astrometric and photometric instrument. If an interval of one day is assigned to a data set, then there will be about 2000 data sets covering each of the order of 500 million observations. As was described above, all reductions for *Gaia* are iterative, and the same data will be processed many times, in particular data obtained near the start of the mission. The iterative processes are essential to ensure the full-sky homogeneity and reliability of the various catalogues.

Consider the very simple case of starting the reductions every 6 months from start again. In that case the reductions will be effectively for 27.5 years of data, or 5.5 full mission lengths. It can reasonably be expected that this is a low rather than a high estimate, and a figure of in total 10^{13} CCD transits to be processed is probably realistic.

The most critical figure is, however, the time required for the final iterations. If the cycle time for the final iterations gets too long, there will be no margin left for re-processing, and convergence of the solution becomes difficult to establish. From the experience with the *Hipparcos* iterations, a maximum time span of 4 weeks for the final iteration runs seems a good target. This is equivalent to 2.4 million seconds, to process 10^{12} CCD transits. Considering time required for I/O and the difficulty to run computers at 100 per cent for 4 weeks, a requirement for the processing of about 1 million CCD transits s^{-1} follows. It then depends on the complexity of the calibration models how much is required in processing for each transit. The current processing of the *Hipparcos* data spends of the order of 30 seconds on the processing of about 50 000 transits, so it should be possible to accommodate the additional factor 100 and an increase in model complexity through extensive parallel processing and faster processors by 2018, when this really matters.

14.7.2 Interdependence of the Data Streams

As will be clear from the preceding sections, the processing of the astrometric, photometric and spectroscopic data are all closely interconnected: the astrometric data will need a colour index from the photometry to eliminate chromaticity effects, the photometry and spectroscopy will need accurate positions and satellite attitude for the reconstruction of parameters for double stars and accidentally disturbed images. Spectroscopic processing will also use photometric magnitudes in a pseudo-pass band compatible to the RVS filter, and these too will improve as the mission and the reduction of its products progresses. The same applies to the initial data treatment: with better positions available, more accurate photometric parameters can be derived from the transits, and better photometry can similarly contribute to improved positional determinations. This effectively amounts to solving for the photometric and astrometric parameters of the images of each object in a global way rather than individually, though the initial determinations will all be individual. These additional sources of information lead to a data processing that consists of internal and external iterative loops, where the internal loop concerns itself with defining the "system", while the external loop works on improvements to the direct analysis of the observations. With all three of the so-called core processing tasks depending on global iterative solutions, that in addition have mutual dependencies, it will be clear that the *Gaia* data processing is going to be a laborious and time and resources consuming task. A variety of data analysis tasks, such as object classification, variability analysis, 2D image reconstruction and analysis of non-singular and extended sources, all contribute information to these iteration cycles, information that will only gradually become available as the mission and its data processing progress.

14.8 Organization of the data processing and analysis

The organization of the complicated and rather substantial data processing tasks described above is structured around specific tasks, covered by coordination units, the activities of which are coordinated by an executive committee. There are 8 coordination units for the *Gaia* data processing, which can be split into core processing, support and object processing, as shown in Fig. 14.6. A ninth unit is planned for data exploration, to start its activities around 2008. Each coordination unit is responsible for ensuring its funding and is, within the requirements set by ESA, relatively free to organize itself according to their own internal agreements. All units, as well as the *Gaia* project team at ESTEC, share a range of facilities for documents and exchange of information, most of which has been fixed by mutual consent through the two main committees managing the scientific aspects of the *Gaia* mission: the *Gaia* Science team and the DPAC Executive.

Data processing will take place at a number of data processing centres: European Space Astronomy Centre (ESAC) for the astrometric solution, Centre National des Etudes Spatial (CNES) for the spectroscopic solution, and Cambridge for the photometric solution. CNES will also take care of the implementation of the object classification and analysis of non-singular objects, while Geneva takes care of the variability analysis. Barcelona and CNES will be responsible for the production of simulated data, and the Barcelona Super Computer is likely to be used for the re-processing of the initial data treatment as part of the external iteration loops. A quality assurance implementation will be implemented in Torino.

Crucially important in the development of the software are the two support units: system architecture and data simulations. The first of these provides the central data base at which raw, partly and fully processed data are archived and made available for reductions. The data flows have been organized such that:

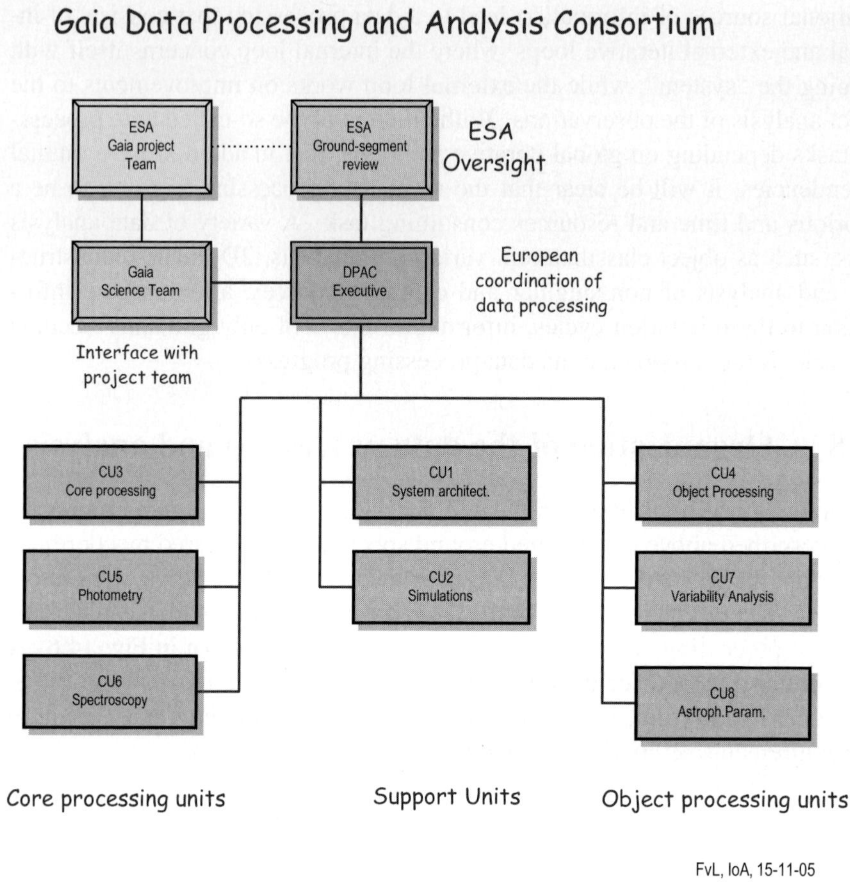

Figure 14.6. The *Gaia* Data Processing and Analysis Consortium, DPAC, showing its structure in coordination units, as well as the committees overseeing the software and processing developments

- Data will only be exchanged between the central data base and the various data processing centres, at a provisional schedule of once every six months;
- There will be no direct data exchanges between other data processing centres;
- The central data base will compile the accumulated reduced data into one or more reference data bases, from which the data processing centres can extract supplementary information for their part of the processing and analysis.

This process is in particular relevant at the start and end of the reductions: at the start most of the supplementary data will still be unavailable, and making it available as soon as possible will speed up the iterative processes. At the end of the reductions it will be important to identify the "convergence" of the different processed data streams, and their dependence on the processing status of other data streams. For example, if the final spectroscopic processing would depend much on object classification, while object classification depends on the final astrometric and photometric processing, then this would define a sequence of final reduction steps to be executed as part of the finalization of the *Gaia* data processing. Without such agreed sequence, inconsistencies may enter the published catalogue.

The second unit, data simulations, provides the input data at various levels of processing, needed for the development and testing of the reduction software. The simulations are produced using three packages, referred to as GIBIS (Babusiaux 2005), Gaia System Simulator (GASS) (Masana et al. 2005) and Gaia Object Generator (GOG), representing decreasing detail of the simulations. GIBIS can simulate image data at the CCD transit level, GASS the telemetry stream, and GOG produces data resulting from various levels of processing. Together they allow most aspects of the data processing software to be tested before the start of the mission. Crucial elements in the simulation are the description of the instrument and of the sky. The first is provided through one of the common user tools for *Gaia*: the *Gaia* parameter data base. The second is part of an ongoing development: a galaxy model, able to predict accurately the distribution of stars on the sky as a function of magnitude, colour, luminosity class and chemical composition (see for example Debray et al. 2006; Marshall et al. 2006)

The current estimates for the time needed to do the data processing put the release of the *Gaia* data at end-of-mission plus three years, which for the nominal 5-year mission will be early 2020. Earlier partial releases of photometric data are considered possible, but less desirable for the astrometric data, which depend on epoch coverage to reach the desired accuracies.

Appendix A
Transformations for heliotropic and Tait-Bryant angles

The Tait-Bryant angles are a way to describe the actual pointing of the satellite relative to a reference nominal pointing. Small corrections to these angles can be evaluated from the inertial rate measurements, and as such they form a useful tool in describing the actual attitude of the satellite. This appendix presents the various relations that have been used in the *Hipparcos* attitude reconstruction involving these Tait-Bryant angles.

We define in the usual way an orthogonal rotation by an angle ϕ around axis i as $\mathsf{R}_i(\phi)$. For $i = 1, 2, 3$:

$$\mathsf{R}_1(\phi) = \begin{bmatrix} 1 & 0 & 0 \\ 0 & \mathrm{c}\phi & -\mathrm{s}\phi \\ 0 & \mathrm{s}\phi & \mathrm{c}\phi \end{bmatrix}, \mathsf{R}_2(\phi) = \begin{bmatrix} \mathrm{c}\phi & 0 & \mathrm{s}\phi \\ 0 & 1 & 0 \\ -\mathrm{s}\phi & 0 & \mathrm{c}\phi \end{bmatrix}, \mathsf{R}_3(\phi) = \begin{bmatrix} \mathrm{c}\phi & -\mathrm{s}\phi & 0 \\ \mathrm{s}\phi & \mathrm{c}\phi & 0 \\ 0 & 0 & 1 \end{bmatrix}, \tag{A.1}$$

where $\mathrm{c}\phi \equiv \cos\phi$ and $\mathrm{s}\phi \equiv \sin\phi$. In each coordinate system the positions are represented by means of unit vectors, relating to the specific angles of the coordinate system through direction cosines as defined in Eq. 2.1 (page 40). Transformations between vectors in different coordinate systems are obtained through a succession of rotations. Inverse transformations are simply obtained by application of the transpose of the transformation matrix,

We start at a position (α, δ) in the equatorial system, for which we have a unit-length vector:

$$\boldsymbol{Q} = \begin{bmatrix} \cos\alpha\cos\delta \\ \sin\alpha\cos\delta \\ \sin\delta \end{bmatrix}. \tag{A.2}$$

The ecliptic-coordinates reference system is rotated with respect to the equatorial system by $\epsilon = 23\overset{\circ}{.}43929111$ around axis $i = 1$:

$$\boldsymbol{E} \equiv \mathsf{R}_1(\epsilon)\boldsymbol{Q}, \tag{A.3}$$

which defines the vector $\boldsymbol{E}$ in ecliptic coordinates.

Converting ecliptic to heliotropic coordinates (see Fig. 1.8, page 22) involves four rotations:

$$\boldsymbol{H} \equiv \mathsf{R}_3(-\Omega)\mathsf{R}_2(\xi - \frac{\pi}{2})\mathsf{R}_1(\frac{\pi}{2} - \nu)\mathsf{R}_3(-\lambda_\odot)\boldsymbol{E}, \tag{A.4}$$

where $\lambda_\odot$ is the longitude of the Sun as described by Eq. 1.2, page 22 for the nominal scanning law. An additional rotation of $\pm\frac{\gamma}{2}$ has to be applied to get positions in the preceding field of view (+) or following (−) field of view. Then, for a star observed in one of the fields of view, $\boldsymbol{H}$ is close to $(1, 0, 0)$.

The addition of the Tait-Bryant angles (ψ, θ, ϕ) has to be applied to $\boldsymbol{H}$:

$$\boldsymbol{H}' = \mathsf{R}_3(\psi)\mathsf{R}_2(\theta)\mathsf{R}_1(\phi)\boldsymbol{H}. \tag{A.5}$$

The Tait-Bryant angles can also be used to define a new set of heliotropic angles (ξ', ν', Ω') for the transformed position. We define the following variables:

$$\begin{aligned} a &= \cos\Omega\cos\phi - \sin\Omega\sin\theta \\ b &= \sin\xi\cdot(\cos\Omega\cos\theta - \sin\Omega\sin\phi\sin\theta) - \cos\xi\cos\phi\sin\theta \\ c &= \sin\xi\sin\Omega\cos\phi - \cos\xi\sin\phi \\ d &= \sqrt{b^2 + c^2} \\ e &= (c\cdot\cos\Omega - b\cdot\sin\Omega)/d. \end{aligned} \tag{A.6}$$

From these new variables we can derive the transformed heliotropic angles:

$$\begin{aligned} \xi' &= \arcsin(d) \\ \nu' &= \nu + \arcsin(a/d) \\ \Omega' &= \Omega + \psi + \arcsin(c). \end{aligned} \tag{A.7}$$

Given a set of actual heliotropic angles, the Tait-Bryant angles with respect to the nominal attitude can be derived quite easily. We determine for this transformation the rotation matrix A:

$$\mathsf{A} = \mathsf{R}_3(-\Omega')\mathsf{R}_2(\xi' - \frac{\pi}{2})\mathsf{R}_1(\frac{\pi}{2} - \nu')\mathsf{R}_1(\nu - \frac{\pi}{2})\mathsf{R}_2(\frac{\pi}{2} - \xi)\mathsf{R}_3(\Omega). \tag{A.8}$$

The Tait-Bryant angles can be derived from elements of the matrix A:

$$\begin{aligned} \phi &= \arctan(-A_{2,3}/A_{3,3}) \\ \theta &= \arcsin(A_{1,3}) \\ \psi &= \arctan(-A_{1,2}/A_{1,1}). \end{aligned} \tag{A.9}$$

Important in the analysis of the *Hipparcos* data are also the relations between inertial rotation rates and changes in the heliotropic angles. For *Hipparcos* data analysis rate changes at a level of less than a mas s^{-1} are still very relevant. The projection of the changes in the nominal solar longitude is at a level of 40 mas s^{-1}:

$$\boldsymbol{V}_{\lambda_\odot} = \dot{\lambda}\begin{bmatrix} -\cos\xi\sin\nu\cos\Omega - \cos\nu\sin\Omega \\ +\sin\xi\sin\nu\sin\Omega - \cos\nu\cos\Omega \\ +\sin\xi\sin\nu \end{bmatrix}. \tag{A.10}$$

The total inertial rotation as a result of changes in the nominal heliotropic angles is now given by:

$$\boldsymbol{V}_\mathrm{N} = \boldsymbol{V}_\lambda + \begin{bmatrix} -\sin\Omega & \sin\xi\cos\Omega & 0 \\ -\cos\Omega & -\sin\xi\sin\Omega & 0 \\ 0 & \cos\xi & 1 \end{bmatrix}\cdot\begin{bmatrix} \dot{\xi} \\ \dot{\nu} \\ \dot{\Omega} \end{bmatrix}. \tag{A.11}$$

In the integrations for the attitude modelling, the relation between the inertial rates and the accumulation of the Tait-Bryant angles is an important element. Say the inertial rates of the satellite have been determined as $\boldsymbol{V}_I$, the current Tait-Bryant angles are given by (ϕ, θ, ψ), and the inertial rate associated with the nominal scanning law is through Eq. A.11 given by $\boldsymbol{V}_{\mathrm{N}}$. The observed rates are in the actual reference frame of the satellite, and we therefore have to transform $\boldsymbol{V}_N$ to that frame through the usual sequence of rotations:

$$\boldsymbol{V}'_N = \mathsf{R}_3(\psi)\mathsf{R}_2(\theta)\mathsf{R}_1(\phi)\boldsymbol{V}_N. \tag{A.12}$$

As the Tait-Bryant angles are measured with respect to the non-stationary reference frame of the nominal scanning law, we need to correct the observed inertial velocity:

$$\boldsymbol{V}'_I = \boldsymbol{V}_I - \boldsymbol{V}'_N. \tag{A.13}$$

The three rotations, as they are applied in succession, are not orthogonal. To bring the orthogonal rates $\boldsymbol{V}'_I$ in line with the Tait-Bryant angles we apply a further transformation:

$$\begin{bmatrix} \dot{\phi} \\ \dot{\theta} \\ \dot{\psi} \end{bmatrix} = \begin{bmatrix} \cos\psi / \cos\theta & -\sin\psi / \cos\theta & 0 \\ \sin\psi & \cos\psi & 0 \\ -\cos\psi\sin\theta / \cos\theta & \sin\psi\sin\theta / \cos\theta & 1 \end{bmatrix} \cdot \boldsymbol{V}'_I. \tag{A.14}$$

One of the consequences of Eq. A.14 is that starting values for the Tait-Bryant angles and the inertial rates are linked, which needs to be taken care off during the first iteration stage of the attitude modelling. In particular corrections to angular offset ψ are noted clearly in the rate corrections $\dot{\phi}$ and $\dot{\theta}$.

Appendix B
Spline functions

Spline functions have been the backbone of the attitude solution, but have also been implemented to smooth observations of calibration parameters as functions of time. The implementation used here has been somewhat different from the more traditional B-splines, in order to ease the flexibility towards higher orders, needed to represent the underlying torques as third-order splines and solving for rate or positional corrections. The representation presented here fits in well with the least-squares solutions as based on Householder transformations, described in Appendix C, and provides a simple relation between say a fifth order spline and its second derivative, a cubic spline.

Splines are piecewise polynomials which can be fitted to a series of data points. The limits of the piecewise portions are called "knots" or "nodes". A spline $f(x)$ of order n with knots at x_i ($i = 1, ..., N$) is defined as a sequence of polynomials $f_i(x)$ of the kind

$$f_i(x) = a_{i0} + a_{i1}x + a_{i2}x^2 + ... + a_{in}x^n \quad \text{with } x_i \le x \le x_{i+1} \tag{B.1}$$

Restrictions are introduced in order to ensure that the function and all its derivatives up to order $r = n - 1$ are continuous at the knots:

$$f_i(x_i) = f_{(i+1)}(x_i): \quad \sum_{k=0}^{n} a_{ik}x_i^k = \sum_{k=0}^{n} a_{(i+1)\ k}x_i^k \tag{B.2}$$

$$f_i'(x_i) = f_{(i+1)}'(x_i): \quad \sum_{k=1}^{n} k a_{ik}x_i^{k-1} = \sum_{k=1}^{n} k a_{(i+1)\ k}x_i^{k-1} \tag{B.3}$$

$$\cdots \qquad \cdots$$

$$f_i^r(x_i) = f_{(i+1)}^r(x_i): \quad \sum_{k=r}^{n} a_{ik}x_i^{k-r} \prod_{j=k}^{k-r+1} j = \sum_{k=r}^{n} a_{(i+1)k}x_i^{k-r} \prod_{j=k}^{k-r+1} j \tag{B.4}$$

The above set of constraints allows to express the first n coefficients of the polynomial $f_{i+1}(x)$ as a function of $a_{(i+1)\ n}$ (i.e. the last coefficient of $f_{i+1}(x)$) and of a_{i0}, a_{i1}, a_{i2},..., a_{in} (i.e. the coefficients of the polynomial $f_i(x)$ defined over the preceding interval). By introducing such expressions in the definition of $f_{i+1}(x)$, it is possible to rewrite it in the following way:

$$f_{i+1}(x) = f_i(x) + (a_{(i+1)\ n} - a_{in})(x - x_{i+1})^n \tag{B.5}$$

Let us consider the central interval in the given sequence as characterized by index $\bar{i}$ and let us call $f_{\bar{i}}(x)$ "basic" polynomial. Any other polynomial $f_i(x)$ defined over an interval $[x_i, x_{i+1}]$ can be represented as:

$$f_i(x) = f_{\bar{i}}(x) + \begin{cases} \sum_{k=\bar{i}+1}^{i} (a_{kn} - a_{(k-1)\,n})(x - x_k)^n & \text{if } i > \bar{i} \\ \sum_{k=i}^{\bar{i}-1} (a_{(k+1)\,n} - a_{kn})(x - x_{k+1})^n & \text{if } i < \bar{i} \end{cases} \tag{B.6}$$

The continuity of the spline up to its derivative of order $(n-1)$ follows immediately from Eq. B.6. This equation also tells that a spline is completely determined once the coefficients of the basic polynomial and one additional coefficient per interval are given. In other words a spline of order n, covering m intervals and having knots at x_i $(i = 1, ..., m+1)$ requires $(n+1) + (m-1) = (n+m)$ coefficients for its definition, where $(n+1)$ is the number of coefficients in the basic polynomial and $(m-1)$ is the number of intervals - 1 (the central interval), which is also the number of internal knots.

This method also allows to obtain any derivative $f^r(x)$ of order r of the function (provided that $r \le n$). As a matter of fact, by deriving Eq. B.5 r times, one obtains:

$$f_{i+1}^r(x) = f_i^r(x) + (a_{(i+1)\,n} - a_{in})(x - x_{i+1})^{n-r} \prod_{j=n}^{n-r+1} j \tag{B.7}$$

in which

$$f_i^r(x) = \sum_{k=r}^{n} a_{ik} x^{k-r} \prod_{j=k}^{k-r+1} j \tag{B.8}$$

$f^r(x)$ is a spline of order $n-r$. The 2nd derivative of a 5th order spline $g^5(x)$ with knots at x_i $(i = 1, ..., m-1)$ is a 3rd order spline $g^3(x)$. The polynomials $g_i^5(x)$ and $g_i^3(x)$ based on $[x_i, x_{i+1}]$ can be written respectively as:

$$g_i^5(x) = b_{i0} + b_{i1}x + b_{i2}x^2 + b_{i3}x^3 + b_{i4}x^4 + b_{i5}x^5, \tag{B.9}$$

for the fifth-order, and

$$g_i^3(x) = 2b_{i2} + 6b_{i3}x + 12b_{i4}x^2 + 20b_{i5}x^3 \tag{B.10}$$

for its second derivative. With reference to the central interval $\bar{i}$, $g_i^3(x)$ has the following expression:

$$g_i^3(x) = g_{\bar{i}}^3(x) + \begin{cases} \sum_{k=\bar{i}+1}^{i} (b_{k5} - b_{(k-1)\,5})(x - x_k)^3 \cdot 5 \cdot 4 & \text{if } i > \bar{i} \\ \sum_{k=i}^{\bar{i}-1} (b_{(k+1)\,5} - b_{k5})(x - x_{k+1})^3 \cdot 5 \cdot 4 & \text{if } i < \bar{i} \end{cases} \tag{B.11}$$

Eq. B.11 shows that $g^3(x)$ and its first and second derivatives are continuous functions of x. The function $g^3(x)$ contains $3+m$ unknown coefficients, the first 4 represent the central polynomial $g_{\bar{i}}^3(x)$ $(b_{\bar{i}2}, b_{\bar{i}3}, b_{\bar{i}4}, b_{\bar{i}5})$ and the remaining $(m-1)$ (of the kind $(b_{(i+1)\,5} - b_{i5})$) from the additional node intervals.

In fitting the attitude data the thruster firings had to be taken into account. This led to the following situation. The underlying torque remained a continuous function of time, described by a cubic spline. Between two thruster firings the rate corrections are described by a fourth-order, and the angular displacement corrections by a fifth-order spline. Within a solution interval, the different thruster-firing intervals all share the same cubic spline as their second derivative of the actual function fitted, but for interval separate variables are included to account for the local offset and rate corrections, parameters b_{i0} and b_{i1} in Eq. B.9. In order to limit the interaction between the nodes in the spline and the discontinuities in the fitting caused by the thruster firings, nodes were distributed evenly over thruster firing intervals such that the distance between the node nearest to the thruster firing and the instance of that thruster firing was equal to $0.5l/n$, where l is the length of the interval and n the number of nodes fitted in the interval.

Appendix C
Linear Least Squares and Householder Orthogonal Transformations

Least-squares solutions have been applied at all levels of the reductions, and the mechanism to solve has been through Householder orthogonal transformations, providing a numerically stable as well as elegant solution, which is easily adapted to a wide variety of applications, some of which are explored in Appendix D.

Let us consider the linear system

$$\boldsymbol{z} = \mathsf{A}\boldsymbol{x} + \epsilon \tag{C.1}$$

where $\boldsymbol{z}$ is an m vector of observations, $\boldsymbol{x}$ is the n vector of variables to be estimated, $\mathsf{A}(m, n)$ is the coefficient matrix of model parameters used to describe the observations ($m \geq n$) and ϵ is an m vector of observation errors. It is well known that the least squares solution of the system C.1 is the n vector $\tilde{\boldsymbol{x}}$ which minimizes the mean square observation error $J(x)$:

$$J(x) = \| \boldsymbol{z} - \mathsf{A}\boldsymbol{x} \|^2 = (\boldsymbol{z} - \mathsf{A}\boldsymbol{x})^T(\boldsymbol{z} - \mathsf{A}\boldsymbol{x}) \tag{C.2}$$

Following (Bierman 1977), let $\mathsf{T}(m, m)$ be an orthogonal matrix, then the following holds:

$$J(x) \equiv \| \mathsf{I}(\boldsymbol{z} - \mathsf{A}\boldsymbol{x}) \|^2 = \| \mathsf{T}\boldsymbol{z} - (\mathsf{TA})\boldsymbol{x} \|^2 \tag{C.3}$$

and in particular the minimum of $J(x)$ is independent of T. This can be exploited by choosing T so that TA has a computationally attractive form. Householder orthogonal transformations allow choosing T in such a way that

$$\mathsf{TA} = \begin{bmatrix} \mathsf{U} \\ \mathbf{0} \end{bmatrix} \begin{matrix} \}n \\ \}m-n \end{matrix} \tag{C.4}$$

where U is an upper triangular matrix. The Householder orthogonal transformation is based on the concept of elementary reflection: given a non-zero vector $\boldsymbol{u}$ the matrix $\mathsf{T}_u = \mathsf{I} - \beta\boldsymbol{u}\boldsymbol{u}^T$, where $\beta = 2/(\boldsymbol{u}^T\boldsymbol{u})$, transforms a vector $\boldsymbol{y}$ into its reflection $\boldsymbol{y}_r$ in the plane orthogonal to $\boldsymbol{u}$:

$$\boldsymbol{y}_r = \left(\mathsf{I} - \beta\boldsymbol{u}\boldsymbol{u}^T\right)\boldsymbol{y} \tag{C.5}$$

In particular the vector $\boldsymbol{u}$, and hence T_u, can be chosen in such a way that $\boldsymbol{y}_r$ be parallel to the unit vector $\boldsymbol{e}_1 = (1, 0, 0..., 0)^T$:

$$\mathsf{T}_u\boldsymbol{y} = \sigma\boldsymbol{e}_1 \tag{C.6}$$

This can be obtained by defining $\boldsymbol{u}$ and σ in the following way:

$$u_i = \begin{cases} A_{11} - \sigma & \text{for } i = 1 \\ A_{i1} & \forall\ i = 2, ..., m \end{cases} \tag{C.7}$$

$$\sigma = (\boldsymbol{y}^T \boldsymbol{y})^{1/2} \tag{C.8}$$

Application of Eq. C.6 to matrix A (with $\boldsymbol{y}$ as the first column of A) results in the fundamental step to matrix triangularization, i.e. annihilating the first column:

$$\mathsf{T}_u \mathsf{A} = \mathsf{T}_1 \mathsf{A} = \begin{bmatrix} \sigma & \\ 0 & \tilde{\mathsf{A}} \\ \cdots & \\ 0 & \end{bmatrix} \tag{C.9}$$

where the scalar σ is computed as prescribed by Eq. C.8 and the sub-matrix $\tilde{\mathsf{A}}(m, n-1)$ results from application of T_u to the columns of A to the right of the first. By further applying elementary Householder matrices to annihilate the remaining columns of A, the jth transformation T_j zeros the jth column of A below row j and determines the elements of the jth row that are to the right of the jth column with σ_j as the jth diagonal element without altering in any way the rows above the jth. The triangularization of the matrix A is thus described by:

$$\mathsf{T} = \begin{bmatrix} \mathsf{I}_{j-1} & \mathbf{0} \\ \mathbf{0} & \mathsf{T}_j \end{bmatrix} \cdots \begin{bmatrix} \mathsf{I}_1 & \mathbf{0} \\ \mathbf{0} & \mathsf{T}_2 \end{bmatrix} \mathsf{T}_1 \quad j \leq n \tag{C.10}$$

I_j being the identity matrix of order j. Thus this algorithm can be used to orthogonally transform a matrix into a partially triangular form. The same result can be obtained by applying a classical Gauss elimination and then dividing each row by the square root of the corresponding diagonal element. However the algorithm based on Householder transformations is characterized by higher numerical stability. Besides it requires no additional computer storage other than used for A because A and $\tilde{\mathsf{A}}$ share common storage and the matrices T_u are not explicitly computed.

According to Eqs. C.3 and C.4 the solution $\tilde{\boldsymbol{x}}$ to the least squares problem C.1 must satisfy

$$\mathsf{U}\boldsymbol{x} = \boldsymbol{z}_1 \tag{C.11}$$

where $\boldsymbol{z}_1$ is obtained by partitioning the matrix $\mathsf{T}\boldsymbol{z}$ in the following way:

$$\mathsf{T}\boldsymbol{z} = \begin{bmatrix} \boldsymbol{z}_1 \\ \boldsymbol{z}_2 \end{bmatrix} \begin{matrix} \}\, n \\ \}\, m - n \end{matrix} \tag{C.12}$$

Thus Householder transformations provide the solution to the least squares problem. The reader is referred to (Bierman 1977) for details. They are particularly useful in the case of least squares requiring inclusion of additional information in the form of an *a priori* information matrix $\tilde{\Lambda}$. An *a priori* estimate-covariance information can be interpreted as providing additional observations to the original problem, thus resulting in the following augmented system:

$$\begin{bmatrix} \tilde{\boldsymbol{z}} \\ \boldsymbol{z} \end{bmatrix} = \begin{bmatrix} \tilde{\mathsf{U}} \\ \mathsf{A} \end{bmatrix} \boldsymbol{x} + \begin{bmatrix} \tilde{\epsilon} \\ \epsilon \end{bmatrix}, \tag{C.13}$$

where $\tilde{\mathsf{U}}$ is the Householder triangularization of $\tilde{\Lambda}$. The solution to Eq. C.13 is obtained through Householder triangularization of the matrix $\begin{bmatrix} \tilde{\mathsf{U}} \\ \mathsf{A} \end{bmatrix}$. Systems of the kind C.13 will appear recursively as the current estimate and covariance become the *a priori* and are combined with new data to form an updated estimate and covariance.

The elements in the upper-triangular matrix can be stored and processed as a vector. In the case of the solution for the five astrometric parameters:

$$\mathsf{U} = \begin{bmatrix} U_1 & U_2 & U_4 & U_7 & U_{11} \\ 0 & U_3 & U_5 & U_8 & U_{12} \\ 0 & 0 & U_6 & U_9 & U_{13} \\ 0 & 0 & 0 & U_{10} & U_{14} \\ 0 & 0 & 0 & 0 & U_{15} \end{bmatrix} \tag{C.14}$$

The upper-triangular matrix U is a square root of the covariance matrix of the model parameters:

$$\mathsf{A}^T\mathsf{A} = \mathsf{U}^T\mathsf{U} = \mathsf{R}, \tag{C.15}$$

and we can therefore derive the inverse of the covariance matrix directly from U:

$$\mathsf{R}^{-1} = \mathsf{C} = (\mathsf{A}^T\mathsf{A})^{-1} = (\mathsf{U}^T\mathsf{U})^{-1} = \mathsf{U}^{-1}(\mathsf{U}^{-1})^T, \tag{C.16}$$

where the matrix U^{-1} is again upper triangular. The relations between the elements of the n by n upper triangular arrays U and U^{-1} are given by:

$$\sum_{i=k}^{l} U_{k,i}U_{i,l}^{-1} = \begin{cases} 1 & \text{for } k = l \\ 0 & \text{for } k \neq l \end{cases}, \tag{C.17}$$

for $(k = 1 \ldots n)$ and $(l = k \ldots n)$. Starting at element (n, n), Eq. C.17 translates into a simple algorithm for the inversion of an upper-triangular matrix.

The formal error imposed by our model on a fitted value z_i for a given model vector $\boldsymbol{x}_i$ is given by (assuming a solution variance equal to the degrees of freedom):

$$\sigma(z) = \sqrt{\boldsymbol{x}_i^T \mathsf{C} \boldsymbol{x}_i}, \tag{C.18}$$

which can be rewritten as:

$$\sigma_z = |\boldsymbol{x}^T \mathsf{U}^{-1}|, \tag{C.19}$$

which simplifies the required calculations. Similarly, the weight of a predicted observation based on the solution, given a model vector $\boldsymbol{x}$ is given by:

$$w_z = \sigma_z^{-1} = |\mathsf{U}\boldsymbol{x}|. \tag{C.20}$$

Thus, using Eq. C.14 as the matrix U for the astrometric-parameter solution, the formal error on the parallax is given by:

$$\sigma_\varpi = (U_4^2 + U_5^2 + U_6^2)^{-0.5}. \tag{C.21}$$

When combining for example the proper motion and parallax information of different stars (as will be the case when studying a star cluster), a physical model $\boldsymbol{F}$ is used to describe the observed quantities $\boldsymbol{z}$, and the observed quantities are used to fit parameters a of the model:

$$\boldsymbol{F}(a_i) = \boldsymbol{z}_i + \mathsf{U}_i^{-1}\boldsymbol{\epsilon}, \tag{C.22}$$

where each element of $\boldsymbol{\epsilon}$ has an expectation value zero and sigma of one. In this case $\boldsymbol{z}_i = (0, 0, \varpi_i, \mu_{\alpha*,i}, \mu_{\delta,i})$, representing the parallax and proper motion measurements for star i. Multiplying left and right by U_i normalizes the observations:

$$\mathsf{U}_i\boldsymbol{F}(a_i) = \mathsf{U}_i\boldsymbol{z}_i + \boldsymbol{\epsilon}, \tag{C.23}$$

and makes it possible to combine data for different stars in one and the same solution. The matrix U is provided in the catalogue files for each solution.

Appendix D
Chain solutions, running solutions, and common parameters

Solving a least-squares problem by means of Householder orthogonal transformations provides simple means to accommodate situations that were frequently encountered in the *Hipparcos* data reductions, and in particular when applied to calibrations. They all concern variations on the same theme: the distinction between instantaneous model parameters and slowly evolving or fixed instrument parameters. A typical example is the attitude modelling, where the instantaneous parameters represent the ever changing attitude of the satellite, which require the calibration of the fixed (for the duration of a solution) instrument model to represent the relation between the actual observations and suitable calibrated input parameters for the attitude model. A different kind of problem was encountered in the star mapper photometric reductions, where the instrument model was slowly evolving, and where a single data set was generally insufficient to determine some of the instrument parameters with sufficient accuracy. Here we adopted what we call a running solution, which effectively is a form of a Kalman filter.

All solution types described here are variations on Eq. C.13, repeated here:

$$\begin{bmatrix} \tilde{z} \\ z \end{bmatrix} = \begin{bmatrix} \tilde{\mathsf{U}} \\ \mathsf{A} \end{bmatrix} x + \begin{bmatrix} \tilde{\epsilon} \\ \epsilon \end{bmatrix}. \tag{D.1}$$

The upper part of this equation,

$$\tilde{z} = \tilde{\mathsf{U}} x + \tilde{\epsilon}, \tag{D.2}$$

is the equivalent of one observation for each of the model parameters, with weights based on the accumulated information for their calibration. When we consider a running solution for calibration parameters, we need to be able to apply different "decay times" to different parameters, representing variations at different timescales, and we can do that by adjusting the "weights" as given by $\tilde{\mathsf{U}}$:

$$\begin{aligned} \tilde{\mathsf{U}}' &= \mathsf{W} \cdot \tilde{\mathsf{U}}, \\ \tilde{z}' &= \mathsf{W} \cdot \tilde{z}, \end{aligned} \tag{D.3}$$

where W is a diagonal matrix with elements between zero and one. A value of zero represents the "no memory" situation, each calibration will determine this parameter on its own. A value of one is the other extreme, where all data contribute to the calibration of a parameter fixed in value over the mission. The effect of this is that of an individual Kalman filter applied to each model parameter, each filter adjusted to the noise level and long-term behaviour of a single parameter.

At an early stage of the reductions the possibility was considered of fitting the attitude model across thruster firings, using the calibration of the rotation-rate changes as a function of thruster-firing lengths (see Section 13.3.2). The attitude reconstruction across the firing is given by a continuous function of time in its second derivative (in this case a cubic spline), with superimposed a constant and a linear time parameter for each interval between two thruster-firings. The earlier model, used for some time in the production of the published catalogue, used individual polynomial fits per thruster-firing interval, and linked these across the thruster firings using the calibration information referred to above. In that case we have for each interval i a set of equations as given by Eq. D.2, connected by cross-boundary conditions. The procedure is as follows. Information on the first interval is collected in a set of equations like Eq. D.2. We now construct a new set of equations for the parameters of the first and those of the second interval:

$$\begin{bmatrix} \tilde{\boldsymbol{z}}_1 \\ \boldsymbol{z}_2 \\ \boldsymbol{z}_{1,2} \end{bmatrix} = \begin{bmatrix} \tilde{\mathsf{U}}_1 & \mathsf{D}_{1,2} \\ 0 & \mathsf{U}_2 \\ \mathsf{A}_1 & \mathsf{A}_2 \end{bmatrix} \begin{bmatrix} \boldsymbol{x}_1 \\ \boldsymbol{x}_2 \end{bmatrix} + \begin{bmatrix} \tilde{\boldsymbol{\epsilon}} \\ \boldsymbol{\epsilon}_2 \\ \boldsymbol{\epsilon}_{1,2} \end{bmatrix}. \tag{D.4}$$

The meaning of this construction is as follows. At the end of the first interval information has been gathered for the parameters $\boldsymbol{x}_1$, but no information is available yet for $\boldsymbol{x}_2$. This reflects in $\boldsymbol{z}_2$ and ϵ_2 being zero. The information to be added concerns both the first and the second interval through the equation:

$$\boldsymbol{z}_{1,2} = \begin{bmatrix} \mathsf{A}_1 & \mathsf{A}_2 \end{bmatrix} \begin{bmatrix} \boldsymbol{x}_1 \\ \boldsymbol{x}_2 \end{bmatrix} + \boldsymbol{\epsilon}_{1,2}, \tag{D.5}$$

and by applying it, *a priori* information is transferred from the first to the second interval. In the further accumulation of information for the second interval, only the system

$$\tilde{\boldsymbol{z}}_2 = \tilde{\mathsf{U}}_2 \boldsymbol{x}_2 + \tilde{\boldsymbol{\epsilon}}_2 \tag{D.6}$$

needs to be considered. This process is repeated until the last interval has been added. The system is then solved starting from the last interval, which then incorporates the boundary conditions back to the first interval. Storage for only the upper triangular arrays U_i and the rectangular arrays $\mathsf{D}_{i,i+1}$ are required.

A special case of the chain solution described above is the closed chain, a form of which has been used in the calibration of the Walraven VBLUW photometric standards (Lub and Pel 1977). As all good photometric systems, this system was defined through the responses of a selection of standard stars spread over the sky. To ensure that data obtained at different times of the year are on the same system, it is essential to establish the responses of all standards with respect to a chosen reference standard, in this case the B1V star HD 144470 (HIP 78933). The response of this reference standard is further established through spectro-photometric measurements. System-calibration observations are obtained from measurements of standard stars at nearly the same air mass and time of observation. For such measurements the differential extinction correction is very small and can safely be approximated by average extinction conditions. These observations describe a network of connections for the response differences of the primary standard stars, including the reference standard. The fact that the ring is closed is what makes this calibration stable and reliable. This was confirmed through a comparison with the *Hipparcos* Hp photometry, where, contrary to some other photometric systems, no need for seasonal correction of the reduced photometric data was detected (see Vol. 3, Chapter 21 of ESA 1997).

A different kind of application is the common-parameter solution, where we encounter sets of equations similar in form to Eq. D.4:

$$\begin{bmatrix} 0 \\ \tilde{\boldsymbol{z}}_c \\ \boldsymbol{z}_{i,c} \end{bmatrix} = \begin{bmatrix} \tilde{\mathsf{U}}_i & \mathsf{D}_{i,c} \\ 0 & \mathsf{U}_c \\ \mathsf{A}_i & \mathsf{A}_c \end{bmatrix} \begin{bmatrix} \boldsymbol{x}_i \\ \boldsymbol{x}_c \end{bmatrix} + \begin{bmatrix} 0 \\ \tilde{\boldsymbol{\epsilon}}_c \\ \boldsymbol{\epsilon}_i \end{bmatrix}. \tag{D.7}$$

Each observation j in interval i, $z_{i,j}$, is described by a set of local model parameters, $\boldsymbol{A}_i$, and a set of calibration parameters, $\boldsymbol{A}_c$. The latter are shared with other intervals. The *a priori* information transferred between intervals only concerns the accumulated information on the calibration parameters. Once all intervals have been added to the solution, the calibration parameters are recovered from:

$$\boldsymbol{z}_c = \mathsf{U}_c \boldsymbol{x}_c + \boldsymbol{\epsilon}_c, \tag{D.8}$$

and applied to the individual intervals:

$$\boldsymbol{z}_i - \mathsf{D}_{c,i} \boldsymbol{x}_c = \mathsf{U}_i \boldsymbol{x}_i + \boldsymbol{\epsilon}_i. \tag{D.9}$$

Solution of the system only requires the storage of the matrices and vectors as indicated.

In the context of the *Hipparcos* data processing, this procedure was used in the attitude reconstructions for both the star mapper and the main-grid data, to accommodate the calibration of instrument parameters in data sets for which the solution of the local parameters had to be split over a number of individual solutions due to gaps in the data coverage.

Appendix E
Orbit parameters for binary stars

For the description of the orbit parameters of binary stars the presentation by Bond and Allman (1996) is followed in detail, and where also a much more detailed derivation of the basic equations can be found. A further extension to issues specific for the *Hipparcos* data is given: the definition of the reference point relative to which the orbital elements are given, which depends on the separation of the components, their magnitude difference and the length of the orbital period. The description of the two-body problem presented here is valid for separated point sources, and is effectively applicable to separated stars.

The following definition are used throughout: a vector is given as $\boldsymbol{a}$; its length by $a = \sqrt{\boldsymbol{a} \cdot \boldsymbol{a}}$; the unit vector in the direction of $\boldsymbol{a}$ is $\hat{\boldsymbol{a}} = \boldsymbol{a}/a$; the first and second derivatives in time are given by $\dot{\boldsymbol{a}}$ and $\ddot{\boldsymbol{a}}$ respectively.

Assume 2 point sources with masses m_1 and m_2, at relative position $\boldsymbol{r} = \boldsymbol{r}_2 - \boldsymbol{r}_1$. Newton's second and third law:

$$\begin{aligned} m_2 \ddot{\boldsymbol{r}}_2 &= -\boldsymbol{F}, \\ m_1 \ddot{\boldsymbol{r}}_1 &= \boldsymbol{F}, \\ \boldsymbol{F} &= F \hat{\boldsymbol{r}} \end{aligned} \tag{E.1}$$

and Newton's universal law of gravitation,

$$F = G \frac{m_1 m_2}{r^2}, \tag{E.2}$$

lead to the differential equation of motion for the two-body problem:

$$\ddot{\boldsymbol{r}} + \frac{\mu}{r^3} \boldsymbol{r} = 0, \tag{E.3}$$

where $\mu = G(m_1 + m_2)$ and G is the Universal Gravitational constant. Equation E.3 has the same form as Newton's second law, and describes in a reference system with a fixed origin the motions of a unit mass attracted by mass $(m_1 + m_2)$ located at a distance r.

Equation E.3 requires 6 integrals of motion for a complete analytic solution. An integral of motion for this system is a first-order differential equation of the kind:

$$Q(\boldsymbol{r}, \dot{\boldsymbol{r}}, t) = C, \tag{E.4}$$

where C is constant. These integrals of motion can be derived from Eq. E.3 through relatively simple manipulations. The first is obtained by taking the cross-product of $\boldsymbol{r}$ with Eq. E.3, to obtain (as $\boldsymbol{r} \times \boldsymbol{r} = 0$):

$$\boldsymbol{r} \times \ddot{\boldsymbol{r}} = 0, \tag{E.5}$$

which is equivalent to:

$$\frac{\mathrm{d}}{\mathrm{d}t}(\boldsymbol{r} \times \dot{\boldsymbol{r}}) = 0, \tag{E.6}$$

from which the first three integrals of motion ar obtained:

$$\boldsymbol{r} \times \dot{\boldsymbol{r}} = \boldsymbol{c}, \tag{E.7}$$

where $\boldsymbol{c}$ is a constant known as the angular momentum vector, which, by being perpendicular to both the position vector and the velocity vector, proves that the orbit of a two body system is restricted to the plane described at any time by the relative position vector $\boldsymbol{r}$ and its time derivative $\dot{\boldsymbol{r}}$.

The next integral is obtained by taking the dot-product of the velocity vector $\dot{\boldsymbol{r}}$ and Eq. E.3:

$$\dot{\boldsymbol{r}} \cdot \left(\ddot{\boldsymbol{r}} + \frac{\mu}{r^3}\boldsymbol{r} \right) = 0, \tag{E.8}$$

which can be re-written in the form:

$$\frac{\mathrm{d}}{\mathrm{d}t}\left(\frac{1}{2}(\dot{\boldsymbol{r}} \cdot \dot{\boldsymbol{r}}) - \frac{\mu}{r} \right) = 0, \tag{E.9}$$

from which the fourth integral of motion is derived:

$$\frac{1}{2}(\dot{\boldsymbol{r}} \cdot \dot{\boldsymbol{r}}) - \frac{\mu}{r} = h, \tag{E.10}$$

where the constant h represents the total energy of the system, and the two terms on the left representing the kinetic and potential energy respectively.

Next we derive the Laplacian integral P. This is obtained by taking the cross-product of the angular momentum vector and Eq. E.3:

$$\boldsymbol{c} \times \ddot{\boldsymbol{r}} + \frac{\mu}{r^3}(\boldsymbol{r} \times \dot{\boldsymbol{r}}) \times \boldsymbol{r} = 0, \tag{E.11}$$

which can be reorganized to give:

$$\boldsymbol{c} \times \ddot{\boldsymbol{r}} + \frac{\mu}{r^3}\big((\boldsymbol{r} \cdot \boldsymbol{r})\dot{\boldsymbol{r}} - (\dot{\boldsymbol{r}} \cdot \boldsymbol{r})\boldsymbol{r}\big) = 0. \tag{E.12}$$

This, considering

$$\frac{\mu}{r}\boldsymbol{r} = \frac{\mu}{\sqrt{\boldsymbol{r} \cdot \boldsymbol{r}}}\boldsymbol{r}, \tag{E.13}$$

is equivalent to:

$$\frac{\mathrm{d}}{\mathrm{d}t}\left(\boldsymbol{c} \times \dot{\boldsymbol{r}} + \frac{\mu}{r}\boldsymbol{r} \right) = 0, \tag{E.14}$$

from which three further integrals are derived:

$$\boldsymbol{c} \times \dot{\boldsymbol{r}} + \frac{\mu}{r}\boldsymbol{r} = -\boldsymbol{P}, \tag{E.15}$$

where $\boldsymbol{P}$ is a vector constant called the Laplace or eccentricity vector. The Laplace vector is normal to the angular momentum vector:

$$\boldsymbol{c} \cdot (\boldsymbol{c} \times \boldsymbol{r}) + \boldsymbol{c} \cdot \boldsymbol{r}\frac{\mu}{r} = -\boldsymbol{c} \cdot \boldsymbol{P} = 0. \tag{E.16}$$

The vector $\boldsymbol{P}$ is therefore in the orbital plane. A further relationship between $\boldsymbol{c}$, h and $\boldsymbol{P}$ is obtained when the dot-product $\boldsymbol{P} \cdot \boldsymbol{P}$ is evaluated:

$$P^2 = (\boldsymbol{c} \times \dot{\boldsymbol{r}}) \cdot (\boldsymbol{c} \times \dot{\boldsymbol{r}}) + \frac{2\mu}{r} \boldsymbol{r} \cdot (\boldsymbol{c} \times \dot{\boldsymbol{r}}) + \frac{\mu^2}{r^2} (\boldsymbol{r} \cdot \boldsymbol{r}). \tag{E.17}$$

This reduces simply to:

$$P^2 = c^2 \left(\dot{\boldsymbol{r}} \cdot \dot{\boldsymbol{r}} - \frac{2\mu}{r} \right) + \mu^2, \tag{E.18}$$

and using Eq. E.10:

$$P^2 = 2hc^2 + \mu^2, \tag{E.19}$$

which for further use is written in the form:

$$\frac{c^2}{\mu} = -\frac{\mu}{2h} \left(1 - \frac{P^2}{\mu^2} \right). \tag{E.20}$$

The seven integrals of motion, for $\boldsymbol{c}$, h and $\boldsymbol{P}$, are reduced to 5 through the two scalar relations (Eq. E.17 and Eq. E.20) described above. Thus, one more integral is needed to fully describe the system. This integral is known as Kepler's equation, and describes the relation between time t and separation r. It is derived by the dot-product of the angular momentum vector with itself:

$$\boldsymbol{c} \cdot \boldsymbol{c} = c^2 = (\boldsymbol{r} \times \dot{\boldsymbol{r}}) \cdot (\boldsymbol{r} \times \dot{\boldsymbol{r}}), \tag{E.21}$$

which is reorganized to give:

$$c^2 = (\dot{\boldsymbol{r}} \cdot \dot{\boldsymbol{r}})(\boldsymbol{r} \cdot \boldsymbol{r}) - (\dot{\boldsymbol{r}} \cdot \boldsymbol{r})(\dot{\boldsymbol{r}} \cdot \boldsymbol{r}). \tag{E.22}$$

Relating the first term on the right to the total energy h gives:

$$c^2 = r^2 \left(2h + \frac{2\mu}{r} \right) - r^2 \dot{r}^2. \tag{E.23}$$

From here on only bound orbits ($h < 0$) are considered. Then we can derive from the equation above:

$$r\dot{r} = \pm\sqrt{-2h} \sqrt{-\left(r^2 + \frac{\mu}{h} \right) + \frac{c^2}{2h}}, \tag{E.24}$$

which, using Eq. E.20, can be written as:

$$r\frac{\mathrm{d}r}{\mathrm{d}t} = \pm\sqrt{-2h} \sqrt{\left(\frac{P}{2h} \right)^2 - \left(r + \frac{\mu}{2h} \right)^2} \tag{E.25}$$

After substituting $z = r + \mu/2h$, and some reorganizing, a relation of the form $\mathrm{d}t = f(r)\mathrm{d}r$ is obtained:

$$\pm\sqrt{-2h}\, \mathrm{d}t = \frac{z - (\mu/2h)}{\sqrt{\left(\frac{P}{2h}\right)^2 - z^2}}\, \mathrm{d}z. \tag{E.26}$$

This equation has the following solution:

$$\pm\sqrt{-2h}\, (t + t_0) = -\frac{P}{2h} \sin E + \frac{\mu}{2h} E, \tag{E.27}$$

where t_0 is the integration constant (the final constraint for the solution of the differential equation of motion), and:

$$\begin{aligned} \cos E &= \frac{2h}{P} z, \\ \sin E &= \frac{2h}{P} \sqrt{\left(\frac{P}{2h}\right)^2 - z^2}. \end{aligned} \tag{E.28}$$

The angle E is referred to as the eccentric anomaly. Multiplying Eq. E.27 by $2h/\mu$ gives the final integral, Keppler's equation:

$$\pm \frac{2h}{\mu} \sqrt{-2h}\,(t + t_0) = E - \frac{P}{\mu} \sin E. \tag{E.29}$$

The separation distance between the two components of the system can now be derived as:

$$r = -\frac{\mu}{2h} \left(1 - \frac{P}{\mu} \cos E\right) \tag{E.30}$$

Relating the above equations to Keppler's three laws shows the relations between the physical constants and the traditional orbital parameters, the semi-major axis a, eccentricity e and the true anomaly ϕ. The first law, stating that the orbits of planets describe ellipses with the Sun in one focus, can be derived by taking the dot product between the position vector $\boldsymbol{r}$ and the Laplacian:

$$\boldsymbol{r} \cdot \boldsymbol{c} \times \dot{\boldsymbol{r}} + \frac{\mu}{r} \boldsymbol{r} \cdot \boldsymbol{r} = -\boldsymbol{r} \cdot \boldsymbol{P}, \tag{E.31}$$

which reduces to:

$$-c^2 + \mu r = -rP \cos \phi. \tag{E.32}$$

Solving this equation for the distance r gives:

$$r = \frac{c^2/\mu}{1 + (P/\mu) \cos \phi}, \tag{E.33}$$

which has the form of the standard equation of a conic section:

$$r = \frac{p}{1 + e \cos \phi} = \frac{a(1 - e^2)}{1 + e \cos \phi}. \tag{E.34}$$

This leads to two important relations for the eccentricity and semi-major axis:

$$\begin{aligned} e &= \frac{P}{\mu}, \\ a &= -\frac{\mu}{2h}. \end{aligned} \tag{E.35}$$

The first relation shows why the Laplace vector is also referred to as the eccentricity vector. The second of these relations states that given two masses, the semi-major axis of the orbits they describe relative to each other is only determined by the total energy in the system. Furthermore, the true anomaly ϕ is defined to measure the angle between the Laplace vector $\boldsymbol{P}$ and the position vector $\boldsymbol{r}$, aligning $\boldsymbol{P}$ such that it points towards the perigee of the orbit, as seen from the centre of attraction.

Kepler's second law derives directly from the first integral of motion, the area integral expressed by Eq E.7, which defined the angular momentum vector as constant. Describing the position and velocity vectors in local coordinates:

$$\begin{aligned} \boldsymbol{r} &= r\hat{\boldsymbol{r}}, \\ \dot{\boldsymbol{r}} &= \dot{r}\hat{\boldsymbol{r}} + r\dot{\phi}\hat{\boldsymbol{\phi}}, \end{aligned} \tag{E.36}$$

where:

$$\begin{aligned} \hat{\boldsymbol{\phi}} &\equiv \hat{\boldsymbol{c}} \times \hat{\boldsymbol{r}}, \\ \hat{\boldsymbol{c}} &\equiv \boldsymbol{c}/\,c. \end{aligned} \tag{E.37}$$

Substituting these relations in Eq. E.7 gives:

$$\boldsymbol{c} = r^2\dot{\phi}\,\hat{\boldsymbol{c}}, \tag{E.38}$$

from which follows that the magnitude of the angular momentum, $c = r^2\dot{\phi}$, is constant. The area swept per unit of time is given by:

$$\frac{\mathrm{d}A}{\mathrm{d}t} = c/2. \tag{E.39}$$

Kepler's third law (the period of an orbit is proportional to the cube of the orbit's semi-major axis) can be derived from Eq. E.30, which, after substituting the relations for a and e as given in Eq. E.35 becomes:

$$r = a\,(1 - e\cos E). \tag{E.40}$$

Substituting this relation and those of Eq. E.35 into Eq. E.25, the following relation can be obtained:

$$\sqrt{\frac{\mu}{a}}\,\mathrm{d}t = a\,(1 - e\cos E)\mathrm{d}E. \tag{E.41}$$

Integration over a full cycle in E gives the rotation period T:

$$T = 2\pi\sqrt{\frac{a^3}{\mu\varpi^3}}. \tag{E.42}$$

The additional factor ϖ represents the parallax of the system, and provides a transformation from observed to physical coordinates for the semi-major axis.

Two directions of the orbit, $\boldsymbol{c}$ defining the plane of the orbit and $\boldsymbol{P}$ the direction of perigee, can be used to describe the positions of the two masses. An orthogonal coordinate system is created from the directions of these vectors:

$$\hat{\boldsymbol{Q}} \equiv \hat{\boldsymbol{c}} \times \hat{\boldsymbol{P}}, \tag{E.43}$$

which defines the position vector in the orbit-coordinate system as:

$$\boldsymbol{r} = r\cos\phi\,\hat{\boldsymbol{P}} + r\sin\phi\,\hat{\boldsymbol{Q}}, \tag{E.44}$$

and the velocity $\dot{\boldsymbol{r}}$ as:

$$\dot{\boldsymbol{r}} = (\dot{r}\cos\phi - r\dot{\phi}\sin\phi)\,\hat{\boldsymbol{P}} + (\dot{r}\sin\phi + r\dot{\phi}\cos\phi)\,\hat{\boldsymbol{Q}} \tag{E.45}$$

The relation between the true anomaly ϕ and eccentric anomaly E is given by Bond and Allman (1996):

$$\cos\phi = \frac{\cos E - e}{1 - e\cos E} \tag{E.46}$$
$$\sin\phi = \frac{\sqrt{1-e^2}\,\sin E}{1 - e\cos E};$$

the inverse relations are:

$$\cos E = \frac{\cos\phi + e}{1 + e\cos\phi}, \tag{E.47}$$
$$\sin E = \frac{\sqrt{1-e^2}\,\sin\phi}{1 + e\cos\phi}.$$

Using Eq. E.40 the following relations are obtained:

$$\boldsymbol{r} = a\big[(\cos E - e)\,\hat{\boldsymbol{P}} + \sqrt{1-e^2}\sin E\,\hat{\boldsymbol{Q}}\big], \tag{E.48}$$
$$\dot{\boldsymbol{r}} = \frac{\sqrt{\mu a}}{r}\big[-\sin E\,\hat{\boldsymbol{P}} + \sqrt{1-e^2}\cos E\,\hat{\boldsymbol{Q}}\big].$$

The position vector can also be written as:

$$\boldsymbol{r} = a\begin{bmatrix}(\cos E - e)\\ \sqrt{1-e^2}\sin E\\ 0\end{bmatrix}, \tag{E.49}$$

and the velocity vector as:

$$\dot{\boldsymbol{r}} = \frac{\sqrt{\mu a}}{r}\begin{bmatrix}-\sin E\\ \sqrt{1-e^2}\cos E\\ 0\end{bmatrix}. \tag{E.50}$$

As verification we can see that the first integrals of motion (Eq. E.7) are verified:

$$\boldsymbol{r}\times\dot{\boldsymbol{r}} = \sqrt{\mu a}\,\sqrt{1-e^2}\,\frac{a\,(1-e\cos E)}{r}\,\hat{\boldsymbol{c}} = \sqrt{\mu a}\,\sqrt{1-e^2}\,\hat{\boldsymbol{c}} = \boldsymbol{c}. \tag{E.51}$$

The orientation of the $(\hat{\boldsymbol{P}}, \hat{\boldsymbol{Q}}, \hat{\boldsymbol{c}})$ reference system with respect to a set of observational coordinates (ξ, ν, τ) is defined by three angles:

- The ascending node Ω, which defines the line along which the orbital plane is tilted with respect to the observational plane (the node vector);
- The inclination i, which defines the tilt of the orbital plane with respect to the observation plane;
- The argument of perigee ω, which defines the orientation of the orbit in the orbital plane as measured from the node vector.

Thus are defined the three rotations to be applied to obtain the orbital plane with respect to the observational plane. The latter will generally be the plane perpendicular to the line of sight. Using the definitions in Eq. A.1, the rotation matrix to be applied to the observed coordinates to represent them in the obit coordinates is given by:

$$\mathsf{A} = \mathsf{R}_3(\Omega)\mathsf{R}_1(i)\mathsf{R}_3(\omega), \tag{E.52}$$

from which:

$$\mathsf{A} = \begin{bmatrix} \mathrm{c}\Omega\,\mathrm{c}\omega - \mathrm{s}\Omega\,\mathrm{s}\omega\,\mathrm{c}i & -\mathrm{c}\Omega\,\mathrm{s}\omega - \mathrm{s}\Omega\,\mathrm{c}\omega\,\mathrm{c}i & -\mathrm{s}\Omega\,\mathrm{s}i \\ \mathrm{s}\Omega\,\mathrm{c}\omega + \mathrm{c}\Omega\,\mathrm{s}\omega\,\mathrm{c}i & -\mathrm{s}\Omega\,\mathrm{s}\omega + \mathrm{c}\Omega\,\mathrm{c}\omega\,\mathrm{c}i & \mathrm{c}\Omega\,\mathrm{s}i \\ \mathrm{s}\omega\,\mathrm{s}i & -\mathrm{c}\omega\,\mathrm{s}i & \mathrm{c}i \end{bmatrix}, \tag{E.53}$$

where $\mathrm{c}\Omega = \cos\Omega$, $\mathrm{s}\Omega = \sin\Omega$ etc. The matrix A is a fixed quantity for each binary-star system.

Applying matrix A to $\boldsymbol{r}$ as defined in Eq. E.49 gives the following relations for the displacements in the observed coordinates ξ and ν:

$$\begin{aligned} \Delta\xi &= B\,X(t) + G\,Y(t), \\ \Delta\nu &= A\,X(t) + F\,Y(t), \end{aligned} \tag{E.54}$$

where:

$$\begin{aligned} X(t) &= \cos E - e, \\ Y(t) &= \sqrt{1-e^2}\sin E, \end{aligned} \tag{E.55}$$

and:

$$\begin{aligned} A &= a\,(\ \ \cos\Omega\cos\omega - \sin\Omega\sin\omega\cos i), \\ B &= a\,(\ \ \sin\Omega\cos\omega + \cos\Omega\sin\omega\cos i), \\ F &= a\,(-\cos\Omega\sin\omega - \sin\Omega\cos\omega\cos i), \\ G &= a\,(-\sin\Omega\sin\omega + \cos\Omega\cos\omega\cos i), \end{aligned} \tag{E.56}$$

are derived from the relevant elements of array A, and are referred to as the Thiele-Innes elements, which can be used to calculate predicted positions from orbital elements.

Similarly, the observed radial-velocity variation is given by:

$$V_R = -\frac{\sqrt{\mu a/\varpi}}{r(E)}\sin i\left[\sin E\sin\omega + \sqrt{1-e^2}\cos E\cos\omega\right], \tag{E.57}$$

from which it is clear that the radial velocity is, as would be expected, unable to determine the rotation Ω around the line-of-sight. The extra factor $1/\varpi$ (the inverse of the parallax of the system in the same units as a) has been introduced to transform the semi-major axis a into AU. The radial velocities scale linearly with $\sin i$, which therefore remains undetermined when measuring radial velocities only. The extremes in the radial velocity are found at $\partial V_R/\partial E = 0$. Applied to Eq. E.57 gives:

$$(\cos E - e)\sin\omega - \sqrt{1-e^2}\sin E\cos\omega = 0. \tag{E.58}$$

Using Eq. E.46 for the relations between E and ϕ shows that the extremes are found at $\sin(\omega - \phi) = 0$, or $\phi = (\omega, \omega + \pi)$. This can be substituted in Eqs. E.40 and E.47 to give:

$$\begin{aligned} \cos E &= \frac{\pm\cos\omega + e}{1 \pm e\cos\omega}, \\ \sin E &= \frac{\pm\sqrt{1-e^2}\sin\omega}{1 \pm e\cos\omega}, \\ r &= \frac{a}{\varpi}\,\frac{1-e^2}{1 \pm e\cos\omega}. \end{aligned} \tag{E.59}$$

Substituting this in Eq. E.57:

$$V_R(\max, \min) = -\sqrt{\mu\varpi/a}\, \sin i\, \frac{\pm 1 + e\cos\omega}{\sqrt{1-e^2}}. \tag{E.60}$$

The amplitude for the radial-velocity variations, usually indicated by K, is:

$$K = |V_R(\max) - V_R(\min)|/2 = \sqrt{\mu\varpi/a}\, \sin i\, \frac{1}{\sqrt{1-e^2}}, \tag{E.61}$$

or, using Eq. E.42 for the orbital period T:

$$K = 2\pi\, \frac{a\, \sin i}{\varpi T\, \sqrt{1-e^2}}, \tag{E.62}$$

stating that the amplitude of the radial velocity variations is independent of the argument of perigee ω, which does need to be used, however, to calculate the actual radial velocity of the system. For conversion, 1 AU yr^{-1} = 4.74047 km s^{-1}.

The equations presented above for the astrometric measurements provide the orbital description of the primary with respect to the secondary star. Although this is generally the case for ground-based differential data, it often does not apply to the *Hipparcos* data. For the *Hipparcos* data we can distinguish three different situations:

1 Long-period binaries ($P \gtrsim 25$ Yr) with good on-ground determined orbital parameters, but only one component visible for *Hipparcos*; the *Hipparcos* data on its own cannot be used to examine the orbit; the reference-epoch position is that of the primary if no orbital corrections are applied, and positions can be obtained for both components when these corrections are applied; for large separation binaries the position on the grid may not always have been known to sufficient accuracy, leaving in the reference grid line associated with an observation and the possibility of a grid-step ambiguities in the derived position;

2 Short-period binaries ($P \lesssim 25$ Yr) with good, ground-based orbit determinations and a single or both components visible for *Hipparcos*; the *Hipparcos* data contain (some) information on the orbit, which may allow for a determination of the astrometry of the centre of mass of the system and the determination of the individual masses of the components through incorporating scaling of the semi-major axis a in the astrometric solution;

3 Short-period binaries with good orbit determinations, but for which the components are too close for *Hipparcos* to separate; the orbit as observed by *Hipparcos* is that of the photo centre of the system within the positional reference frame, and is obtained by appropriately scaling the semi-major axis a in the solution for the astrometric parameters.

Incorporating *Hipparcos* observations with ground-based observations in orbit determinations is thus usually not a straightforward procedure, with the character of the *Hipparcos* data depending on orbital period, separation and magnitude difference.

Appendix F
Reference orbital parameters

Since the publication of the *Hipparcos* data, parameters describing orbital systems have in many cases been re-determined using additional ground-based data, while at the same time new orbital systems have been established (some after indications from the *Hipparcos* data). The new reduction has tried to take the new orbital parameters into account. For this purpose a list has been compiled of the most recent determinations of parameters for orbital systems in the *Hipparcos* catalogue. In the compilation of this list, given here in Table F.1, the most recent determination was always assumed superior, primarily because in nearly all cases the latest solutions incorporated the most recent additional data. The addition of new orbital determinations removed some stars from the lists of accelerated and stochastic solutions. Marginal detections of brown dwarfs and planetary companions have not been included.

The main inputs used for the current compilations are:

- Table DMCA_O in ESA (1997), providing a compilation of orbital parameters for systems that are well separated in magnitude and/or position, and for which the observation could generally be treated as direct measurements of transits;
- The studies by the Nice group (Martin et al. 1997; Martin and Mignard 1998; Martin et al. 1998) on primarily short period double stars and the influences on the measured transit times caused by the proximity on the sky and in magnitude of the components of the system;
- A similar study by Soederhjelm (1999), more focussed on incorporating new external data, and covering a wider range of systems;
- Two studies on detection of astrometric orbits for spectroscopic binaries (Pourbaix and Jorissen 2000; Pourbaix and Boffin 2003);
- Two papers on newly discovered systems by Balega et al. (2005, 2006);
- A summary of the CHARA project, a systematic Speckle survey that took place over some 20 years (Mason et al. 1999; Hartkopf et al. 2000);
- Some results from the WIYN telescope (Horch et al. 2004).

Magnitude differences have been more difficult to compile than orbital elements. In most cases these have been derived from the DMSA_C table in ESA (1997), giving the details of the resolved binary systems. For stars in the DMSA_O table the magnitude differences were assumed to be large enough to be ignored.

Differences between different studies often reflect a 180 degrees ambiguity in determining some of the projection angles, and are for that same reason irrelevant for the application in the astrometric solutions. In other cases there is a factor-two ambiguity on the orbital period as based on the *Hipparcos* data alone. These cases have been kept in the Table to allow checks with the new reduction results.

Columns 11, 12 and 13 in Table F.1 provide some further reference. The number in column 11 refers to the paper from which the data have been extracted; from 0 to 8 these are: ESA (1997) (0); Martin et al. (1998) Table 3 (1) and Table 4 (2); Soederhjelm (1999) Table 1 (3), Table 3(4) and Table 5 (5); Pourbaix and Jorissen (2000), Pourbaix and Boffin (2003) (6); Balega et al. (2005, 2006) (7); Mason et al. (1999) (8); Hartkopf et al. (2000) (9). In column 12, 0 refers to the basic solution, 1 to a possible alternative. In column 13 the solution types used in the published data are given, 5, G, X, C, O refer to standard 5-parameter solution, an accelerated solution, a stochastic solution, a resolved double star, and an orbital solution respectively.

Table F.1. Orbital parameters used in the new reduction

HIP	HD	dmag	a	e	i	Ω	ω	T	P	r	s	H
			mas			degrees		Years				
2	224690		14.3	0.000	118.1	77.3	0.0	1991.380	1.368	0	0	O
171	224930	3.12	830.0	0.380	49.0	290.0	96.0	1989.400	26.280	4	0	X
443	28		4.6	0.272	36.0	66.3	337.7	1991.167	0.200	0	0	O
518	123	0.94	1440.0	0.450	45.0	221.0	277.0	1943.100	106.700	4	0	C
677	358		6.5	0.527	105.7	104.2	77.3	1988.580	0.265	0	0	O
999	0		18.4	0.000	81.3	118.0	0.0	1991.608	3.426	0	0	O
1242	0	3.00	350.0	0.080	133.0	96.0	45.0	1991.000	4.800	5	0	X
1349	1273		19.9	0.567	80.5	352.6	4.7	1990.965	1.126	0	0	O
1674	1624	0.37	1000.0	0.890	88.0	87.0	202.0	1941.000	161.000	5	0	C
2170	2343		6.0	0.000	47.0	300.0	0.0	1975.320	2.563	6	0	C
2237	2475	0.08	146.0	0.660	64.0	134.0	134.0	1994.100	5.650	3	0	C
2237	2475	0.07	211.0	0.060	72.0	119.0	45.0	1993.000	11.330	3	1	C
2552	0	3.25	450.0	0.000	27.0	24.0	0.0	1989.000	15.400	5	0	C
2762	3196	1.27	230.0	0.760	47.0	149.0	283.0	1994.050	6.890	3	0	C
2912	3369		1.8	0.542	103.0	94.7	170.7	1989.519	0.393	0	0	O
2941	3443	0.32	669.0	0.210	78.0	111.0	143.0	1999.200	25.040	4	0	C
3504	4180		7.2	0.132	100.8	89.0	99.7	1991.450	2.828	0	0	O
4809	6009	0.20	128.0	0.393	58.0	57.1	106.9	1998.620	16.410	7	0	C
4849	6101	1.70	465.0	0.678	144.0	92.6	205.4	2002.680	28.990	7	0	C
5249	6586	0.00	131.0	0.232	23.9	104.1	97.1	1997.490	33.720	9	0	C
5300	6767	1.35	250.0	0.460	66.0	143.0	329.0	1979.000	29.000	5	0	C
5336	6582		188.5	0.620	110.0	227.3	157.2	1975.797	21.399	0	0	O
5531	6840	0.80	87.0	0.720	54.0	151.6	219.1	1995.710	7.300	7	0	C
5685	7275		15.1	0.000	116.7	113.3	0.0	1990.369	2.281	0	0	O
5778	7374		2.4	0.310	152.8	313.4	201.0	1911.618	2.193	0	0	O
5842	7693	0.55	1140.0	0.040	35.0	142.0	141.0	1919.000	85.200	4	0	C
6486	8272	0.21	830.0	0.930	99.0	137.0	130.0	1984.800	222.000	5	0	5
6564	8556	0.46	199.0	0.930	117.0	29.0	348.0	1988.840	16.110	5	0	C
6867	9053		4.9	0.000	50.3	25.3	0.0	1991.514	0.531	0	0	O
7078	9021		6.2	0.310	88.0	160.2	188.2	1938.274	0.367	0	0	O
7213	9525		8.1	0.516	25.1	49.3	96.6	1990.454	1.674	0	0	O
7372	9770	0.18	172.0	0.330	22.0	158.0	307.0	1991.860	4.560	3	0	C
7580	10009	1.03	312.0	0.780	96.0	160.0	250.0	1989.800	29.000	3	0	G
7918	10307	6.00	580.0	0.430	105.0	33.0	22.0	1997.100	19.500	5	0	G
7981	10476		4.6	0.000	89.0	12.7	0.0	1990.878	0.567	0	0	O
8514	11262		11.9	0.000	106.3	79.0	0.0	1991.391	3.073	0	0	O
8833	11559		13.1	0.180	65.8	270.1	71.0	1961.960	4.579	0	0	O
8882	11753		5.9	0.000	93.5	161.9	0.0	1991.906	2.403	0	0	O
8903	11636	3.18	36.0	0.903	44.7	79.1	209.1	1980.098	0.290	1	0	O
8922	11613		8.1	0.000	24.0	155.0	0.0	1978.029	2.294	6	0	X
9236	12311		21.7	0.643	42.5	85.1	234.0	1991.897	1.659	0	0	O
9480	12111	2.13	628.0	0.386	19.4	188.4	244.0	1965.800	60.550	8	0	C
9500	12376	0.27	150.0	0.404	67.0	191.4	295.1	1989.060	12.940	9	0	C

Continued on next page

Table F.1 – continued from previous page

HIP	HD	dmag	a	e	i	Ω	ω	T	P	r	s	H
			mas			degrees			years			
9727	12230		39.3	0.000	148.3	64.6	0.0	1990.208	4.425	0	0	O
10064	13161		-1.6	0.440	129.9	64.9	298.1	1989.551	0.086	0	0	O
10324	13611		-4.5	0.000	106.5	71.2	0.0	1954.661	4.496	0	0	O
10340	13520		3.0	0.340	127.0	6.0	358.0	1962.602	2.048	6	0	5
10366	13530		16.8	0.750	115.6	168.3	270.0	1988.117	4.517	0	0	O
10403	13594	0.65	896.0	0.270	63.0	99.0	319.0	1898.000	144.000	5	0	C
10438	13475	0.00	78.0	0.267	29.7	203.9	291.0	1989.760	32.800	9	0	5
10514	13738		5.4	0.064	74.1	316.4	63.0	1973.814	3.792	0	0	O
10535	13872	0.08	234.0	0.680	104.0	236.0	84.0	1986.220	23.700	5	0	C
10542	14001	0.75	2000.0	0.140	41.0	170.0	19.0	1803.000	225.000	5	0	C
10644	13974		2.8	0.020	167.0	15.0	301.0	1990.613	0.027	0	0	O
10723	14214		5.6	0.445	118.3	176.2	100.9	1991.109	0.256	0	0	O
11231	15064		5.6	0.295	24.9	234.3	192.8	1991.435	0.390	0	0	O
11352	15013	0.20	100.0	0.284	50.0	15.1	4.4	1995.120	6.850	7	0	C
11452	15285	0.16	580.0	0.220	74.0	109.0	223.0	1987.700	25.300	4	0	C
11569	15089	4.03	700.0	0.300	106.0	175.0	156.0	1980.000	52.000	5	0	C
11840	15755		6.3	0.000	85.2	17.8	0.0	1989.686	1.653	0	0	O
12153	16234	0.20	119.0	0.884	101.0	171.3	256.9	1986.830	1.920	1	0	5
12390	16620	0.76	107.0	0.230	24.0	86.0	46.0	1991.120	2.650	3	0	C
12421	16619	1.30	137.0	0.469	40.5	62.8	348.2	1973.760	17.940	9	0	5
12623	16739	0.30	53.0	0.656	126.8	230.6	89.5	1995.148	0.906	2	0	O
12709	16909		31.2	0.521	76.5	27.4	117.0	1979.849	3.324	0	0	O
12717	17326	0.37	290.0	0.850	94.0	36.0	15.0	1999.000	36.000	5	0	C
12719	16908		6.5	0.140	155.0	66.5	320.0	1991.884	1.342	0	0	O
13055	16458		10.8	0.099	87.1	280.5	113.1	1985.758	5.525	6	0	G
13531	17879		8.8	0.734	95.0	101.0	234.6	1901.842	4.149	0	0	O
14075	18774	0.00	111.0	0.474	50.0	165.5	184.9	1998.320	13.890	7	0	C
14230	18940	1.70	110.0	0.876	68.0	21.4	144.1	1999.310	5.910	7	0	C
14328	18925		21.5	0.720	88.0	242.6	353.2	1947.208	14.648	0	0	O
14576	19356		19.0	0.225	84.0	132.3	130.3	1987.367	1.862	0	0	O
14669	0	1.70	569.0	0.138	96.0	13.2	165.6	2009.160	28.310	7	0	C
14879	20010	3.04	4000.0	0.730	81.0	117.0	43.0	1947.000	269.000	5	0	C
14913	20121	0.71	410.0	0.900	165.0	110.0	118.0	1977.500	45.200	5	0	C
15134	0	0.00	442.0	0.975	126.0	227.0	286.0	1991.300	100.500	9	0	5
15799	21175	4.14	1800.0	0.200	32.0	49.0	348.0	1982.000	111.000	5	0	X
16134	21531		19.4	0.324	35.0	115.1	235.1	1991.233	1.378	0	0	O
16369	21754		4.3	0.397	36.9	207.6	326.3	1989.051	2.628	0	0	O
16602	21794	0.24	120.0	0.350	45.0	236.0	235.0	1990.140	14.000	5	0	5
16628	22262	0.95	230.0	0.360	87.0	141.0	8.0	1997.000	19.200	5	0	C
16900	22418		8.8	0.000	135.1	14.3	0.0	1990.757	1.963	0	0	O
17138	22921		3.5	0.000	138.0	115.9	0.0	1991.836	1.934	0	0	O
17296	22649		4.5	0.088	114.0	217.0	361.0	1976.040	1.632	6	0	O
17336	23052		32.1	0.000	78.3	79.5	0.0	1990.359	2.740	0	0	O
17440	23817		27.1	0.210	82.9	22.4	13.8	1992.681	5.233	0	0	O
17694	23610		5.9	0.000	54.8	137.6	0.0	1991.756	2.684	0	0	O
17846	23728		9.6	0.000	82.0	161.8	0.0	1991.286	3.609	0	0	O
17847	23850		4.2	0.000	106.7	111.5	0.0	1991.129	0.796	0	0	O
17932	23838		7.4	0.724	93.3	238.8	108.8	1974.654	2.636	0	0	O
17954	23985	0.80	420.0	0.620	84.0	25.0	353.0	2000.000	62.000	5	0	C
19209	0	1.40	223.0	0.686	122.0	38.2	255.6	1996.770	21.330	7	0	G
19719	26690	0.91	134.0	0.330	65.0	144.0	305.0	1990.680	7.200	3	0	O
19758	27019	1.28	250.0	0.790	57.0	84.0	252.0	1987.100	18.200	5	0	C
19758	27019	1.29	290.0	0.000	58.0	132.0	180.0	1975.900	36.100	5	1	C
20070	26961		5.1	0.235	83.5	115.3	263.0	1968.782	1.921	0	0	O
20087	27176	2.03	133.0	0.170	125.0	352.0	344.0	1989.200	11.320	3	0	O
20215	27383	1.31	590.0	0.600	53.0	62.0	134.0	1987.000	89.700	5	0	C
20347	27710	0.25	490.0	0.610	36.0	359.0	106.0	1924.000	80.000	5	0	C
20482	284414		7.6	0.638	26.8	11.0	303.1	1978.007	1.617	0	0	O
20661	27991	0.15	105.0	0.730	122.0	215.0	271.0	1988.700	6.280	5	0	C
20686	27989	0.85	259.0	0.040	97.0	23.0	235.0	1996.000	27.400	5	0	C
20885	28307	3.04	219.0	0.570	92.0	355.0	250.0	1998.500	16.260	5	0	5
20916	28363	0.70	390.0	0.330	92.0	78.0	311.0	1960.000	40.700	5	0	C
20935	28394		10.8	0.242	19.9	306.7	127.0	1977.420	0.654	0	0	O
21123	28634		13.4	0.148	52.8	212.4	325.0	1980.473	2.312	0	0	O

Continued on next page

Table F.1 – continued from previous page

HIP	HD	dmag	a	e	i	Ω	ω	T	P	r	s	H
			mas			degrees		years				
21273	28910		2.4	0.000	96.2	60.8	0.0	1991.321	0.288	0	0	O
21280	285931	1.29	230.0	0.660	70.0	80.0	88.0	1989.900	13.000	5	0	C
21281	29305	0.94	180.0	0.800	31.0	140.0	193.0	1986.000	12.100	5	0	C
21402	29140	1.77	220.0	0.034	67.3	147.2	293.0	1996.000	16.250	9	0	C
21594	29503	2.76	750.0	0.550	67.0	171.0	24.0	1975.000	77.000	5	0	C
21698	29803	0.97	290.0	0.110	77.0	80.0	308.0	1999.000	32.000	5	0	C
22500	30997		12.8	0.000	126.0	158.8	0.0	1991.455	3.133	0	0	O
22505	30712	0.95	96.0	0.000	71.0	132.0	0.0	1994.300	7.500	5	0	5
22550	30810	0.24	188.0	0.460	17.0	147.0	257.0	1988.390	16.280	3	0	C
22607	30869	2.00	710.0	0.590	48.0	143.0	310.0	1982.000	95.200	5	0	C
23166	31925	1.52	440.0	0.880	118.0	141.0	0.0	1979.000	55.000	5	0	C
23395	32092	1.94	260.0	0.325	21.0	106.0	96.0	1949.690	25.140	8	0	C
23416	31964		22.4	0.070	87.0	264.0	0.0	1950.247	27.069	0	0	O
23452	32450	2.14	1800.0	0.890	76.0	63.0	94.0	1996.000	44.000	5	0	C
23453	32069		4.2	0.406	99.3	145.4	336.0	1953.566	2.662	0	0	O
23662	32662		6.9	0.000	82.7	162.7	0.0	1990.911	1.828	0	0	O
23786	32850		10.1	0.140	29.4	153.0	293.1	1991.510	0.560	0	0	O
23922	237354		8.5	0.000	68.8	132.7	0.0	1991.775	1.840	0	0	O
24526	34540		19.7	0.109	50.3	154.9	221.0	1990.069	1.876	0	0	O
24608	34029		2.2	0.000	137.2	220.8	0.0	1989.001	0.285	0	0	O
24727	34334		7.2	0.100	52.9	56.2	40.0	1991.245	1.190	0	0	O
25092	35155		2.0	0.071	115.0	297.0	306.0	1990.545	1.754	6	0	5
25119	35112	2.79	1100.0	0.130	114.0	164.0	153.0	1950.000	93.000	5	0	C
26001	37297		4.6	0.509	44.6	104.4	332.9	1991.559	0.495	0	0	O
26563	37507		1.8	0.549	87.4	326.9	200.7	1990.791	1.220	0	0	O
26926	38089	0.60	210.0	0.400	72.0	95.0	273.0	1999.000	20.300	5	0	C
28311	40705		8.4	0.000	136.4	11.7	0.0	1990.674	2.753	0	0	O
28360	40183		-0.5	0.000	76.0	115.4	0.0	1988.755	0.011	0	0	O
28442	40887	0.96	900.0	0.450	103.0	125.0	279.0	1998.000	68.000	5	0	X
28614	40932	1.93	270.0	0.750	96.0	25.0	217.0	1985.100	18.500	5	0	C
28734	41116	0.45	208.0	0.340	62.0	178.0	190.0	1995.400	13.200	5	0	C
29234	42443	0.02	207.0	0.330	46.0	265.0	299.0	1997.800	18.200	5	0	C
29982	43821		12.2	0.437	110.9	264.8	38.6	1981.686	3.628	0	0	O
30060	43378		8.5	0.298	59.8	52.5	298.3	1990.732	2.245	0	0	O
30277	44762		9.6	0.695	115.1	282.1	117.1	1991.991	2.379	0	0	O
30501	44780		2.9	0.240	120.7	246.1	85.8	1978.317	1.581	0	0	O
30920	0	2.68	1100.0	0.380	54.0	30.0	229.0	1983.500	16.500	5	0	G
30953	46273	0.05	510.0	0.230	33.0	112.0	17.0	1905.000	101.000	5	0	C
31150	47121		7.2	0.000	80.2	59.5	0.0	1990.452	1.962	0	0	O
31205	46407		5.2	0.013	85.0	297.0	118.0	1989.409	1.252	6	0	O
31509	47230	1.27	300.0	0.470	100.0	156.0	286.0	1993.100	28.900	5	0	C
32104	48097		8.8	0.000	65.8	98.8	0.0	1990.950	1.321	0	0	O
32349	48915		2490.4	0.592	136.5	44.9	327.3	1894.129	50.090	0	0	O
32578	49293		6.6	0.400	96.3	47.6	172.0	1973.708	4.821	0	0	O
32761	50337		2.2	0.000	101.3	141.2	0.0	1989.881	0.535	0	0	O
32768	50310		7.9	0.088	79.9	194.4	64.0	1991.052	2.919	0	0	O
32894	50264		12.0	0.098	109.0	69.2	225.1	1986.982	2.498	6	0	X
33142	0	1.44	400.0	0.300	109.0	101.0	73.0	1996.000	12.000	5	0	5
33449	50522	0.91	1000.0	0.660	79.0	42.0	100.0	1993.000	190.000	5	0	5
33451	51825	0.42	210.0	0.410	29.0	126.0	240.0	1992.300	16.800	3	0	C
34047	51708		13.4	0.000	126.8	103.4	0.0	1991.538	2.850	0	0	O
34608	54563		4.2	0.400	89.2	77.7	103.4	1980.778	0.310	0	0	O
34860	55130	0.08	800.0	0.930	141.0	19.0	244.0	1920.700	119.000	5	0	C
35550	56986		14.3	0.353	92.4	70.0	214.6	1901.220	6.129	0	0	O
36042	58368		4.9	0.221	101.0	295.8	75.2	1983.768	1.842	6	0	5
36238	58728	1.87	87.0	0.480	97.0	172.0	51.0	1990.400	2.030	3	0	C
36377	59717		8.2	0.170	68.1	332.0	349.3	1991.630	0.706	0	0	O
36832	61033		17.5	0.257	140.2	29.7	7.0	1991.269	0.761	0	0	O
36890	60766		10.0	0.360	132.7	38.3	236.6	1990.519	2.437	0	0	O
37279	61421		1179.0	0.365	31.9	284.8	88.8	1967.853	40.378	0	0	O
38052	62522	0.38	227.0	0.480	122.0	71.0	108.0	1990.720	18.700	3	0	C
38300	63799	0.00	179.0	0.603	45.9	91.9	294.8	1992.100	106.900	9	0	5
38382	64096	0.88	600.0	0.750	80.0	102.0	72.0	1985.800	23.300	5	0	C
38474	64235	1.04	330.0	0.660	77.0	185.0	122.0	1989.300	32.000	5	0	C

Continued on next page

Table F.1 – continued from previous page

HIP	HD	dmag	a	e	i	Ω	ω	T	P	r	s	H
			mas			degrees			years			
38538	64145		6.0	0.000	95.6	19.7	0.0	1989.802	1.593	0	0	O
40167	68257	0.35	862.0	0.320	167.0	13.0	187.0	1989.190	59.560	4	0	C
40239	0	0.58	870.0	0.680	51.0	142.0	167.0	1998.700	62.900	5	0	C
40326	69142		10.9	0.400	136.3	251.0	140.0	1991.823	2.546	0	0	O
41261	71386	0.64	290.0	0.250	36.0	130.0	231.0	1975.200	25.800	5	0	C
41426	71805	0.89	160.0	0.320	62.0	103.0	177.0	1997.000	14.600	5	0	C
41820	71974	1.62	410.0	0.710	120.0	143.0	243.0	1984.000	31.300	5	0	C
42075	72954	0.38	130.0	0.590	19.0	79.0	230.0	1996.000	17.300	5	0	C
42430	73752	1.46	1690.0	0.320	83.0	211.0	124.0	1986.000	127.000	4	0	C
42455	73900	1.22	500.0	0.800	105.0	58.0	355.0	1992.000	67.000	5	0	O
42805	74556		5.3	0.000	134.9	355.1	0.0	1990.744	2.235	0	0	O
43109	74874	0.84	257.0	0.650	50.0	108.0	265.0	1991.260	15.070	5	0	C
43671	76360	1.05	90.0	0.280	15.0	0.0	220.0	1993.400	7.250	5	0	O
44248	76943	2.34	644.0	0.150	131.0	204.0	33.0	1993.800	21.800	3	0	C
44471	77327	0.39	182.0	0.567	109.4	106.0	356.8	1996.880	35.590	9	0	C
44676	78549		19.0	0.362	40.5	123.3	5.7	1992.864	3.115	0	0	O
45075	78362		11.3	0.478	67.1	116.1	349.4	1929.297	2.909	0	0	O
45170	79096	0.19	116.0	0.430	124.0	317.0	170.0	1993.510	2.700	3	0	5
45527	79910		10.6	0.293	92.6	282.8	92.3	1992.477	2.524	0	0	O
45571	80671	0.39	90.0	0.500	141.0	125.0	102.0	1992.800	3.070	5	0	C
45617	79969	0.01	680.0	0.320	76.0	24.0	311.0	1981.000	34.170	4	0	5
46396	81919		8.0	0.571	116.4	123.7	107.9	1992.214	1.975	0	0	O
46404	81809	1.51	422.0	0.310	85.0	151.0	150.0	1975.100	32.000	5	0	C
46454	81858	0.65	840.0	0.560	64.0	326.0	302.0	1959.400	117.600	5	0	C
46651	82434	1.21	800.0	0.440	57.0	287.0	48.0	1970.000	34.200	4	0	C
46706	0	0.19	630.0	0.290	143.0	48.0	285.0	1983.900	18.400	5	0	C
46893	82674		4.8	0.149	135.0	5.0	261.0	1976.929	2.274	6	0	X
47479	84121	0.37	130.0	0.320	129.0	91.0	20.0	1995.100	10.600	3	0	C
48348	85563		17.0	0.350	49.2	5.6	272.5	1991.475	3.405	0	0	O
49166	87080		3.4	0.177	109.1	338.1	99.7	1991.313	0.749	6	0	5
49841	88284		16.8	0.138	79.5	249.6	238.9	1992.111	4.342	0	0	O
50805	89948		15.6	0.117	100.6	343.8	146.3	1987.330	1.828	6	0	O
51233	90537	1.32	300.0	0.690	79.0	42.0	24.0	1998.000	39.000	5	0	C
51384	89571		14.2	0.258	148.9	163.5	11.8	1990.645	2.204	0	0	O
51885	91881	1.10	870.0	0.750	128.0	147.0	40.0	1949.000	159.000	5	0	C
51986	92139	1.62	300.0	0.730	131.0	41.0	295.0	1986.500	16.600	5	0	C
52085	92214		5.9	0.100	146.7	300.7	270.0	1991.553	3.285	0	0	O
52271	92626		2.0	0.000	76.1	35.2	49.3	1993.435	2.514	6	0	5
52419	93030		3.4	0.000	25.8	147.3	0.0	1990.628	2.242	0	0	O
53240	94363		8.8	0.375	123.8	290.1	301.0	1976.333	3.192	0	0	O
53423	94672	3.05	800.0	0.670	121.0	42.0	143.0	1919.000	134.000	5	0	C
53763	0		4.9	0.000	68.7	254.2	103.5	1986.301	0.899	6	0	5
54061	95689	2.98	590.0	0.390	180.0	0.0	222.0	1958.000	44.500	4	0	C
54155	96064	0.00	350.0	0.080	65.0	70.0	105.0	1995.000	23.300	5	0	C
54204	96202	0.06	138.8	0.349	96.5	224.4	343.0	1983.455	7.553	8	0	O
54632	96511		2.0	0.282	122.0	254.0	333.0	1922.268	0.052	6	0	5
54677	97233		38.7	0.773	89.1	88.5	258.8	1990.370	2.572	0	0	O
54977	97961		11.3	0.459	39.5	15.6	123.6	1991.984	1.919	0	0	O
55016	97907	2.83	46.0	0.415	49.1	283.3	156.5	1974.015	8.110	1	0	G
55203	98230	0.52	2530.0	0.410	121.0	100.0	126.0	1995.040	59.950	4	0	5
55266	98353	0.33	93.0	0.120	57.0	121.0	68.0	1992.600	5.090	5	0	O
55425	98718	1.49	259.0	0.788	54.8	290.7	35.0	1935.530	38.710	8	0	C
55642	99028	2.75	1910.0	0.530	128.0	235.0	325.0	1948.800	186.000	5	0	C
56290	100203	1.56	790.0	0.400	46.0	80.0	132.0	1981.800	72.700	4	0	C
56528	0		4.3	0.000	53.5	345.8	0.0	1991.395	1.370	0	0	O
56675	101132		7.9	0.324	104.9	160.3	253.1	1991.456	0.611	0	0	O
56731	101013		7.9	0.195	77.6	347.8	308.8	1979.160	4.684	6	0	O
57363	102249		6.3	0.299	134.0	29.4	251.5	1991.101	1.239	0	0	O
57565	102509		0.8	0.000	50.1	138.0	0.0	1989.314	0.196	0	0	O
57791	102928		7.7	0.309	87.1	107.6	125.1	1974.831	1.333	0	0	O
57994	103246	1.05	410.0	0.500	39.0	22.0	247.0	1987.300	76.700	5	0	C
58113	103501		6.4	0.000	119.3	154.7	0.0	1992.896	2.750	0	0	O
58590	104321		3.5	0.265	62.7	149.3	312.0	1991.063	0.774	0	0	O
58799	104747	0.54	660.0	0.570	159.0	49.0	123.0	1913.000	110.000	5	0	C

Continued on next page

Table F.1 – continued from previous page

HIP	HD	dmag	a	e	i	Ω	ω	T	P	r	s	H
			mas			degrees		years				
59459	105982		4.7	0.297	63.2	20.2	238.0	1980.432	3.710	0	0	O
59468	105981		4.8	0.169	69.3	320.4	235.3	1990.998	1.267	0	0	O
59750	106516		29.0	0.081	73.9	13.3	52.7	1991.704	2.335	0	0	O
59780	106549	1.11	800.0	0.210	54.0	109.0	239.0	1984.000	63.000	5	0	C
59816	0	0.48	3500.0	0.890	26.0	2.0	275.0	1973.670	1100.000	5	0	C
59856	106760		6.6	0.426	120.1	77.9	303.6	1972.410	3.598	0	0	O
60129	107259	2.18	135.0	0.080	50.0	173.0	4.0	1990.200	13.100	3	0	G
60299	107574		9.9	0.081	123.3	264.1	124.1	1985.753	3.696	6	0	G
60994	108799	3.04	1400.0	0.710	24.0	94.0	75.0	1944.000	151.000	5	0	C
61724	110024		9.9	0.590	84.5	139.7	102.5	1977.435	2.662	0	0	O
61880	110314		21.5	0.249	122.9	77.5	36.2	1991.815	3.098	0	0	O
61932	110304	0.10	860.0	0.790	114.0	2.0	186.0	1931.200	83.000	5	0	C
61941	110379	0.05	3680.0	0.890	148.0	37.0	257.0	1836.400	168.900	4	0	C
62124	110743		22.1	0.253	49.1	87.2	205.5	1990.401	2.188	0	0	O
62145	110833		7.2	0.000	48.9	154.9	0.0	1991.391	0.740	0	0	O
62371	111096		8.2	0.000	88.5	147.8	0.0	1991.651	1.546	0	0	O
62409	0		6.7	0.140	94.2	165.1	168.2	1985.613	4.917	6	0	5
63406	112914		16.5	0.661	80.9	98.1	26.5	1991.505	2.017	0	0	O
63503	113139	3.21	1260.0	0.380	47.0	269.0	295.0	1920.000	105.000	4	0	C
63613	112985		11.7	0.492	120.1	59.3	316.4	1990.149	1.155	0	0	O
63742	113449		11.2	0.512	126.5	33.3	42.0	1991.305	0.633	0	0	O
64241	114378	0.02	678.0	0.530	90.0	12.0	99.0	1989.240	26.500	4	0	X
65026	115953	1.12	1507.4	0.229	94.4	91.1	73.6	1968.550	48.910	8	0	C
65135	116127		8.8	0.640	133.4	31.8	311.9	1990.911	1.492	0	0	O
65203	116114		4.9	0.000	48.3	173.1	0.0	1992.381	2.777	0	0	O
65417	116594		10.3	0.193	60.8	120.2	166.0	1983.441	3.742	0	0	O
65420	116568		25.2	0.663	68.5	69.3	285.6	1990.177	2.479	0	0	O
65783	117025		10.3	0.316	148.8	47.9	142.9	1992.561	2.356	0	0	O
66438	118261	0.40	450.0	0.780	117.0	71.0	94.0	1967.500	35.000	5	0	C
66458	118623	2.03	1020.0	0.800	147.0	87.0	159.0	1864.000	228.000	5	0	C
66640	118889	0.12	199.8	0.545	43.5	34.6	359.9	1929.850	22.460	8	0	C
67234	119834		6.4	0.134	62.4	197.3	58.6	1991.787	1.196	0	0	O
67422	120476	0.44	2440.0	0.450	47.0	155.0	199.0	1916.700	157.000	4	0	C
67483	120600		8.2	0.309	66.4	23.4	217.2	1991.779	1.269	0	0	O
67927	121370		35.4	0.257	116.6	74.5	326.3	1991.384	1.353	0	0	O
68682	122742	4.30	330.0	0.547	93.5	252.3	189.0	1951.960	9.900	1	0	O
68756	123299		1.5	0.400	131.8	241.1	23.2	1982.400	0.141	0	0	O
69112	124547		5.7	0.137	138.5	143.7	311.8	1991.984	1.659	0	0	O
69176	123585		4.8	0.062	53.9	8.5	203.5	1990.859	1.253	6	0	5
69283	124138		10.2	0.000	119.9	32.2	0.0	1991.577	3.212	0	0	O
69879	125351		3.7	0.574	88.2	345.4	224.9	1969.172	0.581	0	0	O
70327	126129	0.66	240.0	0.250	42.0	46.0	141.0	1998.000	40.000	5	0	C
70857	128642		13.1	0.148	54.8	80.4	158.6	1991.283	0.492	0	0	O
70973	127352	0.32	230.0	0.540	52.0	177.0	163.0	1994.200	20.600	5	0	C
71094	127726	0.47	200.0	0.160	158.0	23.3	57.0	1983.850	29.930	2	0	C
71141	127743		11.2	0.000	100.5	344.7	0.0	1990.247	3.106	0	0	O
71469	128429		21.3	0.000	127.6	88.7	0.0	1990.557	2.974	0	0	O
71510	128563	0.29	136.0	0.839	50.4	134.6	37.5	1994.520	21.500	9	0	5
71683	128620	1.25	17600.0	0.520	79.0	204.0	232.0	1955.610	79.850	4	0	C
71729	129132	1.00	73.0	0.040	106.4	78.0	241.9	1984.852	9.260	1	0	5
71914	0	0.23	640.0	0.090	120.0	178.0	204.0	1982.000	51.700	5	0	C
72217	129980	0.00	292.0	0.030	58.0	137.0	67.0	1993.000	25.800	5	0	C
72479	130669	0.02	122.0	0.510	40.0	143.0	338.0	1988.200	9.970	5	0	C
72659	131156	2.27	4940.0	0.510	139.0	347.0	203.0	1909.300	151.600	4	0	C
72848	131511		14.3	0.512	99.4	247.4	221.9	1974.012	0.343	0	0	O
73182	131976	1.77	140.0	0.760	109.0	14.0	308.0	1991.450	0.847	5	0	C
73199	132813		7.1	0.130	63.8	228.8	212.0	1990.681	2.050	0	0	O
73507	133388		6.8	0.000	126.1	13.1	0.0	1990.453	1.892	0	0	O
73695	133640	0.78	3800.0	0.550	84.0	57.0	45.0	2013.000	206.000	4	0	C
73787	133412		11.4	0.000	74.4	33.6	0.0	1991.249	1.428	0	0	O
74392	134759		129.4	0.247	153.6	177.4	343.9	1971.098	23.469	8	0	5
74893	136176	0.05	1220.0	0.650	58.0	62.0	52.0	1941.400	200.000	5	0	C
75312	137107	0.29	867.6	0.262	59.0	203.2	38.4	1933.721	41.585	8	0	C
75379	137052		8.4	0.680	49.2	145.1	339.5	1991.226	0.621	0	0	O

Continued on next page

Table F.1 – continued from previous page

HIP	HD	dmag	a	e	i	Ω	ω	T	P	r	s	H
			mas			degrees			years			
75389	137687		7.7	0.200	45.9	14.2	246.1	1991.223	1.093	0	0	O
75411	137391	0.37	102.0	0.280	131.0	134.0	51.0	1991.390	3.750	4	0	G
75415	137392	0.60	1470.0	0.580	134.0	174.0	336.0	1864.200	257.000	4	0	C
75508	137664		6.0	0.000	62.6	137.0	0.0	1992.345	2.068	0	0	O
75695	137909		56.4	0.538	111.1	327.2	178.1	1980.469	10.551	0	0	O
75695	137909	2.01	203.0	0.550	111.0	327.0	357.0	1990.980	10.550	3	0	O
75949	138439	0.50	285.0	0.603	23.2	123.0	305.0	1989.210	217.000	9	0	C
76031	138369		17.0	0.271	145.2	336.7	203.9	1990.898	1.687	0	0	O
76267	139006		1.8	0.370	88.2	330.4	311.0	1959.496	0.048	0	0	O
76382	139341	0.19	794.0	0.590	63.0	0.0	23.0	1993.990	55.600	4	0	C
76466	139905	0.23	410.0	0.630	121.0	130.0	232.0	1985.000	88.000	5	0	C
76734	139312		6.7	0.000	70.4	28.9	0.0	1989.015	2.447	0	0	O
76752	140139		7.3	0.000	110.2	121.1	0.0	1991.946	1.934	0	0	O
76852	140159	0.14	209.0	0.070	83.0	70.0	72.0	1984.800	21.900	5	0	C
76952	140436	1.56	750.0	0.510	94.0	111.0	103.0	1931.000	92.700	5	0	C
77541	141194		4.0	0.000	95.0	321.9	0.0	1991.064	0.890	0	0	O
77760	142373		1.0	0.000	131.7	51.7	0.0	1991.248	0.140	0	0	O
78459	143761		2.3	0.000	103.8	91.5	0.0	1991.308	0.214	0	0	O
78662	143474	0.26	320.0	0.520	159.0	114.0	33.0	1991.000	26.900	5	0	C
78727	144069		671.2	0.713	42.3	204.5	344.6	1951.694	45.648	8	0	5
78918	144294		7.3	0.000	76.9	4.6	0.0	1988.631	2.825	0	0	O
79101	145389		9.1	0.470	36.2	187.6	357.0	1991.475	1.535	0	0	O
80166	147508		2.6	0.371	123.0	19.0	230.0	1975.041	2.526	0	0	O
80346	0		51.9	0.580	147.4	289.3	254.7	1965.725	3.733	0	0	O
80725	148653	0.18	2210.0	0.750	108.0	94.0	130.0	1921.100	224.000	4	0	C
80816	148856		11.0	0.550	46.4	17.8	24.6	1991.368	1.124	0	0	O
81023	149162		9.9	0.282	109.5	320.6	17.6	1982.851	0.618	0	0	O
81126	149630	1.60	74.0	0.533	108.7	13.2	175.7	1982.560	7.480	1	0	5
81693	150680	2.66	1330.0	0.460	131.0	50.0	111.0	1967.700	34.450	4	0	C
81726	150710		13.6	0.095	83.0	95.8	101.1	1990.532	2.518	0	0	O
81754	150453		11.5	0.353	63.6	164.1	290.1	1991.342	1.353	0	0	O
82020	151613		39.3	0.720	112.4	249.0	99.0	1991.981	3.794	0	0	O
82817	152751	0.11	230.0	0.060	161.0	147.0	104.0	1991.600	1.710	3	0	C
82860	153597		6.7	0.210	62.4	224.4	339.0	1968.344	0.143	0	0	O
83575	154732		8.8	0.217	59.7	19.4	348.0	1987.025	2.165	0	0	O
83838	155103	0.18	110.0	0.550	120.0	130.0	55.0	1991.700	8.110	5	0	O
83895	155763	1.04	67.0	0.000	0.0	0.0	0.0	1980.760	6.090	1	0	G
84012	155125	0.50	1300.0	0.940	95.0	39.0	275.0	1936.800	88.000	5	0	C
84123	0	0.02	780.0	0.190	14.0	98.0	284.0	1989.000	34.000	5	0	5
84140	155876	0.42	770.0	0.750	146.0	158.0	97.0	1991.040	12.960	3	0	C
84179	156558		11.4	0.122	158.7	139.4	202.4	1990.657	2.418	0	0	O
84425	155826	1.57	280.0	0.500	123.0	4.0	349.0	1986.500	23.300	5	0	C
84709	156384	1.00	1810.0	0.580	128.0	313.0	247.0	1975.900	42.150	4	0	X
84924	156643		13.7	0.000	81.4	138.8	0.0	1989.560	3.937	0	0	O
84949	157482	0.78	79.0	0.690	55.0	318.0	47.0	1991.940	5.525	5	0	X
85019	157060		7.3	0.000	118.0	10.8	0.0	1991.552	0.547	0	0	O
85106	157821		10.0	0.000	47.3	117.2	0.0	1992.838	3.497	0	0	O
85141	157498	0.65	146.0	0.570	22.0	82.0	1.0	1990.600	15.000	5	0	G
85333	157979		5.5	0.445	91.5	252.4	222.3	1990.860	3.203	0	0	O
85582	0	0.40	650.0	0.190	81.0	66.0	64.0	1972.000	59.000	5	0	C
85667	158614	0.07	970.0	0.170	99.0	151.0	326.0	1962.400	46.300	4	0	C
85683	158806		14.6	0.000	152.5	156.0	0.0	1991.019	3.111	0	0	O
85727	158094		6.7	0.000	90.9	130.3	0.0	1989.893	2.856	0	0	O
85749	158837		3.3	0.203	76.3	10.3	301.4	1978.584	1.145	0	0	O
85846	159304	0.29	160.0	0.260	70.0	157.0	205.0	1992.000	23.300	5	0	C
86036	160269	3.36	1530.0	0.180	104.0	151.0	307.0	1947.000	76.100	4	0	C
86221	0	0.61	290.0	0.210	165.0	140.0	190.0	1985.300	24.000	5	0	C
86254	160181	0.00	143.0	0.856	99.7	259.6	80.7	1984.880	9.790	9	0	5
86373	160365		5.1	0.000	57.9	141.5	0.0	1991.477	1.321	0	0	O
86400	160346		13.8	0.302	44.7	72.1	33.9	1991.404	0.230	0	0	O
86722	161198	3.96	174.0	0.936	42.8	309.2	129.6	1994.189	7.000	1	0	G
87204	162338	0.05	179.0	0.620	110.0	135.0	81.0	1988.770	20.800	5	0	C
87428	162596		5.3	0.000	149.0	163.0	0.0	1975.193	1.279	6	0	5
87655	163151	1.80	88.0	0.320	157.0	169.0	233.0	1989.200	8.900	3	0	G

Continued on next page

Table F.1 – continued from previous page

HIP	HD	dmag	a	e	i	Ω	ω	T	P	r	s	H
			mas			degrees		years				
87861	163642		24.5	0.000	88.4	102.3	0.0	1989.051	2.705	0	0	O
87895	163840	2.34	83.0	0.400	68.0	177.0	317.0	1991.940	2.410	3	0	O
87899	163750		8.6	0.000	42.3	90.4	0.0	1988.998	2.744	0	0	O
88404	164765	0.67	1400.0	0.770	52.0	60.0	42.0	1829.000	257.000	5	0	C
88436	164967		9.3	0.000	95.8	168.8	0.0	1990.172	2.755	0	0	O
88601	165341	1.85	4560.0	0.500	120.0	301.0	13.0	1984.320	88.340	4	0	C
88637	165590	0.79	256.0	0.960	77.0	271.0	359.0	1998.300	19.940	5	0	C
88715	166067		6.3	0.000	120.6	118.1	0.0	1990.085	2.861	0	0	O
88728	165670		14.3	0.000	119.9	12.2	0.0	1988.935	2.859	0	0	O
88745	165908	3.71	1000.0	0.750	34.0	216.0	301.0	1998.000	56.400	5	0	C
88932	165893	0.32	160.0	0.520	55.0	69.0	60.0	1988.000	9.400	5	0	C
88964	166233	1.33	1190.0	0.610	103.0	71.0	307.0	1912.800	294.000	5	0	C
89507	167096		8.1	0.000	74.0	50.6	0.0	1989.686	1.812	0	0	O
89808	167954		4.0	0.428	85.8	188.3	134.4	1941.020	0.329	0	0	O
89937	170153	2.40	124.0	0.450	75.0	230.0	297.0	1990.970	0.760	3	0	O
90659	170737		9.4	0.501	140.1	355.3	56.0	1976.399	3.515	0	0	O
91009	234677		3.1	0.000	113.2	147.3	0.0	1991.338	0.312	0	0	O
91196	172569		9.4	0.050	96.4	15.6	166.7	1990.498	1.381	0	0	O
91394	172088	0.48	147.0	0.250	124.0	173.0	258.0	1995.000	12.130	5	0	C
91395	173084	0.69	650.0	0.600	88.0	129.0	353.0	1868.000	135.000	5	0	C
91751	172831		6.1	0.209	56.7	296.6	78.0	1976.407	1.329	0	0	O
92122	173950	0.72	220.0	0.560	131.0	75.0	285.0	1995.700	14.500	5	0	C
92175	173764		5.1	0.350	116.2	114.2	33.9	1920.425	2.283	0	0	O
92512	175306		3.1	0.114	96.6	34.4	274.3	1911.601	0.379	0	0	O
92677	176523		4.4	0.000	38.2	116.9	0.0	1990.934	0.794	0	0	O
92757	175292		7.1	0.000	110.7	132.8	0.0	1991.743	2.639	0	0	O
92818	175492		2.4	0.102	40.2	250.1	186.7	1989.701	0.672	0	0	O
93017	176051	2.61	1250.0	0.250	114.0	48.0	279.0	1971.800	61.000	4	0	C
93244	176411		13.7	0.272	87.5	59.7	82.0	1973.094	3.479	0	0	O
93383	176650		11.6	0.000	43.8	120.7	0.0	1989.673	3.696	0	0	O
93506	176687	0.20	489.2	0.204	110.8	74.3	7.0	1921.689	21.075	8	0	C
93574	175986	0.40	160.0	0.340	152.0	135.0	227.0	1990.100	14.100	5	0	C
93864	177716		8.9	0.000	62.5	128.1	0.0	1991.371	0.190	0	0	O
93995	178593		16.3	0.648	59.9	146.9	41.8	1992.033	2.456	0	0	O
94076	178911	1.45	78.0	0.624	142.2	107.0	270.0	1997.260	3.540	9	0	C
94252	179484	0.38	420.0	0.460	110.0	93.0	190.0	1957.000	63.000	5	0	C
94349	0		34.7	0.000	126.3	26.1	0.0	1990.608	2.284	0	0	O
94521	179799		6.7	0.000	50.3	27.7	0.0	1992.166	1.927	0	0	O
94739	179930	0.07	260.0	0.320	117.0	20.0	69.0	1995.300	7.700	5	0	C
95066	181391		7.7	0.833	71.4	132.8	152.6	1950.375	0.730	0	0	O
95477	182369		23.4	0.000	90.0	85.1	0.0	1958.296	10.801	0	0	O
95501	182640		55.7	0.360	150.0	337.0	191.0	1954.578	3.426	0	0	O
95769	183536		11.3	0.789	147.7	173.5	289.5	1990.360	2.260	0	0	O
95995	184467	0.31	83.0	0.380	142.0	66.0	355.0	1990.660	1.351	4	0	C
96302	184760	0.64	30.0	0.820	114.6	29.3	45.5	1985.560	4.560	1	0	5
96683	185734		1.6	0.561	78.4	70.4	210.5	1988.465	1.188	0	0	O
96807	185762	0.00	85.0	0.067	163.0	175.0	272.0	1993.600	21.700	9	0	G
97016	186158		11.6	0.199	142.2	128.2	271.9	1991.615	2.460	0	0	O
97222	186858	0.16	2070.0	0.770	156.0	91.0	128.0	1945.300	232.000	4	0	C
97237	0	0.67	2800.0	0.500	78.0	67.0	179.0	1961.000	228.000	5	0	5
97837	188307		10.4	0.432	142.0	188.4	345.8	1991.005	1.402	0	0	O
98001	188753	0.75	270.0	0.500	34.0	43.0	236.0	1988.200	25.700	5	0	C
98416	189340	1.50	236.0	0.020	54.0	111.0	326.0	1989.000	9.700	3	0	C
99312	191589		2.4	0.253	122.5	255.1	82.4	1992.603	1.033	6	0	5
99376	191854	0.54	460.0	0.480	115.0	142.0	339.0	1970.100	85.300	5	0	C
99473	191692		1.3	0.600	143.5	99.0	215.0	1989.749	0.047	0	0	O
99668	191889		11.3	0.000	105.9	119.2	0.0	1989.318	2.020	0	0	O
99675	192577		16.2	0.222	88.8	316.5	201.1	1960.641	10.361	0	0	O
99848	192910		5.3	0.301	64.2	226.6	218.2	1949.613	3.143	0	0	O
99965	193216		16.1	0.103	98.2	130.7	312.5	1991.463	1.115	0	0	O
100259	193554		14.9	0.326	36.3	97.9	146.6	1991.864	2.278	0	0	O
100345	193496		28.8	0.432	84.0	212.8	120.1	1917.800	3.762	0	0	O
101093	195725		14.3	0.030	102.5	60.6	83.7	1993.307	2.301	0	0	O
101235	195483	0.60	407.0	0.862	63.7	177.6	232.1	1998.450	272.000	9	0	C

Continued on next page

Table F.1 – continued from previous page

HIP	HD	dmag	a	e	i	Ω	ω	T	P	r	s	H
			mas			degrees			years			
101382	195987		5.2	0.306	89.5	327.7	356.8	1977.500	0.157	0	0	O
101750	197433	2.72	500.0	0.580	27.0	202.0	226.0	1998.000	32.000	5	0	C
101769	196524	0.90	440.0	0.360	61.0	177.0	349.0	1989.500	26.660	4	0	C
101955	196795	1.45	850.0	0.100	85.0	128.0	143.0	1967.000	39.000	5	0	C
101958	196867	2.55	158.0	0.470	160.0	129.0	99.0	1983.800	17.000	5	0	C
102782	0	0.20	880.0	0.260	123.0	31.0	296.0	1958.000	159.000	5	0	C
103519	199870		7.3	0.441	34.6	307.7	148.1	1966.161	1.739	0	0	O
103546	199939		1.4	0.237	45.8	185.1	132.9	1984.921	8.819	6	0	5
103655	0	2.06	700.0	0.550	54.0	67.0	207.0	1979.000	29.000	5	0	C
104019	200499	2.30	265.5	0.392	161.5	152.0	219.0	1974.460	28.240	8	0	C
104486	201626		7.5	0.000	64.8	104.4	57.7	1984.429	1.115	6	0	G
104788	202582	0.24	600.0	0.820	82.0	62.0	37.0	1921.000	79.000	5	0	C
104858	202275	0.08	231.0	0.440	99.0	203.0	3.0	1992.850	5.710	3	0	C
104887	202444	2.90	910.0	0.240	133.0	159.0	118.0	1989.000	49.600	4	0	C
105200	202908	1.48	520.0	0.870	99.0	253.0	169.0	1987.000	79.000	5	0	C
105431	203345	0.43	97.0	0.860	135.0	288.0	23.0	1991.900	6.030	5	0	O
105712	203244		31.8	0.287	110.3	70.7	11.9	1990.752	2.904	0	0	O
105881	204075		12.3	0.282	112.0	240.0	261.0	1984.806	6.511	6	0	5
105947	204236	1.50	168.0	0.354	57.0	149.1	129.7	1993.960	18.790	7	0	C
105969	204613		12.1	0.130	133.1	277.5	233.0	1988.866	2.404	6	0	X
106711	205835		8.6	0.535	102.1	18.9	306.6	1991.879	1.573	0	0	O
106811	0	3.63	1500.0	0.220	153.0	64.0	184.0	1945.000	68.000	5	0	C
106972	0	1.20	268.0	0.186	71.0	144.7	42.2	2005.230	18.570	7	0	C
106985	206088		12.9	0.000	68.1	63.2	0.0	1992.141	2.192	0	0	O
107354	206901	0.04	237.0	0.310	108.0	290.0	305.0	1990.810	11.590	3	0	C
107522	206804	0.13	205.0	0.590	129.0	103.0	347.0	1988.900	6.210	3	0	C
107522	206804	0.13	324.0	0.000	113.0	109.0	0.0	1986.000	12.460	3	1	C
107788	207652	1.17	368.0	0.240	70.0	231.0	248.0	1989.800	26.600	4	0	C
107818	207585		3.4	0.030	107.3	175.7	302.1	1988.428	1.836	6	0	5
108084	208174		5.3	0.000	35.6	67.4	0.0	1990.779	2.276	0	0	O
108431	208450	0.51	100.0	0.860	32.0	178.0	100.0	1996.300	6.110	3	0	C
108431	208450	0.68	160.0	0.060	71.0	94.0	258.0	1990.000	12.200	3	1	C
108478	208496		15.0	0.432	68.8	129.4	273.5	1991.414	2.028	0	0	O
109554	210647		17.1	0.904	88.0	300.1	237.7	1983.638	7.734	0	0	O
110102	211332		14.7	0.000	117.0	323.7	0.0	1988.502	3.598	0	0	O
110108	211594		5.3	0.058	85.1	128.2	172.3	1991.766	2.790	6	0	5
110130	211416		32.3	0.380	113.0	262.0	76.0	1989.917	11.493	0	0	O
110893	239960	1.62	2420.0	0.410	172.0	152.0	209.0	1970.300	44.640	4	0	C
111062	213235	1.55	720.0	0.500	89.0	116.0	214.0	1911.000	125.000	5	0	C
111170	213429		23.5	0.351	59.5	83.5	188.2	1991.496	1.757	0	0	O
111200	213464		7.8	0.000	95.7	338.3	0.0	1990.960	2.272	0	0	O
111314	213974	0.36	740.0	0.300	79.0	262.0	185.0	1973.000	224.000	5	0	C
111528	214222	0.70	140.0	0.370	60.0	111.0	145.0	1985.300	22.300	5	0	5
111685	0	2.00	330.0	0.256	55.0	69.0	118.7	1991.780	16.770	7	0	C
111805	214608	0.46	330.0	0.320	88.0	154.0	82.0	1979.800	29.900	5	0	C
111965	214810	1.03	230.0	0.740	85.0	128.0	206.0	1982.700	28.000	5	0	C
111965	214810	1.03	383.0	0.000	87.0	130.0	0.0	1994.900	54.000	5	1	C
111974	214850	0.81	287.0	0.740	141.0	253.0	24.0	1983.560	20.830	4	0	C
112158	215182		13.6	0.155	70.6	203.6	5.6	1990.260	2.240	0	0	O
112915	216448	0.85	1100.0	0.840	104.0	20.0	68.0	1932.000	107.000	5	0	C
113048	216608	1.78	600.0	0.590	38.0	26.0	13.0	1940.000	105.000	5	0	C
113445	217166	0.58	350.0	0.500	89.0	166.0	262.0	1978.100	26.000	4	0	5
113718	217580		11.1	0.000	58.9	292.7	0.0	1990.201	1.173	0	0	O
113860	217792		9.3	0.529	75.8	304.6	2.6	1955.576	0.488	0	0	O
113996	218060	0.30	203.0	0.390	47.0	204.0	82.0	1983.040	21.840	5	0	C
114222	218658	2.32	810.0	0.610	30.0	81.0	98.0	1934.000	168.000	5	0	C
114273	218527		10.5	0.000	80.8	62.7	0.0	1990.469	2.138	0	0	O
114421	218670		6.2	0.656	122.8	127.6	240.8	1902.997	1.121	0	0	O
114922	0	0.20	219.0	0.440	117.0	28.6	324.8	1999.810	19.720	7	0	C
115031	219678		7.2	0.000	142.2	285.0	0.0	1991.028	0.268	0	0	O
115126	219834	3.27	189.0	0.180	45.0	157.0	42.0	1993.500	6.300	3	0	X
116233	221531		11.1	0.165	23.0	312.0	201.9	1987.984	3.877	6	0	G
116436	221839	1.84	600.0	0.490	51.0	69.0	108.0	1968.000	78.000	5	0	C
116849	222516	0.04	148.0	0.290	50.0	105.0	112.0	1989.300	20.700	5	0	C

Continued on next page

Table F.1 – continued from previous page

HIP	HD	dmag	a	e	i	Ω	ω	T	P	r	s	H
			mas			degrees			years			
117570	223551	0.27	830.0	0.280	152.0	51.0	210.0	1978.000	120.000	5	0	C
117607	223617		3.4	0.061	82.3	173.3	154.6	1988.312	3.542	6	0	G
117666	223688	0.14	172.7	0.303	150.8	21.6	353.9	1960.240	30.300	8	0	C
118169	224464		8.3	0.303	155.0	172.9	66.2	1992.660	1.783	0	0	O
118266	224621		2.3	0.048	60.1	18.5	205.2	1993.975	0.843	6	0	5

Appendix G
The data disk

The enclosed DVD contains various data files, additional diagnostic figures and the colour-versions (with extended captions) of figures from the book for which colour is an important aspect.

G.1 The colour figures

For the figures in Table G.1 colour versions have been included on the DVD in the directory *Figures*. Files are presented in the PS (original) and PDF formats.

Table G.1. Summary of the colour figures available on the DVD. Column 1: File name; column 2: Chapter; column 3: Figure number; column 4: Brief description

Name	Ch.	Fig.	Description
Ch01F01	1	1.1	The Carlsberg Meridian Telescope
Ch01F08	1	1.8	Heliotropic angles
Ch01F12	1	1.12	Scan-phase discontinuities
Ch01F13	1	1.13	An external hit
Ch01F14	1	1.14	Basic angle stability
Ch02F01	2	2.1	Heliotropic reference frame
Ch02F02	2	2.2	Distribution of parallax factor with ecliptic latitude
Ch02F08	2	2.8	IDT: first-order modulation amplitude
Ch02F09	2	2.9	Normalized formal errors on abscissae
Ch02F10	2	2.10	Correlated errors in orbit 237
Ch02F11	2	2.11	Abscissa-error correlations
Ch02F12	2	2.12	Evolution of the mean scale correction
Ch02F13	2	2.13	Medium-scale distortion maps
Ch02F14	2	2.14	Small-scale grid distortions
Ch02F15	2	2.15	Distribution of orientation angles
Ch02F17	2	2.17	Error corrections with colour and ordinate
Ch02F18	2	2.18	Chromaticity corrections
Ch02F19	2	2.19	Parallax accuracies
Ch03F01	3	3.1	Modulation scale factor for phase errors

Continued on next page

Table G.1 – continued from previous page

Name	Ch.	Fig.	Description
Ch03F03	3	3.3	Distribution abscissa residuals
Ch03F04	3	3.4	Negative parallax distribution
Ch03F05	3	3.5	Parallax update correlations
Ch03F07	3	3.7	Parallax update dispersions
Ch03F09	3	3.9	The Lutz-Kelker effect
Ch03F10	3	3.10	Magnitude to parallax dispersion
Ch03F11	3	3.11	The HR diagram G8V and K0V stars
Ch03F12	3	3.12	Relative-parallax distribution G8V & K0V stars
Ch03F13	3	3.13	Parallax calibration NGC 752 through photometric boxes
Ch03F18	3	3.18	LMC and SMC proper motion diagrams
Ch03F19	3	3.19	Statistics for 7-parameter solutions
Ch03F20	3	3.20	The light curve of HIP 117054
Ch03F21	3	3.21	Cosmic dispersion for stochastic solutions
Ch04F03	4	4.3	AC and DC magnitudes for HIP 25
Ch04F04	4	4.4	Fitting double-star parameters
Ch04F05	4	4.5	Parameter variations for narrow double star
Ch04F06	4	4.6	Differences between differential double star parameters
Ch04F08	4	4.8	Formal errors as function of ρ for double-star parameters
Ch04F09	4	4.9	Formal errors as function of dMag for double-star parameters
Ch04F13	4	4.13	The quadruple system HIP 26220, 26221, 26224
Ch04F14	4	4.14	Periods and semi-major axes for orbital binaries
Ch05F01	5	5.1	Hyades radial velocities comparison
Ch05F02	5	5.2	Hyades proper motion dispersion
Ch05F04	5	5.4	The Hyades HR-diagram
Ch05F05	5	5.5	HR Diagram for the nearby clusters
Ch05F06	5	5.6	Parallaxes for members of Coma Ber and Pleiades
Ch05F07	5	5.7	Proper motions for the Cep OB6 cluster
Ch05F08	5	5.8	The HR diagram for Open Clusters and field stars
Ch05F09	5	5.9	The Orion OB1 association
Ch05F10	5	5.10	Estimated distance moduli versus measured parallaxes
Ch05F11	5	5.11	Orion OB1 association: proper motion diagrams
Ch05F12	5	5.12	Orion OB1 association: maps of proper motion members
Ch05F14	5	5.14	The HR diagram in the δ Sct region
Ch05F15	5	5.15	Period-Luminosity diagram for A-type variables
Ch05F16	5	5.16	RR Lyrae absolute magnitude calibration
Ch05F17	5	5.17	Sub dwarfs in the HR diagram
Ch05F19	5	5.19	PL calibration for 100 Cepheids
Ch06F03	6	6.3	Solar motion and velocity dispersion
Ch06F04	6	6.4	Asymmetric drift
Ch06F06	6	6.6	Galactic rotation in Cepheid proper motions
Ch06F07	6	6.7	HR Diagram for the A and F main sequence
Ch06F08	6	6.8	Parallax accuracies for an A and F star sample
Ch06F10	6	6.10	Inverse-density profiles for A & F stars
Ch06F12	6	6.12	Velocity distributions for A & F stars
Ch07F02	7	7.2	Response changes in IDT photometry
Ch07F10	7	7.10	Periodicity analysis for HIP 8163

Continued on next page

Table G.1 – continued from previous page

Name	Ch.	Fig.	Description
Ch08F01	8	8.1	Position Chevron slits
Ch08F03	8	8.3	Gyro response calibrations
Ch08F04	8	8.4	Example of gravity gradient torques over an orbit
Ch08F05	8	8.5	Magnetic field contributions
Ch08F06	8	8.6	Example of the local magnetic field variations
Ch08F07	8	8.7	Amplitudes of scan-phase jumps
Ch08F08	8	8.8	Phase distribution of scan-phase jumps
Ch09F01	9	9.1	Solar radiation torques
Ch09F02	9	9.2	Solar radiation torque components
Ch09F03	9	9.3	Constant torque on y-axis
Ch09F04	9	9.4	Solar aspect angle as function of rotation phase
Ch09F05	9	9.5	High-frequency torque component on z axis
Ch09F10	9	9.10	Systematics in torque residuals
Ch10F01	10	10.1	Error-angles update from ground
Ch10F02	10	10.2	Gyro-drift in orbit 320
Ch10F03	10	10.3	Gyro-drift modulation in orbit 1264
Ch10F04	10	10.4	Attitude modelling: gyro data
Ch10F06	10	10.6	Star mapper V-channel background in orbit 75
Ch10F08	10	10.8	Attitude modelling: star mapper data
Ch10F10	10	10.10	Attitude modelling: IDT data
Ch10F14	10	10.14	Distribution of scan-phase jumps
Ch10F17	10	10.17	Occurrences of external hits
Ch10F19	10	10.19	Instrument parameters: Mean rotation
Ch11F02	11	11.2	Gyro-4 breakdown
Ch11F03	11	11.3	Thermal control failure
Ch11F06	11	11.6	Effect of grid rotation correction
Ch12F01	12	12.1	OTF: evolution of mean values
Ch12F03	12	12.3	OTF: evolution of colour terms
Ch12F04	12	12.4	Grid rotation
Ch12F11	12	12.11	IDT Photometry: Along-scan gradient
Ch12F12	12	12.12	IDT Photometry: Across-scan gradient
Ch12F15	12	12.15	IDT Photometry: Colour terms in preceding FOV
Ch12F16	12	12.16	IDT Photometry: pass band variation along scan
Ch12F17	12	12.17	SM Photometry: single-slit response functions
Ch12F18	12	12.18	SM Photometry: FWHM evolution of SSRFs
Ch12F19	12	12.19	SM Photometry: Response changes over the mission
Ch12F20	12	12.20	SM Photometry: Pass band corrections
Ch12F22	12	12.22	SM Photometry: Colour and ordinate gradient
Ch13F01	13	13.1	On-board clock drift orbit 968
Ch13F03	13	13.3	Mean on-board clock drift over the mission
Ch13F04	13	13.4	Spin-synchronous on-board clock drift
Ch13F05	13	13.5	On-board clock drift, harmonics, amplitude and phase evolution
Ch13F06	13	13.6	Gyro drift and noise over the mission
Ch13F08	13	13.8	Spin-synchronous gyro drifts
Ch13F11	13	13.11	Thruster response calibration
Ch13F12	13	13.12	Thruster calibrations over the mission

Continued on next page

Table G.1 – continued from previous page

Name	Ch.	Fig.	Description
Ch13F13	13	13.13	Position Centre of Gravity
Ch14F01	14	14.1	The Gaia focal plane
Ch14F02	14	14.2	The Gaia BP/RP response
Ch14F04	14	14.4	The Gaia BP/RP dispersion spectra

G.2 The science data

Table G.2. Summary of the catalogue tables. Column 1: File name; column 2: Folder; column 3: Record length; column 4: Number of records; column 5: File length in Kbytes; column 6: Description

1	2	3	4	5	6
MainCat.d	data/Catalogues	278	117 955	32 023	Main astrometric catalogue
MCIndex.d	data/Catalogues	27	117 955	3 111	Index for main catalogue
SevenPCat.d	data/Catalogues	130	1679	214	7-par.supplement
NinePCat.d	data/Catalogues	275	91	25	9-par.supplement
VIMCat.d	data/Catalogues	130	41	6	VIM supplement

G.2.1 The Astrometric Catalogue

The astrometric catalogues are provided in ASCII files of fixed record length. The summary information on these files is given in Table G.2. The astrometric catalogue is presented as a main catalogue with references to supplementary catalogues for complex solutions. These are the 7- and 9-parameter solutions and the VIM solutions. The solution type as implemented is given by isol_n. This parameter consists of two parts, as $\text{isol}_n = 10 \times d + s$. The values for s are 5, 7, and 9 for the astrometric 5, 7 and 9 parameter solutions, 3 for VIM solutions, and 1 for stochastic solutions. A value of $s = 0$ is used when no new solution is available, and the published solution is presented instead. In the case of s equal to 3, 7 or 9, the variable "ic" provides the entry for the supplementary information in the relevant table, and is also found in Table G.4. Only when $s = 1$, the value of "var" will be non-zero, and a measure of the cosmic noise added to the solution. The value for "d" is zero for single stars, and starts with 1 for double stars. If there is a variable star in the system, with amplitude above 0.2 mag., "d" becomes 2. If the astrometry for the double system refers to the photo centre, 4 is added to "d", and if the measurements concern the secondary (fainter) star in a double system, 8 is added to "d". The solution type in the published catalogue is given by isol_o, and is equal to 0 for standard 5-parameter solutions, 1 for 7- or 9-parameter solutions, 2 for stochastic solutions, 3 for double and multiple stars, 4 for orbital binaries as resolved in the published catalogue, and 5 for VIM solutions. The upper triangular matrix U and its use in applying weights when combining data is explained in Appendix C. When 7- or 9- parameter solutions were applied, the first 15 elements, as defined in Eq. C.14, are given in the main catalogue file, remaining elements in the supplementary files.

Table G.3. Description of the contents of the astrometric catalogue. Notes: isol_n, isol_o see text; ic: entry in the relevant table for supplementary information; VarAnn: 1 for periodic, 2 for unresolved variables, 0 for non-variables

Variable	Start	Type	Units	Notes
HIP	1	I6		Hipparcos identifier
isol_n	8	I3		Sol. type new reduction
isol_o	12	I1		Sol. type old reduction
ncomp	14	I1		Number of components
α	16	F13.10	rad.	Right Ascension in ICRS
δ	30	F13.10	rad.	Declination in ICRS
ϖ	44	F7.2	mas	Parallax
$\mu_{\alpha,*}$	52	F8.2	mas yr^{-1}	Proper motion in Right Ascension
μ_δ	61	F8.2	mas yr^{-1}	Proper motion in Declination
$\sigma_{\alpha*}$	70	F6.2	mas	Formal error on α
σ_δ	77	F6.2	mas	Formal error on δ
σ_ϖ	84	F6.2	mas	Formal error on ϖ
$\sigma_{\mu_{\alpha,*}}$	91	F6.2	mas yr^{-1}	Formal error on $\mu_{\alpha,*}$
σ_{μ_δ}	98	F6.2	mas yr^{-1}	Formal error on μ_δ
nobs	105	I3		Number of field transits used
gof	109	F5.2		Goodness of fit
prej	115	I2		Percentage rejected data
var	118	F6.1		Cosmic dispersion added
ic	125	I4		Entry in one of the suppl.catalogues
Hp	130	F7.4	mag.	Hp magnitude
ϵ(Hp)	138	F6.4	mag.	Error on mean Hp
σ(Hp)	145	F5.3	mag.	Scatter of Hp
VarAnn	151	I1		Reference to variability annex
V − B	153	F6.3	mag.	Colour index
σ(V − B)	160	F5.3	mag.	Formal error on colour index
V − I	166	F6.3	mag.	Colour index
$U_1 \ldots U_{15}$	172	15(1X,F6.2)		Upper-triangular weight matrix

Table G.4. Index table for the astrometric catalogues, giving record numbers for each Hipparcos entry

Param.	Start	Type	Note
HIP	1	I6	Hipparcos number
CM	8	I6	Main-catalogue entry
iol_n	15	I1	Solution type
ic	17	I4	Supplement-catalogue entry

Table G.5 presents the supplementary data for the 7-parameter solutions, the detection statistic, acceleration terms and formal errors, and the additional elements of the upper-triangular weight matrix U.

Table G.5. Description of the supplementary data provided for 7-parameter solutions.

Variable	Start	Type	Units	Notes
HIP	1	I6		Hipparcos identifier
Fg	8	F5.2		Detection statistic
$\dot{\mu}_{\alpha,*}$	14	F6.2	mas yr^{-2}	Acceleration in Right Ascension
$\dot{\mu}_{\delta}$	21	F6.2	mas yr^{-2}	Acceleration in Declination
$\sigma_{\dot{\mu}_{\alpha,*}}$	28	F5.2	mas yr^{-2}	Formal error on $\mu_{\alpha,*}$
$\sigma_{\dot{\mu}_{\delta}}$	34	F5.2	mas yr^{-2}	Formal error on μ_{δ}
$U_{16} \ldots U_{28}$	39	13(1X,F6.2)		Upper-triangular weight matrix

Table G.6. Description of the supplementary data provided for 9-parameter solutions.

Variable	Start	Type	Units	Notes
HIP	1	I6		Hipparcos identifier
Fg	8	F5.2		Detection statistic
$\dot{\mu}_{\alpha,*}$	14	F6.2	mas yr^{-2}	Acceleration in $\alpha*$
$\dot{\mu}_{\delta}$	21	F6.2	mas yr^{-2}	Acceleration in δ
$\ddot{\mu}_{\alpha,*}$	28	F6.2	mas yr^{-3}	Change in acceleration in $\alpha*$
$\ddot{\mu}_{\delta}$	35	F6.2	mas yr^{-3}	Change in acceleration in δ
$\sigma_{\dot{\mu}_{\alpha,*}}$	42	F5.2	mas yr^{-2}	Formal error on $\dot{\mu}_{\alpha,*}$
$\sigma_{\dot{\mu}_{\delta}}$	48	F5.2	mas yr^{-2}	Formal error on $\dot{\mu}_{\delta}$
$\sigma_{\ddot{\mu}_{\alpha,*}}$	54	F5.2	mas yr^{-3}	Formal error on $\ddot{\mu}_{\alpha,*}$
$\sigma_{\ddot{\mu}_{\delta}}$	60	F5.2	mas yr^{-3}	Formal error on $\ddot{\mu}_{\delta}$
$U_{16} \ldots U_{45}$	65	30(1X,F6.2)		Upper-triangular weight matrix

Table G.7. Description of the supplementary data provided for VIM solutions.

Variable	Start	Type	Units	Notes
HIP	1	I6		Hipparcos identifier
Fg	8	F5.2		Detection statistic
$v_{\alpha,*}$	14	F6.2	mas	Proper motion in Right Ascension
v_{δ}	21	F6.2	mas	Proper motion in Declination
$\sigma_{v_{\alpha,*}}$	28	F5.2	mas	Formal error on $v_{\alpha,*}$
$\sigma_{v_{\delta}}$	34	F5.2	mas	Formal error on v_{δ}
$U_{16} \ldots U_{28}$	39	13(1X,F6.2)		Upper-triangular weight matrix

G.2.2 The Intermediate Astrometric Data

The intermediate data are stored in the folder *Intermediate data*, which itself contains two folders, *absrec* and *resrec*.

The intermediate astrometric data is presented in two ways. The simple ACSII format gives the basic information as defined by the final catalogue solution (in the *resrec* folder). The full format gives all information that was used to derive the final catalogue, i.e. the one-but-last iteration (number 14) and all the modulation, amplitude and supplementary information for each field-of-view transit (in the *absrec* folder). The latter files, which are the only binary files on the disk, can be used to disentangle double stars for example. In each case there is a single data file per star, in folders based on the star number divided by 1000. Thus, the residual data for star HIP 12345 is found in the folder *012* in folder *resrec* as file *HIP012345.d*. A total of 117 955 files is present in each case.

In the simple files (Table G.8), the abscissa residuals and formal errors are given for each field transit, as well as the orbit number, the epoch of observation, the parallax factor and the sine and cosine of the scan orientation. A header record provides the necessary references to the solution applied and where the solution details can be found in the main catalogue, and when relevant, in one of the supplementary catalogues. The record length is fixed at 50 bytes (including the header record).

Table G.8. The abscissa residuals per star as determined for the final solution

Variable	Start	Type	Units	Notes
		The header record		
HIP	1	I6		Hipparcos identifier
MCE	8	I6		Main-catalogue entry
NOB	15	I3		Number of observations
NC	19	I1		Number of components
isol_n	21	I3		Solution type
SCE	25	I4		Supplement-catalogue entry
F2	30	F6.2		Goodness of fit
NR	37	I2		Percentage rejected
		"NOB" data records		
IORB	1	I4		Orbit number
EPOCH	6	F6.3	Yr-1991.25	Epoch
PARF	13	F6.3		Parallax factor
CPSI	20	F7.4		$\cos\psi$
SPSI	28	F7.4		$\sin\psi$
RES	36	F7.2	mas	Abscissa residual
SRES	44	F6.2	mas	Formal error on abscissa residual

The detailed field-transit data are presented in Table G.9 as binary files with byte-order as used on PCs. For use on Solaris hardware byte swapping has to be applied. Data come either as two-byte integer **short**, or as four-byte integer **int**, and has been scaled to fit these formats as indicated in the table. As the **int** format is insufficient to store the stellar coordinates with sufficient accuracy (at least 0.1 mas is required), those data are presented in two parts. Detailed information on the parameters contained in these tables can be found in Chapter 2.

Table G.9. The detailed field-transit information.

Variable	Start	Type	Units	Notes
		The header record (32 bytes)		
HIP	0	int		Identifier
NOB	4	int		Number of observations
dRA	8	int	0.01 mas	decimal part of α
dDEC	12	int	0.01 mas	decimal part of $\delta + 90$

Continued on next page

Table G.9 – continued from previous page

Variable	Start	Type	Units	Notes
PAR	16	int	0.01 mas	Reference parallax
MUA	20	int	0.01 mas Yr^{-1}	Reference proper motion in $\alpha*$
MUD	24	int	0.01 mas Yr^{-1}	Reference proper motion in δ
iRA	28	short	deg.	Integer part of α
iDEC	30	short	deg.	Integer part of $\delta + 90$
"NOB" data records (of 64 bytes)				
da	0	int	mas$\times$100	Abscissa residual
β_4	4	int	$\times 10^5$	Normalize β_4
β_5	8	int	mas$\times$100	Normalized β_5
BCJD	12	int	$\times 10^5$	Baricentric Julian Day - 2 440 000
IFRM	16	int		Mean observing frame
ψ	20	int	rad$\times 10^5$	Scan-orientation angle
NTR	24	short		Number of field transits used
EPOCH	26	short	Yr $\times 10^4$	Epoch of observation-1991.25
PARF	28	short	$\times 10^4$	parallax factor
$\sigma_{\mathrm{d}a}$	30	short	mas$\times$100	Formal error on abscissa residual
SSC	32	short		$\log I_{\mathrm{tot}} = (SSC + 10000)/5000$
χ^2	34	short	$\times$100	Mean χ^2 for frame transits
F2	36	short	$\times$100	Goodness of fit for frame transits
$\langle\zeta\rangle$	38	short	$\times$30000	Ordinate of transit
dη	40	short	$\times 10^4$	Abscissa correction for light-bending
NSLT	42	short		Total number of slots in field transit
IORB	44	short		Orbit number
dummy	46	short		
dummy	48	short		
dx	50	short	arcsec	IFOV offset along scan
dy	52	short	arcsec	IFOV offset across scan
σ_{β_4}	54	short	$\times 10^5$	Formal error on β_4
σ_{β_5}	56	short	mas$\times$100	Formal error on β_5
R(OTF)	58	short	$\times 10^3$	Nominal relative amplitude of 2nd harmonic
V $-$ I	60	short	$\times 10^3$	Colour index used
M_1	62	short	$\times 10^3$	Nominal amplitude first harmonic

G.3 Payload calibration data

In the description of the payload calibration data the field coordinates are given by $g = \eta/w$ and $h = \zeta/w$, where (η, ζ) are the coordinates of a transit as input to the calibration, and w the half-width of the field of view in the same units as (η, ζ). The coordinates (g, h) therefore range from -1 to +1. Colour dependencies are described as a function of the colour index $c = (\mathrm{V} - \mathrm{I}) - 0.7$. Orbits are identified through the orbit number and the time associated with an orbit is expressed in days since 1989.0

Table G.10. Summary of the data tables for payload calibrations. Column 1: File name; column 2: Folder; column 3: Record length; column 4: Number of records; column 5: File length in Kbytes; column 6: Description

1	2	3	4	5	6
OTFb4p.d	data/Science/OTF	239	2297	537	β_4, PFOV
OTFb4f.d	data/Science/OTF	239	2297	537	β_4, FFOV
OTFb5p.d	data/Science/OTF	239	2297	537	β_5, PFOV
OTFb5f.d	data/Science/OTF	239	2297	537	β_5, FFOV
IDTdcp.d	data/Science/IDT	214	2292	506	IDT DC-phot., PFOV
IDTdcf.d	data/Science/IDT	214	2292	506	IDT DC-phot., FFOV
IDTacp.d	data /Science/IDT	201	2292	475	IDT AC-phot., PFOV
IDTacf.d	data/Science/IDT	201	2292	475	IDT AC-phot., FFOV
InParBAC	data/Science/IDT	32	2304	72	Basic angle and chrom.
InParG.d	data/Science/IDT	105	2304	237	Mean instr.par.
InParH.d	data/Science/IDT	105	2304	237	Diff.instr.par.
IDTmsd.d	data/Science/IDT	49	1932	93	IDT MSD
IDTssd.d	data/Science/IDT	55	1100	69	IDT SSD
IDTssvi.d	data/Science/IDT	62	28	2	IDT $V - I$ corr.
SMBp.d	data/Science/SM	205	329	66	SM B-Phot., PFOV
SMBf.d	data/Science/SM	205	329	66	SM B-Phot., FFOV
SMVp.d	data/Science/SM	205	329	66	SM V-Phot., PFOV
SMVf.d	data/Science/SM	205	329	66	SM V-Phot., FFOV
SMGeom.d	data/Science/SM	68	2306	153	SM Geometry

G.3.1 The OTF Calibration Files

The OTF calibration files describe the calibration results for the relative amplitude and phase of the second harmonic in the modulated signal for the IDT measurements. These calibrations are used to help recognize non-single stars, and to prepare the modulation parameters of those stars for further analysis on the components. The third-order positional parameters as well as the cross terms between position and colour were fixed in these calibrations. For those parameters no formal errors are given.

Table G.11. Contents of the OTF calibration files (**OTFb4p.d**, **OTFb4f.d**, **OTFb5p.d**, **OTFb5f.d**)

Depend.	Start	Type	Start	Type
	Value		sigma	
orbit	1	I4		
time	6	F7.2		
observ.	14	I5		
1	20	F9.5	30	F9.5
g	40	F9.5	50	F9.5
h	60	F9.5	70	F9.5
g^2	80	F9.5	90	F9.5
gh	100	F9.5	110	F9.5
h^2	120	F9.5	130	F9.5
c	140	F9.5	150	F9.5

Continued on next page

Table G.11 – continued from previous page

Depend.	Start	Type	Start	Type
	Value		sigma	
c^2	160	F9.5	170	F9.5
g^3	180	F9.5		
g^2h	190	F9.5		
gh^2	200	F9.5		
h^3	210	F9.5		
cg	220	F9.5		
ch	230	F9.5		

G.3.2 IDT Photometry

The photometric calibration files provide the coefficients for the geometric parameters only, as the colour parameters were replaced by an *a posteriori* re-calibration. The background estimate is only relevant and given for the DC calibrations. Part of the photometric calibration model consisted of the representation of a circular-symmetric response feature, represented by a linear spline as a function of the distance from the centre of this feature. These are coefficients X_9 to X_{14}.

Table G.12. Contents of the IDT-photometry calibration files (**IDTdcp.d**, **IDTdcf.d**, **IDTacp.d**, **IDTacf.d**)

Depend.	Start	Type	Start	Type
	Value		sigma	
orbit	1	I4		
time	6	F7.2		
1	14	F9.4	24	F7.4
g	32	F8.4	41	F7.4
h	49	F8.4	58	F7.4
gc	66	F8.4	75	F7.4
hc	83	F8.4	92	F7.4
X_9	100	F8.4	109	F7.4
X_{10}	117	F8.4	126	F7.4
X_{11}	134	F8.4	143	F7.4
X_{12}	151	F8.4	160	F7.4
X_{13}	168	F8.4	177	F7.4
X_{14}	185	F8.4	194	F7.4
DC backgr.	202	F6.2	209	F5.2

G.3.3 IDT Geometric Calibrations

This section covers all calibrations of systematic offsets in the abscissa measurements, from the large-scale instrument parameters, through the medium- and small-scale distortions to the detailed colour dependencies. All offsets ($\mathrm{d}a$) and related accuracies ($\sigma_{\mathrm{d}a}$) are given in mas, noise figures refer to unit-weight measurements, indicated by w, and should ideally be equal to 1.0. The subscript p or f is used to indicate the PFOV or FFOV where such data are in the same table.

Table G.13. The basic angle and chromaticity evolution over the mission (file **InParBAC.d**)

Depend.	Start	Type	Description
orbit	1	I4	Orbit number
time	6	F7.2	Reference time (days)
$\Delta\gamma/2$	14	F6.2	Basic-angle correction (mas)
$\sigma_{\Delta\gamma/2}$	21	F6.4	Formal error on $\Delta\gamma/2$
c	27	F5.2	Chromaticity coefficient (mas/mag.)

Table G.14. Contents of the instrument calibration files **InParG.d** and **InParH.d**

Depend.	Start	Type	Start	Type
	Value		sigma	
orbit	1	I4		
time	6	F7.2		
g	14	F8.2	23	F4.2
h	28	F8.2	37	F4.2
g^2	42	F6.2	49	F4.2
gh	54	F6.2	61	F4.2
h^2	66	F6.2	73	F4.2
g^3	78	F6.2		
g^2h	85	F6.2		
gh^2	92	F6.2		
h^3	99	F6.2		

The grid coordinates for the medium-scale distortion are given by (with $hfow$ half the width of the field of view):

$$ig = 21 \cdot (\eta/hfow + 1.0) \tag{G.1}$$

$$ih = 23 \cdot (\zeta/hfow + 1.0), \tag{G.2}$$

as a result of which ig covers groups of four scan-fields along scan, and ih effectively counts the rows of scan-fields across scan.

Table G.15. The medium-scale distortion data as accumulated over the mission (file **IDTmsd.d**)

Depend.	Start	Type	Description
ig	1	I2	Along-scan coordinate
ih	4	I2	Across-scan coordinate
observ.	7	I5	Number of observations
w	14	F5.2	Unit-weight dispersion
$\mathrm{d}a_p$	20	F7.3	Mean offset in PFOV (mas)
$\sigma_{\mathrm{d}a,p}$	28	F6.3	Formal error on $\mathrm{d}a_p$
$\mathrm{d}a_f$	35	F7.3	Mean offset in FFOV (mas)
$\sigma_{\mathrm{d}a,f}$	43	F6.3	Formal error on $\mathrm{d}a_f$

The small-scale distortions have been accumulated at a resolution of 1/24 points per scan-field, a total 1100 points across scan. The four most extreme positions are not covered, as these tend to be affected by few observations and high noise. The across-scan coordinate h is given as the fractional scanfield.

Table G.16. The small-scale distortion data as accumulated over the mission (file **IDTssd.d**)

Depend.	Start	Type	Description
h	1	F7.3	Across-scan coordinate
da_p	9	F6.3	Mean offset in PFOV (mas)
$\sigma_{da,p}$	18	F5.3	Formal error on da_p
observ.	27	I4	Observations in PFOV
da_f	33	F6.3	Mean offset in FFOV (mas)
$\sigma_{da,f}$	42	F5.3	Formal error on da_f
observ.	51	I4	Observations in FFOV

The high-resolution colour dependence of the abscissa residuals represents the systematic effects as a function of V – I, averaged over the mission, after subtracting for each orbit the linear colour dependence. The data have been binned with variable bin-width, to obtain both good resolution and coverage.

Table G.17. The high-resolution colour dependence of the abscissa residuals as accumulated over the mission (file **IDTssvi.d**)

Depend.	Start	Type	Description
$\langle \mathrm{V-I} \rangle_p$	2	F6.3	Mean V – I in PFOV
da_p	9	F6.3	Mean offset in PFOV (mas)
$\sigma_{da,p}$	18	F5.3	Formal error on da_p
observ.	25	I6	Observations in PFOV
$\langle \mathrm{V-I} \rangle_f$	33	F6.3	Mean V – I in FFOV
da_f	40	F6.3	Mean offset in FFOV (mas)
$\sigma_{da,f}$	49	F5.3	Formal error on da_f
observ.	56	I6	Observations in FFOV

G.3.4 The Star Mapper Photometric Calibration Files

The star mapper photometric calibration records cover each a range of orbits, the first of which is given in each record, and the last of which is defined by the orbit number of the next record. There are eight independent calibrations, covering the B and V bands, the preceding and following fields of view, and the inclined and vertical slit groups. The calibration provides the factor with which to multiply the pseudo intensity to obtain the expected intensity-scaling factor for the fitting of the star mapper signals with the SSRFs. The pseudo intensity for a star with magnitude B is given by:

$$I_B = 10^{-0.4\cdot(B-10)}. \tag{G.3}$$

The zero-level is therefore equivalent to the peak count rate for a star of magn. 10. The colour index used in the star mapper photometric calibration is (B – V) – 0.7. In Table G.18 the slit groups are indicated by index v for vertical or i for inclined.

Table G.18. Contents of the star mapper photometric calibration files (**SMBp.d**, **SMBf.d**, **SMVp.d** and **SMVf.d**)

Depend.	Start	Type	Start	Type
	Value		sigma	
orbit	1	I4		
time	6	F7.2		
1_v	14	F8.4	23	F6.4
c_v	30	F8.4	39	F6.4
c_v^2	46	F8.4	55	F6.4
ch_v	62	F8.4	71	F6.4
ch_v^2	78	F8.4	87	F6.4
$c^2 z_v$	94	F8.4	103	F6.4
1_i	110	F8.4	119	F6.4
c_i	126	F8.4	135	F6.4
c_i^2	142	F8.4	151	F6.4
ch_i	158	F8.4	167	F6.4
ch_i^2	174	F8.4	183	F6.4
$c^2 z_i$	190	F8.4	199	F6.4

G.3.5 The Star Mapper Geometry Calibration Data

The Star Mapper Geometric calibration covers the differential and the global aspects of the positions of the star mapper slits. The differential aspects concerns the relative position of the vertical slits with respect to the inclined slits, the global aspects cover the effective basic angle for the star mapper slits, the rotation relative to the scan circle, and their position relative to the centre of the main (IDT) grid. Other parameters were initially also calibrated, such as the differential rotation between the two fields of view and relative tilt of the vertical slits, but these parameters showed only marginally measurable variation over the mission, and were given fixed values based on the accumulated data. Not all of the remaining four parameters were always included in the calibration model. If the volume of data was small, predicted values based on fitted curves over the mission data were used instead. In Table G.19 these cases can be recognized by formal errors equal to zero.

Table G.19. Contents of the SM geometric-calibration file (**SMGeom.d**)

Depend.	Start	Type	Start	Type	units
	Value		sigma		
orbit	1	I4			
time	6	F7.2			days
$\mathrm{d}\gamma/2$	14	F7.2	22	F5.2	mas
$\mathrm{d}p$	28	F7,2	36	F5.2	mas
h	42	F7.2	49	F5.2	arcsec
g^2	55	F7.2	63	F5.2	mas

G.4 Spacecraft calibration data

Table G.20. Summary of the data tables for spacecraft calibration. Column 1: File name; column 2: Folder; column 3: Record length; column 4: Number of records; column 5: File length in Kbytes; column 6: Description

1	2	3	4	5	6
GyroDrif.d	data/Spacecr	43	2092	88	Gyro drift
GyroOrie.d	data/Spacecr	175	2090	358	Gyro orientation
ClockCal.d	data/Spacecr	90	2014	178	On-board clock drift

G.4.1 The Gyro-calibration Data

The gyro-calibration data show a range of peculiarities, in particular concerning the way different units on-board the spacecraft interacted. They also provide a record on stability and temperature sensitivity, although the detailed data (resolved on orbital phase) is not released here, but can be obtained from the author.

Table G.21. Contents of the Gyro drift-calibration files (**GyroDrif.d**)

Depend.	Start	Type	Description
orbit	1	I4	Orbit number
time	6	F7.2	Time in days
NOBS	14	I4	Number of observations
D_1	19	F7.1	Drift Gyro 1, mas s^{-1}
D_2	27	F7.1	Drift Gyro 2, mas s^{-1}
$D_{4/5}$	35	F7.1	Drift Gyro 4 or 5, mas s^{-1}

The gyro-orientation calibration provides a record of the relations between the gyro readings and the reconstructed inertial scan rates of the satellite. These relations may be affected by thermal variations, but appear not to be. There are, however, other less easy to explain effects on the orientations, such as the influence of the operations of redundant gyros (so-called de-spin).

Table G.22. Contents of the gyro-orientation-calibration files (**GyroOrie.d**)

Depend.	Start	Type	Description
orbit	1	I4	Orbit number
time	6	F7.2	Time in days
G_1, x	14	F7.4	x component of Gyro 1
$\sigma_{G1,x}$	22	F7.4	Formal error on G_1, x
G_1, y	30	F7.4	y component of Gyro 1
$\sigma_{G1,y}$	38	F7.4	Formal error on G_1, y
G_1, z	46	F7.4	z component of Gyro 1
$\sigma_{G1,z}$	54	F7.4	Formal error on G_1, z
w_1	62	F5.2	Unit-weight dispersion Gyro 1 solution
G_2, x	68	F7.4	x component of Gyro 2
$\sigma_{G2,x}$	76	F7.4	Formal error on G_2, x

Continued on next page

Table G.22 – continued from previous page

Depend.	Start	Type	Description
G_2, y	84	F7.4	y component of Gyro 2
$\sigma_{G2,y}$	92	F7.4	Formal error on G_2, y
G_2, z	100	F7.4	z component of Gyro 2
$\sigma_{G2,z}$	108	F7.4	Formal error on G_2, z
w_2	116	F5.2	Unit-weight dispersion Gyro 2 solution
$G_{4/5}, x$	122	F7.4	x component of Gyro 4/5
$\sigma_{G45,x}$	130	F7.4	Formal error on $G_{4/5}, x$
$G_{4/5}, y$	138	F7.4	y component of Gyro 4/5
$\sigma_{G45,y}$	146	F7.4	Formal error on $G_{4/5}, y$
$G_{4/5}, z$	154	F7.4	z component of Gyro 4/5
$\sigma_{G45,z}$	162	F7.4	Formal error on $G_{4/5}, z$
$w_{4/5}$	170	F5.2	Unit-weight dispersion Gyro 4/5 solution

G.4.2 The On-board Clock Calibration File

The on-board clock has provided a unique record on the temperature variations in the spacecraft, with a range of peculiarities still unresolved. The data provided here concern the measurements around perigee of the zero point, drift, second derivative and the spin-synchronous modulation coefficients, all with their formal errors. Data on the on-board clock was available, independent of whether science data could be obtained. The spin-synchronous modulation coefficients are zero for Sun-pointing mode data sets.

Table G.23. Contents of the gyro-orientation-calibration files (**GyroOrie.d**)

Depend.	Start	Type	Description
orbit	1	I4	Orbit number
time	6	F7.2	Time in days
ν	14	F5.3	Precession phase of spin axis in radians
$\Delta\tau$	20	F9.5	Time offset on-board clock in seconds
$\mathrm{d}\Delta\tau/\mathrm{d}t$	30	F9.5	First derivative of time offsetin $10^{-6}s/s$
$\sigma_{\mathrm{d}\Delta\tau/\mathrm{d}t}$	40	F7.5	Formal error on $\mathrm{d}\Delta\tau/\mathrm{d}t$ in $10^{-6}s/s$
$\mathrm{d}^2\Delta\tau/\mathrm{d}t^2$	48	F8.4	Second derivative of time offset in $10^{-9}s/s^2$
$\sigma_{G1,z}$	57	F7.4	Formal error on $\mathrm{d}^2\Delta\tau/\mathrm{d}t^2$ in $10^{-9}s/s^2$
Fc	65	F6.2	Modulation amplitude of $\cos\Omega$ in $10^{-6}s/s$
$\sigma(Fc)$	72	F5.2	Formal error on Fc in $10^{-6}s/s$
Fs	78	F6.2	Modulation amplitude of $\sin\Omega$ in $10^{-6}s/s$
$\sigma(Fs)$	85	F5.2	Formal error on Fs in $10^{-6}s/s$

Bibliography

Alcock, C., Allsman, R. A., Axelrod, T. S., Bennett, D. P., Cook, K. H., Freeman, K. C., Griest, K., Marshall, S. L., Peterson, B. A., Pratt, M. R., Quinn, P. J., Reimann, J., Rodgers, A. W., Stubbs, C. W., Sutherland, W., and Welch, D. L. (1995). The MACHO project LMC variable star inventory. 1: Beat Cepheids-conclusive evidence for the excitation of the second overtone in classical Cepheids. *AJ*, 109:1653–1662.

Ambronn, l. (1900). *Handbuch der Astronomischen Instrumentkunde. Erster Band.* Springer, Berlin.

Antonello, E. and Mantegazza, L. (1997). Luminosity and related parameters of δ Scuti stars from HIPPARCOS parallaxes. General properties of luminosity. *A&A*, 327:240–244.

Arenou, F., Lindegren, L., Frœschlé, M., Gómez, A. E., Turon, C., Perryman, M. A. C., and Wielen, R. (1995). Zero-point and external errors of the hipparcos parallaxes. *A&A*, 304:52–60.

Arias, E. F., Charlot, P., Feissel, M., and Lestrade, J.-F. (1995). The extragalactic reference system of the International Earth Rotation Service, ICRS. *A&A*, 303:604–608.

Babusiaux, C. (2005). The Gaia Instrument and Basic Image Simulator. In Turon, C., O'Flaherty, K. S., and Perryman, M. A. C., editors, *ESA SP-576: The Three-Dimensional Universe with Gaia*, pages 417–420.

Babusiaux, C. and Gilmore, G. (2005). Red clump distances to the inner Galactic structures. *ArXiv Astrophysics e-prints*.

Bailer-Jones, C. A. L. (2005). Microarcsecond astrometry with Gaia: the solar system, the Galaxy and beyond. In Kurtz, D. W., editor, *IAU Colloq. 196:*

Transits of Venus: New Views of the Solar System and Galaxy, pages 429–443.

Balega, I. I., Balega, Y. Y., Hofmann, K.-H., Malogolovets, E. V., Schertl, D., Shkhagosheva, Z. U., and Weigelt, G. (2006). Orbits of new Hipparcos binaries. II. *A&A*, 448:703–707.

Balega, I. I., Balega, Y. Y., Hofmann, K.-H., Pluzhnik, E. A., Schertl, D., Shkhagosheva, Z. U., and Weigelt, G. (2005). Orbits of new Hipparcos binaries. I. *A&A*, 433:591–596.

Barraclough, D. R. (1985). International Geomagnetic Reference Field revision 1985. *Pure Appl.Geophysics*, 123:641–645.

Baumjohann, W., Haerendel, G., Treumann, R. A., Bauer, R. A., Rustenbach, T. M., Georgescu, J., Auster, E., Fornacon, U., Glassmeier, K. H., and Lühr, K. (1999). First elf wave measurements with the equator-s magnetometer. *Advances in Space Research*, 24:77–80.

Bernstein, H.-H. (1997). Astrometric Indications of Brown Dwarfs based on HIPPARCOS Data. In *ESA SP-402: Hipparcos - Venice '97*, pages 705–708.

Bernstein, H.-H. (1999). Derivation of Orbital Parameters of Very Low-Mass Companions in Double Stars from Radial Velocities and Observations of Space-Astrometry Missions like HIPPARCOS, DIVA and GAIA. In Hearnshaw, J. B. and Scarfe, C. D., editors, *ASP Conf. Ser. 185: IAU Colloq. 170: Precise Stellar Radial Velocities*, pages 410–415.

Bertin, G. and Mark, J. W.-K. (1978). Density wave theory for spiral galaxies - The regime of finite spiral arm inclination in stellar dynamics. *A&A*, 64:389–397.

Bessell, M. S. (2000). The Hipparcos and Tycho Photometric System Passbands. *PASP*, 112:961–965.

Bierman, G. J. (1977). *Factorization Methods for discrete Sequential Estimation*. Academic Press, New York.

Binney, J. and Tremaine, S. (1987). *Galactic dynamics*. Princeton, NJ, Princeton University Press, 1987, 747 p.

Binney, J. J. (1999). Dynamics of the Solar Neighborhood. In Merritt, D. R., Valluri, M., and Sellwood, J. A., editors, *ASP Conf. Ser. 182: Galaxy Dynamics - A Rutgers Symposium*, pages 285–296.

Blaauw, A. (1964). The O Associations in the Solar Neighborhood. *ARA&A*, 2:213–246.

Bois, E. (1986). First-order theory of satellite attitude motion application to HIPPARCOS. *Celestial Mechanics*, 39:309–327.

Bois, E. (1987). A Second-Order Theory of the Rotation of an Artificial Satellite. *Celestial Mechanics*, 42:141–168.

Bond, V. R. and Allman, M. C. (1996). *Modern Astrodynamics: Fundamentals and perturbation methods*. Princeton University Press, Princeton.

Brown, A. G. A., de Geus, E. J., and de Zeeuw, P. T. (1994). The Orion OB1 association. 1: Stellar content. *A&A*, 289:101–120.

Brown, A. G. A., Hartmann, D., and Burton, W. B. (1995). The Orion OB1 association. II. The Orion-Eridanus Bubble. *A&A*, 300:903–922.

Butkevich, A. G., Berdyugin, A. V., and Teerikorpi, P. (2005a). Statistical biases in stellar astronomy: the Malmquist bias revisited. *MNRAS*, 362:321–330.

Butkevich, A. G., Berdyugin, A. V., and Teerikorpi, P. (2005b). The absolute magnitude of K0V stars from Hipparcos data using an analytical treatment of the Malmquist bias. *A&A*, 435:949–954.

Butler, R. P., Wright, J. T., Marcy, G. W., Fischer, D. A., Vogt, S. S., Tinney, C. G., Jones, H. R. A., Carter, B. D., Johnson, J. A., McCarthy, C., and Penny, A. J. (2006). Catalog of Nearby Exoplanets. *ApJ*, 646:505–522.

Caldwell, J. A. R. and Laney, C. D. (1991). Cepheids in the Magellanic Clouds. In Haynes, R. and Milne, D., editors, *IAU Symp. 148: The Magellanic Clouds*, pages 249–257.

Chereul, E., Crézé, M., and Bienayme, O. (1998). The distribution of nearby stars in phase space mapped by HIPPARCOS. II. Inhomogeneities among A-F type stars. *A&A*, 340:384–396.

Creze, M., Chereul, E., Bienayme, O., and Pichon, C. (1998). The distribution of nearby stars in phase space mapped by Hipparcos. I. The potential well and local dynamical mass. *A&A*, 329:920–936.

Dalla Torre, A. and van Leeuwen, F. (2003). A detailed analysis of the operational orbit of the hipparcos satellite. *Space Sci. Rev.*, 108:451–470.

Daly, E. J., van Leeuwen, F., Evans, H. D. R., and Perryman, M. A. C. (1994). Radiation-belt and transient solar-magnetospheric effects on hipparcos radiation background. *IEEE Trans.nucl.science*, 41(6):2376–2382.

de Cat, P., Eyer, L., Cuypers, J., Aerts, C., Vandenbussche, B., Uytterhoeven, K., Reyniers, K., Kolenberg, K., Groenewegen, M., Raskin, G., Maas, T., and Jankov, S. (2006). A spectroscopic study of southern (candidate) γ Doradus stars. I. Time series analysis. *A&A*, 449:281–292.

De Simone, R., Wu, X., and Tremaine, S. (2004). The stellar velocity distribution in the solar neighbourhood. *MNRAS*, 350:627–643.

De Vries, C. P. (1986). Optical and infrared observations of high latitude dust clouds. *Ph.D. Thesis*.

de Zeeuw, P. T., Hoogerwerf, R., de Bruijne, J. H. J., Brown, A. G. A., and Blaauw, A. (1999). A HIPPARCOS Census of the Nearby OB Associations. *AJ*, 117:354–399.

Debray, B., Robin, A. C., Reylé, C., and Schultheis, M. (2006). The Besançon Model of our Galaxy: a simulation tool towards the Virtual Observatory. In Gabriel, C., Arviset, C., Ponz, D., and Enrique, S., editors, *ASP Conf. Ser. 351: Astronomical Data Analysis Software and Systems XV*, pages 224–227.

Dehnen, W. and Binney, J. J. (1998). Local stellar kinematics from HIPPARCOS data. *MNRAS*, 298:387–394.

Donati, F. and Sechi, G. (1992). Method of comparison between determinations of the hipparcos attitude. *A&A*, 258:46–52.

Dravins, D., Lindegren, L., and Madsen, S. (1999). Astrometric radial velocities. I. Non-spectroscopic methods for measuring stellar radial velocity. *A&A*, 348:1040–1051.

Dravins, D., Lindegren, L., Madsen, S., and Holmberg, J. (1997). Astrometric radial velocities from hipparcos. In Perryman, M.A.C. and P.L.Bernacca, editors, *Hipparcos Venice'97*, number 402 in ESA-SP, pages 733–738.

Drew, D. (2000). *The lost chronicles of the Maya Kings*. Phoenix, Orion House, London.

ESA, editor (1992). *The Hipparcos Input Catalogue*. Number 1136 in SP. ESA.

ESA, editor (1997). *The Hipparcos and Tycho Catalogues*. Number 1200 in SP. ESA.

Evans, D. W. (2000). Clarification on the Hipparcos numbering in the Trapezium. *The Observatory*, 120:402–403.

Evans, D. W. (2001). The Carlsberg Meridian Telescope: an astrometric robotic telescope. *Astronomische Nachrichten*, 322:347–351.

Evans, D. W. (2003). The Carlsberg Meridian Telescope. In *The Future of Small Telescopes In The New Millennium. Volume II - The Telescopes We Use*, pages 49–61.

Evans, D. W., Irwin, M. J., and Helmer, L. (2002). The Carlsberg Meridian Telescope CCD drift scan survey. *A&A*, 395:347–356.

Evans, D. W., van Leeuwen, F., Penston, M.J., Ramamani, N., and Hoeg, E. (1992). Hipparcos photometry: NDAC reductions. *A&A*, 258:149–156.

Eyer, L. and Mignard, F. (2005). Rate of correct detection of periodic signal with the Gaia satellite. *MNRAS*, 361:1136–1144.

Fabricius, C., Hoeg, E., Makarov, V. V., Mason, B. D., Wycoff, G. L., and Urban, S. E. (2002). The Tycho double star catalogue. *A&A*, 384:180–189.

Fabricius, C. and Makarov, V. V. (2000). Two-colour photometry for 9473 components of close Hipparcos double and multiple stars. *A&A*, 356:141–145.

Famaey, B., Jorissen, A., Luri, X., Mayor, M., Udry, S., Dejonghe, H., and Turon, C. (2005). Local kinematics of K and M giants from CORAVEL/Hipparcos/Tycho-2 data. Revisiting the concept of superclusters. *A&A*, 430:165–186.

Fantino, E. (2000). *Attitude dynamics of the ESA Hipparcos satellite and application to the German astrometry mission DIVA*. PhD thesis, Universita' degli studi di Padova.

Fantino, E. and van Leeuwen, F. (2003). Modelling the torques affecting the hipparcos satellite. *Space Sci.Rev.*, 108:499–535.

Feast, M. (1999). Cepheids as Distance Indicators. *PASP*, 111:775–793.

Feast, M. and Whitelock, P. (1997). Galactic kinematics of Cepheids from HIPPARCOS proper motions. *MNRAS*, 291:683–693.

Feast, M. W. and Catchpole, R. M. (1997). The cepheid period-luminosity zero-point from hipparcos trigonometrical parallaxes. *MNRAS*, 286:L1–L5.

Feast, M. W., Glass, I. S., Whitelock, P. A., and Catchpole, R. M. (1989). A period-luminosity-colour relation for Mira variables. *MNRAS*, 241:375–392.

Feissel, M. and Mignard, F. (1998). The adoption of ICRS on 1 January 1998: meaning and consequences. *A&A*, 331:L33–L36.

Fernley, J., Barnes, T. G., Skillen, I., Hawley, S. L., Hanley, C. J., Evans, D. W., Solano, E., and Garrido, R. (1998). The absolute magnitudes of RR Lyraes from HIPPARCOS parallaxes and proper motions. *A&A*, 330:515–520.

Flynn, C. and Fuchs, B. (1994). Density of dark matter in the Galactic disk. *MNRAS*, 270:471–479.

Fricke, W., Schwan, H., Corbin, T., Bastian, U., Bien, R., Cole, C., Jackson, E., Jährling, R., Jahreiß, H., Lederle, T., and Röser, S. (1991). Fifth fundamental catalogue. Part 2: The FK5 extension - new fundamental stars. *Veroeffentlichungen des Astronomischen Rechen-Instituts Heidelberg*, 33:1–143.

Fricke, W., Schwan, H., Lederle, T., Bastian, U., Bien, R., Burkhardt, G., Du Mont, B., Hering, R., Jährling, R., Jahreiß, H., Röser, S., Schwerdtfeger, H.-M., and Walter, H. G. (1988). Fifth fundamental catalogue (FK5). Part 1: The basic fundamental stars. *Veroeffentlichungen des Astronomischen Rechen-Instituts Heidelberg*, 32:1–106.

Froeschle, M., Mignard, F., and Arenou, F. (1997). Determination of the PPN Parameter gamma with the HIPPARCOS Data. In *ESA SP-402: Hipparcos - Venice '97*, pages 49–52.

Fuchs, B. and Wielen, R. (1993). Kinematical Constraints on the Dynamically Determined Local Mass Density of the Galaxy. In Holt, S. S. and Verter, F., editors, *AIP Conf. Proc. 278: Back to the Galaxy*, pages 580–583.

Gatewood, G. (1995). Map-based trigonometric parallaxes of open clusters: Coma. *ApJ*, 445:712–715.

Gatewood, G., Castelaz, M., Han, I., Persinger, T., Stein, J., Stephenson, B., and Tangren, W. (1990). Map-based trigonometric parallaxes of open clusters - The Pleiades. *ApJ*, 364:114–117.

Gatewood, G. and de Jonge, J. K. (1994). Map-based trigonometric parallaxes of open clusters: The Praesepe. *ApJ*, 428:166–169.

Gatewood, G., de Jonge, J. Kiewiet, and Han, I. (2000). The pleiades, map-based trigonometric parallaxes of open clusters. *ApJ*, 533:938–943.

Glass, I. S. (1989). Beginnings of Astronomical Photography at the Cape. *Monthly Notes of the Astronomical Society of South Africa*, 48:117–122.

Gliese, W. and Upgren, A. R. (1990). Parallaxes - History survey and outlook. *Astronomisches Rechen-Institut Heidelberg, Mitteilungen, Serie A, no. 225*, 40:7–18.

Goldstein, H. (1980). *Classical Mechanics*. Addison-Wesley, Reading, MA.

Gratton, R. G., Fusi Pecci, F., Carretta, E., Clementini, G., Corsi, C. E., and Lattanzi, M. (1997). Ages of Globular Clusters from HIPPARCOS Parallaxes of Local Subdwarfs. *ApJ*, 491:749–771.

Grenon, M., Mermilliod, M., and Mermilliod, J. C. (1992). The HIPPARCOS Input Catalogue. III - Photometry. *A&A*, 258:88–93.

Hammersley, P. L., Garzon, F., Mahoney, T., and Calbet, X. (1995). The tilted old Galactic disc and the position of the Sun. *MNRAS*, 273:206–214.

Hartkopf, W. I., Mason, B. D., and McAlister, H. A. (1996). Binary Star Orbits From Speckle Interferometry. VIII. Orbits of 37 Close Visual System. *AJ*, 111:370–392.

Hartkopf, W. I., Mason, B. D., McAlister, H. A., Roberts, Jr., L. C., Turner, N. H., ten Brummelaar, T. A., Prieto, C. M., Ling, J. F., and Franz, O. G. (2000). ICCD Speckle Observations of Binary Stars. XXIII. Measurements during 1982-1997 from Six Telescopes, with 14 New Orbits. *AJ*, 119:3084–3111.

Hartkopf, W. I., McAlister, H. A., and Franz, O. G. (1989). Binary star orbits from speckle interferometry. II - Combined visual-speckle orbits of 28 close systems. *AJ*, 98:1014–1039.

Hawley, S. L., Jefferys, W. H., Barnes, T. G., and Lai, W. (1986). Absolute magnitudes and kinematic properties of RR Lyrae stars. *ApJ*, 302:626–631.

Helmer, L. and Morrison, L. V. (1985). Carlsberg Automatic Meridian Circle. *Vistas in Astronomy*, 28:505–518.

Hemenway, P. D., Duncombe, R. L., Bozyan, E. P., Lalich, A. M., Argue, A. N., Franz, O. G., McArthur, B., Nelan, E., Taylor, D., White, G., Benedict, G. F., Crifo, F., Fredrick, L. W., Jefferys, W. H., Johnston, K., Kovalevsky, J., Kristian, J., Perryman, M. A. C., Preston, R., Shelus, P. J., Turon, C., and van Altena, W. (1997). The Program to Link the HIPPARCOS Reference Frame to an Extragalactic Reference System Using the Fine Guidance Sensors of the Hubble Space Telescope. *AJ*, 114:2796–2810.

Hentschel, K. (2005). Einstein and gravitational redshift. *Acta Historica Astronomiae*, 27:12–43.

Hirshfeld, A. W. (2001). *Parallax, the race to measure the Cosmos*. W.H.Freeman and Company, New York.

Hoeg, E. (1968). Refraction anomalies: The mean power spectrum of star image motion. *ZAp*, 69:313–325.

Hoeg, E. (1970). A theory of a photoelectric multislit micrometer. *A&A*, 4:89–95.

Hoeg, E. (1972). The photelectric meridian circle of bergedorf/perth. *A&A*, 19:27–40.

Hoeg, E., Bassgen, G., Bastian, U., Egret, D., Fabricius, C., Grossmann, V., Halbwachs, J. L., Makarov, V. V., Perryman, M. A. C., Schwekendiek, P., Wagner, K., and Wicenec, A. (1997). The TYCHO Catalogue. *A&A*, 323:L57–L60.

Hoeg, E., Fabricius, C., Makarov, V. V., Bastian, U., Schwekendiek, P., Wicenec, A., Urban, S., Corbin, T., and Wycoff, G. (2000a). Construction and verification of the Tycho-2 Catalogue. *A&A*, 357:367–386.

Hoeg, E., Fabricius, C., Makarov, V. V., Urban, S., Corbin, T., Wycoff, G., Bastian, U., Schwekendiek, P., and Wicenec, A. (2000b). The Tycho-2 catalogue of the 2.5 million brightest stars. *A&A*, 355:L27–L30.

Hoeg, E., Kuzmin, A., Bastian, U., Fabricius, C., Kuimov, K., Lindegren, L., Makarov, V. V., and Roeser, S. (1998). The TYCHO Reference Catalogue. *A&A*, 335:L65–L68.

Hoeg, E. and Petersen, J. O. (1997). HIPPARCOS parallaxes and the nature of δ Scuti stars. *A&A*, 323:827–830.

Horch, E. P., Meyer, R. D., and van Altena, W. F. (2004). Speckle Observations of Binary Stars with the WIYN Telescope. IV. Differential Photometry. *AJ*, 127:1727–1735.

Hoskin, M. (1997). The astrolabe. In Hoskin, M., editor, *Cambridge Illustrated History: Astronomy*, pages 64–67, Cambridge. Cambridge University Press.

Houk, N. and Swift, C. (1999). *Michigan catalogue of two-dimensional spectral types for the HD Stars ; vol. 5.* Department of Astronomy, University of Michigan.

Humphreys, R. M. (1970). The space distribution and kinematics of supergiants. *AJ*, 75:602–623.

Jahreiss, H., Requieme, Y., Argue, A. N., Dommanget, J., Rousseau, M., Lederle, T., Le Poole, R. S., Mazurier, J. M., Morrison, L. V., and Nys, O. (1992). The HIPPARCOS Input Catalogue. II - Astrometric data. *A&A*, 258:82–87.

Jenkins, J. F. (1952). *General Catalogue of Trigonometric Stellar Parallaxes.* Yale University Observatory.

Jenkins, J. F. (1963). *Supplement to the General Catalogue of Trigonometric Stellar Parallaxes.* Yale University Observatory.

Jones, B. F. and Walker, M. F. (1988). Proper motions and variabilities of stars near the Orion Nebula. *AJ*, 95:1755–1782.

Jordi, C., Høg, E., Brown, A. G. A., Lindegren, L., Bailer-Jones, C. A. L., Carrasco, J. M., Knude, J., Straižys, V., de Bruijne, J. H. J., Claeskens, J.-F., Drimmel, R., Figueras, F., Grenon, M., Kolka, I., Perryman, M. A. C., Tautvaišienė, G., Vansevičius, V., Willemsen, P. G., Bridžius, A., Evans, D. W., Fabricius, C., Fiorucci, M., Heiter, U., Kaempf, T. A., Kazlauskas, A., Kučinskas, A., Malyuto, V., Munari, U., Reylé, C., Torra, J., Vallenari, A., Zdanavičius, K., Korakitis, R., Malkov, O., and Smette, A. (2006). The design and performance of the Gaia photometric system. *MNRAS*, 367:290–314.

Jorissen, A., Jancart, S., and Pourbaix, D. (2004). Binaries in the Hipparcos data: Keep digging, I. Search for binaries without 'a priori' knowledge of their orbital elements: Application to barium stars. In Hidlitch, R. W., Hensberge, H., and Pavlovski, K., editors, *ASP Conf. Ser. 318: Spectroscopically and Spatially Resolving the Components of the Close Binary Stars*, pages 141–143.

Kallinger, T. and Weiss, W. W. (2002). Detecting low amplitude periodicities with HIPPARCOS. *A&A*, 385:533–536.

Kaye, A. B., Handler, G., Krisciunas, K., Poretti, E., and Zerbi, F. M. (1999). Gamma Doradus Stars: Defining a New Class of Pulsating Variables. *PASP*, 111:840–844.

Kharchenko, N. V., Piskunov, A. E., Röser, S., Schilbach, E., and Scholz, R.-D. (2005). Astrophysical parameters of Galactic open clusters. *A&A*, 438:1163–1173.

Klemola, A. R., Robinson, L. B., and Vasilevskis, S. (1974). Computer control of the Lick-Gaertner automatic measuring system. *PASP*, 86:820–825.

Knapp, G., Pourbaix, D., and Jorissen, A. (2001). Reprocessing the Hipparcos data for evolved giant stars II. Absolute magnitudes for the R-type carbon stars. *A&A*, 371:222–232.

Knapp, G. R., Pourbaix, D., Platais, I., and Jorissen, A. (2003). Reprocessing the Hipparcos data of evolved stars. III. Revised Hipparcos period-luminosity relationship for galactic long-period variable stars. *A&A*, 403:993–1002.

Koen, C. (2001). Multiperiodicities from the Hipparcos epoch photometry and possible pulsation in early A-type stars. *MNRAS*, 321:44–56.

Koen, C. and Eyer, L. (2002). New periodic variables from the Hipparcos epoch photometry. *MNRAS*, 331:45–59.

Kovalevsky, J., Falin, J. L., Pieplu, J. L., Bernacca, P. L., Donati, F., Froeschle, M., Galligani, I., Mignard, F., Morando, B., Perryman, M. A. C., Schrijver, H., van Daalen, D. T., van der Marel, H., Villenave, M., Walter, H. G., Badiali, M., Borriello, L., Brouw, W. N., Canuto, E., Guerry, A., Hering, R., Huc, C., Iorio-Fili, D., Lacroute, P., Lattanzi, M., Le Poole, R. S., Murgolo, F. P., Preston, R. A., Röser, S., Sanso, F., Wielen, R., Belforte, P., Bernstein, H.-H., Bucciarelli, B., Cardini, D., Emanuele, A., Fassino, B., Lenhardt, H., Lestrade, J. F., Prezioso, G., and Tommasini Montanari, T. (1992). The FAST HIPPARCOS Data Reduction Consortium: Overview of the Adopted Reduction Software. *A&A*, 258:7–17.

Kovalevsky, J., Lindegren, L., Perryman, M. A. C., Hemenway, P. D., Johnston, K. J., Kislyuk, V. S., Lestrade, J. F., Morrison, L. V., Platais, I., Röser, S., Schilbach, E., Tucholke, H.-J., de Vegt, C., Vondrak, J., Arias, F., Gontier, A. M., Arenou, F., Brosche, P., Florkowski, D. R., Garrington, S. T., Kozhurina-Platais, V., Preston, R. A., Ron, C., Rybka, S. P., Scholz, R.-D., and Zacharias, N. (1997). The HIPPARCOS catalogue as a realisation of the extragalactic reference system. *A&A*, 323:620–633.

Kroupa, P., Aarseth, S., and Hurley, J. (2001). The formation of a bound star cluster: from the Orion nebula cluster to the Pleiades. *MNRAS*, 321:699–712.

Lacroute, P. (1982). Histoire du projet d'astrometrie spatiale. In Perryman, M.A.C. and Guyenne, T.D., editors, *Scientific aspects of the Hipparcos space astrometry mission*, number 177 in ESA-SP, pages 3–12.

Laney, C. D. and Stobie, R. S. (1994). Cepheid Period / Luminosity Relations in K H J and V. *MNRAS*, 266:441–454.

Lee, T. A. (1968). Interstellar extinction in the Orion association. *ApJ*, 152:913–941.

Lestrade, J.-F., Jones, D. L., Preston, R. A., Phillips, R. B., Titus, M. A., Kovalevsky, J., Lindegren, L., Hering, R., Froeschle, M., Falin, J. L., Mignard, F., Jacobs, C. S., Sovers, O. J., Eubanks, M., and Gabuzda, D. (1995).

Preliminary link of the HIPPARCOS and VLBI reference frames. *A&A*, 304:182–188.

Lestrade, J.-F., Preston, R. A., Jones, D. L., Phillips, R. B., Rogers, A. E. E., Titus, M. A., Rioja, M. J., and Gabuzda, D. C. (1999). High-precision VLBI astrometry of radio-emitting stars. *A&A*, 344:1014–1026.

Lin, C. C., Yuan, C., and Shu, F. H. (1969). On the Spiral Structure of Disk Galaxies. III. Comparison with Observations. *ApJ*, 155:721–746.

Lindegren, L. (1995). Estimating the external accuracy of hipparcos parallaxes by deconvolution. *A&A*, 304:61–68.

Lindegren, L., Hoeg, E., van Leeuwen, F., Murray, C. A., Evans, D. W., Penston, M. J., Perryman, M. A. C., Petersen, C., Ramamani, N., and Snijders, M. A. J. (1992). The ndac hipparcos data analysis consortium, overview of the reduction methods. *A&A*, 258:18–30.

Lindegren, L., Le Poole, R. S., Perryman, M. A. C., and Petersen, C. (1992). Geometrical stability and evolution of the HIPPARCOS telescope. *A&A*, 258:35–40.

Lindegren, L., Madsen, S., and Dravins, D. (2000). Astrometric radial velocities. ii. maximum-likelihood estimation of radial velocities in moving clusters. *A&A*, 356:1119–1135.

Lindegren, L., Mignard, F., Söderhjelm, S., Badiali, M., Bernstein, H.-H., Lampens, P., Pannunzio, R., Arenou, F., Bernacca, P. L., Falin, J. L., Froeschlé, M., Kovalevsky, J., Martin, C., Perryman, M. A. C., and Wielen, R. (1997). Double star data in the HIPPARCOS Catalogue. *A&A*, 323:L53–L56.

Lub, J. and Pel, J. W. (1977). Properties of the Walraven VBLUW photometric system. *A&A*, 54:137–158.

Lucy, L. B. (1974). An iterative technique for the rectification of observed distributions. *AJ*, 79:745–754.

Luri, X., Ansari, S., Torra, J., Figueras, F., Jordi, C., Masana, E., and Llimona, P. (2004). Gaia: understanding our galaxy. In Ochsenbein, F., Allen, M. G., and Egret, D., editors, *ASP Conf. Ser. 314: Astronomical Data Analysis Software and Systems (ADASS) XIII*, pages 653–660.

Luri, X., Gomez, A. E., Torra, J., Figueras, F., and Mennessier, M. O. (1998). The LMC distance modulus from HIPPARCOS RR Lyrae and classical Cepheid data. *A&A*, 335:L81–L84.

Luri, X., Mennessier, M. O., Torra, J., and Figueras, F. (1996). A new maximum likelihood method for luminosity calibrations. *A&AS*, 117:405–415.

Lutz, T. E. and Kelker, D. E. (1973). On the use of trigonometric parallaxes for the calibration of luminosity systems : theory. *PASP*, 85:573–578.

Madsen, S., Lindegren, L., and Dravins, D. (2001). The Velocity Dispersion of the Hyades as a Function of Mass and Radius. In Deiters, S., Fuchs, B., Just, A., Spurzem, R., and Wielen, R., editors, *ASP Conf. Ser. 228: Dynamics of Star Clusters and the Milky Way*, pages 506–508.

Makarov, V. V. (2003). Improved Hipparcos Parallaxes of Coma Berenices and NGC 6231. *AJ*, 126:2408–2410.

Makarov, V. V., Odenkirchen, M., and Urban, S. (2000). Internal velocity dispersion in the Hyades as a test for Tycho-2 proper motions. *A&A*, 358:923–928.

Marshall, D. J., Robin, A. C., Reylé, C., Schultheis, M., and Picaud, S. (2006). Modelling the Galactic interstellar extinction distribution in three dimensions. *A&A*, 453:635–651.

Martin, C. and Mignard, F. (1998). Mass determination of astrometric binaries with Hipparcos. II. Selection of candidates and results. *A&A*, 330:585–599.

Martin, C., Mignard, F., and Froeschle, M. (1997). Mass determination of astrometric binaries with Hipparcos. I. Theory and simulation. *A&AS*, 122:571–580.

Martin, C., Mignard, F., Hartkopf, W. I., and McAlister, H. A. (1998). Mass determination of astrometric binaries with Hipparcos. III. New results for 28 systems. *A&AS*, 133:149–162.

Masana, E., Luri, X., Anglada-Escudé, G., and Llimona, P. (2005). The Gaia System Simulator. In Turon, C., O'Flaherty, K. S., and Perryman, M. A. C., editors, *ESA SP-576: The Three-Dimensional Universe with Gaia*, pages 457–460.

Mason, B. D., Douglass, G. G., and Hartkopf, W. I. (1999). Binary Star Orbits from Speckle Interferometry. I. Improved Orbital Elements of 22 Visual Systems. *AJ*, 117:1023–1036.

Mazeh, T., Zucker, S., dalla Torre, A., and van Leeuwen, F. (1999). Analysis of the HIPPARCOS Measurements of upsilon Andromedae: A Mass Estimate of Its Outermost Known Planetary Companion. *ApJ*, 522:L149–L151.

McCarthy, D. D. (2005). Precision time and the rotation of the Earth. In Kurtz, D. W., editor, *IAU Colloq. 196: Transits of Venus: New Views of the Solar System and Galaxy*, pages 180–197.

McNamara, B. J. (1976). Proper motion membership analysis of the ORI nebula cluster. *AJ*, 81:375–382.

McNamara, B. J., Hack, W. J., Olson, R. W., and Mathieu, R. D. (1989). A proper-motion membership analysis of stars in the vicinity of the Orion Nebula. *AJ*, 97:1427–1439.

Mignard, F. and Froeschle, M. (2000). Global and local bias in the FK5 from the Hipparcos data. *A&A*, 354:732–739.

Morrison, L. V., Argyle, R. W., Requieme, Y., Helmer, L., Fabricius, C., Einicke, O. H., Buotempo, M. E., Muinos, J. L., and Rapaport, M. (1990). Comparison of FK5 with Bordeaux and Carlsberg meridian circle observations. *A&A*, 240:173–177.

Munari, U., Dallaporta, S., Siviero, A., Soubiran, C., Fiorucci, M., and Girard, P. (2004). The distance to the Pleiades from orbital solution of the double-lined eclipsing binary HD 23642. *A&A*, 418:L31–L34.

Murray, C. A. and Nicholson, W. (1975). The Use of the "galaxy" Machine at the Royal Greenwich Observatory. In de Jager, C. and Nieuwenhuijzen, H., editors, *ASSL Vol. 54: Image Processing Techniques in Astronomy*, pages 171–184.

Myllari, A., Flynn, C., and Orlov, V. (2001). Stellar Moving Groups in HIPPARCOS. In Deiters, S., Fuchs, B., Just, A., Spurzem, R., and Wielen, R., editors, *ASP Conf. Ser. 228: Dynamics of Star Clusters and the Milky Way*, pages 329–334.

Narayanan, V. K. and Gould, A. (1999). Correlated errors in hipparcos parallaxes towards the pleiades and the hyades. *ApJ*, 523:328–339.

Nicolet, B. (1981). Geneva photometric boxes. III - Distances and reddenings for 43 open clusters. *A&A*, 104:185–197.

Noel, F. (1997). Systematic differences Astrolabe-FK5 derived from observations at 60 deg zenith distance. *A&AS*, 124:153–155.

Nordstroem, B., Mayor, M., Andersen, J., Holmberg, J., Pont, F., Jørgensen, B. R., Olsen, E. H., Udry, S., and Mowlavi, N. (2004). The Geneva-Copenhagen survey of the Solar neighbourhood. Ages, metallicities, and kinematic properties of $\sim$14 000 F and G dwarfs. *A&A*, 418:989–1019.

Oort, J. H. (1927a). Investigations concerning the rotational motion of the galactic system together with new determinations of secular parallaxes, precession and motion of the equinox (Errata: 4 94). *Bull. Astron. Inst. Netherlands*, 4:79–89.

Oort, J. H. (1927b). Observational evidence confirming Lindblad's hypothesis of a rotation of the galactic system. *Bull. Astron. Inst. Netherlands*, 3:275–282.

Orchiston, W. (2005). James Cook's 1769 transit of Venus expedition to Tahiti. In Kurtz, D. W., editor, *IAU Colloq. 196: Transits of Venus: New Views of the Solar System and Galaxy*, pages 52–66.

Oudmaijer, R. D., Groenewegen, M. A. T., and Schrijver, H. (1998). The Lutz-Kelker bias in trigonometric parallaxes. *MNRAS*, 294:L41–L46.

Pan, X., Shao, M., and Kulkarni, S. R. (2004). A distance of 133-137 parsecs to the pleuades star cluster. *Nature*, 427:326–328.

Papoulis, A. (1991). *Probability, Random Variables, and Stochastic Processes*. McGraw-Hill, New York.

Percy, J. R., Hosick, J., Kincaide, H., and Pang, C. (2002). Autocorrelation Analysis of Hipparcos Photometry of Short-Period Be Stars. *PASP*, 114:551–558.

Perryman, M. A. C., Brown, A. G. A., Lebreton, Y., Gomez, A., Turon, C., de Strobel, G. C., Mermilliod, J. C., Robichon, N., Kovalevsky, J., and Crifo, F. (1998). The Hyades: distance, structure, dynamics, and age. *A&A*, 331:81–120.

Perryman, M. A. C., Lindegren, L., Kovalevsky, J., Hoeg, E., Bastian, U., Bernacca, P. L., Crézé, M., Donati, F., Grenon, M., van Leeuwen, F., van der Marel, H., Mignard, F., Murray, C. A., Le Poole, R. S., Schrijver, H., Turon, C., Arenou, F., Froeschlé, M., and Petersen, C. S. (1997). The HIPPARCOS Catalogue. *A&A*, 323:L49–L52.

Petersen, J. O. and Hoeg, E. (1998). HIPPARCOS parallaxes and period-luminosity relations of high-amplitude delta Scuti stars. *A&A*, 331:989–994.

Pinsonneault, M. H., Staufer, J., Soderblom, D. R., King, J. R., and Hanson, R. B. (1998). The problem of hipparcos distances to open clusters. i. constraints from multicolor main-sequence fitting. *ApJ*, 504:170–191.

Platais, I., Kozhurina-Platais, V., Barnes, S., Girard, T. M., Demarque, P., van Altena, W. F., Deliyannis, C. P., and Horch, E. (2001). WIYN Open Cluster

Study. VII. NGC 2451A and the Hipparcos Distance Scale. *AJ*, 122:1486–1499.

Platais, I., Pourbaix, D., Jorissen, A., Makarov, V. V., Berdnikov, L. N., Samus, N. N., Lloyd Evans, T., Lebzelter, T., and Sperauskas, J. (2003). Hipparcos red stars in the HpV_T2 and V I_C systems. *A&A*, 397:997–1010.

Plett, M. (1978). The earth's magnetic field. In Wertz, J. R., editor, *Spacecraft attitude determination and control*, volume 73 of *Astrophysics and Space Science Library*, pages 113–123. D. Reidel, Dordrecht.

Pont, F. (1999). The Cepheid Distance Scale after Hipparcos. In Egret, D. and Heck, A., editors, *ASP Conf. Ser. 167: Harmonizing Cosmic Distance Scales in a Post-HIPPARCOS Era*, pages 113–128.

Pont, F., Mayor, M., and Burki, G. (1994). New radial velocities for classical cepheids. Local galactic rotation revisited. *A&A*, 285:415–439.

Pourbaix, D. (2004). Orbits from Hipparcos. In Hidlitch, R. W., Hensberge, H., and Pavlovski, K., editors, *ASP Conf. Ser. 318: Spectroscopically and Spatially Resolving the Components of the Close Binary Stars*, pages 132–140.

Pourbaix, D. and Boffin, H. M. J. (2003). Reprocessing the Hipparcos Intermediate Astrometric Data of spectroscopic binaries. II. Systems with a giant component. *A&A*, 398:1163–1177.

Pourbaix, D. and Jorissen, A. (2000). Re-processing the Hipparcos Transit Data and Intermediate Astrometric Data of spectroscopic binaries. I. Ba, CH and Tc-poor S stars. *A&AS*, 145:161–183.

Pourbaix, D., Platais, I., Detournay, S., Jorissen, A., Knapp, G., and Makarov, V. V. (2003). How many Hipparcos Variability-Induced Movers are genuine binaries? *A&A*, 399:1167–1175.

Reffert, S. and Quirrenbach, A. (2006). Hipparcos astrometric orbits for two brown dwarf companions: HD 38529 and HD 168443. *A&A*, 449:699–702.

Richardson, W. H. (1972). Bayesian-based iterative method of image restoration. *Optical Society of America Journal A*, 62:55–59.

Roberts, Jr., W. W. (1972). Application of the Density-Wave Theory of Spiral Structure: Shock Formation Along the Perseus Arm. *ApJ*, 173:259–283.

Robichon, N., Arenou, F., Mermilliod, J. C., and Turon, C. (1999). Open clusters with hipparcos. i. mean astrometric parameters. *A&A*, 345:471–484.

Robin, A. C. (2005). Gaia Census and Completeness. In Turon, C., O'Flaherty, K. S., and Perryman, M. A. C., editors, *ESA SP-576: The Three-Dimensional Universe with Gaia*, pages 83–88.

Robin, A. C., Reylé, C., Derrière, S., and Picaud, S. (2003). A synthetic view on structure and evolution of the Milky Way. *A&A*, 409:523–540.

Robin, A. C., Reylé, C., Derrière, S., and Picaud, S. (2004). Erratum: A synthetic view on structure and evolution of the Milky Way. *A&A*, 416:157.

Rodriguez, E. and Breger, M. (2001). delta Scuti and related stars: Analysis of the R00 Catalogue. *A&A*, 366:178–196.

Roeser, S. and Bastian, U. (1989). P P M : Positions and proper motions of 181731 stars north of -2. 5 degrees declination for equinox and epoch J2000. 0. *Bulletin d'Information du Centre de Donnees Stellaires*, 37:153–170.

Scargle, J. D. (1982). Studies in astronomical time series analysis. II - Statistical aspects of spectral analysis of unevenly spaced data. *ApJ*, 263:835–853.

Scargle, J. D. (1989). Studies in astronomical time series analysis. III - Fourier transforms, autocorrelation functions, and cross-correlation functions of unevenly spaced data. *ApJ*, 343:874–887.

Schaefer, B. E. (2004). Discovery of the Lost Star Catalog of Hipparchus on the Farnese Atlas. In *Bulletin of the American Astronomical Society*, page 1413.

Schrijver, H. and van der Marel, H. (1992). Geometrical calibration and assessment of the stability of the hipparcos payload. *A&A*, 258:31–34.

Schwan, H. (2001). An analytical representation of the systematic differences HIPPARCOS-FK5. *A&A*, 367:1078–1086.

Schwarzenberg-Czerny, A. (1989). On the advantage of using analysis of variance for period search. *MNRAS*, 241:153–165.

Schwarzenberg-Czerny, A. (1997). The Correct Probability Distribution for the Phase Dispersion Minimization Periodogram. *ApJ*, 489:941–945.

Smith, H. (2003). Is there really a Lutz-Kelker bias? Reconsidering calibration with trigonometric parallaxes. *MNRAS*, 338:891–902.

Soderblom, D. R., King, J. R., Hanson, R. B., Jones, B. F., Fischer, D., and Stauffer, J. (1998). The problem of hipparcos distances to open clusters. ii. constraints from nearby field stars. *ApJ*, 504:192–199.

Soderblom, D. R., Nelan, E., Benedict, G. F., McArthur, B., Ramirez, I., Spiesman, W., and Jones, B. F. (2005). Confirmation of Errors in Hipparcos Parallaxes from Hubble Space Telescope Fine Guidance Sensor Astrometry of the Pleiades. *AJ*, 129:1616–1624.

Soederhjelm, S. (1999). Visual binary orbits and masses POST HIPPARCOS. *A&A*, 341:121–140.

Spence, C. B. (1978). Environmental torques. In Wertz, J. R., editor, *Spacecraft attitude determination and control*, volume 73 of *Astrophysics and Space Science Library*, pages 566–576. D. Reidel, Dordrecht.

Standish, E. M. (2005). The Astronomical Unit now. In Kurtz, D. W., editor, *IAU Colloq. 196: Transits of Venus: New Views of the Solar System and Galaxy*, pages 163–179.

Stephenson, F. R. (1997). *Historical eclipses and earth's rotation*. Cambridge University Press, Cambridge.

Stephenson, F. R. and Morrison, L. V. (1982). History of the Earth's Rotation Since 700 B.C. In *Tidal Friction and the Earth's Rotation II*, pages 29–50.

Stone, R. C. (1997). Systematic Errors in the FK5 Catalog as Derived from CCD Observations in the Extragalactic Reference Frame. *AJ*, 114:850–858.

Tammann, G. A., Sandage, A., and Reindl, B. (2003). New Period-Luminosity and Period-Color relations of classical Cepheids: I. Cepheids in the Galaxy. *A&A*, 404:423–448.

Templeton, M., Basu, S., and Demarque, P. (2002). High-Amplitude δ Scuti and SX Phoenicis Stars: The Effects of Chemical Composition on Pulsations and the Period-Luminosity Relation. *ApJ*, 576:963–975.

ten Brummelaar, T., Mason, B. D., McAlister, H. A., Roberts, Jr., L. C., Turner, N. H., Hartkopf, W. I., and Bagnuolo, Jr., W. G. (2000). Binary Star Differential Photometry Using the Adaptive Optics System at Mount Wilson Observatory. *AJ*, 119:2403–2414.

Terlevich, E. (1987). Evolution of n-body open clusters. *MNRAS*, 224:193–225.

Tian, K. P., van Leeuwen, F., Zhao, J. L., and Su, C. G. (1996). Proper motions of stars in the region of the Orion Nebula cluster (C 0532-054). *A&AS*, 118:503–515.

Tsyganenko, N. A. (1990). Quantitative models of the magnetospheric magnetic field - methods and results. *Space Sci. Rev.*, 54:75–186.

Tsyganenko, N. A. (1996). Effects of the solar wind conditions on the global magnetospheric configuration as deduced from data-based models. *ESA SP*, 389:181–185.

Turon, C., Gomez, A., Crifo, F., Creze, M., Perryman, M. A. C., Morin, D., Arenou, F., Nicolet, B., Chareton, M., and Egret, D. (1992). The HIPPARCOS Input Catalogue. I - Star selection. *A&A*, 258:74–81.

Upgren, A. R. (1977). Internal and external errors in trigonometric parallaxes. *Vistas in Astr.*, 21:241–264.

van Altena, W. F., Lee, J. T., and Hoffleit, E. D. (1995). *The general catalogue of trigonometric [stellar] paralaxes*. Yale University Observatory, New Haven, CT.

van Altena, W. F., Lee, J. T., Lee, J.-F., Lu, P. K., and Upgren, A. R. (1988). The velocity dispersion of the Orion Nebula cluster. *AJ*, 95:1744–1754.

van der Marel, H. (1988). *On the "Great-Circle Reduction" in the data analysis for the astrometric satellite Hipparcos*. PhD thesis, Technische Universiteit Delft.

van der Marel, H. and Petersen, C. S. (1992). Hipparcos great-circle reduction. theory, results and intercomparisons. *A&A*, 258:60–69.

van Herk, G. and van Woerkom, A. J. J. (1961). Problems in meridian astronomy. *AJ*, 66:87–95.

van Leeuwen, F. (1980). Mass and luminosity function of the Pleiades. In *IAU Symp. 85: Star Formation*, pages 157–162.

van Leeuwen, F. (1994). Measuring velocity dispersions in open clusters. In Morisson, L. V. and Gilmore, G., editors, *Galactic and Solar System Optical Astrometry*, pages 223–229. Cambridge University Press.

van Leeuwen, F. (1997). The hipparcos mission. *Space Sci. Rev.*, 81(3–4):201–412.

van Leeuwen, F. (1999a). Hipparcos distance calibrations for 9 open clusters. *A&A*, 341:L71–L74.

van Leeuwen, F. (1999b). Open cluster distances from hipparcos parallaxes. In Egret, D. and Heck, A., editors, *Harmonizing cosmic distance scales in a post-Hipparcos era*, volume 167, pages 52–71. PASPC.

van Leeuwen, F. (2005a). Rights and wrongs of the Hipparcos data. A critical quality assessment of the Hipparcos catalogue. *A&A*, 439:805–822.

van Leeuwen, F. (2005b). The Pleiades question, the definition of the zero-age main sequence, and implications. In *IAU Colloq. 196: Transits of Venus: New Views of the Solar System and Galaxy*, pages 347–360.

van Leeuwen, F., Alphenaar, P., and Brand, J. (1986). A VBLUW photometric survey of the Pleiades cluster. *A&AS*, 65:309–347.

van Leeuwen, F., Challinor, A. D., Mortlock, D. J., Ashdown, M. A. J., Hobson, M. P., Lasenby, A. N., Efstathiou, G. P., Shellard, E. P. S., Munshi, D., and Stolyarov, V. (2002). Harmonic analysis of cosmic microwave background data - I. Ring reductions and point-source catalogue. *MNRAS*, 331:975–993.

van Leeuwen, F. and Evans, D. W. (1998). On the use of the hipparcos intermediate astrometric data. *A&A*, 323:157–172.

van Leeuwen, F., Evans, D. W., Grenon, M., Grossmann, V., Mignard, F., and Perryman, M. A. C. (1997a). The HIPPARCOS mission: photometric data. *A&A*, 323:L61–L64.

van Leeuwen, F., Evans, D. W., Lindegren, L., Penston, M. J., and Ramamani, N. (1992). Early improvements to the HIPPARCOS Input Catalogue through the accumulation of data from the satellite - Including the NDAC attitude reconstruction description. *A&A*, 258:119–124.

van Leeuwen, F., Evans, D. W., and van Leeuwen-Toczko, M. B. (1997b). Statistical Aspects of the Hipparcos Photometric Data. In *Statistical Challenges in Modern Astronomy II*, pages 259–280.

van Leeuwen, F. and Fantino, E. (2003). Dynamic modelling of the hipparcos attitude. *Space Sci. Rev.*, 108:537–576.

van Leeuwen, F. and Fantino, E. (2005). A new reduction of the raw Hipparcos data. *A&A*, 439:791–803.

van Leeuwen, F., Feast, M. W., Whitelock, P. A., and Yudin, B. (1997c). First results from HIPPARCOS trigonometrical parallaxes of Mira-type variables. *MNRAS*, 287:955–960.

van Leeuwen, F. and Penston, M. J. (2003). The operational environment of the hipparcos mission. *Space Sci. Rev.*, 108:471–497.

van Leeuwen, F. and van Genderen, A. M. (1997). The discovery of a new massive O-type close binary: tau CMa (HD 57061), based on HIPPARCOS and Walraven photometry. *A&A*, 327:1070–1076.

Varma, B. P. and Ghosh, C. (1973). Some optical properties of a multialkali (S20) photocathode and the processing parameters . *Journal of Physics D Applied Physics*, 6:628–632.

Warner, P. B., Kaye, A. B., and Guzik, J. A. (2003). A Theoretical γ Doradus Instability Strip. *ApJ*, 593:1049–1055.

Warren, W. H. and Hesser, J. E. (1977a). A photometric study of the ORI OB1 association. II. Photometric analysis. *ApJS*, 34:207–231.

Warren, W. H. and Hesser, J. E. (1977b). A photometric study of the Orion OB 1 association. I - Observational data. *ApJS*, 34:115–206.

Warren, W. H. and Hesser, J. E. (1978). A photometric study of the Orion OB 1 association. III - Subgroup analyses. *ApJS*, 36:497–572.

Wicenec, A. and van Leeuwen, F. (1995). The tycho star mapper background analysis. *A&A*, 304:160–167.

Wielen, R., Schwan, H., Dettbarn, C., Jahreiß, H., and Lenhardt, H. (1997). Statistical astrometry based on a comparison of individual proper motions and positions of stars in the FK5 and in the HIPPARCOS Catalogue. In *ESA SP-402: Hipparcos - Venice '97*, pages 727–732.

Yuan, C. (1969). Application of the Density-Wave Theory to the Spiral Structure of the Milky way System. II. Migration of Stars. *ApJ*, 158:889–898.

Zucker, S. and Mazeh, T. (2000). Analysis of the Hipparcos Measurements of HD 10697: A Mass Determination of a Brown Dwarf Secondary. *ApJ*, 531:L67–L69.

Zucker, S. and Mazeh, T. (2001). Analysis of the Hipparcos Observations of the Extrasolar Planets and the Brown Dwarf Candidates. *ApJ*, 562:549–557.

Zwahlen, N., North, P., Debernardi, Y., Eyer, L., Galland, F., Groenewegen, M. A. T., and Hummel, C. A. (2004). A purely geometric distance to the binary star Atlas, a member of the Pleiades. *A&A*, 425:L45–L48.

Subject index

Object index